FEEDBACK CONTROL SYSTEMS

Third Edition

JOHN VAN DE VEGTE
Department of Mechanical Engineering
University of Toronto

Prentice Hall
Englewood Cliffs, New Jersey 07632

Library of Congress Cataloging-in-Publication Data

Van de Vegte, J. (John)
    Feedback control systems / John Van de Vegte.—3rd ed.
        p.   cm.
    Includes bibliographical references and index.
    ISBN 0-13-016379-1
    1. Feedback control systems.   I. Title.
TJ217.V29   1994
629.8′3—dc20   93-10418
                                                          CIP

Acquisitions editor: *Don Fowley*
Editorial/production supervision: *Raeia Maes*
Cover designer: *Maureen Eide*
Prepress buyer: *Linda Behrens*
Manufacturing buyer: *Dave Dickey*

The author and publisher of this book have used their best efforts in preparing this book. These efforts include the development, research, and testing of the theories and programs to determine their effectiveness. The author and publisher make no warranty of any kind, expressed or implied, with regard to these programs or the documentation contained in this book. The author and publisher shall not be liable in any event for incidental or consequential damages in connection with, or arising out of, the furnishing, performance, or use of these programs.

Printed in the United States of America

10   9   8   7   6   5   4   3

ISBN 0-13-016379-1

Prentice-Hall International (UK) Limited, *London*
Prentice-Hall of Australia Pty, Limited, *Sydney*
Prentice-Hall Canada, Inc., *Toronto*
Prentice-Hall Hispanoamericana, S.A., *Mexico*
Prentice-Hall of India Private Limited, *New Delhi*
Prentice-Hall of Japan, Inc., *Tokyo*
Simon & Schuster Asia Pte. Ltd., *Singapore*
Editora Prentice-Hall do Brasil, Ltda., *Rio de Janeiro*

To
Maria

and to
Joyce
John
David
Michael

Contents

Contents

Preface

This book, like its previous editions, is intended to serve as a text for a first course in control systems for third- or fourth-year engineering students. Material in later chapters beyond what can be covered in a first course is used by the author in an elective second course at the senior undergraduate/graduate level. The book has been written to be suitable also for students with a more remote or less complete educational background and for self-study.

The text has grown out of many years of experience in teaching the subject to students in mechanical engineering, industrial engineering, engineering science, and correspondence courses and to students from industry in night courses. However, to show the generality and power of the subject, a first course should not be directed toward a particular department, and the book has been written to be suitable in all branches of engineering.

In this third edition, again many smaller changes have been made throughout to improve clarity by modifying or adding words or sentences. However, the heavy emphasis has been on revisions in the first eight chapters, on classical analysis and design. Numerous rearrangements of material were made among chapters to permit a more focused treatment of key topics. The s-plane is introduced in Chapter 1, permitting transient response calculation to be concentrated here.

An expanded and formalized treatment of block diagram reduction has been included in Chapter 3, following the derivation of such diagrams for physical systems. Signal flow graphs and Mason's Gain Formula, omitted from editions 1 and 2, have been included and are discussed in a separate section. With these changes, Chapter 4 can focus on the performance of systems with feedback and Chapter 5 on how this performance may be changed, that is, on introducing control system design. New here is, for example, the discussion of controller tuning.

However, the most significant feature of this edition is a greatly expanded treatment of the classical root locus and frequency response design techniques in Chapters 6 and 8. Instead of being discussed rather briefly and largely only in the context of examples, they are now presented as stepwise procedures, followed by examples and preceded by illustrative examples to motivate the procedures. In each case, phase-lag, phase-lead, and PID control design are treated in separate sections, and new related problems have been added to these chapters.

It is emphasized, however, that the fundamental approach of the book is unchanged. To develop insight, concepts are explained in the simplest possible mathematical framework, and concepts of design are developed in parallel with those of analysis. Thus Chapter 1 immediately identifies the two questions to be answered, that is, how dynamic systems behave and how this behavior may be changed by the use of feedback. This chapter also introduces the basic compromise between stability and accuracy that underlies all feedback system design. And Chapter 5 allows a focus on a physical explanation of the basic actions of dynamic controllers, unencumbered by the intricacies of the root locus and frequency response techniques discussed in Chapters 6 through 8.

A text must accommodate a wide spectrum of preferences on the extent to which computer aids are incorporated in a course. To achieve this, the text and its problems have been made independent of such aids, and programs for interactive computer-aided analysis and design with graphics are collected in Appendix B, with examples of their use. Reference to these aids is made where appropriate. The graphics no longer requires access to any special graphics software program package or to a plotter for hardcopy output. The author is pleased to acknowledge the work of then student now Dr. Philip W. P. Cheng, who developed the original programs for use in the Department of Mechanical Engineering of the University of Toronto, and of student Tony Paikeday, who prepared the first version of the present package.

Finally, the author expresses his gratitude to his wife, who was as pleased as he was that a third edition was called for and is at least as pleased as he is that the associated assault on his free time is nearing an end.

John Van de Vegte
Toronto, Canada

1

Introduction and Linearized Dynamic Models

1.1 INTRODUCTION

In the first part of this chapter, after a general introduction, the concepts of open-loop and closed-loop control are discussed in the context of a water-level control system. This example is then used to introduce fundamental considerations in control system analysis and design.

In the second part of the chapter, Laplace transforms are discussed and used to define the transfer function of a system. This is a linearized model of the dynamic behavior of the system that will serve as the basis for system analysis and design in most of this book. Such transfer functions will be derived in Chapter 2 for a variety of physical systems.

Transfer functions are used in this chapter to calculate transient responses of the corresponding systems to given input signals. The powerful s-plane, on which the transfer function can be represented in terms of the locations of its so-called poles and zeros, is introduced and applied here to the calculation of transient responses.

1.2 EXAMPLES AND CLASSIFICATIONS OF CONTROL SYSTEMS

Control systems exist in a virtually infinite variety, both in type of application and level of sophistication. The heating system and the water heater in a house are systems in which only the sign of the difference between desired and actual temperatures is used for control. If the temperature drops below a set value, a constant heat source is switched on, to be switched off again when the temperature rises above a set maximum. Variations of such relay or on–off control systems, sometimes quite sophisticated, are very common in practice because of their relatively low cost.

In the nature of such control systems, the controlled variable will oscillate continuously between maximum and minimum limits. For many applications, this control is not sufficiently smooth or accurate. In the power steering of a car, the controlled variable or system output is the angle of the front wheels. It must follow the system input, the angle of the steering wheel, as closely as possible, but at a much higher power level.

One classification of control systems is the following:

1. *Process control or regulator systems:* The controlled variable, or output, must be held as close as possible to a usually constant desired value, or input, despite any disturbances.
2. *Servomechanisms:* The input varies and the output must be made to follow it as closely as possible.

Power steering is one example of the second class, as are the hydraulic servos used for positioning control surfaces on aircraft. Indeed, the control of aircraft, missiles, and satellites in aerospace engineering is an area of often very advanced systems. This applies equally to robotics and automated manufacturing machinery in general. Numerically controlled machine tools use servos extensively for the control of positions or speeds.

In the process industries, including refineries and chemical plants, there are many temperatures and levels to be held to usually constant values in the presence of various disturbances. In an electrical power generation plant, controlled values of voltage and frequency are outputs, but inside such a plant there are again many temperatures, levels, pressures, and other variables to be controlled. Process control in general is a very large area of application of control systems.

Among many other possible classifications of control systems is the distinction between continuous and discrete systems. The latter are inherent in the use of digital computers for control. The classification into linear and nonlinear control systems is also important. Analysis and design are in general much simpler for the former, to which most of this book is devoted. Yet most systems become nonlinear if the variables move over wide enough ranges. The importance in practice of linear techniques relies on linearization based on the assumption that the variables stay close enough to a given *operating point.*

1.3 OPEN-LOOP CONTROL AND CLOSED-LOOP CONTROL

To introduce the subject, it is useful to consider an example. In Fig. 1.1, let it be desired to maintain the actual water level c in the tank as close as possible to a desired level r. The desired level will be called the system *input,* and the actual level the *controlled variable* or system *output.* Water flows from the tank via a valve V_o and enters the tank from a supply via a *control valve* V_c. The control valve is adjustable, either manually or by some type of *actuator.* This may be an electric motor or a hydraulic or pneumatic cylinder. Very often it would be a pneumatic diaphragm actuator, indicated in Fig. 1.2. Increasing the pneumatic pressure above the diaphragm pushes it down against a spring and increases the valve opening.

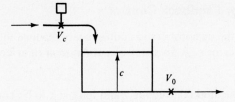

Figure 1.1   Water level control.

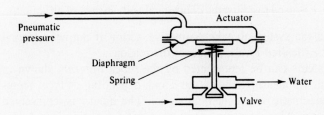

Figure 1.2   Pneumatically actuated valve.

Open-Loop Control

In this form of control, the valve is adjusted to make output c equal to input r, but not readjusted continually to keep the two equal. Open-loop control, with certain safeguards added, is very common, for example, in the context of sequence control, that is, guiding a process through a sequence of predetermined steps. However, for systems such as the one at hand, this form of control will normally not yield high performance. A difference between input and output, a system *error* $e = r - c$ would be expected to develop, due to two major effects:

1. *Disturbances* acting on the system
2. *Parameter variations* of the system

These are prime motivations for the use of feedback control. For the example, pressure variations upstream of V_c and downstream of V_o can be important disturbances affecting inflow and outflow, and hence level. In a steel rolling mill, very large disturbance torques act on the drive motors of the rolls when steel slabs enter or leave, and these may affect speeds.

For the water-level example, a sudden or gradual change of flow resistance of the valves due to foreign matter or valve deposits represents a system parameter variation. In a broader context, not only are the values of the parameters of a process often not precisely known, but they may also change greatly with operating condition.

For an aircraft or a rocket, the effectiveness of control surfaces changes rapidly as the device rises through the atmosphere. In an electrical power plant, parameter values are different at 20% and 100% of full power. In a valve, the relation between pressure drop and flow rate is often nonlinear, and as a result the resistance parameter of the valve changes with flow rate. Even if all parameter variations were known precisely, it would be complex, say in the case of the level example, to schedule the valve opening to follow time-varying desired levels.

Closed-Loop Control or Feedback Control

To improve performance, the operator could continuously readjust the valve based on observation of the system error e. A *feedback control system* in effect automates this action, as follows:

- The output c is measured continuously and fed back to be compared with the input r.
- The error $e = r - c$ is used to adjust the control valve by means of an actuator.

The feedback loop causes the system to take corrective action if output c (actual level) deviates from input r (desired level), whatever the reason.

A broad class of systems can be represented by the *block diagram* shown in Fig. 1.3. The *sensor* in Fig. 1.3 measures the output c and, depending on type, represents it by an electrical, pneumatic, or mechanical signal. The input r is represented by a signal in the same form. The *summing junction* or *error junction* is a device that combines the inputs to it according to the signs associated with the arrows: $e = r - c$.

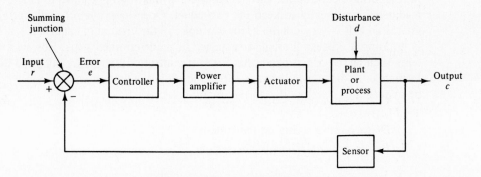

Figure 1.3   System block diagram.

It is important to recognize that if the control system is any good the error e will usually be small, ideally zero. Therefore, it is quite inadequate to operate an actuator. A task of the *controller* is to amplify the error signal. The controller output, however, will still be at a low power level. That is, voltage or pressure has been raised but current or airflow is still small. The *power amplifier* raises power to the levels needed for the actuator.

The *plant* or *process* in the level control example includes the valve characteristics as well as the tank. In part this is related to the identification of a *disturbance d* in Fig. 1.3 as an additional input to the block diagram. For the level control, d could represent supply pressure variations upstream of the control valve.

1.4 CONTROL SYSTEM ANALYSIS AND DESIGN

Control system analysis and design can be summarized in terms of the following two questions:

1. *Analysis:* What is the performance of a given system in response to changes of inputs or disturbances?
2. *Design:* If the performance is unsatisfactory, how can it be improved without changing the process, actuator, and power amplifier blocks?

It is particularly important to note the constraints imposed on the designer. The blocks indicated generally represent relatively or very expensive equipment and must be considered as a fixed part of the system. The power of design techniques that will permit large changes in performance to be achieved by changing only the controller should be appreciated.

The term *performance* summarizes several aspects of the behavior. Assume that in Fig. 1.3 a sudden change of input is applied to a new constant value. A certain period of time will be required for transient response terms to decay and for the output to level off at the new value. One key feature of this transient period is that it should be sufficiently short. Another is that the transient response should not be excessively oscillatory or severely overshoot the final level.

The *steady-state* response, after the transients have decayed, is an equally important aspect of the performance. Any steady-state errors between r and c must be satisfactorily small. To a disturbance input, the output should ideally not respond at all, and in any case the steady-state value of this output should be acceptably small.

The performance of a design is also measured by its success in reducing the dynamic and steady-state effects of parameter variations in the plant on the output.

Disturbances and parameter variations were given as motivations for feedback control. However, the transient response and steady-state error characteristics can also be improved by the use of feedback, and the motivations for feedback can be listed as follows:

1. Reducing the effects of parameter variations
2. Reducing the effects of disturbance inputs
3. Improving transient response characteristics
4. Reducing steady-state errors

In fact, improvements in the first two items are usually achieved in the course of design procedures aimed at the last two. An intuitive idea of the response characteristics can be obtained by assuming the controller in Fig. 1.3 to be an amplifier with *gain K;* that is, the output of this block is K times its input. Larger K means greater amplification of the error signal e. Therefore, the errors for given output values are smaller. Hence large gains are desirable to reduce errors, that is, to improve accuracy. Also, larger gain means a larger change of valve opening for a certain change of error. This suggests a faster change of the output and greater speed of response of the system.

On the other hand, these faster changes of output intuitively suggest an increasing danger of severe overshoot and oscillations of the output following a sudden change of the input. Figure 1.4 shows the large effect on the response to a step change of r, which can easily result if the gain is increased from a rather low to a rather high value. In fact, with a further increase of gain the oscillations may grow instead of decay. The system is then *unstable.*

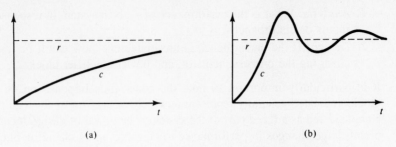

(a)   (b)

Figure 1.4   Step responses for (a) low and (b) high gains.

Stability is always the primary concern in feedback control design. But to be useful a system must also possess adequate *relative stability;* that is, the overshoot of a step response must be acceptably small, and this response must not be unduly oscillatory during the transient period.

Relative stability considerations usually impose an upper limit on gain, and hence on accuracy and speed of response. Much of control system design can be summarized as being concerned with achieving a satisfactory compromise between these features. If this is not possible with only a gain K, controller complexity is increased.

The remainder of this chapter provides a basis for the tools needed to move beyond this intuitive discussion and to answer the questions it raises.

1.5 LINEARIZED DYNAMIC MODELS

The concept of a transfer function will be developed to describe individual blocks and their interconnections in a block diagram. This is a linear model and requires a linearized description of system dynamic behavior.

Figure 1.5 shows a nonlinear relation $y = f(x)$ between the input x and the output y of a block in a block diagram. It could represent the force–deflection relation of a rubber spring or the flow versus valve opening characteristic of a control valve. If the variations of x about x_0 are small enough, the nonlinear curve can be approximated by its tangent at 0. The model is then said to be linearized about the steady-state *operating point* 0. In Fig. 1.5, if $\Delta x = x - x_0$, $\Delta y = y - y_0$, the linearized model of $f(x)$ about x_0 is

$$\Delta y = \left(\frac{df}{dx}\right)_{x_0} \Delta x \tag{1.1}$$

where $(df/dx)_{x_0}$ is the slope of the tangent at 0.

It may also be that the output y of a block is a nonlinear function of two or more inputs. For example, valve flow is a function of valve opening and the pressure drop across the valve. If $y = f(x, z)$, the linearized model for small variations Δx and Δz about an operating point (x_0, z_0) is

$$\Delta y = \left(\frac{\partial f}{\partial x}\right)_{x_0, z_0} \Delta x + \left(\frac{\partial f}{\partial z}\right)_{x_0, z_0} \Delta z \tag{1.2}$$

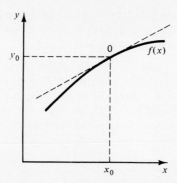

Figure 1.5   Linearization.

where $(\partial f/\partial x)_{x_0, z_0}$ and $(\partial f/\partial z)_{x_0, z_0}$ are partial derivatives. Note that the variables in these linearized models are not the actual values of the variables, but the deviations from those at the steady-state operating point.

It is common practice in these linearized models to redefine the variables to represent the variations. Thus the models (1.1) and (1.2) will usually be written as

$$y = Kx \qquad y = K_1 x + K_2 z \tag{1.3}$$

For many systems the dynamic behavior of the variations about operating-point values can be approximated by an nth-order linear differential equation:

$$\frac{d^n c}{dt^n} + a_{n-1}\frac{d^{n-1}c}{dt^{n-1}} + \cdots + a_1\frac{dc}{dt} + a_0 c = b_m\frac{d^m r}{dt^m} + \cdots + b_1\frac{dr}{dt} + b_0 r \tag{1.4}$$

where the variations $c(t)$ and $r(t)$ of output and input are functions of time t, and a_i and b_i are real constants.

Frequently, it is sufficient to determine the dynamic behavior of the output variations $c(t)$. If the actual variables are required, the steady-state solutions at the operating point must be added, corresponding to x_0 and y_0 in Fig. 1.5. These can be found from the actual nonlinear equations by setting derivatives with respect to time to zero.

Example 1.5.1   Spring–Mass–Damper System

Figure 1.6 shows a spring–mass–damper system subject to an external force $F(t)$. The displacement of mass m from the position of zero spring force is $x(t)$, and x and F are positive in downward direction. By Newton's second law, the acceleration $\ddot{x}$ is given by

$$m\ddot{x} = F + mg - f_c(\dot{x}) - f_k(x)$$

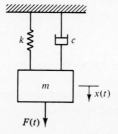

Figure 1.6   Spring–mass–damper system.

Here mg is the weight, and f_c and f_k are upward forces on m due to the damper and spring. Frequently, these forces will not be linear functions of $\dot{x}$ and x. As will be discussed in Section 2.2, $f_c(\dot{x}) = c\dot{x}$ for a linear damper with damping coefficient c. Let the spring be nonlinear, with $f_k(x) = kx^3$. Substituting and rearranging yields the following nonlinear differential equation of motion:

$$m\ddot{x} + c\dot{x} + kx^3 = F + mg$$

At a steady-state operating point where $F = F_0 = $ constant, the solution $x = x_0 = $ constant. Then $\dot{x} = \ddot{x} = 0$, so $kx_0^3 = F_0 + mg$, or $x_0 = [(F_0 + mg)/k]^{1/3}$. To linearize the equation for small variations $\Delta x = x - x_0$, $\Delta F = F - F_0$ about the operating point, the nonlinear term $f_k(x) = kx^3$ is linearized according to (1.1) as

$$kx_0^3 + \left(\frac{df_k}{dx}\right)_{x_0} \Delta x = kx_0^3 + 3kx_0^2\,\Delta x$$

Then, since $(\Delta\dot{x}) = \dot{x} - \dot{x}_0 = \dot{x}$, $(\Delta\ddot{x}) = \ddot{x} - \ddot{x}_0 = \ddot{x}$, the linear model is

$$m(\Delta\ddot{x}) + c(\Delta\dot{x}) + 3kx_0^2(\Delta x) + kx_0^3 = F_0 + \Delta F + mg$$

Here $kx_0^3 = F_0 + mg$, and redefining x and F to represent the variations about x_0 and F_0 gives

$$m\ddot{x} + c\dot{x} + (3kx_0^2)x = F$$

This is the linearized model. If, say to determine physical clearances, the actual positions are needed, x_0 must be added to the solution of this linear differential equation.

For linear systems, the equations for actual variables and deviations are the same. For example, if $\ddot{x} + c\dot{x} + dx = y$, then substituting $x = x_0 + \Delta x$, $y = y_0 + \Delta y$, where x_0 and y_0 are constant, yields

$$\Delta\ddot{x} + c\Delta\dot{x} + d\Delta x + dx_0 = \Delta y + y_0$$

But $dx_0 = y_0$ is the steady-state solution, so the equation in terms of Δx is the same as that for x.

1.6 LAPLACE TRANSFORMS

In Chapter 11 the state-space model will be introduced. It is a description of dynamic systems in terms of a set of first-order differential equations, written compactly in matrix form. This form of model is very powerful for the study of even very large systems. But in classical control theory, systems are commonly described by means of transfer functions. These will be defined by the use of Laplace transforms. Laplace transform theory is quite extensive, and it is therefore fortunate that only a small and isolated part of it is needed. The *Laplace transform F(s)* of a function $f(t)$ is defined by

$$F(s) = L[f(t)] = \int_0^\infty f(t)e^{-st}\,dt \qquad (1.5)$$

For the present it suffices to consider the *Laplace variable s* simply as a complex variable with real part σ and with frequency ω in radians per second as the imaginary part:

$$s = \sigma + j\omega \qquad (1.6)$$

From (1.5), the transform changes a function of time into a function of this new variable s. The advantage will be found to be that differentiation and integration are changed into algebraic operations.

Important examples and theorems are discussed next. The solutions of the integrals, which need not be evaluated in the use that will be made of all these results, may be verified, if desired, from tables of integrals.

1. *Unit step function u(t):* Shown in Fig. 1.7(a), this is a common test input to evaluate the performance of control systems. Its Laplace transform, from substitution of $f(t) = 1$ into (1.5), is

$$L[u(t)] = \int_0^\infty (1)e^{-st}\,dt = \frac{1}{s} \qquad (1.7)$$

For a step of magnitude A, $Au(t)$, the transform is A/s.

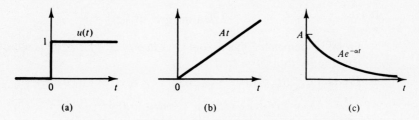

(a)   (b)   (c)

Figure 1.7   (a) Unit step; (b) ramp; (c) decaying exponential.

2. *Ramp function At:* This is also a common test input and is shown in Fig. 1.7(b). With $f(t) = At$, integration by parts in Eq. (1.5) yields

$$L[At] = \int_0^\infty Ate^{-st}\,dt = \frac{A}{s^2} \qquad (1.8)$$

3. *Decaying exponential $Ae^{-\alpha t}$:* In many physical systems, the transient response that follows a change of input or a disturbance decays according to this characteristic, which is shown in Fig. 1.7(c).

$$L[Ae^{-\alpha t}] = \int_0^\infty (Ae^{-\alpha t})e^{-st}\,dt = \frac{A}{s + \alpha} \qquad (1.9)$$

4. *Derivatives $d^n f(t)/dt^n$ if $F(s) = L[f(t)]$:*

$$L\left[\frac{df(t)}{dt}\right] = \int_0^\infty \left[\frac{df(t)}{dt}\right]e^{-st}\,dt$$

$$= e^{-st}f(t)\Big|_0^\infty + s\int_0^\infty f(t)e^{-st}\,dt = sF(s) - f(0) \qquad (1.10)$$

Repeating this procedure yields for the second derivative:

$$L\left[\frac{d^2 f(t)}{dt^2}\right] = s^2 F(s) - sf(0) - \frac{df(0)}{dt} \qquad (1.11)$$

In general, for the nth derivative, if the initial conditions of $f(t)$ and its derivatives are zero,

$$L\left[\frac{d^n f(t)}{dt^n}\right] = s^n F(s) \qquad (1.12)$$

From this remember that *taking a derivative of a function is equivalent to multiplying its transform by s.*

5. *Integrals of f(t) if F(s) = L[f(t)]:* By using integration by parts in the definition (1.5), it can be shown that

$$L\left[\int_{-\infty}^{t} f(\zeta)d\zeta\right] = \frac{1}{s}F(s) + \frac{1}{s}\left[\int_{-\infty}^{t} f(\zeta)d\zeta\right]_{t=0} \qquad (1.13)$$

where $\left[\int_{-\infty}^{t} f(\zeta)\,d\zeta\right]_{t=0}$ is the initial value of the integral. For zero initial conditions,

$$L[n\text{th integral of } f(t)] = \frac{F(s)}{s^n} \qquad (1.14)$$

From this remember that *taking the integral of a function is equivalent to multiplying its transform by 1/s.*

6. *Product $e^{-at}f(t)$ if F(s) = L[f(t)]:*

$$L[e^{-at}f(t)] = \int_{0}^{\infty} e^{-at}f(t)e^{-st}\,dt$$
$$= \int_{0}^{\infty} f(t)e^{-(s+a)t}\,dt = F(s+a) \qquad (1.15)$$

since the last integral is the definition of the transform, with s replaced by $(s + a)$. As an example of the use of this result, let it be desired to determine $L[te^{-at}]$. Then $f(t) = t$ and, from (1.8), $F(s) = 1/s^2$, so $F(s + a) = 1/(s + a)^2$. Hence

$$L[te^{-at}] = \frac{1}{(s + a)^2} \qquad (1.16)$$

7. *Delay theorem:* As shown in Fig. 1.8, the *translated function* or *delayed function* $f(t - t_d)$ equals $f(t)$ translated by t_d along the t-axis and is zero for $t < t_d$. If $F(s) = L[f(t)]$, then, with $\tau = t - t_d$,

$$L[f(t - t_d)] = \int_{0}^{\infty} f(t - t_d)e^{-st}\,dt$$
$$= \int_{t_d}^{\infty} f(t - t_d)e^{-st}\,dt = e^{-st_d}\int_{t_d}^{\infty} f(\tau)e^{-s\tau}\,dt \qquad (1.17)$$
$$= e^{-st_d}\int_{0}^{\infty} f(\tau)e^{-s\tau}\,d\tau = e^{-st_d}F(s)$$

Thus *translation over t_d is equivalent to multiplying the transform by e^{-st_d}.*

8. *Linearity theorem:* If $F_1(s) = L[f_1(t)]$, $F_2(s) = L[f_2(t)]$, and c_1 and c_2 are independent of t and s, then

$$L[c_1 f_1(t) + c_2 f_2(t)] = c_1 F_1(s) + c_2 F_2(s) \qquad (1.18)$$

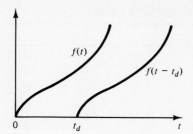

Figure 1.8   Translated function.

9. *Final value theorem:*

$$\lim_{t \to \infty} f(t) = \lim_{s \to 0} sF(s) \tag{1.19}$$

This allows the final or steady-state value of $f(t)$, that is, its value as $t \to \infty$, to be found from $F(s)$. The theorem is valid only if the limit exists, so it cannot be used to determine stability.

Table 1.6.1 is a table of *Laplace transform pairs,* which includes the preceding examples. The additional entries will be considered later. The transform of a unit impulse can be obtained by a limit process in which the width of a rectangular pulse of unit area is allowed to approach zero. The left column shows the *inverse Laplace transforms* of the expressions on the right. The entries in the table will be used to determine the inverse transforms of more complex expressions by expanding these expressions in terms of their partial fractions. This is discussed later.

However, it is desirable to define the concept of a transfer function first to show how such transforms arise in the study of control systems.

1.7 TRANSFER FUNCTIONS, BLOCK DIAGRAMS, AND THE s-PLANE

Figure 1.9(a) shows a block representing a system with input r and output c, which could also be a subsystem in a block diagram. The general linearized differential equation (1.4) will be taken to relate r and c. Based on the linearity theorem in Section 1.6, its Laplace transform can be written simply by transforming each term in turn. If the initial conditions are assumed to be zero, the result is, using (1.12) or Table 1.6.1,

$$(s^n + a_{n-1}s^{n-1} + \cdots + a_1 s + a_0)C(s) = (b_m s^m + \cdots + b_1 s + b_0)R(s) \tag{1.20}$$

where

$$C(s) = L[c(t)] \qquad R(s) = L[r(t)] \tag{1.21}$$

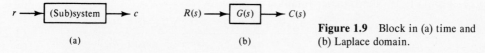

(a)   (b)

Figure 1.9   Block in (a) time and (b) Laplace domain.

Equation (1.20) gives immediately the common system representation of classical control theory.

TABLE 1.6.1 LAPLACE TRANSFORM PAIRS

$f(t)$	$F(s)$
$u(t)$ (unit step)	$\dfrac{1}{s}$
$At;\ At^n$	$\dfrac{A}{s^2};\ \dfrac{An!}{s^{n+1}}$
$Ae^{-\alpha t}$	$\dfrac{A}{s+\alpha}$
$\dfrac{df(t)}{dt}$	$sF(s)-f(0)$
$\dfrac{d^2f(t)}{dt^2}$	$s^2F(s)-sf(0)-\dfrac{df(0)}{dt}$
$\dfrac{d^nf(t)}{dt^n}$ (0 initial conditions)	$s^nF(s)$
$\displaystyle\int_{-\infty}^{t} f(\zeta)\,d\zeta \equiv f^{(-1)}(t)$	$\dfrac{1}{s}F(s)+\dfrac{1}{s}\left[\displaystyle\int_{-\infty}^{t} f(\zeta)\,d\zeta\right]_{t=0}$
$f^{(-n)}(t)$ (0 initial conditions)	$\dfrac{F(s)}{s^n}$
$e^{-at}f(t)$	$F(s+a)$
$te^{-at};\ t^n e^{-at}$	$\dfrac{1}{(s+a)^2};\ \dfrac{n!}{(s+a)^{n+1}}$
Delay theorem: $f(t-t_d), t>t_d;\ 0, t<t_d$	$e^{-st_d}F(s)$
Linearity theorem: $c_1 f_1(t)+c_2 f_2(t)$	$c_1 F_1(s)+c_2 F_2(s)$
Final value theorem: $\lim_{t\to\infty} f(t)$	$\lim_{s\to 0} sF(s)$
$\delta(t)$ (unit impulse)	1
$A\sin\omega t$	$\dfrac{A\omega}{s^2+\omega^2}$
$e^{-\zeta\omega_n t}\sin(\omega_n\sqrt{1-\zeta^2}t),\ \ \zeta<1$	$\dfrac{\omega_n\sqrt{1-\zeta^2}}{s^2+2\zeta\omega_n s+\omega_n^2}$
$1-\dfrac{1}{\sqrt{1-\zeta^2}}e^{-\zeta\omega_n t}\sin(\omega_n\sqrt{1-\zeta^2}t+\phi),\ \ \zeta<1$ $\left(\phi=\tan^{-1}\left[\dfrac{\sqrt{1-\zeta^2}}{\zeta}\right]\right)$	$\dfrac{\omega_n^2}{s(s^2+2\zeta\omega_n s+\omega_n^2)}$

Definition.   The *transfer function* of a (sub)system is the ratio of the Laplace transforms of its output and input, assuming zero initial conditions:

$$G(s)=\frac{C(s)}{R(s)} \qquad (1.22)$$

From (1.20), the transfer function is given by

$$G(s)=\frac{b_m s^m + b_{m-1}s^{m-1}+\cdots+b_1 s+b_0}{s^n+a_{n-1}s^{n-1}+\cdots+a_1 s+a_0} \qquad (1.23)$$

The representation of Fig. 1.9(a) is now replaced by that of Fig. 1.9(b), in terms of transforms and the transfer function. Equation (1.22) will often be used in the following form:

$$C(s) = G(s)R(s) \tag{1.24}$$

In words, the transform $C(s)$ of the output equals the transfer function $G(s)$ times the transform $R(s)$ of the input.

The assumption of zero initial conditions is very common in system analysis and design and is not a constraint. For linear systems the principle of superposition applies, which means that the total response is the sum of those to the input and to the initial conditions applied separately. Thus the initial conditions can be set to zero in determining the response to input $r(t)$, and $r(t)$ set to zero when calculating the response to initial conditions.

The blocks in a block diagram can now be represented by their transfer functions. Figure 1.10 shows two examples, of which the second is equivalent to that in Fig. 1.3 and is extremely common. Using (1.24), it is seen that for Fig. 1.10(a)

$$C_1 = G_2M \qquad M = G_1R_1 \qquad C_2 = G_3R_2 \tag{1.25}$$

Combining C_1 and C_2 as indicated by the summing junction then yields the overall relation

$$C = G_1G_2R_1 - G_3R_2 \tag{1.26}$$

For Fig. 1.10(b), analogous to (1.24) to (1.26),

$$C = G_2M \qquad M = G_1E \qquad E = R - B \qquad B = HC$$

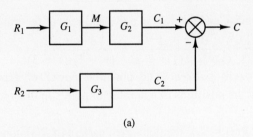

(a)

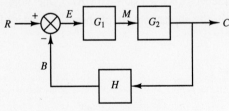

(b) Standard loop

Figure 1.10   Block diagrams.

Combining each pair yields

$$C = G_1G_2E \qquad E = R - HC$$

Eliminating E gives

$$C = G_1G_2R - G_1G_2HC$$

and rearranging this,

$$\textit{Closed-loop transfer function:} \quad \frac{C}{R} = \frac{G_1G_2}{1 + G_1G_2H} \tag{1.27}$$

This important result shows that the transfer function representing the overall closed-loop system can be derived by algebraic manipulations of those of the subsystem blocks. Such manipulations will be discussed in detail in Chapter 3, after the derivation of block diagrams for many physical systems.

The s-Plane, Poles, and Zeros

In the input–output relation $C(s) = G(s)R(s)$, each of C, G, and R is in general a ratio of polynomials in s. Consider now the following definitions:

- *Zeros* of C, G, and R are the roots of their numerator polynomials.
- *Poles* of C, G, and R are the roots of their denominator polynomials.
- *System zeros* and *system poles* are those of the system transfer function $G(s)$.
- *The system characteristic polynomial* is the denominator polynomial of $G(s)$.
- *The system characteristic equation* is obtained if the denominator polynomial of $G(s)$ is equated to zero. Evidently, its roots are the system poles.

Since the polynomials have real coefficients, poles and zeros are either real or occur in complex conjugate pairs. They are values of s and can be plotted on a complex plane called the *s-plane*. Because $s = \sigma + j\omega$, the real axis of this plane is the σ-axis, and frequencies ω are plotted along the imaginery $j\omega$-axis.

It is important to note that angles in the s-plane have great significance, and therefore the scales on both axes must be identical.

Figure 1.11(a) shows this plane and the system poles and zero of the transfer function

$$G(s) = \frac{2K}{3} \frac{0.5s + 1}{(\frac{1}{3})s^2 + (\frac{4}{3})s + 1} = \frac{K(s + 2)}{(s + 1)(s + 3)} \tag{1.28}$$

This is the *system pole–zero pattern*. The potential power of representing all dynamic characteristics by the positions of a number of points in this manner may be appreciated.

For a unit step input $R(s) = 1/s$, the pole–zero pattern of $R(s)$ consists of just a pole at the origin of the s-plane. Since $C(s) = G(s)R(s)$, the pole–zero pattern of $C(s)$ in Fig. 1.11(b) is simply the superposition of those of $G(s)$ and of $R(s)$:

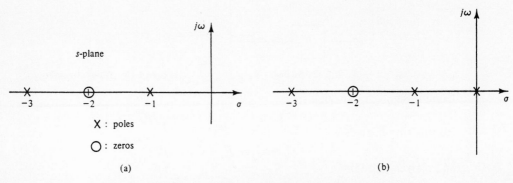

Figure 1.11   Pole–zero patterns of (a) $G(s)$ and (b) $C(s)$.

$$C(s) = G(s)R(s) = \frac{K(s + 2)}{s(s + 1)(s + 3)} \qquad (1.29)$$

These concepts, as well as (1.24) and (1.27), will be used in the next section in the calculation of system transient responses.

1.8 TRANSIENT RESPONSE AND INVERSE LAPLACE TRANSFORMATION

Inverse Laplace transformation of the output $C(s)$ of a block or a block diagram will give the corresponding response in the time domain. For very simple cases these inverses may be available directly in a table of Laplace transform pairs such as Table 1.6.1, but it is evident that a technique is needed to invert more complex expressions. This technique is based on the partial fraction expansion of $C(s)$ and is discussed in this section. This will also help motivate the derivation of transfer functions for physical systems in Chapter 2, by showing at least one of the uses to which they can be put.

The simplest case of the technique occurs when the roots of the denominator of $C(s)$ are real and distinct. For example, let

$$C(s) = \frac{s + 9}{s^2 + 7s + 6}$$

For inverse transformation, the denominator is factored as $(s + 1)(s + 6)$, and $C(s)$ is expanded into partial fractions according to

$$C = \frac{s + 9}{(s + 1)(s + 6)} = \frac{K_1}{s + 1} + \frac{K_2}{s + 6}$$

with a separate term for each real root factor in the denominator of $C(s)$. These terms are of the form $A/(s + \alpha)$, for which Table 1.6.1 gives the inverse $Ae^{-\alpha t}$. The constants K_i are the *residues* at the corresponding roots. To find K_1, multiply both sides of the equation by its denominator $(s + 1)$:

$$(s + 1)C(s) = \frac{s + 9}{s + 6} = K_1 + \frac{K_2(s + 1)}{s + 6}$$

If now s is put equal to the root -1 of $(s + 1)$, the second term on the right disappears, and the residue K_1 is seen to be

$$K_1 = [(s + 1)C(s)]_{s=-1} = \left(\frac{s + 9}{s + 6}\right)_{s=-1} = 1.6$$

Residue K_2 at root -6 is found in the same manner by multiplying through by $(s + 6)$ and then putting s equal to -6:

$$K_2 = [(s + 6)C(s)]_{s=-6} = \left(\frac{s + 9}{s + 1}\right)_{s=-6} = -0.6$$

The partial fraction expansion is now

$$C(s) = \frac{1.6}{s + 1} - \frac{0.6}{s + 6}$$

By the linearity theorem, the inverse is the sum of the inverses of the terms, so the transient response $c(t)$ is the sum of two decaying exponentials:

$$c(t) = 1.6e^{-t} - 0.6e^{-6t}$$

Example 1.8.1   Response to a Decaying Exponential Input

Find the response $c(t)$ of a system with transfer function

$$G(s) = \frac{2(s + 3)}{(s + 1)(s + 6)} \tag{1.30}$$

to a decaying exponential input $r(t) = e^{-2t}$.

From Table 1.6.1, $R(s) = 1/(s + 2)$, so $C(s)$ and its partial fraction expansion can be written as

$$C(s) = G(s)R(s) = \frac{2(s + 3)}{(s + 2)(s + 1)(s + 6)}$$

$$= \frac{K_1}{s + 2} + \frac{K_2}{s + 1} + \frac{K_3}{s + 6}$$

The residues are calculated next:

$$K_1 = [(s + 2)C(s)]_{s=-2} = \left[\frac{2(s + 3)}{(s + 1)(s + 6)}\right]_{s=-2} = -0.5$$

$$K_2 = [(s + 1)C(s)]_{s=-1} = \left[\frac{2(s + 3)}{(s + 2)(s + 6)}\right]_{s=-1} = +0.8$$

$$K_3 = [(s + 6)C(s)]_{s=-6} = \left[\frac{2(s + 3)}{(s + 2)(s + 1)}\right]_{s=-6} = -0.3$$

The partial fraction expansion is

$$C(s) = -\frac{0.5}{s + 2} + \frac{0.8}{s + 1} - \frac{0.3}{s + 6}$$

so the response sought is

$$c(t) = -0.5e^{-2t} + 0.8e^{-t} - 0.3e^{-6t}$$

The next example will show the calculation of the response to initial conditions. As noted in Section 1.7, for linear systems this response can simply be added to the response to an input. For a system

$$\ddot{c} + a_1\dot{c} + a_0c = r \qquad c(0) = c_0, \qquad \dot{c}(0) = \dot{c}_0 \tag{1.31}$$

the response $c_2(t)$ to initial conditions is found by setting $r = 0$ in (1.31) and transforming this equation, by using (1.10) and (1.11) or Table 1.6.1, to

$$[s^2C_2(s) - sc_0 - \dot{c}_0] + a_1[sC_2(s) - c_0] + a_0C_2(s) = 0$$

Rearranging yields the output transform

$$C_2(s) = \frac{(s + a_1)c_0 + \dot{c}_0}{s^2 + a_1s + a_0} \tag{1.32}$$

Example 1.8.2   Response to Initial Conditions

Determine the response to initial conditions $c(0) = 1$, $\dot{c}(0) = 2$ of the system

$$\ddot{c} + 7\dot{c} + 6c = r \tag{1.33}$$

For the response to initial conditions, the input r can be taken to be zero, and the transform is

$$[s^2 C - sc(0) - \dot{c}(0)] + 7[sC - c(0)] + 6C = 0$$

or

$$C = \frac{(s + 7)c(0) + \dot{c}(0)}{s^2 + 7s + 6} = \frac{s + 9}{s^2 + 7s + 6}$$

This is the transform of which the inverse was found earlier in this section.

The next example illustrates calculation of the transient response for a feedback control system of the form of Fig. 1.10(b) and also shows how this response may be modified by changing the controller.

Example 1.8.3   Step Response of a Temperature Control System

In Example 2.5.2 the differential equation and transfer function of a space heating system are derived. If $T(t)$ is the temperature difference with a constant ambient temperature, then, analogous to Ohm's law $v = iR$, it makes sense to take the heat loss to ambient as being proportional to T and to model this loss rate as T/R, where R is constant. If $q_i(t)$ is the heat inflow rate from a space heater, the net heat inflow rate is $(q_i - T/R)$. Clearly, T will rise faster if this is larger, and it is reasonable that $\dot{T}(t)$ should be proportional to net heat inflow rate. This can be expressed as $C\dot{T} = q_i - T/R$, with C a constant. On rearranging, this gives the differential equation

$$RC\dot{T}(t) + T(t) = Rq_i(t)$$

By Laplace transformation, the process transfer function $G_2(s)$ in the standard loop of Fig. 1.10(b) is then seen to be

$$G_2(s) = \frac{T(s)}{Q_i(s)} = \frac{R}{RCs + 1}$$

This *simple lag* form of transfer function will be found to approximate the behavior of many physical systems. In fact, the sensor used to measure the temperature for use in feedback control is often modeled as a simple lag. $H(s)$ in Fig. 1.10(b) is therefore taken to be

$$H(s) = \frac{1}{\tau s + 1}$$

and if the controller is simply an amplifier, then $G_1(s)$ in Fig. 1.10(b) can be taken to be a constant gain K:

$$G_1(s) = K$$

Using (1.27), the closed-loop transfer function $C/R = G_1 G_2/(1 + G_1 G_2 H)$ becomes

$$\frac{C(s)}{R(s)} = \frac{KR(\tau s + 1)}{(\tau s + 1)(RCs + 1) + KR}$$

For a unit step input, $R(s) = 1/s$, and the transform of the system output is

$$C(s) = \frac{KR(\tau s + 1)}{s[\tau RCs^2 + (\tau + RC)s + 1 + KR]}$$

If, in a consistent system of units,

$$\tau = 2 \qquad C = 80 \qquad R = \frac{1}{4}$$

then

$$C(s) = \frac{0.00625 K(2s + 1)}{s(s^2 + 0.55s + 0.025 + 0.00625K)}$$

For $K = 4$, Laplace transform inversion by partial fraction expansion is found to yield

$$c(t) = 0.5 - 0.523e^{-0.115t} + 0.023e^{-0.435t}$$

This unit step response is plotted in Fig. 1.12 and is decidedly unattractive, since in the steady state ($t \to \infty$) the output changes by only 0.5° for a 1° change of desired temperature.

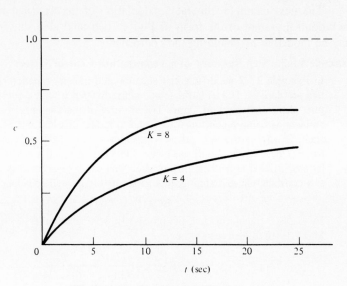

Figure 1.12   Step response, Example 1.8.3.

Doubling the gain to $K = 8$ yields the solution

$$c(t) = 0.667 - 2e^{-0.25t} + 1.333e^{-0.30t}$$

and the response, while the steady-state error is still quite large, is certainly much improved, also in that the response reaches steady state sooner; that is, the speed of response is increased.

It is noted, finally, that the steady-state value c_{ss} of the output could also have been found directly from $C(s)$ and the final value theorem:

$$c_{ss} = \lim_{t \to \infty} c(t) = \lim_{s \to 0} sC(s) = \frac{KR}{1 + KR}$$

Graphical Determination of Residues

Thus far, inverse transformation of $C(s) = G(s)R(s)$ was used to find the response $c(t)$ for the case where the roots of the denominator of $C(s)$ are real and distinct. The s-plane is the basis for a graphical alternative to the analytical method of calculating the residues, which is often preferred for more complicated transforms. Later this alternative will also help to understand the effect of system zeros on the transient response.

In the partial fraction expansion of the output $C(s)$ of (1.29),

$$C(s) = \frac{K(s + 2)}{s(s + 1)(s + 3)} = \frac{K_1}{s} + \frac{K_2}{s + 1} + \frac{K_3}{s + 3} \tag{1.34}$$

the residues are

$$K_1 = \frac{K(s+2)}{(s+1)(s+3)}\bigg|_{s=0} \qquad K_2 = \frac{K(s+2)}{s(s+3)}\bigg|_{s=-1} \qquad K_3 = \frac{K(s+2)}{s(s+1)}\bigg|_{s=-3}$$

$$(1.35)$$

Figure 1.13 helps to interpret the typical factor $(s+a)\big|_{s=b}$ in these expressions in terms of a vector in the s-plane. The factor can be expressed as

$$(s+a)\big|_{s=b} = b + a = b - (-a)$$

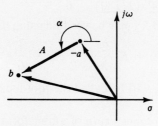

Figure 1.13   The vector $(s+a)$ for $s=b$.

Here b is a vector from the origin to point b in the s-plane, and $-a$ from the origin to point $-a$. The difference of these two vectors is that from $-a$ to b. This vector can be represented by $Ae^{j\alpha}$, where A is the vector length and α the angle, measured positive counterclockwise from the direction of the positive real axis:

$$(s+a)\big|_{s=b} = (\text{vector from } -a \text{ to } b) = Ae^{j\alpha} = A\underline{/\alpha} \qquad (1.36)$$

Here A and α may be found by measurement or calculation from the s-plane. Now a residue of the form of the K_i in (1.35) at a pole $s=-d$ is

$$K_i = \frac{K(s+a)}{(s+b)(s+c)}\bigg|_{s=-d} = \frac{K(A\underline{/\alpha})}{(B\underline{/\beta})(C\underline{/\gamma})} = \frac{KA}{BC}\underline{/\alpha-\beta-\gamma} \qquad (1.37)$$

where the vectors are those drawn to $-d$ from the zero at $-a$ and from the poles at $-b$ and $-c$. To find in (1.35) the residue K_1 at the pole $s=0$, consider the pole–zero pattern of $C(s)$ in Fig. 1.14, repeated from Fig. 1.11(b). The vectors as shown in the upper half-plane are drawn to the pole $s=0$ of $C(s)$ from its zero at -2 and from its other two poles at -1 and -3. These give

$$K_1 = \frac{K(2\underline{/0})}{(1\underline{/0})(3\underline{/0})} = \frac{2}{3}K\underline{/0} = \frac{2}{3}K$$

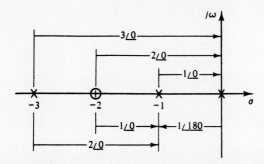

Figure 1.14   Pole–zero pattern of $C(s)$ as in Fig. 1.11(b).

The residue K_2 at the pole -1 is found from the vectors to this point, in the bottom half of Fig. 1.14. The vector from the pole at $s = 0$ to that at -1 is in the direction of the negative real axis, so has angle $\pm 180°$. Then, since an angle of $\pm 180°$ points in the direction of the negative real axis and is therefore equivalent to a minus sign,

$$K_2 = \frac{K(1\underline{/0})}{(1\underline{/180})(2\underline{/0})} = \frac{1}{2}K\underline{/-180} = -\frac{1}{2}K$$

Similarly,

$$K_3 = \frac{K(1\underline{/180})}{(3\underline{/180})(2\underline{/180})} = \frac{1}{6}K\underline{/180 - 360} = -\frac{1}{6}K$$

These values of K_1, K_2, and K_3 may be verified using the analytical method and give

$$c(t) = K\left(\frac{2}{3} - \frac{1}{2}e^{-t} - \frac{1}{6}e^{-3t}\right) \tag{1.38}$$

The gain factor K in (1.34) will be called the root locus gain.

> **Definition:** The *root locus gain* of a transform or a transfer function is that which results if the coefficients of the highest powers of s in the numerator and denominator polynomials are made equal to unity.

For, say,

$$C(s) = 2\frac{0.5s + 1}{(s + 3)(0.1s + 1)} = \frac{2(0.5)}{0.1}\frac{s + 2}{(s + 3)(s + 10)}$$

the root locus gain is $2(0.5/0.1) = 10$.

Reviewing the expressions for K_1, K_2, and K_3 reveals the following general rule.

> **Graphical Residue Rule:** The residue K_i at the pole $-p_i$ of $C(s)$ equals the root locus gain times the product of the vectors from all zeros of $C(s)$ to $-p_i$ divided by the product of the vectors from all other poles of $C(s)$ to $-p_i$.

Example 1.8.4   Two-Tank Level Control

Figure 1.15(a) models a level control system of which the process consists of the two-tank system with control valve in Figs. 2.19 and 2.20, and the level in the second tank is to be controlled. Find the unit step response of the system for $K = 0.025$. The closed-loop transfer function is

$$\frac{C}{R} = \frac{(0.025)5}{(0.5s + 1)(0.2s + 1) + (0.025)5} = \frac{1.25}{s^2 + 7s + 11.25} = \frac{1.25}{(s + 2.5)(s + 4.5)}$$

and with $R(s) = 1/s$,

$$C(s) = \frac{1.25}{s(s + 2.5)(s + 4.5)} = \frac{K_1}{s} + \frac{K_2}{s + 2.5} + \frac{K_3}{s + 4.5}$$

The pole–zero pattern of $C(s)$ in Fig. 1.15(b) and the graphical residue rule now yield

$$K_1 = \frac{1.25}{(2.5\underline{/0})(4.5\underline{/0})} = \frac{1}{9} \qquad K_2 = \frac{1.25}{(2.5\underline{/180})(2\underline{/0})} = -\frac{1}{4}$$

$$K_3 = \frac{1.25}{(4.5\underline{/180})(2\underline{/180})} = \frac{1.25}{9} \qquad c(t) = \frac{1}{9} - \frac{1}{4}e^{-2.5t} + \frac{1.25}{9}e^{-4.5t} \tag{1.39}$$

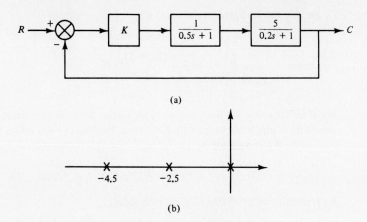

(a)

(b)

Figure 1.15   Example 1.8.4.

1.9 TRANSIENT RESPONSE WITH REPEATED AND COMPLEX POLES

In this section, transient responses are calculated for repeated real poles, a complex conjugate pair of poles, and for combinations of distinct real poles and complex conjugate pairs.

Repeated Real Poles

Suppose that $C(s)$ has m poles at $-p_1$:

$$C(s) = \frac{Z(s)}{(s + p_1)^m(s + p_2)\cdots(s + p_n)} \tag{1.40}$$

Then it can be shown that the partial fraction expansion must be written as

$$C(s) = \frac{K_1}{(s + p_1)^m} + \frac{K_2}{(s + p_1)^{m-1}} + \cdots + \frac{K_m}{s + p_1} + \frac{K_{m+1}}{s + p_2} + \cdots + \frac{K_{m+n-1}}{s + p_n} \tag{1.41}$$

The residues $K_{m+1}, \ldots, K_{m+n-1}$ are found as before, but for the repeated root the equation is

$$K_i = \frac{1}{(i - 1)!} \frac{d^{i-1}}{ds^{i-1}} [C(s)(s + p_1)^m] \bigg|_{s=-p_1} \qquad i = 1, \ldots, m \tag{1.42}$$

Since m rarely exceeds 2, this equation is not difficult to apply. This is illustrated by the next examples, which also imply the derivation of (1.42).

Example 1.9.1   Unit Ramp Response

Find the response to a unit ramp input, $R(s) = 1/s^2$, of the system

$$G(s) = \frac{1}{Ts + 1} \quad \text{(simple lag)}$$

$$C(s) = G(s)R(s) = \frac{1}{s^2(Ts + 1)} = \frac{K_1}{s^2} + \frac{K_2}{s} + \frac{K_3}{s + 1/T} \tag{1.43}$$

K_1 and K_3 can be found in the usual way:

$$K_3 = \left[\left(s + \frac{1}{T} \right) \frac{1}{s^2(Ts + 1)} \right] \Bigg|_{s=-1/T} = \frac{1}{Ts^2} \Bigg|_{s=-1/T} = T$$

$$K_1 = \left[s^2 \frac{1}{s^2(Ts + 1)} \right] \Bigg|_{s=0} = \frac{1}{Ts + 1} \Bigg|_{s=0} = 1$$

But if to determine K_2 the equation is multiplied by its denominator s, the first term becomes K_1/s and tends to infinity for $s \to 0$. Equation (1.42) arises by first multiplying both sides of the equation by s^2.

$$s^2 C(s) = K_1 + K_2 s + \frac{K_3 s^2}{s + 1/T}$$

K_1 is eliminated by differentiating both sides:

$$\frac{d}{ds} [s^2 C(s)] = K_2 + K_3 s \frac{2(s + 1/T) - s}{(s + 1/T)^2}$$

If now $s \to 0$, only K_2 remains on the right, so that, as in (1.42),

$$K_2 = \frac{d}{ds} (s^2 C) \Bigg|_{s=0} = \frac{d}{ds} \left(\frac{1}{Ts + 1} \right) \Bigg|_{s=0} = -T$$

Hence

$$C(s) = \frac{1}{s^2} - \frac{T}{s} + \frac{T}{s + 1/T}$$

Using Table 1.6.1, the transient response is

$$c(t) = t - T + Te^{-t/T}$$

Since the simple lag is so common, this response has been sketched in Fig. 1.16. The value T of the steady-state ($t \to \infty$) difference between r and c could have been found more directly by the final value theorem:

$$\lim_{t \to \infty} [r(t) - c(t)] = \lim_{s \to 0} s[R(s) - C(s)] = \lim_{s \to 0} s \left[\frac{T}{s(Ts + 1)} \right] = T$$

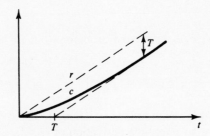

Figure 1.16   Simple lag ramp response.

Example 1.9.2   Figure 1.15 for $K = 0.045$

The response for $K = 0.025$ in Example 1.8.4, with a steady-state output of $\frac{1}{9}$ for a unit step input and so a steady-state error of $\frac{8}{9}$, is really quite poor. For $K = 0.045$ it is found that

$$\frac{C}{R} = \frac{2.25}{s^2 + 7s + 12.25} = \frac{2.25}{(s + 3.5)^2}$$

and for a unit step input

$$C = \frac{2.25}{s(s + 3.5)^2} = \frac{K_1}{s} + \frac{K_2}{(s + 3.5)^2} + \frac{K_3}{s + 3.5}$$

$$K_1 = [sC(s)]\bigg|_{s=0} = \frac{9}{49} = 0.1837$$

$$K_2 = [(s + 3.5)^2 C(s)]\bigg|_{s=-3.5} = \frac{2.25}{s}\bigg|_{s=-3.5} = -0.7429$$

$$K_3 = \frac{d}{ds}\left(\frac{2.25}{s}\right)\bigg|_{s=-3.5} = \frac{-2.25}{s^2}\bigg|_{s=-3.5} = -0.1837$$

Hence

$$c(t) = 0.1837 - 0.7429te^{-3.5t} - 0.1837e^{-3.5t}$$

The steady-state error $1 - 0.1837 = 0.8163$ has been reduced, as expected with an increase of K, but is still quite large.

A Complex Conjugate Pair of Poles

The technique for distinct real poles applies also for distinct complex poles, but the residues are now found to be complex conjugates, and the graphical approach tends to be simpler than the analytical one.

Example 1.9.3   One Complex Conjugate Pair of Poles

Find the inverse transform of an entry in Table 1.6.1:

$$C(s) = \frac{\omega_n^2}{s(s^2 + 2\zeta\omega_n s + \omega_n^2)} \tag{1.44}$$

This could be the unit impulse response of a system $G(s) = C(s)$ or the unit step response of

$$G(s) = \frac{\omega_n^2}{s^2 + 2\zeta\omega_n s + \omega_n^2}$$

For $\zeta < 1$, the poles of $C(s)$ are

$$s_1 = 0 \qquad s_2 = -\zeta\omega_n + j\omega_n\sqrt{1 - \zeta^2} \qquad s_3 = -\zeta\omega_n - j\omega_n\sqrt{1 - \zeta^2} \tag{1.45}$$

The pole–zero pattern of $C(s)$ is shown in Fig. 1.17, and its partial fraction expansion

$$C(s) = \frac{K_1}{s} + \frac{K_2}{s + \zeta\omega_n - j\omega_n\sqrt{1 - \zeta^2}} + \frac{K_3}{s + \zeta\omega_n + j\omega_n\sqrt{1 - \zeta^2}}$$

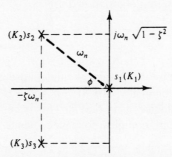

Figure 1.17 $C(s)$ of (1.44).

shows that the response will be of the form

$$c(t) = K_1 + e^{-\zeta\omega_n t}[K_2 \exp(j\omega_n \sqrt{1-\zeta^2}t) + K_3 \exp(-j\omega_n \sqrt{1-\zeta^2}t)]$$

In Fig. 1.17 the distance of s_2 and s_3 to the origin is ω_n and the angle ϕ is given by

$$\tan \phi = \frac{\sqrt{1-\zeta^2}}{\zeta} \qquad (1.46)$$

This may be shown by taking the square root of the sum of the squares of the real and imaginary parts of s_2 and s_3.

The residues can now be expressed. From (1.44), the root locus gain is ω_n^2. Recall that angles are positive counterclockwise from the direction of the positive real axis and that an angle $\phi - 3\pi/2$ is equivalent to $\phi + \pi/2$. The graphical residue rule then gives

$$K_1 = \frac{\omega_n^2}{(\omega_n \underline{/-\phi})(\omega_n \underline{/+\phi})} = 1$$

$$K_2 = \frac{\omega_n^2}{(\omega_n \underline{/\pi - \phi})(2\omega_n \sqrt{1-\zeta^2} \underline{/\pi/2})} = \frac{1}{2\sqrt{1-\zeta^2}} \underline{/\phi + \pi/2}$$

$$K_3 = \frac{\omega_n^2}{(\omega_n \underline{/\pi + \phi})(2\omega_n \sqrt{1-\zeta^2} \underline{/-\pi/2})} = \frac{1}{2\sqrt{1-\zeta^2}} \underline{/-\phi - \pi/2}$$

$$(1.47)$$

Substitution into $c(t)$ yields

$$c(t) = 1 + \frac{1}{\sqrt{1-\zeta^2}} e^{-\zeta\omega_n t} \frac{1}{2} \left\{ \exp\left[j\left(\omega_n \sqrt{1-\zeta^2}t + \phi + \frac{\pi}{2} \right) \right] \right.$$

$$\left. + \exp\left[-j\left(\omega_n \sqrt{1-\zeta^2}t + \phi + \frac{\pi}{2} \right) \right] \right\} \qquad (1.48)$$

or, since $\cos x = \frac{1}{2}(e^{jx} + e^{-jx}) = -\sin(x - \frac{\pi}{2})$

$$c(t) = 1 - \frac{1}{\sqrt{1-\zeta^2}} e^{-\zeta\omega_n t} \sin(\omega_n \sqrt{1-\zeta^2}t + \phi) \qquad (1.49)$$

This verifies the entry in Table 1.6.1. Equation (1.55) will give an alternative form of this result.

Distinct Real Poles and Complex Conjugate Pairs: General Case

For distinct poles, whether real or complex, the partial fraction expansion of $C(s) = G(s)R(s)$ and the corresponding solution $c(t)$ are

$$C(s) = \frac{K_1}{s + p_1} + \frac{K_2}{s + p_2} + \cdots + \frac{K_n}{s + p_n} \qquad (1.50)$$

$$c(t) = K_1 \exp(-p_1 t) + K_2 \exp(-p_2 t) + \cdots + K_n \exp(-p_n t)$$

All residues can be found by the graphical rule. Those corresponding to real poles will be real. For complex conjugate pairs, if $-p_i$ and $-p_{i+1}$ form such a pair, K_i and K_{i+1} are also complex conjugate, as in (1.47). This is because all poles and zeros are real or occur in pairs that are complex conjugate. Hence let

$$p_i = \zeta_i \omega_{ni} - j\omega_{ni} \sqrt{1-\zeta_i^2} \qquad K_i = \mathbf{K}_i e^{j\theta_i} = \mathbf{K}_i \underline{/\theta_i} \qquad (1.51)$$

where $\mathbf{K}_i$ is the magnitude and θ_i the phase. Then

$$p_{i+1} = \zeta_i \omega_{ni} + j\omega_{ni}\sqrt{1 - \zeta_i^2} \qquad K_{i+1} = K_i \underline{/-\theta_i}$$

and the corresponding terms in the inverse are

$$
\begin{aligned}
K_i \exp(-p_i t) &+ K_{i+1} \exp(-p_{i+1} t) \\
&= K_i \exp(-\zeta_i \omega_{ni} t) \{\exp[j(\omega_{ni}\sqrt{1 - \zeta_i^2}\, t + \theta_i)] \\
&\quad + \exp[-j(\omega_{ni}\sqrt{1 - \zeta_i^2}\, t + \theta_i)]\} \\
&= 2K_i \exp(-\zeta_i \omega_{ni} t) \cos(\omega_{ni}\sqrt{1 - \zeta_i^2}\, t + \theta_i)
\end{aligned}
\tag{1.52}
$$

Observe that θ_i in this solution is the phase angle of the residue at the upper half-plane pole. This gives the following general result for distinct poles:

If $C(s)$ has k real poles $-p_i$ and m complex conjugate pairs of the form $-\zeta_i \omega_{ni} \pm j\omega_{ni}\sqrt{1 - \zeta_i^2}$, then the total response is

$$
\begin{aligned}
c(t) = &\sum_{i=1}^{k} K_{ri} \exp(-p_i t) \\
&+ 2\sum_{i=1}^{m} K_i \exp(-\zeta_i \omega_{ni} t) \cos(\omega_{ni}\sqrt{1 - \zeta_i^2}\, t + \theta_i)
\end{aligned}
\tag{1.53}
$$

where the K_{ri} are the residues at the real poles, K_i the magnitudes of the residues at the complex poles, and θ_i their phase angles at the upper half-plane poles.

One application of this result is to obtain an alternative form of the solution (1.49) for the special case of Example 1.9.3. Here $k = m = 1$ and $p_1 = 0$, and, from (1.47),

$$K_{r1} = 1 \qquad K_1 = \frac{1}{2\sqrt{1 - \zeta^2}} \qquad \theta = \phi + \frac{\pi}{2} \tag{1.54}$$

so

$$c(t) = 1 + \frac{1}{\sqrt{1 - \zeta^2}} e^{-\zeta\omega_n t} \cos(\omega_n\sqrt{1 - \zeta^2}\, t + \theta) \tag{1.55}$$

This agrees with (1.49) because of the trigonometric relation shown preceding (1.49).

Example 1.9.4

Find the unit step response of a system with transfer function

$$G(s) = \frac{0.89}{(0.5s + 1)(s^2 + s + 0.89)} \tag{1.56}$$

The roots of the quadratic are $(-0.5 \pm 0.8j)$, and Fig. 1.18 shows the pole–zero pattern of

$$C(s) = \frac{1.78}{s(s + 2)(s + 0.5 - 0.8j)(s + 0.5 + 0.8j)}$$

The root locus gain is 1.78, and Fig. 1.18 shows significant vector lengths and angles for determination of the residues. The residues corresponding to the poles indicated are, from the graphical rule, with phase angles in degrees:

$$K_1 = \frac{1.78}{(2\underline{/0})(0.943\underline{/-58})(0.943\underline{/58})} = 1$$

$$K_2 = \frac{1.78}{(2\underline{/180})(1.7\underline{/208.1})(1.7\underline{/151.9})} = 0.307\underline{/180} = -0.307$$

$$K_3 = \frac{1.78}{(1.7\underline{/28.1})(0.943\underline{/122})(1.6\underline{/90})} = 0.693\underline{/120}$$

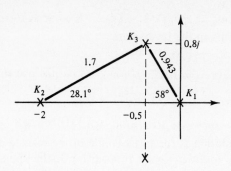

Figure 1.18 Example 1.9.4.

From (1.53) then, the response is

$$c(t) = 1 - 0.307e^{-2t} + 1.386e^{-0.5t}\cos(0.8t + 120°) \qquad (1.57)$$

since, using Fig. 1.17,

$$-\zeta\omega_n = -0.5 = \text{real part of complex poles}$$
$$\omega_n\sqrt{1 - \zeta^2} = 0.8 = \text{imaginary part}$$

Note that $c(0) = 1 - 0.307 + 1.386\cos 120° = 0$.

1.10 CONCLUSION

In this chapter a general introduction has been given first, including physical discussion of some fundamental features of control system behavior. A level control example led to a common block diagram configuration.

Laplace transforms led to the transfer function description of dynamic behavior and to the introduction of the s-plane and the concepts of poles and zeros.

The application of transfer functions and transforms to the calculation of the response $c(t)$ to an input $r(t)$ and initial conditions has been illustrated for a number of examples, including closed-loop control systems. This calculation of transient responses is based on inverse Laplace transformation by partial fraction expansion.

PROBLEMS

1.1. Set up a basic block diagram for a ship's autopilot system. Describe the blocks in the diagram, indicating what hardware might be used to indicate the desired heading and how the actual heading might be made to agree with it.

1.2. Set up a basic block diagram for a position control system in which an electric motor must control angular shaft position. A potentiometer is used to measure the position.

1.3. The output y and input x of a block in a block diagram are related by $y = 2x + 0.5x^3$.

    (a) Find the values of the output for steady-state operation at the operating points

$$\text{(i) } x_0 = 0 \qquad \text{(ii) } x_0 = 1 \qquad \text{(iii) } x_0 = 2$$

    (b) Obtain linearized models for small variations about these operating points. For these models, redefine x and y to represent variations about operating-point values.

1.4. The output z of a block in a block diagram is a function of two inputs, x and y, according to

$$z = 5x - 3y + x^3 + y^3 + x^2y$$

Derive a linearized model for small variations about the operating point $x_0 = 1$, $y_0 = 2$. For this model, redefine x, y, and z to represent only the variations about operating-point values.

1.5. Find the Laplace transform $R(s)$ of a signal $r(t)$ if $r(t) = 2t$, $0 \leqslant t < 1$; $r(t) = 2$, $t \geqslant 1$.

1.6. In a pipeline of length L with fluid velocity v, the temperature T_i of the inflow undergoes a step change of magnitude A at $t = 0$. Neglecting heat loss and with variables representing only changes:
  (a) Describe the outflow temperature T_o and its transform.
  (b) Give the transfer function relating T_i and T_o.

1.7. Use the final value theorem to determine the steady-state value $c_{ss} = \lim_{t \to \infty} c(t)$ of $c(t)$ for the following forms of the transform $C(s)$ of $c(t)$:

  (a) $C(s) = \dfrac{s + 11}{(s + 2)(s + 5)}$ (c) $C(s) = \dfrac{4s^2 + 5s + 6}{s(2s^3 + 7s^2 + 13s + 2)}$

  (b) $C(s) = \dfrac{2}{s(s + 3)}$ (d) $C(s) = \dfrac{2s^2 + 5s + 1}{s^3 + 2s^2 + 3s + 4}$

1.8. For the following differential equations and initial conditions, write the transforms of the solutions:
  (a) $\ddot{c} + 6\dot{c} + 13c = 5u(t)$, $c_0 = 1$, $\dot{c}_0 = 4$
  (b) $\ddot{c} + 3\dot{c} + 4c = 7u(t) + 2t$, $c_0 = 1$, $\dot{c}_0 = 2$
  (c) $\ddot{c} + 3\dot{c} + 4c = 6 \sin \omega t$, $c_0 = 4$, $\dot{c}_0 = 5$
  (d) $\dot{c} + 2c = u(t)$, $c_0 = 1$

1.9. Calculate the unit step response of a system with the transfer function

$$G(s) = \frac{5(1 - 0.4s)}{(s + 1)(0.2s + 1)}$$

1.10. Calculate the response to a decaying exponential input $r(t) = e^{-t}$ of a system with the transfer function

$$G(s) = 3\frac{s^2 + 9s + 18}{s^2 + 6s + 8}$$

1.11. For the systems described by the differential equations with initial conditions

  (i) $\dot{c} + 6c = r$, $c(0) = 1$ (ii) $\ddot{c} + 7\dot{c} + 10c = r$, $c(0) = 1$, $\dot{c}(0) = 4$

determine:
  (a) The transfer functions C/R.
  (b) The transient responses to the initial conditions.

1.12. For the systems shown in Fig. P1.12:
  (a) Find the closed-loop transfer functions.
  (b) Calculate the responses to unit step inputs.
  The types of transfer functions given here and in other problems approximate the behavior of many physical systems.

1.13. Calculate the unit step responses of the systems shown in Fig. P1.13.

1.14. For the systems in Fig. P1.13, determine the transient responses to the initial conditions:

$$\text{(i) } c(0) = 1 \qquad \text{(ii) } c(0) = 1 \qquad \dot{c}(0) = 1$$

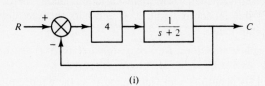

(i)

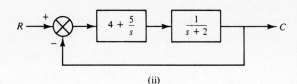

(ii)   **Figure P1.12**

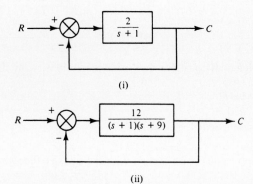

(i)

(ii)   **Figure P1.13**

1.15. Plot the pole–zero patterns of systems described by the following transfer functions.

$$\text{(a) } \frac{K(2s + 1)}{s(4s + 1)(s + 3)} \qquad \text{(b) } \frac{K(s^2 + 2s + 2)}{(s + 2)(s^2 + 2s + 10)}$$

What are the root locus gains of these transfer functions?

1.16. The pole–zero pattern of a system with root locus gain 80 is shown in Fig. P1.16. What is its transfer function?

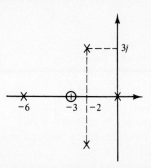

Figure P1.16

1.17. Use graphical calculation of the residues to find the unit step response of the following systems.

$$\text{(a) } G(s) = \frac{4}{(2s + 1)(4s + 8)} \qquad \text{(b) } G(s) = \frac{4(s + 1)}{(2s + 1)(4s + 8)}$$

1.18. For the system

$$G(s) = \frac{2(s^2 + 9s + 19)}{s^2 + 6s + 8}$$

(a) What is the system characteristic polynomial?
(b) What is its characteristic equation?
(c) Find the system poles and zeros.
(d) Plot the system pole–zero pattern.
(e) Plot the pole–zero pattern of $C(s) = G(s)R(s)$ if the input $r(t)$ is a decaying exponential e^{-t}.
(f) Find $c(t)$ by graphical residue determination.

1.19. Find the unit ramp response of the system in Problem 1.17(a).

1.20. Find the unit ramp response for Problem 1.17(b).

1.21. Calculate the response of the system

$$G(s) = \frac{4}{2s^2 + 2s + 8}$$

to a unit ramp input. Note that the graphical residue rule can still be used at poles of $C(s) = G(s)R(s)$ other than the repeated pole, since moving the two poles slightly apart should not affect such results materially.

1.22. Calculate the unit step response of

$$G(s) = \frac{1}{(s + 2)^2(s + 1)}$$

1.23. Calculate the unit step response of the system

$$G(s) = \frac{54}{(2s + 6)(s^2 + 3s + 9)}$$

1.24. Calculate the unit step response for

$$G(s) = \frac{1.5(2s + 6)}{s^2 + 3s + 9}$$

2

Transfer Function Models of Physical Systems

2.1 INTRODUCTION

In this chapter, differential equations are derived to describe the dynamic behavior of mechanical, electrical, thermal, and fluid systems. These are used to obtain transfer functions between selected variables. The same differential equations can also be formulated into state-space models. This alternative, mentioned in Section 1.6, is discussed in Chapters 11 and 12.

The chapter will concentrate on subsystem blocks, with attention to the restrictions that must be observed when separating a system into blocks. The modeling of systems that include feedback control is discussed in Chapter 3. This will show that the block diagram structure can be far from obvious and may need to be derived directly from the system equations. For such systems the precise nature of the feedback may be evident only from this block diagram.

It is useful to note that a model should not be expected to be evident "by inspection" of a schematic diagram. Rather, it usually will emerge gradually from equations written for parts of the system, in rather arbitrary order.

2.2 MECHANICAL SYSTEMS

Figure 2.1 shows common elements of mechanical systems with linear and rotational motion, together with the equations used to describe them. In Fig. 2.1 and the diagrams to follow, the arrows identify the positive directions of the associated variables.

1. *Springs:* These exist in many design configurations and materials. For linearized models the force F (torque T) is taken to be proportional to linear deflection x (angular deflection θ). Note that in linearized models F and x represent the variations of force and deflection about the operating point values. For larger

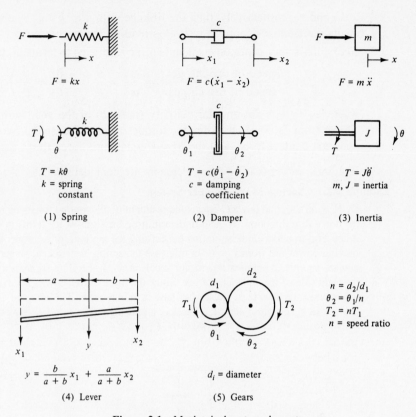

Figure 2.1 Mechanical system elements.

variations of deflections the behavior can sometimes be very nonlinear. The spring constant k for nonlinear springs changes with the operating point.

2. *Dampers or dashpots:* These generate a damping force F (torque T) proportional to the difference $\dot{x}_1 - \dot{x}_2$ $(\dot{\theta}_1 - \dot{\theta}_2)$ of the velocities across the damper, and in the opposite direction. In practice, the friction in mechanical systems may differ greatly from the viscous friction of this linear damper model. For dry friction, or Coulomb friction, the force or torque is opposite the velocity difference, but independent of its magnitude. The approximate linearization of such behavior is a subject in the study of nonlinear systems.

3. *Mass and inertia:* By Newton's second law, force F (torque T) equals mass m (inertia J) times acceleration $\ddot{x}$ $(\ddot{\theta})$.

4. *Lever mechanism:* For small enough angles from horizontal, the total motion y equals the sum of the motion due to x_1 with $x_2 = 0$ and that due to x_2 with $x_1 = 0$. It is useful to observe that the lever is a mechanical implementation of a summing junction in a block diagram and is in fact often used for this purpose. If $a = b$, then $y = 0.5\,(x_1 - x_2)$ if the direction of x_2 is reversed. If the input and output of a system are available in mechanical form and are applied

to x_1 and x_2, respectively, then the linkage with $a = b$ and x_2 opposite implements the feedback loop, as well as providing the system error $e = x_1 - x_2$.

5. *Gears:* This very common element is often identified in terms of its gear ratio n.

$$n = \frac{\text{speed of driving gear}}{\text{speed of driven gear}} = \frac{\omega_1}{\omega_2} = \frac{\dot{\theta}_1}{\dot{\theta}_2} \left(= \frac{\theta_1}{\theta_2} = \frac{\ddot{\theta}_1}{\ddot{\theta}_2} \right) \qquad (2.1)$$

where $\omega_i = \dot{\theta}_i$ is the angular velocity (rad/sec) of the gear with diameter d_i. The relation $T_2 = nT_1$ between the torques arises because the two gears have a common contact force, and the torque equals this force times the gear radius.

Examples incorporating these elements are considered next.

Example 2.2.1   Spring–Mass–Damper System

By Newton's second law, $m\ddot{x}$ equals the resultant of all external forces on m in Fig. 2.2 in the downward direction. To help in determining the signs of the terms in such problems, it is useful to make any assumption concerning the motion, for example, that the mass is moving downward from $x = 0$. In this case the spring is stretched, so spring force kx is upward and hence opposes downward acceleration. It therefore receives a minus sign on the right side of the equation for $m\ddot{x}$. Since the mass moves down, the damping force $c\dot{x}$ is upward, and this term must also have a minus sign. The external force $f(t)$ helps downward acceleration and therefore has a plus sign. The resulting equation is

$$m\ddot{x} = -kx - c\dot{x} + f(t)$$

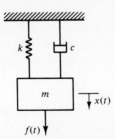

Figure 2.2   Spring–mass–damper system.

Rearranging gives the differential equation of motion in the usual form:

$$m\ddot{x} + c\dot{x} + kx = f(t) \qquad (2.2)$$

Observe that the effects of gravity do not appear, so turning the system upside down will not affect the equation. This is done by choosing $x = 0$ at the position of static equilibrium, where the weight mg is counterbalanced by a spring force. Assume that the transfer function for the response $x(t)$ to $f(t)$ is desired. With zero initial conditions, the transform of (2.2) is

$$(ms^2 + cs + k)X(s) = F(s)$$

and the transfer function of interest is

$$\frac{X(s)}{F(s)} = \frac{1}{ms^2 + cs + k} \qquad (2.3)$$

Example 2.2.2   Two-Mass System

The system in Fig. 2.3 can represent a dynamic absorber, where a relatively small mass m_1 is attached to a main mass m via spring k_1 and damper c_1 to reduce vibrations x due to force f. Assume, say, that m and m_1 both move to the right from the zero positions, m

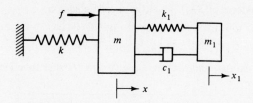

Figure 2.3   Dynamic absorber.

farther and faster than m_1. Then spring force $k_1(x - x_1)$ "opposes" m and "helps" m_1, and damper force $c_1(\dot{x} - \dot{x}_1)$ has the same effect. Hence the differential equations of motion become

$$m\ddot{x} = -kx - k_1(x - x_1) - c_1(\dot{x} - \dot{x}_1) + f$$
$$m_1\ddot{x}_1 = k_1(x - x_1) + c_1(\dot{x} - \dot{x}_1)$$

or

$$m\ddot{x} + c_1\dot{x} + (k + k_1)x = c_1\dot{x}_1 + k_1x_1 + f$$
$$m_1\ddot{x}_1 + c_1\dot{x}_1 + k_1x_1 = c_1\dot{x} + k_1x \tag{2.4}$$

The transfer function showing the effect of f on x is of interest. This requires the elimination of x_1, an algebraic operation if (2.4) are first transformed:

$$(ms^2 + c_1s + k + k_1)X(s) = (c_1s + k_1)X_1(s) + F(s)$$
$$(m_1s^2 + c_1s + k_1)X_1(s) = (c_1s + k_1)X(s) \tag{2.5}$$

Solving $X_1(s)$ from the second equation, substituting it into the first, and rearranging would give the transfer function X/F.

Example 2.2.3   Rotating Drive System

Figure 2.4 indicates a drive system, with c representing a friction coupling and torsion spring k the twisting of a long shaft due to torque. Angle θ_1 is taken to be the input and θ_3 the output. Any other variables may be introduced to facilitate the writing of the equations, such as θ_2, the angle to the right as well as to the left of J_1.

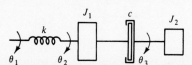

Figure 2.4   Rotating drive system.

The approach of the preceding examples could be used, but an alternative is often convenient. Equations are written in order, starting at the input. To help visualize signs, assume that $\theta_1 > \theta_2, \dot{\theta}_2 > \dot{\theta}_3$. The shaft torque $k(\theta_1 - \theta_2)$ accelerates inertia J_1 and supplies the damping torque:

$$k(\theta_1 - \theta_2) = J_1\ddot{\theta}_2 + c(\dot{\theta}_2 - \dot{\theta}_3)$$

The damping torque in turn accelerates inertia J_2:

$$c(\dot{\theta}_2 - \dot{\theta}_3) = J_2\ddot{\theta}_3$$

Rearranging yields the differential equations

$$J_1\ddot{\theta}_2 + c\dot{\theta}_2 + k\theta_2 = c\dot{\theta}_3 + k\theta_1$$
$$J_2\ddot{\theta}_3 + c\dot{\theta}_3 = c\dot{\theta}_2 \tag{2.6}$$

and the transformed equations

$$(J_1s^2 + cs + k)\theta_2(s) = cs\theta_3(s) + k\theta_1(s)$$
$$(J_2s^2 + cs)\theta_3(s) = cs\theta_2(s)$$

where $\theta_i(s)$ is the transform of $\theta_i(t)$. From the second equation,

$$\theta_2(s) = \left(\frac{J_2}{c}s + 1\right)\theta_3(s)$$

Substituting this into the first and rearranging gives the transfer function

$$\frac{\theta_3(s)}{\theta_1(s)} = \frac{k}{(J_1J_2/c)s^3 + (J_1 + J_2)s^2 + (J_2k/c)s + k} \tag{2.7}$$

Example 2.2.4   Systems with Gears

Although not necessary, it is often convenient for analysis to replace a system with gears by a dynamically equivalent system without gears. Figure 2.5 shows the derivation of the equivalent inertia, spring, and damping, identified as J_e, k_e, and c_e, on the first shaft which can replace J, k, and c on the second shaft. In addition to the relation $T_2 = nT_1$ from Fig. 2.1, (2.1) is used and the fact that $\ddot{\theta}_i = \dot{\omega}_i$. *Note that in each case the equivalent element is obtained by dividing the original element by the square of the speed ratio n.*

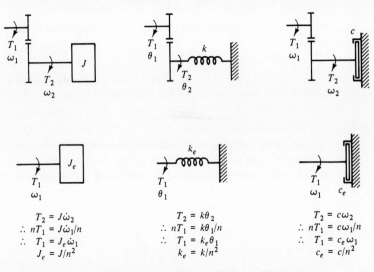

Figure 2.5   Equivalent systems without gears.

Example 2.2.5   Nongeared Equivalent System

Using Fig. 2.5, the system in Fig. 2.6(a) can be replaced immediately by its nongeared equivalent in Fig. 2.6(b), where J_e, k_e, and c_e are as given in Fig. 2.5.

2.3 ELECTRICAL SYSTEMS: CIRCUITS

Figure 2.7 summarizes some important results for the modeling of electrical circuits. Figure 2.7(a) gives the voltage–current relations of the basic elements in the time domain and the Laplace domain, assuming zero initial conditions. In the general transformed relation $V = IZ$, Z is the impedance. In an *ideal* voltage source, the voltage difference across the terminals is independent of the current through them. In a *real* voltage source, the voltage v_s across the terminals decreases with increasing current due to the voltage drop across the internal impedance of the source, which may be a resistance R. In the series circuit shown in Fig. 2.7(b), the same current passes

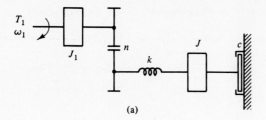

(a)

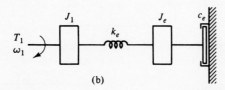

(b)   **Figure 2.6**   Example 2.2.5.

through all three elements, and the total voltage is the sum of the voltage drops across each. In the parallel circuit the voltage drop is the same for each element, and the current is the sum of the individual currents.

The use of these results and of Kirchhoff's laws summarized in Fig. 2.7(c) will be illustrated by application to the examples shown in Fig. 2.8. The transfer functions $E_o(s)/E_i(s)$ between inputs $e_i(t)$ and outputs $e_o(t)$ are desired.

It is important to recognize that to derive these transfer functions the current through the output terminals must be assumed to be negligibly small. Otherwise, the size of this current would affect e_o, so E_o/E_i would not be uniquely defined. In such a case a transfer function can be derived from differential equations written for the combination of the circuit and the so-called "load" connected to its output terminals. This illustrates a key general condition when subdividing a system into subsystem blocks:

The division of systems into blocks must be such that each block does not *load,* that is, affect the output of, the preceding block.

Example 2.3.1   Common Electrical Circuits [Fig. 2.8(a) to (e)]

The characteristics represented by these circuits are used extensively as controllers to improve the performance of feedback control systems. The transfer functions E_o/E_i can be found by use of the results in Fig. 2.7(b). With no current through the output terminals, all are in effect voltage dividers, in which e_o is a fraction of the voltage e_i, determined by the current i through the input terminals caused by e_i. Consider each in turn:

Figure 2.8(a): Simple Lag Network

$$E_i = I\left(R + \frac{1}{Cs}\right)$$

$$\frac{E_o}{E_i} = \frac{1}{RCs + 1} \qquad (2.8a)$$

$$E_o = I\frac{1}{Cs}$$

Figure 2.8(b): Transient-Lead Network

$$E_i = I\left(R + \frac{1}{Cs}\right)$$

$$\frac{E_o}{E_i} = \frac{RCs}{RCs + 1} \qquad (2.8b)$$

$$E_o = IR$$

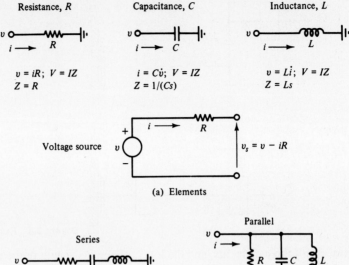

Resistance, R Capacitance, C Inductance, L

$v = iR;\ V = IZ$
$Z = R$

$i = C\dot{v};\ V = IZ$
$Z = 1/(Cs)$

$v = L\dot{i};\ V = IZ$
$Z = Ls$

Voltage source $v_s = v - iR$

(a) Elements

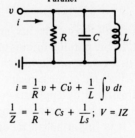

Series

$v = iR + \dfrac{1}{C}\int i\,dt + L\dot{i}$

$Z = R + \dfrac{1}{Cs} + Ls;\ V = IZ$

In general

$Z = Z_1 + Z_2 + Z_3 + \cdots$

Parallel

$i = \dfrac{1}{R}v + C\dot{v} + \dfrac{1}{L}\int v\,dt$

$\dfrac{1}{Z} = \dfrac{1}{R} + Cs + \dfrac{1}{Ls};\ V = IZ$

$\dfrac{1}{Z} = \dfrac{1}{Z_1} + \dfrac{1}{Z_2} + \dfrac{1}{Z_3} + \cdots$

(b) Circuits

Loop method
 Voltage equations:
Sum of voltage drops
around a closed loop = 0

Node method
 Current equations:
Sum of currents at a circuit
node or junction = 0

(c) Kirchhoff's laws

Figure 2.7   Electrical circuit fundamentals.

Figure 2.8(c): Phase-Lag Network

$$E_i = I\left(R_1 + R + \frac{1}{Cs}\right)$$

$$E_o = I\left(R + \frac{1}{Cs}\right)$$

$$\frac{E_o}{E_i} = \frac{\tau s + 1}{(\tau/\alpha)s + 1} \qquad (2.8c)$$

where $\tau = RC \quad \alpha = \dfrac{R}{R_1 + R}$

Figure 2.8(d): Phase-Lead Network

$$E_i = I\left(\frac{1}{(1/R_1) + C_1 s} + R\right)$$

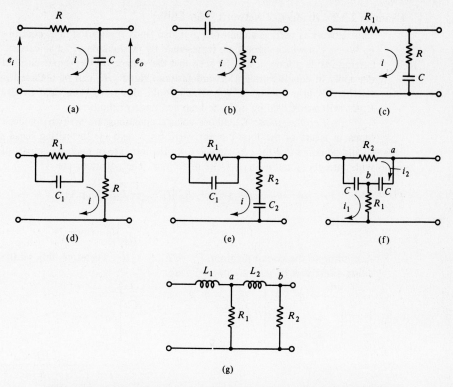

Figure 2.8 Electrical networks: (a) simple lag; (b) transient lead; (c) phase lag; (d) phase lead; (e) lag–lead (notch); (f) bridged-T network; (g) RL ladder network.

since, from Fig 2.7(b) the equivalent impedance Z_1 of the parallel impedances $Z_r = R_1$ and $Z_c = 1/(C_1 s)$ is given by

$$\frac{1}{Z_1} = \frac{1}{Z_r} + \frac{1}{Z_c} \quad \text{or} \quad Z_1 = 1 \left/ \left(\frac{1}{Z_r} + \frac{1}{Z_c} \right) \right.$$

Hence

$$E_i = I \left(\frac{R_1 + R + RR_1 C_1 s}{1 + R_1 C_1 s} \right)$$

$$E_o = IR$$

$$\frac{E_o}{E_i} = \alpha \frac{\tau s + 1}{\alpha \tau s + 1} \quad (2.8d)$$

where $\tau = R_1 C_1 \quad \alpha = \dfrac{R}{R_1 + R}$

Figure 2.8(e): Lag–Lead (Notch) Network

$$E_o = I \left(\frac{1}{(1/R_1) + C_1 s} + R_2 + \frac{1}{C_2 s} \right)$$

$$E_o = I \left(R_2 + \frac{1}{C_2 s} \right)$$

$$\frac{E_o}{E_i} = \frac{(\tau_1 s + 1)(\tau_2 s + 1)}{\tau_1 \tau_2 s^2 + (\tau_1 + \tau_2 + \tau_{12})s + 1} \quad (2.8e)$$

where $\tau_1 = R_1 C_1 \quad \tau_2 = R_2 C_2 \quad \tau_{12} = R_1 C_2$

Example 2.3.2   Bridged-T Network [Fig. 2.8(f)]

This network is used as a controller and also finds application in ac control systems, that is, systems in which signals are represented by modulation of an ac carrier. With e_o occurring inside a loop, it may be seen that the voltage-divider approach used above does not apply in equally straightforward fashion. But E_o/E_i can be obtained by the use of Kirchhoff's laws in Fig. 2.7(c), which could also have been used in Example 2.3.1. E_o/E_i will be derived by both the loop and node methods.

(a) *Loop method:* Kirchhoff voltage equations are written for each of the two loops in terms of the loop current variables i_1 and i_2. In writing these equations, it should be noted that the capacitance C on the left side of node b is a part of both loops, with currents i_1 and i_2 in opposite directions. The equations are

$$E_i = \left(\frac{1}{Cs} + R_1\right)I_1 - \frac{1}{Cs}I_2 \quad \text{(loop 1)}$$

$$0 = -\frac{1}{Cs}I_1 + \left(R_2 + \frac{2}{Cs}\right)I_2 \quad \text{(loop 2)}$$

According to the circuit diagram, $E_o = E_i - I_2 R_2$. Therefore, this set is solved for I_2 either directly or by use of *Cramer's rule:*

$$I_2 = \begin{vmatrix} \frac{1}{Cs} + R_1 & E_i \\ -\frac{1}{Cs} & 0 \end{vmatrix} \Bigg/ \begin{vmatrix} \frac{1}{Cs} + R_1 & -\frac{1}{Cs} \\ -\frac{1}{Cs} & \frac{2}{Cs} + R_2 \end{vmatrix}$$

$$= \frac{E_i Cs}{1 + (2R_1 + R_2)Cs + R_1 R_2 C^2 s^2}$$

Then, since $E_o = E_i - I_2 R_2$, substitution of I_2 yields

$$\frac{E_o}{E_i} = \frac{1 + 2R_1 Cs + R_1 R_2 C^2 s^2}{1 + (2R_1 + R_2)Cs + R_1 R_2 C^2 s^2} \quad (2.8f)$$

(b) *Node method:* Here the Kirchhoff current equations are written in terms of voltage variables at each of the circuit nodes. In Fig. 2.8(f), the unknown node voltages are $E_a(=E_o)$ and E_b, and the equations are

$$\frac{E_a - E_i}{R_2} + \frac{E_a - E_b}{1/(Cs)} = 0 \quad \text{(node } a\text{)}$$

$$\frac{E_b - E_a}{1/(Cs)} + \frac{E_b - E_i}{1/(Cs)} + \frac{E_b}{R_1} = 0 \quad \text{(node } b\text{)}$$

Rearranging yields

$$\left(\frac{1}{R_2} + Cs\right)E_a - CsE_b = \frac{1}{R_2}E_i$$

$$-CsE_a + \left(\frac{1}{R_1} + 2Cs\right)E_b = CsE_i$$

The solution for $E_a(=E_o)$, directly or by Cramer's rule, yields E_o/E_i as in part (a) and, in the case of this example, in a somewhat more direct fashion.

Example 2.3.3   Ladder Network [Fig. 2.8(g)]

Using the node method, current equations are written for the circuit nodes a and b:

$$\frac{E_a - E_i}{L_1 s} + \frac{E_a}{R_1} + \frac{E_a - E_b}{L_2 s} = 0 \qquad \frac{E_b - E_a}{L_2 s} + \frac{E_b}{R_2} = 0$$

The solution for $E_b = E_o$ gives

$$\frac{E_o}{E_i} = \frac{R_1R_2}{L_1L_2s^2 + (L_1R_1 + L_1R_2 + L_2R_1)s + R_1R_2}$$

2.4 ELECTROMECHANICAL SYSTEMS: MOTORS AND GENERATORS

Schematic diagrams of several arrangements of motors and generators are shown in Fig. 2.9. In all cases the motor load is assumed to consist of an inertia J and a damper with damping constant B. Similar to Example 2.2.3, the developed motor torque T accelerates the inertia J and overcomes damping B. Motor shaft position θ and torque T are then related by

$$T(t) = J\ddot{\theta}(t) + B\dot{\theta}(t) \qquad T(s) = s(Js + B)\theta(s) \tag{2.9}$$

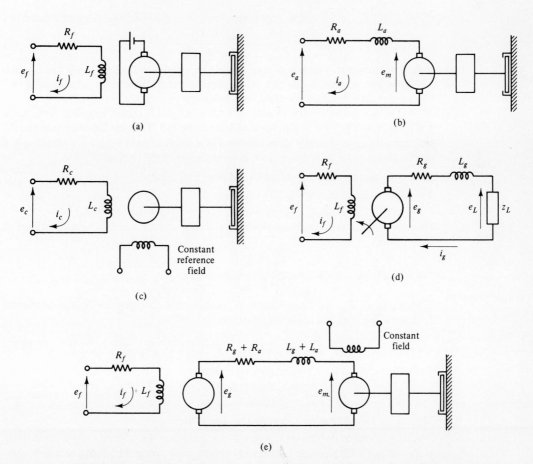

Figure 2.9 Motors and generators: (a) field-controlled dc motor; (b) armature-controlled dc motor; (c) two-phase ac servomotor; (d) dc generator; (e) motor–generator set.

For convenience, the same variables are used in the time and Laplace domains. The identifier (t) or (s) will generally be omitted if it is evident from the context. Each of the systems in Fig. 2.9 will be considered in turn.

Example 2.4.1   Field-Controlled DC Motor [Fig. 2.9(a)]

The equation for the field loop is

$$e_f = R_f i_f + L_f \dot{i}_f \qquad E_f = (R_f + L_f s) I_f$$

With constant armature voltage, the developed motor torque T in (2.9) can be taken to be proportional to field current:

$$T = K_t i_f \qquad T = K_t I_f \qquad K_t = \text{motor torque constant}$$

Eliminating I_f and T between these transformed equations and (2.9) yields the desired transfer function between applied field voltage e_f and shaft position θ:

$$\frac{\theta}{E_f} = \frac{K_t/(R_f B)}{s(T_m s + 1)(T_f s + 1)} \tag{2.10}$$

where $T_m = J/B$ = motor time constant
$T_f = L_f/R_f$ = field time constant

Often $T_f \ll T_m$, and a satisfactory approximation in the operating range of interest is

$$\frac{\theta}{E_f} = \frac{K_t/(R_f B)}{s(T_m s + 1)} \tag{2.11}$$

Note that the transfer function was derived for the combination of the motor and its load. This load affects motor speed (that is, it loads the motor), so a series connection of two blocks with individual transfer functions would be incorrect.

The factor s in the denominator of (2.10) and (2.11) should be noted. From Table 1.6.1, dividing $F(s)$ by s is equivalent to integrating $f(t)$. Thus the factor s represents the fact that a motor is basically an integrator: For a constant input e_f, it has a shaft angle θ that increases at a constant rate, so θ is proportional to the integral of e_f.

Example 2.4.2   Armature-Controlled DC Motor [Fig. 2.9(b)]

The armature loop is described by

$$e_a = R_a i_a + L_a \dot{i}_a + e_m \qquad E_a = (R_a + L_a s) I_a + E_m$$

Here the counter emf (electromotive force) voltage can be taken to be proportional to shaft speed,

$$e_m = K_e \dot{\theta} \qquad E_m = K_e s \theta$$

and the developed torque proportional to current i_a,

$$T = K_t i_a \qquad T = K_t I_a$$

Eliminating I_a, E_m, and T between these equations and (2.9) permits the desired transfer function to be arranged in the common form

$$\frac{\theta}{E_a} = \frac{1/K_e}{s(T_a T_m s^2 + (T_m + \gamma T_a)s + \gamma + 1)} \tag{2.12}$$

where $T_m = JR_a/(K_e K_t)$ = motor time constant
$T_a = L_a/R_a$ = armature time constant
$\gamma = BR_a/(K_e K_t)$ = damping factor

Example 2.4.3   Two-Phase AC Servomotor [Fig. 2.9(c)]

Fixed and variable magnitude ac voltages are applied to the reference and control fields, respectively. A 90° phase shift arranged between these voltages is made positive or negative depending on the desired direction of rotation. The control field is described by

$$e_c = R_c i_c + L_c \dot{i_c} \qquad E_c = (R_c + L_c s) I_c \qquad (2.13)$$

The developed motor torque T can be taken to be proportional to i_c and to decrease proportionally with increasing speed, and it is described by

$$T = K_c i_c - K_\omega \dot\theta \qquad T = K_c I_c - K_\omega s\theta \qquad (2.14)$$

This dependence on speed is assumed to be the same under dynamic conditions as for the steady-state torque–speed motor characteristic curves indicated in Fig. 2.10. Eliminating T and I_c between these equations and (2.9) and rearranging give the transfer function

$$\frac{\theta}{E_c} = \frac{K}{s(T_m s + 1)(T_c s + 1)} \qquad (2.15)$$

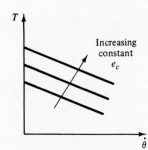

T

Increasing
constant
e_c

$\dot\theta$ **Figure 2.10**   AC motor characteristics.

where $T_m = J/(B + K_\omega)$ = motor time constant
   $T_c = L_c/R_c$ = electrical time constant
   $K = K_c/[R_c(B + K_\omega)]$ = motor constant
Often simplification to the form (2.11) is again satisfactory: $K/[s(T_m s + 1)]$.

Example 2.4.4   DC Generator [Fig. 2.9(d)]

The field loop equation is

$$e_f = R_f i_f + L_f \dot{i_f} \qquad E_f = (R_f + L_f s) I_f$$

The developed generator voltage e_g can be assumed to be proportional to field current:

$$e_g = K_g i_f \qquad E_g = K_g I_f$$

The voltage e_L across the load is given by

$$E_L = Z_L I_g \qquad Z_L = \text{load impedance}$$

and the generator loop is described by

$$e_g = R_g i_g + L_g \dot{i} + e_L \qquad E_g = (L_g s + R_g + Z_L) I_g$$

Hence

$$\frac{E_g}{E_f} = \frac{K_g}{L_f s + R_f} \qquad \frac{E_L}{E_g} = \frac{Z_L}{L_g s + R_g + Z_L} \qquad \frac{E_L}{E_f} = \frac{E_L}{E_g}\frac{E_g}{E_f} \qquad (2.16)$$

Example 2.4.5   Motor–Generator Set [Fig. 2.9(e)]

The generator serves as a rotating power amplifier. $\theta/E_f = (\theta/E_g)(E_g/E_f)$ is obtained directly by appropriate substitutions in (2.12) and (2.16).

2.5 THERMAL SYSTEMS

As in the preceding sections, thermal system elements are discussed first.

Thermal resistance [Fig. 2.11(a)].   A wall of area A separates regions with temperatures T_1 and T_2. The heat flow rate q, in units of heat per unit of time, say Btu's per second, is proportional to the temperature difference $T_1 - T_2$ and to the area A and flows toward the lowest temperature. The constant of proportionality is the heat transfer coefficient h. In the case illustrated, this is an effective coefficient that combines the effects of heat convection at the surfaces and heat conduction through the wall. The equation would also represent heat convection across a single surface.

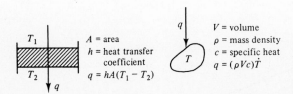

T_1

A = area
h = heat transfer
   coefficient
$q = hA(T_1 - T_2)$

T_2

q

q

T

V = volume
ρ = mass density
c = specific heat
$q = (\rho V c)\dot{T}$

(a) Resistance   (b) Capacitance

Figure 2.11   (a) Thermal resistance; (b) thermal capacitance.

For heat conduction through a wall with surface temperatures T_1 and T_2 and thickness d, the heat flow is proportional to the temperature gradient $(T_1 - T_2)/d$:

$$q = \frac{kA(T_1 - T_2)}{d} = \frac{k}{d} A(T_1 - T_2) \tag{2.17}$$

where k is the thermal conductivity. Hence the equivalent heat transfer coefficient is k/d.

To identify these relations in terms of a thermal equivalent of an electrical resistance, the equation in Fig. 2.11(a) is written as

$$T_1 - T_2 = qR_t \qquad R_t = \frac{1}{hA} \tag{2.18}$$

With $T_1 - T_2$ analogous to voltage drop v, and q to current i, R_t becomes the thermal resistance.

Thermal capacitance [Fig. 2.11(b)].   Let q be the net heat flow rate into a volume V of a material with mass density ρ and specific heat c (= heat required to raise the temperature of a unit mass by $1°$). This net inflow q of heat per second must equal the change per second (that is, the rate of change) of heat stored in V. Since the mass is ρV, the heat required for a $1°$ rise of temperature is $\rho V c$, and hence the heat stored at a temperature T is $\rho V c T$. Assuming ρ, V, and c to be constant, its rate of change is $\rho V c \dot{T}$. Hence follows the equation shown in Fig. 2.11(b).

From the equation $i = C\dot{v}$ in Fig. 2.7 for an electrical capacitance follows immediately the equivalent notion of a thermal capacitance C_t:

$$q = C_t \dot{T} \qquad C_t = \rho V c \tag{2.19}$$

It is seen that C_t is the heat required for a $1°$ temperature rise, since c is this heat per unit mass and the mass is ρV.

Some examples of thermal systems follow.

Example 2.5.1   Process Control

Figure 2.12 shows a tank of volume V filled with an incompressible fluid of mass density ρ and specific heat c. Volume flow rates entering and leaving are f_i and f_o. T_i is the temperature of the inflow. It is assumed that the tank is well stirred so that the outlet temperature equals the tank temperature T.

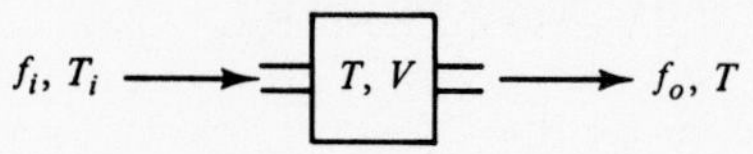

$$f_i, T_i \longrightarrow \boxed{T, V} \longrightarrow f_o, T$$

Figure 2.12   Process flow.

The tank is filled and the fluid incompressible, so $f_i = f_o$, and the mass flow rate entering and leaving is $f_i\rho$. Hence the heat inflow rate is $f_i\rho cT_i$, the outflow rate $f_i\rho cT$, and the net inflow rate $f_i\rho c(T_i - T)$. This is the net inflow of heat per second and so must equal the change per second (that is, the rate of change) of heat stored in the tank. This stored heat is $V\rho cT$, and its rate of change is $V\rho c\dot{T}$. Hence the system is described by the differential equation $V\rho c\dot{T} = f_i\rho c(T_i - T)$, or

$$\frac{V}{f_i}\dot{T} + T = T_i \tag{2.20}$$

The transform is $[(V/f_i)s + 1]T(s) = T_i(s)$, so the following *simple lag* transfer function relates T_i and T:

$$\frac{T(s)}{T_i(s)} = \frac{1}{(V/f_i)s + 1} \tag{2.21}$$

Example 2.5.2   Space Heating

In Fig. 2.13, let T be the difference with a constant ambient temperature. By (2.18), the heat loss q_o to ambient can be modeled by $q_o = T/R_t$, where R_t is the thermal resistance. If q_i is the heat inflow rate from an electrical heater, the net inflow $(q_i - T/R_t)$ per second must equal $C_t\dot{T}$, the change per second or rate of change of stored heat, where C_t is the thermal capacitance. Hence the behavior is modeled by the differential equation

$$R_tC_t\dot{T} + T = R_tq_i \tag{2.22}$$

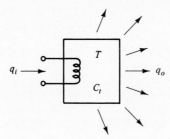

Figure 2.13   Space heating.

Therefore, the effect of heat flow q_i on the temperature T is approximated by the transfer function

$$\frac{T(s)}{Q_i(s)} = \frac{R_t}{R_tC_ts + 1} \tag{2.23}$$

As in Example 2.5.1 and the electrical RC circuit in Example 2.3.1, the behavior is described by a simple lag transfer function.

Example 2.5.3   Three-Capacitance Oven

Figure 2.14 shows an oven in which allowance is made for heat loss to the ambient temperature T_a and for the heating up of an internal material with capacitance C_m and uniform temperature T_m. The heat loss occurs via surface area A_o with heat transfer coefficient h_o and the heating up via surface A_m with heat transfer coefficient h_m. The temperature T_m is measured by a sensor with a significant capacitance C_s, heated via a surface of area A_s and heat transfer coefficient h_s.

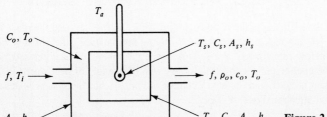

Figure 2.14   Three-capacitance oven.

The approach of the preceding examples is used to equate the net heat flow rate to the rate of change of heat for each capacitance in turn:

$$C_o \dot{T}_o = f\rho_o c_o (T_i - T_o) - A_o h_o (T_o - T_a) - A_m h_m (T_o - T_m)$$
$$C_m \dot{T}_m = A_m h_m (T_o - T_m) \tag{2.24}$$
$$C_s \dot{T}_s = A_s h_s (T_m - T_s)$$

The second of these equations does not include the heat loss term $-A_s h_s (T_m - T_s)$ to the sensor, on the assumption that it is relatively negligible. The last equation immediately gives the transfer function relating T_m and its value as measured by the sensor:

$$\frac{T_s(s)}{T_m(s)} = \frac{1}{\tau_s s + 1} \qquad \tau_s = \frac{C_s}{A_s h_s} \tag{2.25}$$

This is again a simple lag transfer function, as is that relating T_o and T_m from the second of equations (2.24):

$$\frac{T_m(s)}{T_o(s)} = \frac{1}{\tau_m s + 1} \qquad \tau_m = \frac{C_m}{A_m h_m} \tag{2.26}$$

Substituting this for T_m in the transform of the first of equations (2.24) and bringing all terms for $T_o(s)$ to one side give

$$\left[C_o s + (f\rho_o c_o + A_o h_o) + A_m h_m \left(1 - \frac{1}{\tau_m s + 1} \right) \right] T_o(s) = f\rho_o c_o T_i(s) + A_o h_o T_a(s)$$

and some algebraic manipulation then yields

$$T_o(s) = \frac{(\tau_m s + 1)[f\rho_o c_o T_i(s) + A_o h_o T_a(s)]}{C_o \tau_m s^2 + [(f\rho_o c_o + A_o h_o)\tau_m + C_o + C_m]s + f\rho_o c_o + A_o h_o} \tag{2.27}$$

This transform reflects the condition that variations of ambient temperature T_a act as a disturbance input on the system. Thus the system has two inputs, as shown in Fig. 2.15, where G is the transfer function implied by (2.27). As discussed in Section 1.8, the total output is the superposition of the outputs for each input separately with the other set equal to zero.

It should also be observed, however, that if ambient temperature variations are known to be small or slow, $T_a = 0$ can be assumed in the original equations (2.24). This

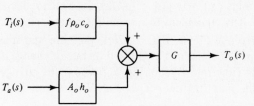

Figure 2.15 Two-input system.

is because in a linearized model the variables represent variations from operating point values, so constant variables are zero.

For $T_a = 0$, the overall transfer function $T_s(s)/T_i(s)$ can now be written from (2.25) to (2.27), or it can be represented by the series connection of blocks in Fig. 2.16, which also identifies the responses $T_o(s)$ and $T_m(s)$.

$T_i(s) \longrightarrow \boxed{f\rho_o c_o G} \xrightarrow{T_o(s)} \boxed{\dfrac{1}{\tau_m s + 1}} \xrightarrow{T_m(s)} \boxed{\dfrac{1}{\tau_s s + 1}} \longrightarrow T_s(s)$ **Figure 2.16** Block diagram for Fig. 2.14.

2.6 FLUID SYSTEMS

Fluid system elements are defined in Fig. 2.17, again in terms of their electrical equivalents.

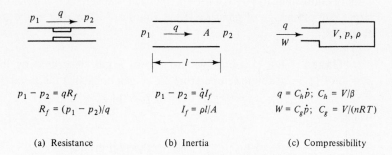

(a) Resistance   (b) Inertia   (c) Compressibility

Figure 2.17 Fluid system elements.

1. *Fluid resistance R_f:* This exists in flow orifices, valves, and fluid lines. With pressure drop $p_1 - p_2$ equivalent to voltage drop and flow rate q to current, R_f is equivalent to electrical resistance. Commonly, the actual relation is nonlinear, and Fig. 2.17(a) shows a linearized model, with $p_1, p_2,$ and q being variations about the values at an operating point. R_f at an operating point may be obtained by calculation or experiment, and the units may yield either volume or mass flow rates.

2. *Fluid inertia I_f:* The mass of fluid of mass density ρ in a line of length l and cross-sectional area A is ρAl. The pressure drop $p_1 - p_2$ generates a force $(p_1 - p_2)A$ to accelerate this mass. Fluid velocity v and volumetric flow rate q are related by $q = Av$, so the acceleration $\dot{v}$ can be expressed as $\dot{v} = \dot{q}/A$.

Newton's law then yields $(p_1 - p_2)A = \rho Al(\dot{q}/A)$, which reduces to the equation in Fig. 2.17(b). It is equivalent to $v = L\dot{i}$ for an electrical inductance.

3. *Fluid compressibility* C_h, C_g: In Fig. 2.17(c), pressure and mass density in volume V are p and ρ. The mass in V is ρV, and its rate of change $d(\rho V)/dt$ must be equal to the mass flow rate W entering V, because the change of mass per second must equal the mass inflow per second. With V constant, therefore,

$$W = V\dot{\rho} \tag{2.28}$$

Liquids and gases are considered in turn.

Liquids.   In high-performance hydraulic systems it is necessary to include the effect of oil compressibility. At constant temperature, near an operating point p_0, ρ_0, a bulk modulus β is defined by

$$\rho - \rho_0 = \frac{\rho_0}{\beta}(p - p_0) \qquad \dot{\rho} = \frac{\rho_0}{\beta}\dot{p} \tag{2.29}$$

A value $\beta \approx 200{,}000$ psi is possible theoretically, but air entrainment usually makes $100{,}000$ psi more realistic. Substituting (2.29) into (2.28) gives

$$W = \frac{V\rho_0}{\beta}\dot{p} \tag{2.30}$$

In hydraulics, volume flow rate $q = W/\rho_0$ is more commonly used:

$$q = C_h\dot{p} \qquad \text{capacitance } C_h = \frac{V}{\beta} \tag{2.31}$$

This is equivalent to $i = C\dot{v}$ for an electrical capacitance. The flow rate q causes a larger rate of change of pressure if the volume V is smaller or the oil stiffer (that is, β larger).

Gases.   For a polytropic process in a gas described by the ideal gas law $p = \rho RT$ ($T = $ absolute temperature; $R = $ gas constant) the p–ρ relation is

$$p = C\rho^n \qquad (\ln p = \ln C + n \ln \rho) \tag{2.32}$$

where

$$n = \begin{cases} 1 & \text{for isothermal processes} \\ k = \dfrac{c_p}{c_v} & \text{for adiabatic frictionless processes} \end{cases}$$

For the latter, where c_p and c_v are the specific heat values at constant pressure and constant volume, there is no heat exchange with the environment. Taking the derivative of (2.32) gives

$$\frac{dp}{p} = \frac{n\,d\rho}{\rho} \quad \text{or} \quad \dot{\rho} = \frac{\rho}{np}\dot{p}$$

and substituting into (2.28) yields

$$W = \frac{V\rho}{np}\dot{p} \tag{2.33}$$

By comparison with (2.30), this shows that

$$\beta = np = \text{bulk modulus for gases} \tag{2.34}$$

Substitution of $p = \rho RT$ gives the form more common for gases:

$$W = C_g \dot{p} \qquad \text{capacitance } C_g = \frac{V}{nRT} \qquad (2.35)$$

($n = 1$ for isothermal; $n = k = c_p/c_v$ for adiabatic).

Example 2.6.1   Hydraulic Tank (Fig. 2.18)

This was the process in the level control system of Section 1.3. The net volumetric inflow rate is $(q_i - q_o)$. This is the volume entering per unit time, so it must equal the change per unit time (that is, the rate of change) of the volume Ah in the tank:

$$A\dot{h} = q_i - q_o \qquad (2.36)$$

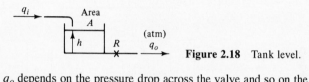

Figure 2.18   Tank level.

The outflow q_o depends on the pressure drop across the valve and so on the head h. The actual relation between h and q_o is nonlinear, but it is approximated by the linearized model, equivalent to Ohm's law $v = iR$,

$$h = q_o R \qquad (2.37)$$

with the variables representing the variations about operating-point values. The valve resistance R is determined as in Fig. 1.5 from the slope of the nonlinear characteristic of h versus q_o at the operating point. Transforming these equations and substituting the second into the first yields

$$(ARs + 1)H(s) = RQ_i(s)$$

and the transfer function

$$\frac{H(s)}{Q_i(s)} = \frac{R}{ARs + 1} \qquad (2.38)$$

Note that this again has the form of a simple lag transfer function, already encountered in electrical and thermal systems.

Example 2.6.2   Two-Tank System with Control Valve (Fig. 2.19)

A control valve V_c with valve opening x controls flow rate q_{i1} into the first tank from a supply with constant pressure P_s. From (2.38), the following transfer functions can be written immediately:

$$\frac{H_1(s)}{Q_{i1}(s)} = \frac{R_1}{A_1 R_1 s + 1} \qquad \frac{H_2(s)}{Q_{i2}(s)} = \frac{R_2}{A_2 R_2 s + 1} \qquad (2.39)$$

To express q_{i1} and q_{i2}, different linearized models may be used, if desired, for R_1 and the control valve:

$$h_1 = q_{i2} R_1 \qquad q_{i1} = K_v x \qquad (2.40)$$

The input to the control valve model is the valve opening x, which in effect controls the valve resistance parameter. The transfer functions corresponding to (2.40) are

$$\frac{Q_{i1}(s)}{X(s)} = K_v \qquad \frac{Q_{i2}(s)}{H_1(s)} = \frac{1}{R_1} \qquad (2.41)$$

Transfer functions (2.39) and (2.41) can be combined into the block diagram shown in Fig. 2.20. It should be noted that this subdivision into blocks would not apply if in Fig. 2.19 the outflow of the first tank fed into the bottom of the second tank. The net

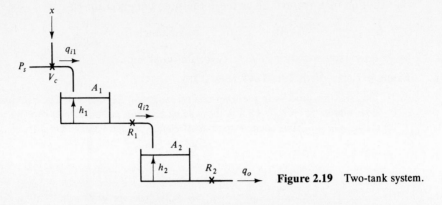

Figure 2.19   Two-tank system.

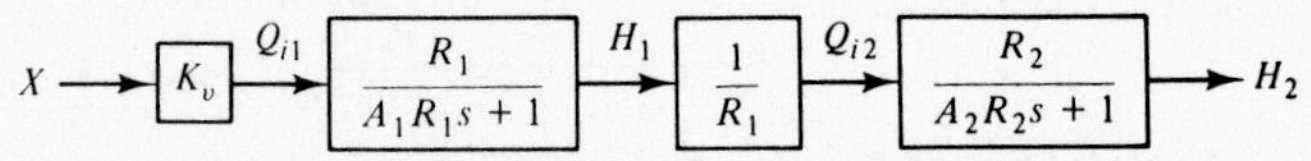

$$X \rightarrow \boxed{K_v} \xrightarrow{Q_{i1}} \boxed{\dfrac{R_1}{A_1 R_1 s + 1}} \xrightarrow{H_1} \boxed{\dfrac{1}{R_1}} \xrightarrow{Q_{i2}} \boxed{\dfrac{R_2}{A_2 R_2 s + 1}} \rightarrow H_2$$

Figure 2.20   Block diagram for system of Fig. 2.19.

head on R_1 would then be $h_1 - h_2$, so the second tank would affect the output of the first. As discussed earlier, this means that the second tank loads the first, and in this case an overall transfer function must be derived directly from the equations for the combined system.

Example 2.6.3   Pneumatic Tank (Fig. 2.21)

The linearized model for subsonic flow through R can be written as

$$p_i - p_o = WR \tag{2.42}$$

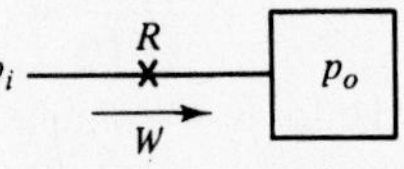

$$p_i \xrightarrow{\quad\overset{R}{\times}\quad} \boxed{p_o} $$
$$\underset{W}{\quad}$$

Figure 2.21   Pneumatic tank.

From Fig. 2.17(c), this flow raises tank pressure p_o according to

$$W = C_g \dot{p}_o \qquad C_g = \text{capacitance} \tag{2.43}$$

Transforming these equations gives

$$P_i(s) - P_o(s) = W(s)R \qquad W(s) = C_g s P_o(s)$$

The transfer function P_o/P_i is again a simple lag:

$$\frac{P_0(s)}{P_i(s)} = \frac{1}{RC_g s + 1} \tag{2.44}$$

Example 2.6.4   Pneumatically Actuated Valve (Fig. 2.22)

This extremely common device is the actuator in the block diagram of Fig. 1.3 for the level control example. The pneumatic line that connects control pressure p_i to pressure p above the diaphragm is represented by the linearized resistance R_g:

$$p_i - p = W_a R_g \qquad P_i(s) - P(s) = W_a(s)R_g \tag{2.45}$$

Diaphragm motion x is so small that the capacitance C_g of the space above it is about constant. Then W_a raises p according to

$$W_a = C_g \dot{p} \qquad W_a(s) = C_g s P(s) \tag{2.46}$$

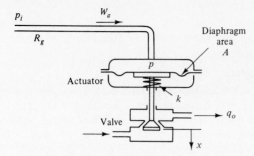

Figure 2.22   Pneumatically actuated valve.

A very simplified model, by no means always acceptable, will be used for diaphragm and valve poppet motion x. The mass and friction forces of the moving parts and the fluid flow forces on the poppet will be neglected. The downward pressure force pA on the diaphragm must then be counterbalanced by the spring force kx:

$$Ap = kx \qquad AP(s) = kX(s) \tag{2.47}$$

Finally, the model for valve flow q_o is taken to be

$$q_o = K_x x + K_p p_d \qquad Q_o(s) = K_x X(s) + K_p P_d(s) \tag{2.48}$$

Here p_d is the pressure drop across the valve, and the linearized model reflects that valve flow q_o will increase with increasing valve pressure drop as well as increasing x. Eliminating W_a between (2.45) and (2.46) gives

$$\frac{P(s)}{P_i(s)} = \frac{1}{R_g C_g s + 1} \tag{2.49}$$

and this with (2.47) and (2.48) gives the block diagram in Fig. 2.23. The overall transfer function relating Q_o and P_i is again a simple lag:

$$\frac{Q_o(s)}{P_i(s)} = \frac{AK_v/k}{R_g C_g s + 1} \tag{2.50}$$

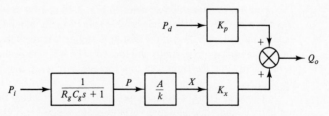

Figure 2.23   Block diagram for Fig. 2.22.

Note how the overall model gradually emerged from equations written for the parts of the system, and not from some form of grand view of the total system.

It is also useful to remark on the signs in (2.47) and (2.48). Remember that the variables represent changes from operating-point values. Figure 2.22 shows that a positive change of p causes a positive change of x, hence the positive signs in (2.47). Also, a positive change of x causes a positive change of q_o, so the corresponding signs in (2.48) must also be positive. If x had been defined as positive in the upward direction, the signs in both equations would be negative on the right or left sides.

2.7 FLUID POWER CONTROL

In this section the modeling of a number of very common fluid power subsystems is discussed by means of examples.

Example 2.7.1   Control Valves

Two common types of valves are shown in Figure 2.24. The supply pressure is p_s and the output pressure p. For $x = 0$ the output port is just blocked off. The linearized valve model (2.48) is used. Valve flow increases with both x and valve pressure drop $(p_s - p)$, and an appropriate linearized valve model is $q = K_x x + K_p(p_s - p)$. In linearized models the variables represent variations about the operating point, so if the supply pressure is constant, the valve model reduces to that shown in Fig. 2.25:

$$q = K_x x - K_p p \tag{2.51}$$

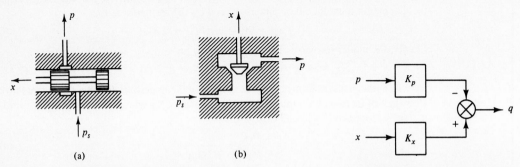

Figure 2.24   Control valves: (a) spool; (b) poppet.

Figure 2.25   Linearized valve model.

The constants K_x and K_p are found from the slopes of the steady-state valve characteristics at the operating point (x_o, p_o).

Example 2.7.2   Hydraulic Cylinder Control

The spool-valve-controlled actuator in Fig. 2.26 is used extensively. For $x = 0$ the valve spool is centered, and the lands on this spool exactly block the ports of fluid lines to the ends of the cylinder so that the piston is stopped. If the valve spool is moved slightly to the left, the ports are partially unblocked. The left side of the cylinder is now connected to the supply and the right side to a low-pressure reservoir. Thus the piston can move to the right.

Consider first the simplest possible model, in which the load connected to the piston is very small and pressure variations are negligible. Oil compressibility can then be ignored, and if the effective area A on both sides of the piston is the same, the flow rate q through both valve ports is also the same and can be modeled as

$$q = K_v x \qquad Q = K_v X \tag{2.52}$$

The change $A\dot{y}$ per second of volume on one side of the piston must equal the flow volume q per second:

$$q = A\dot{y} \qquad Q = AsY \tag{2.53}$$

This is modeled in Fig. 2.27, and the transfer function is

$$\frac{Y(s)}{X(s)} = \frac{K_v}{As} \tag{2.54}$$

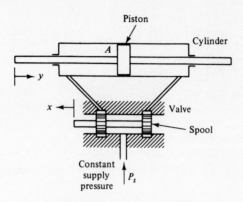

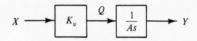

Figure 2.26 Hydraulic cylinder control.

Figure 2.27 Simple model of Fig. 2.26.

The factor s in the denominator represents the fact that, like the electric motor, the cylinder is an integrator, since for a constant flow q the output y increases linearly.

Example 2.7.3 Loaded Hydraulic Cylinder Control

Let the load on the hydraulic cylinder in Fig. 2.26 consist of mass m and damping b. Then if the net pressure on the piston is p, the force balance equation is

$$pA = m\ddot{y} + b\dot{y} \qquad AP(s) = (ms + b)sY(s) \tag{2.55}$$

The valve flow is modeled by (2.51):

$$q = K_x x - K_p p \qquad Q(s) = K_x X(s) - K_p P(s) \tag{2.56}$$

Equation (2.53) must be modified to include the effect of oil compressibility. From Fig. 2.17(c), near a piston position where the volume under pressure p is V, the compressibility flow associated with pressure variations is $(V/\beta)\dot{p}$. The flow q to the cylinder supplies this compressibility flow as well as the flow $A\dot{y}$ corresponding to piston velocity:

$$q = A\dot{y} + \frac{V}{\beta}\dot{p} \qquad Q(s) = s\left[AY(s) + \frac{V}{\beta}P(s)\right] \tag{2.57}$$

From (2.56) and (2.57),

$$K_x X - K_p P = AsY + \frac{V}{\beta}sP \qquad K_x X - AsY = \left(\frac{V}{\beta}s + K_p\right)P$$

Then substituting for P from (2.55) and rearranging yields the following improvement of the model (2.54):

$$\frac{Y(s)}{X(s)} = \frac{K_x}{s\left[\dfrac{mV}{\beta A}s^2 + \dfrac{1}{A}\left(K_p m + \dfrac{V}{\beta}b\right)s + \dfrac{K_p b}{A} + A\right]} \tag{2.58}$$

Example 2.7.4 Hydraulic Motor and Hydrostatic Transmission

Figure 2.28 shows a schematic diagram. A constant-speed hydraulic pump supplies flow to a hydraulic motor. Motor speed can be changed by adjusting pump flow per revolution via a setting ϕ_p. Delivered pump flow is proportional to ϕ_p:

$$q_p = K_p \phi_p = q_l + q_c + q_m \tag{2.59}$$

Of this flow rate, a part q_l is lost in internal leakages, q_c is compressibility flow, and only part q_m causes motor rotation. Let

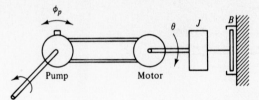

Figure 2.28 Hydrostatic transmission.

p = pressure drop across the motor

V = volume of oil under high pressure p

D_m = motor displacement, the volume of oil needed for 1 rad motor rotation

Then

$q_l = Lp$: leakage flow proportional to p

$q_c = \dfrac{V}{\beta}\, \dot{p}$: compressibility flow [Fig. 2.17(c)]

$q_m = D_m \dot{\theta}$: motor flow ($\dot{\theta}$ = motor speed, rad/sec)

This gives the flow equation:

$$K_p \phi_p = D_m \dot{\theta} + Lp + \frac{V}{\beta}\, \dot{p} \qquad K_p \phi_p(s) = D_m s\theta(s) + \left(L + \frac{V}{\beta}\, s \right) P(s) \tag{2.60}$$

To obtain the load equation, for 100% motor efficiency its mechanical output power equals its hydraulic input power. If the developed motor torque is T, the mechanical output power is $T\dot{\theta}$. The hydraulic input power is $q_m\, p$. This may be verified by thinking of the motor as a cylinder of area A. With flow q_m, the piston velocity is then q_m/A, and piston force is pA. The power is their product, pq_m. Thus $T\dot{\theta} = q_m\, p = D_m \dot{\theta} p$, so

$$T = D_m p \tag{2.61}$$

This torque accelerates inertia J and overcomes damping B, so the load equation becomes

$$T = D_m p = J\ddot{\theta} + B\dot{\theta} \qquad D_m P(s) = s(Js + B)\theta(s) \tag{2.62}$$

Substituting $P(s)$ from this into (2.60) and rearranging yields the transfer function:

$$\frac{\theta(s)}{\phi_p(s)} = \frac{K_p D_m}{s\left[\dfrac{VJ}{\beta} s^2 + \left(\dfrac{VB}{\beta} + LJ \right)s + BL + D_m{}^2 \right]} \tag{2.63}$$

2.8 CONCLUSION

In this chapter, transfer functions were derived for a variety of physical subsystem blocks, including system actuators such as electric motors and the pneumatically actuated valve. A separate section was devoted to the modeling of some fluid power systems. In Chapter 3 the modeling of feedback systems is considered, including cases where the separation into blocks and the precise nature of the feedback may become evident only when all equations for the system have been written.

PROBLEMS

2.1. For the systems shown in Fig. P2.1, write the differential equations and obtain the transfer functions indicated.

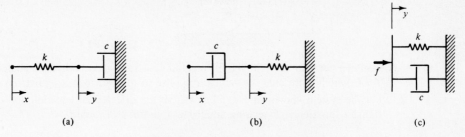

Figure P2.1   (a) Y/X; (b) Y/X; (c) Y/F.

2.2. Figure P2.2 shows a dynamic vibration absorber, often used for the control of mechanical vibrations. A relatively small mass m_2 is attached to the main mass m_1 via spring k_2. For a sinusoidal force f of constant frequency, m_2 and k_2 can be chosen so that the main mass m_1 will not vibrate. Write the system differential equations and obtain the transfer function X_1/F.

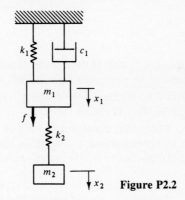

Figure P2.2

2.3. The accelerometer in Fig. P2.3 is mounted on the machine of which the acceleration is to be measured. Under certain conditions, the displacement $(x - y)$ of m relative to the housing is a measure of acceleration. Write the differential equation and the transfer function $(X - Y)/X$.

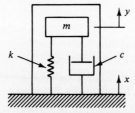

Figure P2.3

2.4. The suspension system of a car is illustrated in Fig. P2.4. On a per wheel basis, the vehicle mass is m_1 and the mass moving with the wheel m_2. The suspension spring and tire are represented by spring constants k_1 and k_2, and the shock absorber by

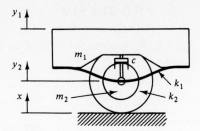

Figure P2.4

damping constant c. Write the differential equations and obtain the transfer function Y_1/X, which represents the vehicle response to road-surface irregularities.

2.5. For the system shown in Fig. P2.5, derive the differential equation and obtain the transfer function X/F relating small mass motions to force f.

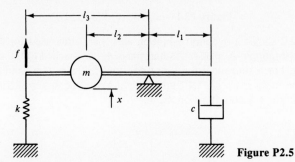

Figure P2.5

2.6. In Example 2.2.3, what is the transfer function relating input and output if the input is redefined to be the torque T_1 at the left end of the shaft instead of the angle?

2.7. In Problem 2.6, determine the transfer function relating this input torque and the shaft angle at the left end of the shaft.

2.8. Write the differential equations and obtain the transfer function relating θ_i and θ_o for the drive system shown in Fig. P2.8, where the springs represent long shafts and damping effects due to bearings and shaft couplings are present.

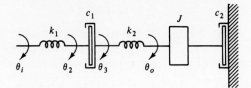

Figure P2.8

2.9. For the drive system in Fig. P2.9, without using equivalent system concepts, obtain the differential equations for T_1 (torque) as input and θ_o as output, and express the corresponding transfer function.

2.10. In Fig. P2.9, without using equivalent system concepts, write the differential equations, and obtain the transfer function between input T_1 and output θ_1.

2.11. Replace the system in Fig. P2.9 by its nongeared equivalent to obtain the transfer function relating T_1 and θ_1 of the output shaft.

2.12. Derive the transfer functions E_o/E_i for the RLC circuits shown in Fig. P2.12.

2.13. Obtain the transfer function E_o/E_i for the circuit shown in Fig. P2.13.

2.14. Obtain the transfer function E_o/E_i for the circuit shown in Fig. P2.14.

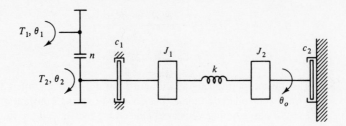

Figure P2.9

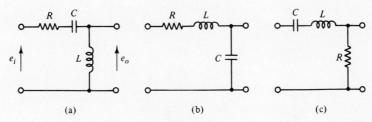

(a)   (b)   (c)

Figure P2.12

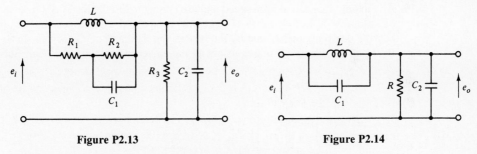

Figure P2.13   **Figure P2.14**

2.15. Derive E_o/E_i for the RC ladder network in Fig. P2.15.
2.16. Derive E_o/E_i for the circuit shown in Fig. P2.16.

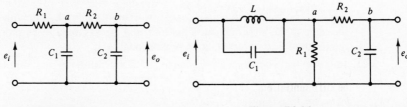

Figure P2.15   **Figure P2.16**

2.17. Show that the transfer function θ/E_f of the motor-generator set in Fig. 2.9(e) can be expressed as

$$\frac{\theta}{E_f} = \frac{K_g/K_e}{s(L_f s + R_f)(T_a T_m s^2 + (T_m + \gamma T_a)s + \gamma + 1)}$$

$$T_m = \frac{J(R_g + R_a)}{K_e K_t}$$

where

$$\gamma = \frac{B(R_g + R_a)}{K_e K_t}$$

$$T_a = \frac{L_g + L_a}{R_g + R_a}$$

2.18. A flexibly supported ring as shown in Fig. P2.18 is mounted on the load inertia J of the field-controlled dc motor. The rotational spring constant of the support is k_r and the rotational damping constant between the inertia and the ring is B_r. Determine the transfer function θ/E_f, where θ is the angle of the motor shaft.

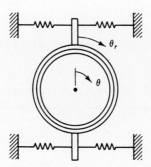

Figure P2.18

2.19. Ac motors often run at high speed and are connected to the load via a gear reduction of ratio $n > 1$. In Fig. P2.19, J represents motor and driving gear inertia, B motor bearing damping, and J_L and B_L the inertia and damping on the driven shaft. Obtain θ_L/E_c if the electrical time constant of the ac servomotor may be neglected.

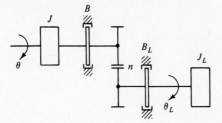

Figure P2.19

2.20. Improve the transfer function model in Example 2.5.1 by allowing for heat loss of the tank to the environment. This loss may be assumed to be proportional to tank temperature T, with constant of proportionality k.

2.21. Improve the equations in Example 2.5.2 by allowing for variations of ambient temperature T_a. Redefine T to be the temperature in the space, and derive the transfer function relating T_a and T.

2.22. A mass M of material of specific heat c and temperature T_i is placed inside an oven at time $t = 0$, and so quickly that the constant oven temperature T_h can be considered as a step input to M. The surface area of M and its coefficient of heat transfer are A and h. Express the transform $T(s)$ of the temperature $T(t)$ of M and solve for this response $T(t)$.

2.23. A constant mass flow rate w of a liquid of specific heat c flows through a tank that contains a mass W of the liquid, and in which a resistance heater adds heat to the flow at rate q_h. As indicated in Fig. P2.23, the inflow temperature is T_i and the outflow (and tank) temperature is T_o. Obtain the transform of T_o that reflects both the effects of varying T_i and varying q_h. What is the transfer function relating q_h and T_o?

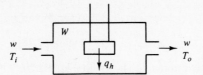

Figure P2.23

2.24. For the mercury thermometer shown in Fig. P2.24, the temperature and the thermal capacitance of the glass and mercury are T_g, C_g and T_m, C_m, respectively. The thermal resistance between ambient temperature T_a and T_g is R_g, and that between the glass and mercury is R_m.
   (a) Obtain the transfer function $T_m(s)/T_a(s)$.
   (b) Also obtain $T_m(s)/T_a(s)$ for a commonly used simplification in which C_g is neglected and an effective thermal resistance R_t is used. What type of transfer function is this?

2.25. In the heat exchanger in Fig. P2.25, the temperature T_h in the outer chamber can be taken to be constant, because of high flow rate through it. The mass flow rate q through the inner chamber is constant. The volume of this chamber is V and its surface area A. Inflow and outflow (and tank) temperatures are T_i and T_o. The density of the fluid is ρ and its specfic heat c. The surface coefficient of heat transfer is h. Express $T_o(s)$, and also show the transfer function T_o/T_i for constant T_h.

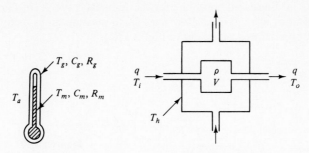

Figure P2.24   **Figure P2.25**

2.26. In Fig. P2.26 a mechanical brake block of mass M is pressed against the drum with force F. The coefficient of friction is μ, so the friction force is μF. The surface velocity between the block and the drum is V. The friction power is converted into heat. The conversion factor that changes mechanical power into heat power is H. Determine the transfer function relating F and the temperature T of the block if all heat power is assumed to enter the block, of which the specific heat is c. The block loses heat to the ambient T_a through a surface area A with heat transfer coefficient h. Note again what type of transfer function results.

2.27. In Example 2.6.2, derive the transfer function relating the level in the second tank to disturbances in the supply pressure.

2.28. For the system in Fig. P2.28, write the linearized differential equations and obtain the transfer function relating volumetric flow rate q_i and level h_2 in the second tank.

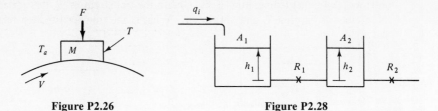

Figure P2.26   **Figure P2.28**

2.29. Part of a pneumatic controller is shown in Fig. P2.29. The opposing bellows are spring centered, and the very small displacement x of the center plate may be taken to be proportional to the difference of the pressures in the bellows. Obtain the transfer function from input pressure p_i to x.

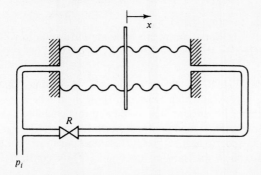

Figure P2.29

2.30. Write all linearized equations for the pneumatic system in Fig. P2.30. The R_i are resistances relating mass flow rates W_i to pressure drops, and C_{g3} and C_{g4} are the tank capacitances according to Fig. 2.17(c).

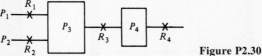

Figure P2.30

2.31. In Problem 2.30, obtain the transfer function $P_4(s)/P_1(s)$.

2.32. Figure P2.32 shows a flapper-nozzle valve, very common in pneumatic and hydraulic systems. A linearized model is analogous to that of other control valves, with flow depending on both the change x of flapper-to-nozzle distance and the change of pressure drop across the nozzle. Obtain the transfer function $P(s)/X(s)$, introducing parameters as needed, if the supply pressure p_s and the drain pressure are constant.

2.33. In Fig. P2.33 the flapper-nozzle valve of Problem 2.32 is used to control piston position y. Neglecting inertia, friction, and oil compressibility, write the linearized system equations, derive a block diagram, and obtain $Y(s)/X(s)$.

2.34. Repeat Problem 2.33 without the simplifying assumptions. Piston mass, damping, and spring constants are m, b, and k, and piston area is A. Write the linearized

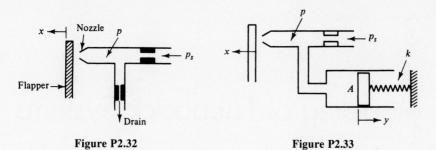

Figure P2.32   **Figure P2.33**

equations and obtain the block diagram and $Y(s)/X(s)$ if the volume under pressure p is V and the bulk modulus is β.

2.35. Obtain the linearized equations, a block diagram, and the transfer function $Y(s)/X(s)$ in Fig. P2.35 if oil compressibility, piston mass, and damping are negligible.

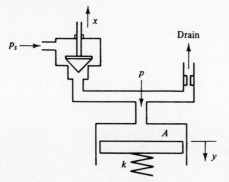

Figure P2.35

2.36. Repeat Problem 2.35 without the simplifying assumptions. Piston mass and damping are m and b, and the volume of oil under pressure p is V, with bulk modulus β.

3

Modeling of Feedback Systems and Controllers

3.1 INTRODUCTION

In the preceding chapter, transfer functions were derived for a variety of physical subsystem blocks. The first part of the present chapter is concerned with the modeling of feedback systems built up from such blocks. As suggested earlier, while the block diagram structure of a system is in the majority of cases more or less immediately evident from the system schematic diagram, there are many occasions where the nature and even the existence of feedback may be rather difficult to see by inspection. In the latter case in particular, the derivation of a "good" block diagram, which clearly identifies the feedback, is an important aid in system analysis and design. Both types are considered through examples.

The next part of the chapter is devoted to the modeling of common feedback system controllers, including electronic controllers based on the use of operational amplifiers. The section on pneumatic PID controllers is limited to their modeling by block diagrams and a physical discussion of the operation.

After the construction of block diagrams it is appropriate to consider their reduction. *Block diagram reduction* involves manipulations that in effect reduce the block diagram to a single block, for example, to determine the transient response of the overall closed-loop system. *Signal flow graphs* are an alternative to block diagrams and are discussed also.

3.2 FEEDBACK SYSTEM MODEL EXAMPLES

First, examples are given in which the system structure is rather evident from the schematic diagram. In subsequent sections, numerous additional examples of this kind will be presented.

Example 3.2.1   Water-Level Control System

For a first example, it is appropriate to return to the level control system in Chapter 1. It would probably operate as a process control or regulator system; that is, the desired level is usually constant and the actual level must be held near it despite disturbances. The model should therefore allow for water supply pressure variations, probably the main disturbance.

A simplified schematic diagram is shown in Fig. 3.1. The level c is measured by means of a float, and a lever is used as a summing junction to determine a measure e of the error with the desired level r. From this mechanical input, the controller and amplifier set a pneumatic output pressure P_o of sufficient power to operate the pneumatic actuator, which adjusts the control valve opening x to control inflow q of the tank.

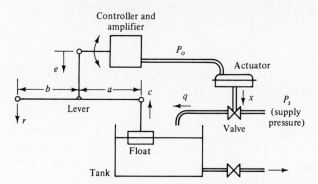

Figure 3.1   Water-level control.

All subsystem transfer functions needed except that of the controller have been found in Chapter 2. The mechanical lever is shown in Fig. 2.1, and the tank and outflow valve are given by (2.38). The pneumatically actuated control valve is modeled in Fig. 2.23, with $P_d = P_s$ to make allowance for disturbances in supply pressure P_s. With these transfer functions and blocks, the translation of Fig. 3.1 to the block diagram in Fig. 3.2 is virtually immediate.

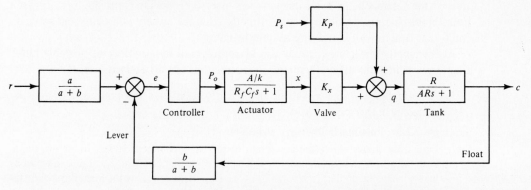

Figure 3.2   Water-level control block diagram.

Example 3.2.2   Hydraulic Servo with Mechanical Feedback: Simple Model

In Fig. 3.3(a) a mechanical lever has been added to a variation of the hydraulic cylinder control in Fig. 2.26. This lever acts as feedback, because it causes piston motion c to affect valve position x. If, say, input r is moved to the right initially, the valve moves

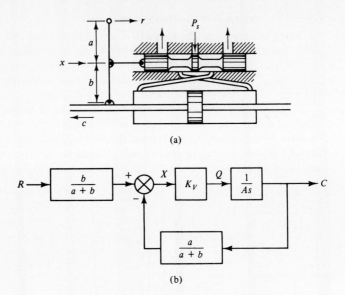

Figure 3.3   Hydraulic servo and block diagram.

right, causing the piston to move left until the valve is again centered. Use of the simple valve–cylinder model in Fig. 2.27, which assumes small loads and neglects oil compressibility, now readily leads to the block diagram in Fig. 3.3(b).

In these examples the feedback was realized by the lever mechanism, and the structure of the block diagram was rather easy to perceive. This applies even more to the motor position and speed control systems and the aerospace examples in the next two sections. However, on many occasions the feedback is generated by the use of signals or physical elements that are an intrinsic part of the system. The precise nature of the feedback, or even its presence, may then be far from obvious. In such cases block diagrams can be derived directly from the system equations and serve an important function in clarifying system behavior.

Frequently, system equations can be represented by a variety of possible block diagrams, which are all mathematically correct but not all equally useful.

A *good block diagram* is one that clearly identifies the components and parameters in the feedback loop.

Example 3.2.3   Pneumatic Pressure Regulator

Figure 3.4 shows a schematic diagram of this very common device. Its purpose is to keep the pressure P_l to the load serviced by the controller constant, equal to a value set by manual adjustment, despite variations of the flow W_l required by the load. Physically, the action is that a reduction of P_l reduces the pressure against the bottom of the diaphragm. This permits the spring force to push it downward to increase valve opening x, and hence increase valve flow from a supply with constant pressure P_s. This increase serves to raise P_l back toward the set value.

To obtain a block diagram, the system equations are written first, similar to Example 2.6.4, for the parts of the system, taken in rather arbitrary order. Only then is consideration given to their combination into a good block diagram model on the basis

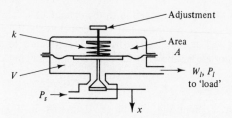

Figure 3.4   Pressure regulator.

of the physical operation of the system. For the weight flow W_v through the valve, the linearized model (2.48) is used in the following form:

$$W_v = K_x x - K_p P_l \qquad W_v(s) = K_x X(s) - K_p P_l(s) \tag{3.1}$$

The pressure drop across the valve is $P_d = P_s - P_l$, but in a linearized model the variables are deviations from operating point values. Hence the constant supply pressure will not appear, and $P_d = -P_l$ in (2.48). The net flow entering the volume below the diaphragm is $(W_v - W_l)$. Frequently, the diaphragm motion is very small, and the volume under pressure P_l can be taken to be constant, equal to V at the operating point. Then

$$W_v - W_l = C_g \dot{P}_l \qquad W_v(s) - W_l(s) = C_g s P_l(s) \tag{3.2}$$

Here $C_g = V/(nRT)$ is the capacitance of V according to Fig. 2.17, and $C_g \dot{P}_l$ the compressibility flow. Force equilibrium on the moving parts gives the equation

$$kx = -AP_l \qquad kX(s) = -AP_l(s) \tag{3.3}$$

This expresses the balance of spring force kx and the pressure force AP_l on the bottom of the diaphragm. The minus sign is needed because the pressure force acts in the direction of negative x. This model is approximate because it neglects friction and mass effects as well as the flow forces on the poppet. To combine (3.1) to (3.3) into a block diagram, it is noted first that the output is the controlled variable P_l, and the input is the disturbance W_l, the unknown flow to the load. Thus, following convention, it is desirable to show W_l at the left and P_l at the right in the diagram. The feedback should then show how changes of P_l are used to make valve flow W_v "follow" W_l. Figure 3.5(a) shows the block diagram constructed from (3.1) to (3.3). Pressure P_l determines x via (3.3), and x and P_l together determine W_v via (3.1). In Fig. 3.5(b) an alternative arrangement is shown, obtained by eliminating $X(s)$ between (3.1) and (3.3).

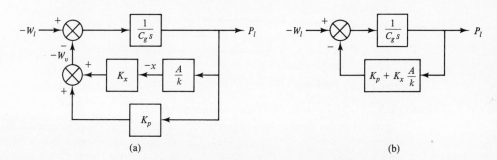

(a)   (b)

Figure 3.5   Block diagrams for Fig. 3.4.

Example 3.2.4   Pneumatic Pressure Regulator: Improved Model

A refinement of (3.2) that allows for a change of V is given by

$$W_v - W_l = C_g \dot{P}_l - \rho A \dot{x} \qquad W_v(s) - W_l(s) = C_g s P_l(s) - \rho A s X(s) \tag{3.2a}$$

The term $\rho A\dot{x}$ is the flow rate corresponding to the change of V. This flow per second equals density times the change per second of volume below the diaphragm. The minus sign arises because Fig. 3.4 defines x as positive in the direction of decreasing volume below the diaphragm. For (3.3), allowing for mass m and damping coefficient b gives the more refined model

$$m\ddot{x} + b\dot{x} + kx = -AP_l \qquad (ms^2 + bs + k)X(s) = -AP_l(s) \qquad (3.3a)$$

These refinements complicate the block diagrams in Fig. 3.5 appreciably. Figure 3.6(a) shows how Fig. 3.5(a) changes. Combining the feedback loops as was done for Fig. 3.5(b) yields a rather complex form. An alternative arrangement, which could also have been used in Fig. 3.5(b), is shown in Fig. 3.6(b). This follows by eliminating W_v between (3.1) and (3.2a):

$$K_xX - K_pP_l - W_l = C_gsP_l - \rho AsX$$

or

$$(C_gs + K_p)P_l = -W_l + (\rho As + K_x)X \qquad (3.4)$$

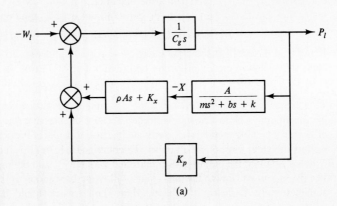

(a)

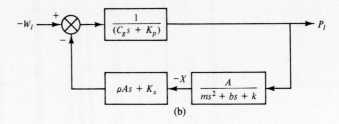

(b)

Figure 3.6   Improved block diagrams for Fig. 3.4.

Example 3.2.5   Motor with *IR*-Drop Compensation

Figure 3.7 shows the armature-controlled dc motor of Fig. 2.9(b) (Example 2.4.2) with a resistance R added in the armature loop. The voltage across this resistance is fed back as shown to the input of the power amplifier. A block diagram should clarify the nature of this feedback and allow its effect to be studied. The transfer function θ/E_a given by (2.12), with R_a replaced by $R_a + R$, does not show I_a, so it is necessary to return to the equations from which it was derived. These are repeated here, with R_a replaced by $R_a + R$:

$$E_a = (R_a + R + L_as)I_a + E_m \qquad E_m = K_es\theta \qquad T = K_TI_a = s(Js + B)\theta$$

$$(3.5)$$

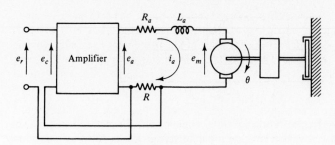

Figure 3.7   Motor with *IR*-drop compensation.

The additional equations in Fig. 3.7 are

$$E_c = E_r - I_a R \qquad E_a = K_a E_c \qquad (3.6)$$

The block diagram in Fig. 3.8 is readily obtained from these equations and allows the effect of the *IR*-drop compensation to be studied.

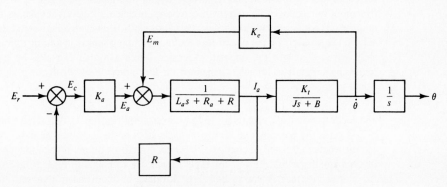

Figure 3.8   Block diagram for Fig. 3.7.

3.3 FEEDBACK SYSTEMS FOR MOTOR POSITION AND SPEED CONTROL

Electric motor-driven servomechanisms for position control or speed control are used in many areas of engineering. Robotic manipulators use position servos for control of the individual joints, similar to the control of the degrees of freedom of a machine tool. Low-power servos are found in such applications as indicating instruments and drafting machines. The servos on a steel rolling mill provide a heavy-duty example of speed control. The following examples illustrate features such as the use of velocity feedback to improve the damping of position servos, the representation of load disturbance torques, and the modeling of load resonance effects.

Example 3.3.1   Motor Position Servos for Robots, Machine Tools, Tracking Radar Antennas, and so on

A simple motor position servo is shown in Fig. 3.9. The motor and its load are represented by the transfer function (2.11) for a field-controlled dc motor. The time constant $T_m = J/B$, where J is the inertia and B the damping constant. A potentiometer can be used to represent shaft position by a voltage, but a rotary variable-differential transformer is often preferred, because it avoids the wear and friction associated with the sliding contact. An operational amplifier, discussed later, can serve as a summing junction to determine the error voltage $E = R - C$, with the voltage R representing desired

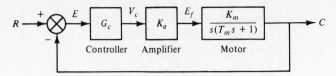

Figure 3.9   Motor position servo.

position. Alternatively, electrical bridge circuits could be used. The controller G_c may be a simple amplifier and generates a low-power output voltage V_c. Its power is raised in a power amplifier of which the output is applied to the motor.

Example 3.3.2   Servo with Velocity Feedback (Minor Loop Feedback)

In speed or position control servomechanisms such as in Example 3.3.1, design for satisfactory performance is often complicated by a lack of adequate inherent damping in motor and load. Then the difficulty of positioning a large inertia J rapidly without severe overshoot in response to a step input can be appreciated intuitively. One possible solution is to install a mechanical damper on the motor shaft. However, a better and more elegant solution is possible by the use of feedback. A damping torque is a torque proportional to shaft speed $\dot{c}$ and in the opposite direction. Such a torque can also be generated by mounting a small tachometer–generator on the motor shaft to obtain a signal proportional to speed,

$$b = K_g\dot{c} \qquad B(s) = K_g sC(s) \tag{3.7}$$

and feeding this back negatively to the power amplifier input. This direct measurement of $\dot{c}$ is generally preferable to taking the derivative of the measured signal c. A derivative circuit reacts to the rate of change of its input signal. Its output is therefore very sensitive to even small irregularities of the input signal due to noise. In effect, the circuit amplifies noise. Figure 3.10 shows how Fig. 3.9 must be modified to represent this velocity feedback. This is an example of *minor loop feedback*. The importance of velocity feedback may be judged from the availability of motors with integrally mounted tachometers on the shaft.

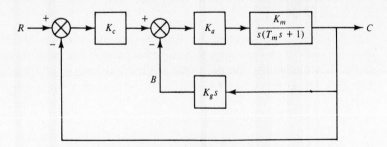

Figure 3.10   Servo with velocity feedback.

Example 3.3.3   Motor Position Servo with Load Disturbance Torques

To improve the model in Fig. 3.9 for a field-controlled dc motor position servo, allowance must be made for load disturbance torques T_l acting on the motor shaft. Severe disturbance torques can arise, for example, in tracking radar antennas due to wind or in steel rolling mill speed controls when slabs enter and leave the rolls. It is necessary to return to the motor equations in Example 2.4.1. Field voltage E_f and motor developed torque T are related to field current I_f by

$$E_f = (R_f + L_f s)I_f \qquad T = K_t I_f \tag{3.8}$$

where R_f and L_f are field resistance and inductance and K_t is the motor torque constant. If T_l is taken as positive in a direction opposite that of T, a net torque $(T - T_l)$ is available to accelerate the motor and load inertia J and overcome their damping B:

$$T - T_l = J\ddot{c} + B\dot{c} \qquad T(s) - T_l(s) = s(Js + B)C(s)$$

Hence

$$\frac{C}{T - T_l} = \frac{1}{s(Js + B)} = \frac{1/B}{s(T_m s + 1)} \qquad T_m = \frac{J}{B} \tag{3.9}$$

Figure 3.9 is now modified to the diagram in Fig. 3.11(a). As was noted, often the field time constant $T_f = L_f/R_f \ll T_m$ and can be neglected. The factor $1/R_f$ can then be considered to be incorporated into K_a and $E_f = I_f$ assumed in the diagram for purposes of analysis. Figure 3.11(a) shows T_l as a second input, a disturbance input, to the block diagram. This is typical of the way in which disturbances are represented in control system block diagrams.

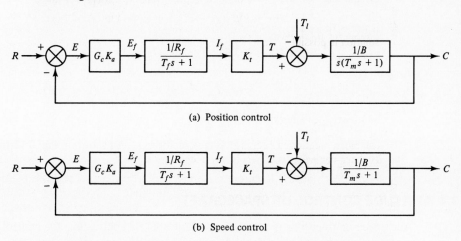

(a) Position control

(b) Speed control

Figure 3.11   Position and speed control servos with disturbance torque.

Example 3.3.4   Speed Control System with Load Disturbance Torques

Figure 3.11(b) shows how Fig. 3.11(a) is modified when speed rather than position is the variable to be controlled. The only apparent difference is that the factor s in the denominator of the block representing the motor and load is removed. This factor represents an integration, and without it the output c is shaft speed instead of position. Correspondingly, a feedback sensor must be used that measures speed instead of position. This could be a tachometer providing a voltage proportional to speed.

Example 3.3.5   Position Servo with Load Resonance

So far, one inertia was used to represent motor and load. However, a common situation can be approximated by the model indicated in Fig. 3.12(a). Separate motor and load inertias J_m and J_l are identified, and the connecting structure is modeled by a shaft with spring constant k and damping coefficient b. A load disturbance torque T_l acts on J_l opposite the developed motor torque T. T accelerates J_m and supplies the torque transmitted by the shaft. This shaft torque, in turn, accelerates J_l and supplies T_l. This leads to the equations

$$T = J_m \ddot{\theta}_m + b(\dot{\theta}_m - \dot{\theta}_l) + k(\theta_m - \theta_l)$$

$$b(\dot{\theta}_m - \dot{\theta}_l) + k(\theta_m - \theta_l) = J_l \ddot{\theta}_l + T_l$$

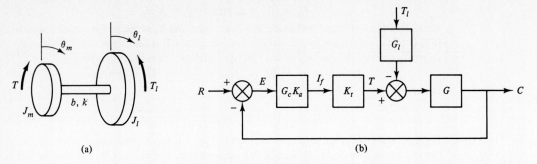

(a)   (b)

Figure 3.12 Position servo with load resonance.

and, on rearranging and Laplace transformation, to

$$(J_m s^2 + bs + k)\theta_m(s) - (bs + k)\theta_1(s) = T(s)$$
$$(bs + k)\theta_m(s) - (J_1 s^2 + bs + k)\theta_1(s) = T_l(s) \tag{3.10}$$

Solving for $\theta_l(s)$ yields

$$\theta_l(s) = G(s)[T(s) - G_l(s)T_l(s)]$$

$$G(s) = \frac{bs + k}{s^2[J_m J_1 s^2 + (J_m + J_1)bs + (J_m + J_1)k]} \tag{3.11}$$

$$G_l(s) = \frac{J_m s^2 + bs + k}{bs + k}$$

The block diagram of Fig. 3.11 is now modified to that shown in Fig. 3.12(b), where T_f has been neglected and $1/R_f$ incorporated into K_a.

3.4 ATTITUDE CONTROL OF SPACECRAFT

Aerospace engineering is a very important area of application for control systems, ranging from the use of hydraulic servos to position control surfaces on aircraft to controls that account for resonances due to the flexibility of a rocket.

Three examples of attitude control systems are given in this section to serve as an introduction to some aspects of aerospace controls.

Example 3.4.1 Satellite Attitude Control

Figure 3.13 shows one axis of a three-axis system for controlling the orientation of a spacecraft. An attitude sensor measures the deviation angle θ from the desired direction, so the reference input is shown as zero. A rate gyro may be available to provide a direct measurement of $\dot{\theta}$. Like velocity feedback in a motor position servo, this direct measurement avoids problems associated with taking the derivative of a noisy signal θ. Also, as in the case of a motor, the rate feedback with gain K_r helps to stabilize the system and permits a simpler controller G_c. Opposing pairs of thrustor jets produce a control torque T on the satellite inertia J. A disturbance torque T_d allows for solar wind and other disturbances. The dynamics are approximated by

$$J\ddot{\theta} = T - T_d \qquad \frac{\theta(s)}{T(s) - T_d(s)} = \frac{1}{Js^2} \tag{3.12}$$

to complete the diagram. Note that the feedback from $\dot{\theta}$ could also be shown as a feedback $K_r s$ from θ, because $s\theta$ is the transform of $\dot{\theta}$.

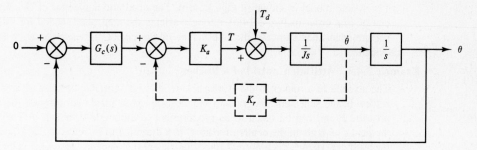

Figure 3.13   Satellite attitude control.

Example 3.4.2   Attitude Control of Spacecraft with Structural Resonance

In the preceding example the satellite is modeled as a simple inertia. But in fact it is often necessary to allow for flexibility in the structure and the presence of one or more mechanical resonances, for example, due to attached solar panels. Tall rockets, robotic manipulators, and airplane wings are other structures with continuous mass and flexibility that theoretically have an infinite number of modes of vibration. Modes with resonant frequencies sufficiently above the frequency range of interest in control can be omitted from the model, as will be discussed later. In the simplest case it may be sufficient to include one resonance. Then the model of Fig. 3.12(a) can serve for rotation and its linear equivalent for translation.

For the satellite, if the thrusters are attached to the part of the structure identified as J_m and the sensors to the part identified as J_l, then $G(s)$ of (3.11) represents the satellite and Fig. 3.14(a) models the control system, with the rate feedback and disturbance input in Fig. 3.13 omitted. This follows because $G(s)$ of (3.11) also relates load position $\theta_l(s)$ to torque T.

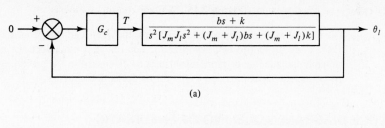

(a)

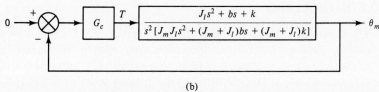

(b)

Figure 3.14   Attitude control of flexible satellites.

A significant change occurs if the sensors as well as the thrusters are attached to J_m. Then the satellite transfer function $G(s)$ is $\theta_m(s)/T(s)$ and is obtained by solving (3.10) for $\theta_m(s)$:

$$G(s) = \frac{\theta_m(s)}{T(s)} = \frac{J_l s^2 + bs + k}{s^2[J_m J_l s^2 + (J_m + J_l)bs + (J_m + J_l)k]} \qquad (3.13)$$

The system model is shown in Fig. 3.14(b). The quadratic numerator of G, compared to that of first order in Fig. 3.14(a), strongly affects the controller needed for satisfactory performance, thus showing the importance of the choice of sensor location in system design.

Example 3.4.3 Attitude Control of a Rocket

The attitude of a rocket in the atmosphere tends to be unstable because of aerodynamic forces. In the schematic diagram in Fig. 3.15(a), these forces act through the center of pressure P and can be taken to be proportional to the angle of attach θ. They cause a moment $C_n l_1 \theta$ about the center of mass C in a direction to increase θ (that is, destabilize the attitude). Here C_n is a normal force coefficient that depends on rocket velocity v and other factors. Feedback control of the deflection angle δ of the rocket thrust T is needed to make $\theta = 0$ and ensure stability. For small δ, the moment about C due to the thrust is $T l_2 \sin \delta \approx T l_2 \delta$. If the moment of inertia about C is J, the equation of motion and the transfer function of the rocket are approximated by

$$J\ddot{\theta} = C_n l_1 \theta + T l_2 \delta \qquad G(s) = \frac{\theta(s)}{\delta(s)} = \frac{T l_2/J}{s^2 - C_n l_1/J} \qquad (3.14)$$

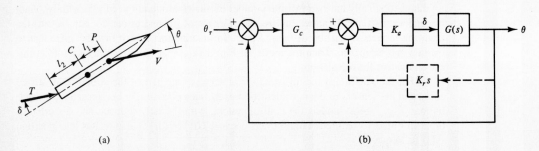

Figure 3.15 Attitude control of a rocket.

The control system, shown in Fig. 3.15(b), includes feedback from a rate gyro, which is likely to be present to help stabilize the control. In a more accurate model, $G(s)$ would probably include one or more resonances as discussed in the preceding example, and K_a might include dynamics due to the thrust deflection actuator.

3.5 BLOCK DIAGRAM MODELING OF PNEUMATIC PID CONTROLLERS

The controllers shown in block diagrams earlier in this chapter must be chosen such that the dynamic behavior of the closed-loop system will be satisfactory. PID controllers are very common, and their proportional (P), integral (I), and derivative (D) actions are basic to all controllers. Among many possible implementations, pneumatic PID controllers continue to be widely used and are available in a large variety of designs and makes. The operation of a generic design is discussed in this section and the behavior modeled by means of block diagrams.

The design may be seen as a possible implementation of the controller block identified in Fig. 3.1 for the water-level control system. This block, shown in Fig. 3.16(a), has a mechanical position input e, which is a measure of the error in the control system, and a pneumatic pressure output P_o, which, after a power amplification stage, which is not indicated, operates the actuator to adjust the valve.

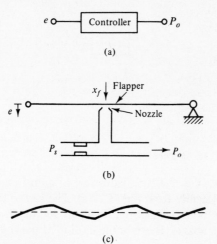

(a)

(b)

(c)   **Figure 3.16** On–off control.

The simplest possible configuration to control P_o from e is indicated in Fig. 3.16(b) and is not PID control, but *on–off control*. The operation is based on the flapper–nozzle device. Air is supplied through a constant restriction from a source with constant pressure P_s. When the flapper is moved closer to the nozzle, the resistance to airflow out of the nozzle is increased, and the back pressure P_o, which is also the output pressure, rises. Linearized models will be used, assuming small variations about an operating point. P_o and x_f in Fig. 3.16(b) represent changes from operating-point values of output pressure and flapper–nozzle distance, with x_f positive in the direction shown. Hence, if x_f increases, P_o increases, and a logical linearized model is

$$P_o = K_f x_f \qquad (3.15)$$

The problem is, however, that the gain K_f of the flapper–nozzle amplifier is extremely large. This means that only a very small change of e will already cause P_o to change from its minimum to its maximum value, or vice versa. Hence the control valve in the level control in Fig. 3.1 switches back and forth between fully open and fully closed for only very small variations of e. Hence the name "on–off control," and as a result the level will fluctuate about the desired value as suggested by Fig. 3.16(c). Proportional control is the first step toward improved performance.

Proportional Control

A P controller is shown in Fig. 3.17(a). The right end of the flapper is controlled by the pressure P_o via a bellows that expands proportional to pressure:

$$x_b = K_b P_o \qquad (3.16)$$

The mechanical lever is modeled as in Fig. 3.3 for the hydraulic servo:

$$x_f = \frac{a}{a+b}\, e - \frac{b}{a+b}\, x_b \qquad (3.17)$$

With input e at the left and output P_o at the right, these equations immediately yield the block diagram in Fig. 3.17(b). It identifies the lever as the summing junction and the bellows and lever as feedback elements.

As will be seen later, the system characteristics can be examined analytically

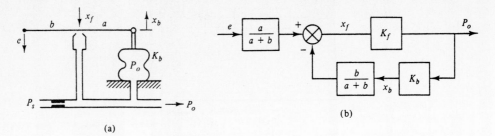

(a)

(b)

Figure 3.17 P controller.

from the block diagram in Fig. 3.17(b). However, to gain insight, these properties will here be explained physically from Fig. 3.17(a). The key to this understanding is that, as noted earlier, the motion x_f is extremely small and the lever essentially pivots around a fixed point at the nozzle. Because the motion of the right end of the lever is proportional to P_o and the left end represents the error in the level control system, two conclusions follow immediately:

1. This is indeed P control, because the ratio P_o/e of output to input is constant. This ratio is called the *gain* and can be set to a desired value by adjusting the lever ratio.
2. With P control the system will have a steady-state error. This follows because the lever implies a change of e if P_o must be changed to a new steady-state value to obtain a different valve flow.

Zero steady-state error would require the possibility for P_o to have any value without a change of the steady-state position of the left end of the lever. This can be realized by expanding the P controller to PI control.

Proportional Plus Integral (or Reset) Control

A PI controller is shown in Fig. 3.18(a). A bellows connected to P_o via a severe resistance R_i is added to oppose the proportional bellows. This resistance is so severe that P_i rises only very slowly after a step increase of e, causing the system to operate initially much like a P controller and causing x_b to rise. When this transient operation is largely complete, P_i is still increasing, pushing x_b down. Finally, in the steady state P_i equals P_o, balancing the two bellows and ensuring that $x_b = 0$. Since, as discussed earlier, the changes of x_f needed to produce the range of output pressures P_o is extremely small, this means that in the steady state the left end of the lever is in virtually unchanged position, $e = 0$. This physical explanation of the zero steady-state errors will be confirmed mathematically later, when this will be shown to be due to the presence of integral control. Physically, this is reflected in the fact that the system produces a nonzero steady-state change of output P_o for a virtually zero steady-state change of input e.

The equations for derivation of a block diagram are, similarly to P control,

$$P_o = K_f x_f \qquad x_f = \frac{a}{a+b}e - \frac{b}{a+b}x_b \qquad (3.18)$$

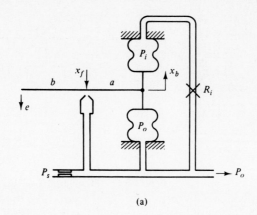

(a)

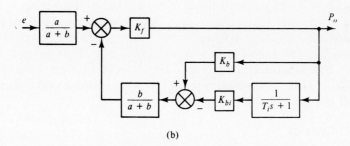

(b)

Figure 3.18   PI controller.

$$x_b = K_b P_o - K_{bi} P_i \tag{3.19}$$

The model for the pressure P_i is analogous to that for the pneumatic tank in Fig. 2.21. The airflow rate through resistance R_i is

$$q = \frac{P_o - P_i}{R_i} \tag{3.20}$$

Assuming small bellows motion, the capacitance C_i is about constant, so q raises pressure P_i according to

$$q = C_i \dot{P}_i \tag{3.21}$$

Eliminating q now yields the model

$$R_i C_i \dot{P}_i + P_i = P_o \qquad \frac{P_i(s)}{P_o(s)} = \frac{1}{T_i s + 1} \qquad T_i = R_i C_i \tag{3.22}$$

These equations immediately provide the block diagram in Fig. 3.18(b), which shows that both bellows and the lever operate as feedback elements.

Proportional Plus Derivative Control

A PD controller is shown in Fig. 3.19(a). It is seen to be identical to the P controller in Fig. 3.17(a) except that a resistance R_d is added in the line to the bellows. In P control, if a step increase of e is applied, x_b rises immediately to counteract its effect on x_f. In PD control, the presence of R_d delays the rise of P_d, and hence of x_b. If e changes faster, P_d lags more behind P_o, and the flapper is correspondingly closer to the nozzle.

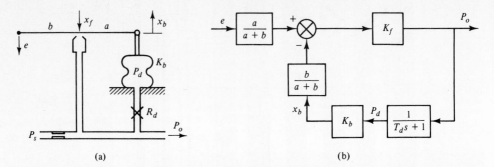

Figure 3.19   PD controller.

This represents PD control because P_o contains a component proportional to the rate of change $\dot{e}$, and so proportional to the derivative of the error.

Physically, this controller provides a stronger control signal if the error changes faster. Thus it anticipates large errors and takes corrective action before they occur. Therefore, PD control tends to have a stabilizing effect on the dynamic behavior.

The block diagram in Fig. 3.19(b) can be verified in a manner similar to that for PI control. Mathematical proof that this controller provides derivative action as well as P control will be given later.

PID Control

The PI controller in Fig. 3.18(a) becomes a PID controller when a resistance is added in the line to the proportional bellows. The I and D actions do not interfere with each other because the integral action is made to occur much more slowly by making resistance R_i much larger than R_d.

3.6 ELECTRONIC CONTROLLERS AND SYSTEM SIMULATION USING OPERATIONAL AMPLIFIERS

The operational amplifier is a very important general-purpose device, used extensively in many applications. It is the basic element of general-purpose analog computers, which may contain hundreds of such amplifiers. These computers can simulate the blocks or the differential equations of systems with or without feedback and show the dynamics of system variables in terms of analog voltages. Operational amplifiers can serve as summing junctions and to realize dynamic compensators and have many other applications. In this section their use in dynamic compensators is emphasized, but analog computers are also discussed briefly.

The operational amplifier is a dc amplifier with very high gain, on the order of 10^5 to 10^8, and very high input impedance. It is shown symbolically in Fig. 3.20(a), where the ground connections have been ignored. The amplifier changes the sign, so amplifier output e_o and input e_g are related by the equation

$$e_o = -Ae_g \qquad A \approx 10^5 \text{ to } 10^8 \tag{3.23}$$

Because A is so large and e_o is limited to a maximum of 100 volts, 10 volts, or less, depending on the amplifier, the input voltage e_g is virtually zero. The amplifier input

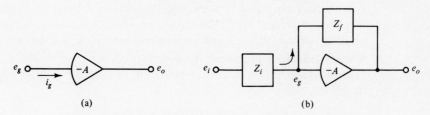

Figure 3.20 (a) Operational amplifier; (b) system building block.

current i_g is also virtually zero, because the input impedance is quite large. Hence, in the application of the amplifier, the following can be assumed:

$$e_g = 0 \qquad i_g = 0 \tag{3.24}$$

Figure 3.20(b) shows the basic building block in which the operational amplifier is used. An input impedance Z_i is connected to the input, and a feedback impedance Z_f is between output and input.

From the equations for the elements of electrical systems in Fig. 2.7(a), voltage v and current i are related by $v = iR$ for a resistance R and by $i = C\dot{v}$ for a capacitance C. The impedance Z can be expressed as the ratio of the Laplace transforms $V(s)$ and $I(s)$:

$$Z = \frac{V}{I} \qquad \begin{array}{ll} \text{resistance: } V = IR; & Z = R \\[2mm] \text{capacitance: } I = CsV; & Z = \dfrac{1}{Cs} \end{array} \tag{3.25}$$

Since $e_g = 0$, the current through Z_i in Fig. 3.20(b) is E_i/Z_i and that through Z_f is $-E_o/Z_f$, where E_i and E_o are the transforms of e_i and e_o. Also, since $i_g = 0$, these two currents must be the same. Hence

$$\frac{E_i}{Z_i} = -\frac{E_o}{Z_f} \qquad \frac{E_o}{E_i} = -\frac{Z_f}{Z_i} \tag{3.26}$$

By choosing Z_i and Z_f, operational amplifiers can be used to realize active dynamic compensation transfer functions for use in control systems, for example, to replace the passive RC networks in Fig. 2.8. They can also serve as summing junctions or as P controllers.

The most common cases are listed next.

1. *Constant gain (P controller):*

$$Z_i = R_i \qquad Z_f = R_f$$
$$\frac{e_o}{e_i} = -\frac{R_f}{R_i}$$

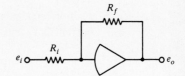

For a gain $a \leq 1$, a potentiometer can be used:

$$a = \frac{R_p}{R_t} \qquad e_o = ae_i$$

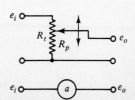

R_f/R_i is only adjustable in steps, usually between 0.1 and 10, limited by available resistor values. To realize a gain of, say, 3.45, a series connection of an amplifier with $R_f/R_i = 5$ and a potentiometer with $a = 0.69$ can be used.

2. *Summer (summing junction):* Analogous to (3.26), the current $-e_o/R_f$ through R_f must equal the sum of the currents e_j/R_j through the parallel input resistors. Hence

$$e_o = -\frac{R_f}{R_1}e_1 - \frac{R_f}{R_2}e_2 - \frac{R_f}{R_3}e_3$$

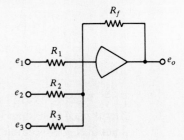

3. *Integrator:* From (3.25), $Z_i = R_i$ and $Z_f = 1/(C_f s)$, so

$$\frac{E_o}{E_i} = \frac{-1}{R_i C_f s} \qquad e_o = \frac{-1}{R_i C_f}\int_0^t e_i\, dt + e_o(0)$$

An initial condition $e_o(0)$ can be applied as an initial voltage across the capacitor.

4. *Differentiator:*

$$Z_i = \frac{1}{C_i s} \qquad Z_f = R_f$$

$$\frac{E_o}{E_i} = -R_f C_i s \qquad e_o = -R_f C_i \dot{e}_i$$

This element is very sensitive to noise that may be superimposed on the input signal, because the derivative represents the slopes of the irregularities in the output.

5. *Proportional plus integral (PI) controller:* $Z_i = R_i$. For the feedback, if the current is i_f, the voltage is

$$R_f i_f + \frac{1}{C_f}\int i_f\, dt$$

so the impedance is $Z_f = R_f + 1/(C_f s)$, and

$$\frac{E_o}{E_i} = -\left[\frac{R_f}{R_i} + \frac{1/(R_i C_f)}{s}\right] = -\frac{K(\tau s + 1)}{s}$$

with $K = 1/(R_iC_f)$ and $\tau = R_fC_f$

6. *Proportional plus derivative (PD) controller:* The sum $[(e_i/R_i) + C_i\dot{e}_i]$ of the parallel currents through R_i and C_i must equal the current $-e_o/R_f$ through R_f:

$$\left(\frac{1}{R_i} + C_is\right)E_i = -\frac{1}{R_f}E_o$$

$$\frac{E_o}{E_i} = -\frac{R_f}{R_i}(R_iC_is + 1)$$

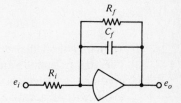

7. *Simple lag:* The current e_i/R_i must equal the sum $-[(e_o/R_f) + C_f\dot{e}_o]$ of the parallel currents through the feedback elements:

$$\frac{1}{R_i}E_i = -\left(\frac{1}{R_f} + C_fs\right)E_o$$

$$\frac{E_o}{E_i} = -\frac{R_f/R_i}{R_fC_fs + 1}$$

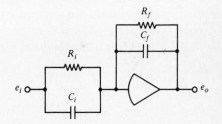

8. *Phase-lead or phase-lag compensators:* Analogous to the above:

$$\left(\frac{1}{R_i} + C_is\right)E_i = -\left(\frac{1}{R_f} + C_fs\right)E_o$$

$$\frac{E_o}{E_i} = -\frac{R_f}{R_i}\frac{R_iC_is + 1}{R_fC_fs + 1}$$

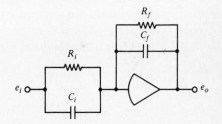

9. *Lead compensator (high-pass filter):* Virtually by inspection, after the preceding examples,

$$\frac{E_o}{E_i} = -\frac{R_fC_is}{R_fC_fs + 1}$$

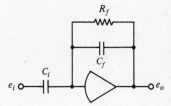

Analog Computer Simulation

Block diagram simulation.   In this approach for simulating a feedback system on an analog computer, the individual blocks are modeled using the preceding examples. Figure 3.21 shows an example of PI control of a process with two simple lags. The numbers in the simulation refer to one of the cases listed previously, of which the equations would be used to determine the resistor and capacitor values required.

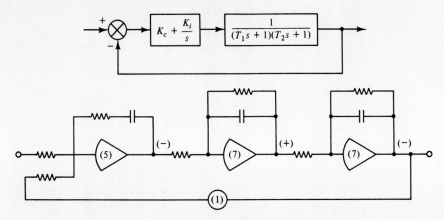

Figure 3.21 Block simulation using operational amplifiers.

It is likely that at least one potentiometer will be needed to match the loop gain. The sign inversions are indicated in parentheses. If the last one had been positive, a constant-gain element $R_f/R_i = 1$ would have to be added in the feedback to ensure that it will be negative.

General-purpose simulation. The preceding approach is natural and convenient for control system simulation. General-purpose analog simulation, however, is usually concerned with one or more differential equations and uses only the elements 1, 2, and 3 of those listed previously.

To explain the idea, Fig. 3.22 shows a simulation of the differential equation

$$\ddot{x} + 6\dot{x} + 22x = r \tag{3.27}$$

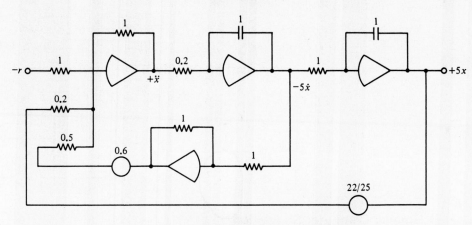

Figure 3.22 Simulation of (3.27).

In Fig. 3.22, the values shown with resistors are the resistances in megohms, and those with capacitors are the capacitances in microfarads. If both are 1 for an integrator, its multiplication factor is

$$-\frac{1}{R_i C_f} = -\frac{1}{10^6 \times 10^{-6}} = -1$$

Equation (3.27) is first solved for $\ddot{x}$:

$$\ddot{x} = -6\dot{x} - 22x + r \tag{3.28}$$

The idea of the simulation is that $\ddot{x}$ is integrated once to obtain $\dot{x}$ and then again to generate x. These variables are then fed back to generate $\ddot{x}$ in a summer amplifier according to (3.28). Because of the sign inversion in the summer amplifier, it is necessary to add an "inverting" amplifier in the feedback from $\dot{x}$. The reason for generating $-5\dot{x}$ instead of $-\dot{x}$ (another choice could have been made) is to obtain a reasonable distribution of gains and avoid, for example, the need for amplifiers in the feedback from x.

   This is only a brief introduction into the subject of analog computers. Discussion of the problems of amplitude scaling, to ensure that output voltages do not exceed permissible levels, and of time scaling, to simulate slow processes in less than real time, has been omitted. Analog computers can be used as well to simulate and study the behavior of multivariable systems, which have more than one input and output and are described by sets of differential equations. Function generators are available to include the effect of system nonlinearities in such simulations. The proliferation of digital computers and the availability of techniques for digital simulation, however, have had great impact on this area.

   Time scaling, which permits a simulation to be speeded up or slowed down as desired, cannot be used freely if real-time simulation is required. For example, sometimes a model is used in the actual control system, or it is desired to test a computer control algorithm under realistic constraints on available computing time. Then the simulation must satisfy time-of-the-clock or real-time constraints, which can complicate matters considerably.

3.7 BLOCK DIAGRAM REDUCTION

Block diagram reduction involves algebraic manipulations of the transfer functions of the subsystems or blocks, which in effect reduce the diagram to a single block. This gives the overall transfer function relating the input r and output c in a block diagram and hence permits, for example, calculation of system transient responses.

   Consider the *cascade* or series connection of two blocks in Fig. 3.23. By definition,

$$C = G_2 M \qquad M = G_1 R$$

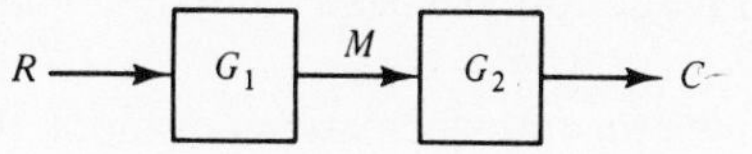

$R \longrightarrow \boxed{G_1} \overset{M}{\longrightarrow} \boxed{G_2} \longrightarrow C$

Figure 3.23   Two-block cascade.

Hence, substituting the second into the first yields

$$C = GR \qquad G = G_1 G_2 \tag{3.29}$$

By direct extension it follows that

   The overall transfer function of a series of blocks equals the product of the individual transfer functions.

The configuration in Fig. 3.24 is extremely common. It is equivalent to that in Fig. 1.3 and to many of the other block diagrams considered earlier. By definition,

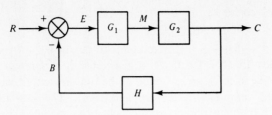

Figure 3.24 Standard feedback loop.

$$C = G_2 M \qquad M = G_1 E \qquad E = R - B \qquad B = HC$$

Combining each pair yields

$$C = G_1 G_2 E \qquad E = R - HC$$

Eliminating E gives

$$C = G_1 G_2 R - G_1 G_2 HC$$

and, rearranging this, the

$$\text{closed-loop transfer function} \quad \frac{C}{R} = \frac{G_1 G_2}{1 + G_1 G_2 H} \tag{3.30}$$

In words, and in somewhat generalized form, this may be stated as follows:

> The *closed-loop transfer function* of the standard loop equals the product of the transfer functions in the forward path divided by the sum of 1 and the loop gain function.

> The *loop gain function* is defined as the product of the transfer functions around the loop.

For the present system the

$$\text{loop gain function} = G_1 G_2 H \tag{3.31}$$

If $H = 1$, then $E = R - C$ is the system error, as in Fig. 1.3, and E/R is the input-to-error transfer function. It will permit the error response for a given input $r(t)$ to be found directly. Since $C = G_1 G_2 E$, (3.30) shows that the

$$\text{input-to-error transfer function} \quad \frac{E}{R} = \frac{1}{1 + G_1 G_2 H} \tag{3.32}$$

The transfer function relating the input to any variable of interest can be found similarly. The derivation of the closed-loop transfer function (3.30) is in effect an example of block diagram reduction, because the block diagram has been reduced to a single block.

Figure 3.25 shows rules for transformations of block diagrams. Note that while blocks in series are multiplied, as in (3.29), parallel loops are added. The following examples show how these rules, and in particular the use of (3.30), can facilitate the reduction of block diagrams.

Example 3.7.1 Minor Loop Feedback

The configuration in Fig. 3.26(a), which includes a *minor feedback loop,* is very common in servomechanisms. The system with velocity feedback in Fig. 3.10 is an example. Derivation of C/R by the approach used to obtain (3.30) would be laborious. However, the reduction becomes simple if the result (3.30) is used directly. It is applied first to

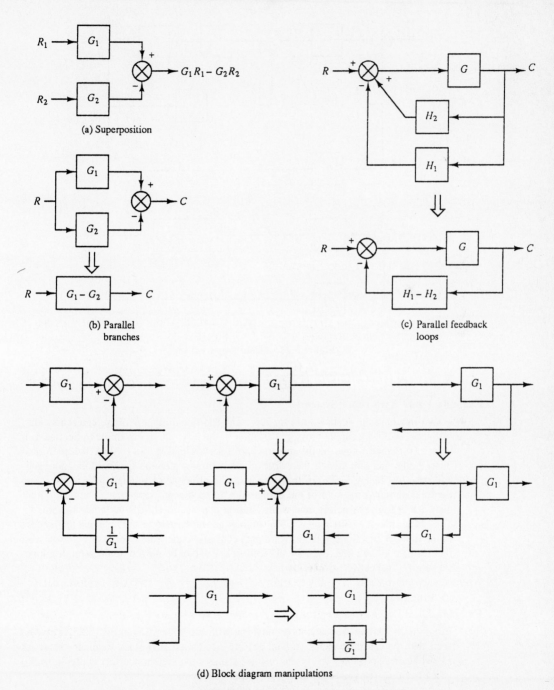

(a) Superposition

(b) Parallel branches

(c) Parallel feedback loops

(d) Block diagram manipulations

Figure 3.25   Rules for block diagram reduction.

reduce the minor feedback loop C/M to a single block, as shown in Fig. 3.26(b). But (3.30) applies again to this new loop and now yields the closed-loop transfer function

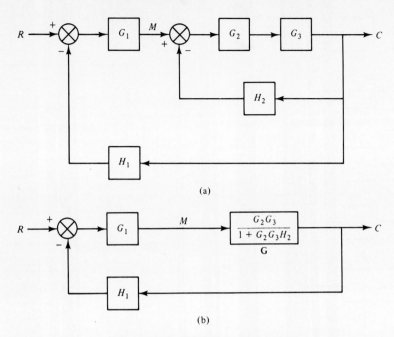

Figure 3.26   Minor loop feedback.

$$\frac{C}{R} = \frac{G_1 G}{1 + G_1 G H_1} = \frac{G_1 G_2 G_3}{1 + G_2 G_3 H_2 + G_1 G_2 G_3 H_1} \qquad (3.33)$$

Example 3.7.2   Two-Input System

A two-input system is shown in Fig. 3.27. The additional input D often represents a disturbance, such as a supply pressure variation in the level control example in Section 1.3, or a disturbance torque in the motor control servos in Fig. 3.11. With the additional block L, the diagram models the effect of the disturbance on the system in the more general case. As in Fig. 3.25, for linear systems the principle of superposition applies, and the total output is the sum of the outputs due to each input separately. Thus the output due to R is found as before, and while finding that due to D, R is put equal to zero.

The rule of (3.30) applies when finding the response to D, but note that the product of the transfer functions in the forward path consists, aside from L, only of G_2. Note also that for $R = 0$ the minus sign for the feedback at R can be moved to the summing junction for D. Inspection now yields

$$\frac{C}{D} = \frac{G_2 L}{1 + G_1 G_2 H} \qquad (3.34)$$

Example 3.7.3

In Fig. 3.28 the two feedback loops interfere with each other. Using Fig. 3.25, it may be seen that the rearrangements (a) and (b) are alternative first steps to make the result (3.30) again applicable. Verify that neither changes the system and that applying (3.30) twice to (a) or (b) yields the closed-loop transfer function

$$\frac{C}{R} = \frac{G_1 G_2}{1 + G_1 H_2 + G_2 H_1} \qquad (3.35)$$

The next example will, in addition to other features in Fig. 3.25, illustrate the transformation of parallel loops, both in the forward and feedback paths.

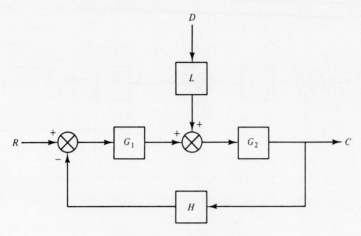

Figure 3.27   Two-input system.

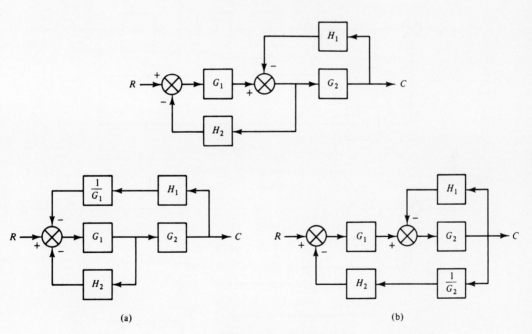

(a)   (b)

Figure 3.28   Example 3.7.3.

Example 3.7.4

Figure 3.29 shows the successive stages of reduction for a system with parallel branches. Other manipulations indicated in Fig. 3.25 are also used, together with the standard equation (3.30). Note that in the reduction from (c) to (d) one of the parallel branches in the forward path in effect contains a block with transfer function unity.

The last example in particular suggests that block diagram reduction can become laborious. In that case the use of signal flow graphs presents an alternative.

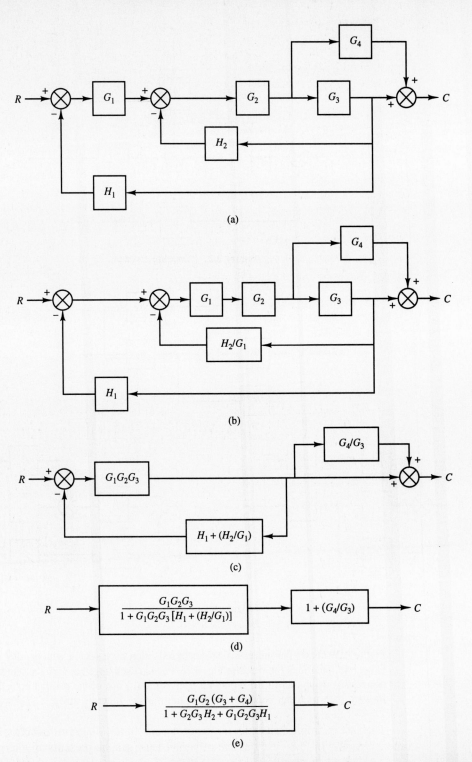

Figure 3.29   Example 3.7.4.

3.8 SIGNAL FLOW GRAPHS AND MASON'S GAIN FORMULA

Signal flow graphs are an alternative to block diagrams. For complex systems, signal flow graphs have the advantage that Mason's gain formula is available to determine the overall transfer function without the need for successive reductions. This formula applies also to block diagrams, but it is often recommended that an equivalent flow graph be constructed first to reduce the possibility of errors. Indeed, even when this is done, the probability of mistakes is considerable for the more complex systems where the formula is needed most, and the presence of such mistakes is not obvious.

For a first course in control systems, where the block diagrams are seldom very complex, the block diagram is preferable because it is closer to the physical reality of the system. Block diagrams are used elsewhere in this book, so study of this section is not needed for material to follow. However, flow graphs are used extensively in control engineering, so their treatment is very appropriate.

Figure 3.30 shows the flow graphs that are equivalent to the block diagrams in the figures indicated. As may be seen from the flow graph for the series connection of two blocks in (a):

- Variables are shown as *nodes* instead of as line segments.
- Line segments between nodes are called *branches,* and the associated arrows and transfer functions, here generally called *gains,* represent the relations between variables.

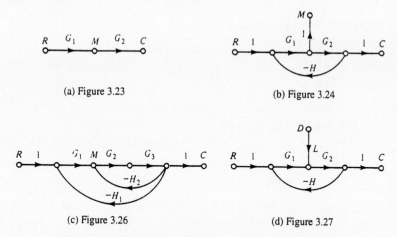

(a) Figure 3.23

(b) Figure 3.24

(c) Figure 3.26

(d) Figure 3.27

Figure 3.30   Signal flow graphs.

Some important definitions and properties related to nodes are the following:

1. The signal at a node is the sum of the signals on all incoming branches, and this signal is transmitted to all outgoing branches. (Note the difference with a summing junction, which has only one "outgoing branch.")
2. An *input node* has only outgoing branches.
3. An *output node* has only incoming branches.

Except for these, all the usual rules for block diagrams apply, and the flow graphs in Fig. 3.30 may be verified to be equivalent to the corresponding block diagrams. The signal at each node is equal to the sum of the signals at all nodes connected to it by incoming branches, each multiplied by the corresponding branch gain. Note how branches with unity gains have been introduced to satisfy the definitions of input and output nodes. In the same way, any node can be considered as an output node by the introduction of a unity gain outgoing branch. For illustration, this has been done for the actuating signal M in Fig. 3.30(b).

Example 3.8.1 Signal Flow Graphs for Figs. 3.28 and 3.29

Figure 3.31 shows the signal flow graphs that correspond to the block diagrams in Figs. 3.28 and 3.29. Figure 3.31(b) is almost the same as Fig. 3.30(c), except for the forward branch with gain G_4 to the output node. This node has only incoming branches, so a unity gain branch to the right of it, as in Fig. 3.31(a), is not needed. However, the unity gain branch to the left of it is necessary, since without it the feedback signals to H_1 and H_2 would be incorrect.

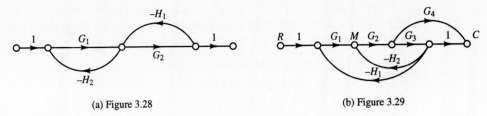

(a) Figure 3.28   (b) Figure 3.29

Figure 3.31   Signal flow graphs for Figs. 3.28 and 3.29.

Mason's Gain Formula

Mason's gain formula gives the overall transfer function or gain T between selected input and output nodes. To express this formula, some additional definitions are needed first.

- A *path* is a sequence of branches between selected nodes.
- A *loop* is a closed path.
- The *path gain* is the product of the branch gains along the path.
- The *loop gain* is the path gain along a loop.
- *Nontouching* loops are loops that do not have any common nodes.

For example, examination of Fig. 3.31(b) shows two loops and two possible forward paths from input node R to output node C. The path gains P_1 and P_2 and the loop gains L_1 and L_2 are

$$\begin{aligned} P_1 &= G_1 G_2 G_3 & P_2 &= G_1 G_2 G_4 \\ L_1 &= -G_1 G_2 G_3 H_1 & L_2 &= -G_2 G_3 H_2 \end{aligned} \tag{3.36}$$

Mason's formula for the overall gain is

$$T = \frac{1}{\Delta} \sum_k P_k \Delta_k \tag{3.37}$$

where $\Delta = 1 -$ (sum of all individual loop gains) + (sum of products of gains of all possible combinations of two nontouching loops) $-$ (sum of products of gains of all possible combinations of three nontouching loops) $+ \cdots$

P_k = path gain of kth forward path

Δ_k = cofactor of kth forward path, obtained from Δ by removing the loops that touch this path.

In the special case where all forward paths and all loops touch each other, it may be seen that the cofactors are all unity and that the gain formula reduces to the following:

The overall gain is equal to the sum of the gains of the forward paths divided by the difference between 1 and the sum of the loop gains.

Example 3.8.2   Gains for Flow Graphs in Fig. 3.31

In Fig. 3.31(a) there is only one forward path from R to C, and it touches both feedback loops, which also touch each other. Hence the system satisfies the conditions for the special case, and the gain is

$$\frac{C}{R} = \frac{G_1 G_2}{1 + G_1 H_2 + G_2 H_1}$$

In Fig. 3.31(b) there are two forward paths and two loops, with gains as given by (3.36). Here also, both loops touch each other as well as both forward paths, so the gain is again obtained from the formula for the special case:

$$\frac{C}{R} = \frac{G_1 G_2 (G_3 + G_4)}{1 + G_2 G_3 H_2 + G_1 G_2 G_3 H_1}$$

Note that these results agree with those obtained by block diagram reduction in (3.35) and Fig. 3.29.

Example 3.8.3   Gain for Flow Graph in Fig. 3.32

Figure 3.32 shows two forward paths added to the block diagram of Fig. 3.28 and how this changes the flow graph of Fig. 3.31(a). There are now three forward paths from R to C and three loops, with gains

$$P_1 = G_1 G_2 \qquad P_2 = G_3 \qquad P_3 = G_4$$
$$L_1 = -G_1 H_2 \qquad L_2 = -G_2 H_1 \qquad L_3 = G_3 H_1 H_2$$

All three loops touch, so Δ is 1 minus the sum of the loop gains:

$$\Delta = 1 + G_1 H_2 + G_2 H_1 - G_3 H_1 H_2$$

The forward paths P_1 and P_2 touch all three loops, so the corresponding cofactors Δ_1 and Δ_2 are unity. But loop L_1 does not touch path P_3, so

$$\Delta_3 = 1 + G_1 H_2$$

Substitution into the gain formula now yields the closed-loop transfer function:

$$\frac{C}{R} = \frac{G_1 G_2 + G_3 + G_4 (1 + G_1 H_2)}{1 + G_1 H_2 + G_2 H_1 - G_3 H_1 H_2} \tag{3.38}$$

3.9 CONCLUSION. HIGHER-ORDER SYSTEMS

Following Chapter 2 on transfer functions of physical subsystems, in this chapter the modeling of systems with feedback was considered. The examples include those in which the structure of the system and the nature of the feedback are clear from the schematic diagram, as well as cases in which even the existence of feedback may not be obvious. In particular in the latter case, a good block diagram is an important aid in system analysis and design. The physical realization and modeling of common system controllers using pneumatics or active circuits based on operational amplifiers

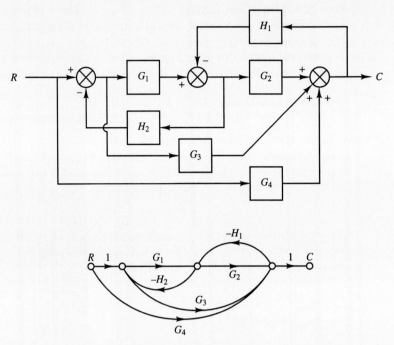

Figure 3.32   Example 3.8.3.

were treated. This modeling also provides additional examples of the derivation of block diagrams for systems with feedback. The application of the operational amplifier in system simulation was touched on as well. Active circuits based on operational amplifiers are an alternative to the passive electrical networks in Fig. 2.8. Digital computer realization of controllers in the form of computer control algorithms will be discussed later.

The reduction of block diagrams such as those derived in this chapter to determine overall system behavior was considered next. Signal flow graphs and Mason's gain formula were included as a good alternative for complex block diagrams.

The examples considered in this chapter and the preceding one have been of relatively low order. Higher-order systems arise all too easily. For example, in mechanical drives an electric or hydraulic motor may operate a rotating system that must be modeled by a number of inertias and interconnecting torsion springs, as in Example 2.2.3. Each added spring–inertia combination in effect adds two poles. An analogous situation exists for multimass translating, instead of rotating, systems along the lines of Example 2.2.2. Temperature control with multiple thermal capacitances as in Example 2.5.3 augmenting the dynamics of the heat source can also involve high-order transfer functions.

Although high-order dynamics are most often caused by the process, other elements in the loop also contribute. For example, a third-order transfer function was derived in Example 2.7.3 to model a valve-controlled hydraulic cylinder with allowance for the effect of oil compressibility. In this model the mechanical displacement of the spool valve was the input. However, in high-pressure hydraulic servos the forces required for positioning this valve are quite large and are generated instead by

applying hydraulic pressures to the ends of the spool. These pressures are generated
in a first stage of hydraulic power amplification, with the valve itself acting as the
second stage. Two-stage electrohydraulic servos are used extensively in many areas
of engineering for high-performance positioning of heavy loads. In these servos the
first-stage amplifier often involves an electromagnetic torque motor, and the system
input is an electrical signal. For high-performance design it may be necessary to allow
for the effects of motor inertia and valve spool mass as well as oil compressibility.
The transfer function from input to valve spool position can then be of sixth order,
raising that from input to output from order 3 to 9.

PROBLEMS

3.1. An armature-controlled dc motor is used in a speed control system shown in
Fig. P3.1. The voltage r representing desired speed is obtained by the setting of a
potentiometer to which a constant voltage V is applied. Actual speed is represented
by the voltage from a tachometer–generator on the motor shaft. The difference of
these voltages, the speed error signal, is applied to a voltage amplifier of gain K_a, of
which the output is raised in power in the power amplifier K_p. Using (2.12) to represent the motor and load, obtain a block diagram and express the steady-state speed
error for step demands.

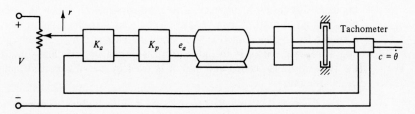

Figure P3.1

3.2. A field-controlled dc motor is used in the position control system indicated in
Fig. P3.2. Using the results of Fig. 3.11(a) for Example 3.3.3 where appropriate,
write the modified equations necessary and obtain a block diagram representation.
Shaft position θ_o is measured by means of a potentiometer. The torsion spring on
the output shaft has spring constant k.

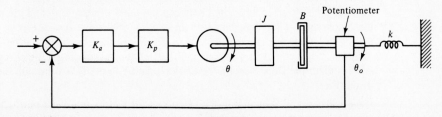

Figure P3.2

3.3. Figure P3.3 shows a schematic diagram of a temperature control system. The fluid
is incompressible, with specific heat c, the mass in the tank is W, and the mass flow
rate through the tank is w. The temperature sensor is described by a simple lag

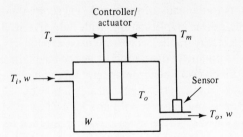

Figure P3.3

transfer function. The desired temperature T_s is the input to a controller with transfer function G_c, and heat flow rate q from the heater is proportional to controller output.

Write the equations and obtain a block diagram from which the effect on T_o of each of T_s, T_i, and w could be studied. Remember that for small variations x and y about x_0 and y_0 the expression $(x_0 + x)(y_0 + y)$ is linearized to $(x_0 y + y_0 x)$, considering only the changes from operating-point values.

3.4. Obtain a block diagram for study of the dynamic behavior of a ship stabilizer from the following description. The differential equation for the roll angle θ due to a disturbance torque T_d on the ship in rough seas is $J\ddot{\theta} + b\dot{\theta} + k\theta = T_d$, where J is the inertia, b a generally small damping constant, and k a spring constant representing the self-righting effect due to ship geometry. To minimize roll, many ships are equipped with stabilizer fins on the sides, of which the angles can be adjusted in opposite directions to produce a restoring torque T_r, much like the flaps on the wings of an aircraft. T_r can be taken to be related to the input of the fin actuator by a simple lag transfer function. Roll angle is measured by a vertical gyro roll sensor, and a roll rate sensor is included to measure the rate of change of θ. Explain why the rate sensor is needed and how it is used in the system, and include this in the block diagram.

3.5. In the pressure regulator of Fig. 3.4, let a resistance R separate the load pressure from the small space under the diaphragm, where the pressure is P. Obtain a block diagram that shows the effect of load pressure disturbances on P. Neglect the capacitance of the volume under the diaphragm, as well as the inertia and friction and the flow corresponding to diaphragm motion. The supply pressure is constant.

3.6. Amend the block diagram for the pressure regulator in Fig. 3.5(a) and the corresponding equations to allow for disturbances in the supply pressure.

3.7. The pneumatic pressure regulator in Fig. P3.7 is a modification of that in Fig. 3.4 discussed in Examples 3.2.3 and 3.2.4. A seal with an orifice of resistance R through it has been incorporated between the output pressure p_l and the space under the diaphragm. The capacitance of this space, where the pressure is p, is C_d, and that of the volume under pressure p_l is C_g. Write the equations and derive a block diagram. Neglect inertia, friction, and the flow corresponding to diaphragm motion. Also neglect the flow through the orifice relative to the flows through the valve and to the load.

3.8. Repeat Problem 3.7 including the effects of all factors that were omitted. Introduce additional parameters as needed.

3.9. A form of water-level control is suggested in Fig. P3.9. Here the valve–cylinder modeled in Fig. 2.27 is used to control the water valve opening. The oil and water supply pressures are constant. Write the necessary equations, introducing parameters as needed, and obtain a block diagram model.

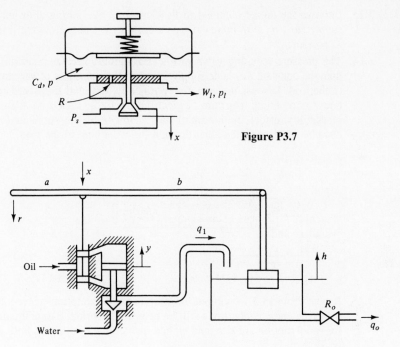

Figure P3.7

Figure P3.9

3.10. Write the linearized equations and derive a block diagram for the hydraulic servo in Fig. P3.10. Introduce parameters as needed and assume the load on the cylinder to be small.

3.11. Refine the hydraulic servo model for Fig. P3.10 derived in Problem 3.10 by assuming the mass m to be so large that the loading on the cylinder can no longer be neglected and oil compressibility must be included. Introduce parameters as needed, write all linearized equations, and obtain a block diagram representation.

3.12. The hydraulic pressure controller in Fig. P3.12 must keep the pressure p_l to the load constant, regardless of variations of load flow q_l. Spring, damping, and mass constants for the valve spool are k, b, and m, and its surface area is A. The flow q_p from a pump may be taken to be constant. The volume of oil under pressure p_l is V and its bulk modulus β. Discuss the operation physically. Write all linearized equations and set up a good diagram that clearly identifies the feedback.

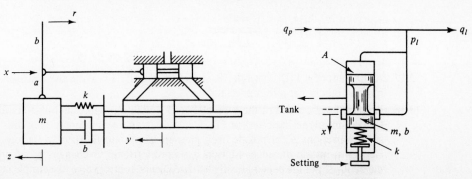

Figure P3.10   Figure P3.12

3.13. Improve the model obtained in Problem 3.12 by making allowance for a drop in supply flow q_p with increasing pressure due to leakage, taking it to be proportional to pressure.

3.14. The pressure-reducing valve in Fig. P3.14 must keep p_l constant, at a value lower than the constant available supply pressure P_s, regardless of variations of the flow q_l to the load. Discuss the operation physically, write all linearized equations, and obtain a good block diagram. Neglect the inertia and friction. Also neglect the compressibility effect in the small end chamber of the spool and the flow to this chamber relative to the flow through the valve and to the load.

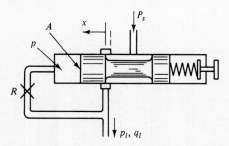

Figure P3.14

3.15. Repeat Problem 3.14 without the simplifying assumptions, except that compressibility in the end chamber may still be neglected. Spring, damping, and mass constants for spool motion are k, b, and m, and the volume of oil, with bulk modulus β, under pressure p_l is V.

3.16. A pressure-compensated pump system is indicated in Fig. P3.16. It uses a variable-displacement pump, of which the stroke, and hence the delivery per revolution, can be varied by a mechanical adjustment x. Variations q_p of pump delivery can be taken to be proportional to the adjustments x. The equivalent spring, damping, and mass load on the piston are represented by the constants k, b, and m, and the volume of oil under pressure p_l is V. How does the system operate? Introducing parameters as needed, write the linearized model equations, with allowance for leakage, and derive the block diagram.

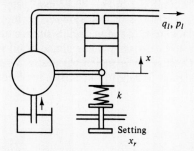

Figure P3.16

3.17. A diagram of a PID controller is shown in Fig. P3.17. Write the equations and derive the block diagram representation.

3.18. A water-level control of the type of Fig. 3.1 with a proportional pneumatic controller (P control) is indicated in Fig. P3.18. Air and water supply pressures are constant. Variations p_o of nozzle back pressure are proportional to variations x_f of the flapper-to-nozzle distance. Motion x_b is proportional to changes p_o of pressure in the bellows. For the pneumatically actuated water control valve, the simple lag transfer

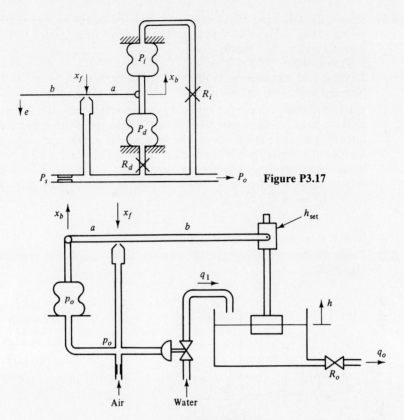

Figure P3.17

Figure P3.18

function $K/(Ts + 1)$ can be assumed. Introducing parameters as needed, write the necessary equations and represent the system by a block diagram. The desired level is set by screw adjustment on the float post.

3.19. The water-level control system of Fig. P3.18 has been extended in Fig. P3.19 by an additional lever and bellows. Write all linearized equations and represent the system by a block diagram.

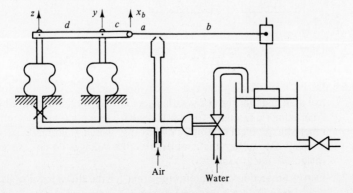

Figure P3.19

3.20. Realize the following dynamic compensators by means of operational amplifiers, giving values for resistors and capacitors in terms of megohms ($= 10^6$ ohm) and microfarads ($= 10^{-6}$ farad):

1. PI controller: $-(10 + 20/s)$.

2. Phase-lead compensator: $-2(0.5s + 1)/(0.1s + 1)$.

Note that capacitors and resistors are not available in odd sizes.

3.21. Velocity or rate feedback is also used to improve damping in systems of which the plant does not include a motor and the rate of change of the output cannot be measured by a tachometer–generator. The derivative of a measured output must then be found, and a simple filter is often added to reduce the effect of high-frequency noise in the measured output. The desired compensator is then a transient rate or transient velocity minor loop feedback transfer function of the form of the following example:

$$-\frac{5s}{0.1s + 1}$$

Realize this compensator using operational amplifiers.

3.22. Derive the transfer functions corresponding to the operational amplifier circuits in Fig. P3.22.

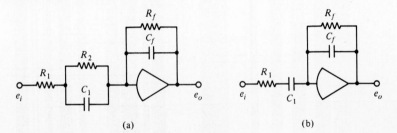

(a)   (b)

Figure P3.22

3.23. Show the general form of an operational amplifier block diagram type of simulation of the motor position control system in Fig. P3.23. The position sensor is described by a simple lag transfer function, and a phase-lag or phase-lead compensator is used.

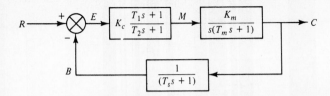

Figure P3.23

3.24. Derive a block diagram type of analog computer simulation for a temperature control system as in Fig. 3.21 with $T_1 = 1$, $T_2 = 0.2$, and with PI control $13(1 + 2/s)$. Give numerical values for the resistors, capacitors, and potentiometers present, but ignore the problem of amplitude scaling, which is normally part of analog computer simulation to ensure that amplifier output voltages do not exceed permissible levels, inside the linear range.

3.25. Derive an analog computer simulation for the differential equation

$$\ddot{x} + 3.5\dot{x} + 9.25x = r$$

(a) Of the type of Fig. 3.22.

(b) An alternative using one less operational amplifier, by using the integrator that generates $\dot{x}$ also as a summing amplifier.

As in Problem 3.24, ignore amplitude scaling and also time scaling, which can be used to simulate slow processes in less than real time.

3.26. For the system in Fig. P3.26, express the transfer function E/R and find the steady-state values of the error following unit step and unit ramp inputs.

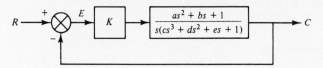

Figure P3.26

3.27. Reduce the block diagrams in Fig. P3.27 and find the closed-loop transfer functions.

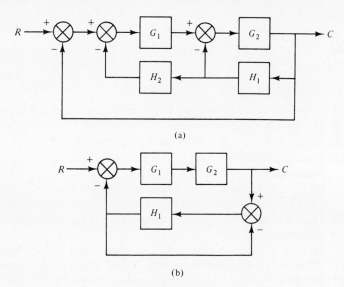

(a)

(b)

Figure P3.27

3.28. Find the transfer functions C/R and C/D for the block diagram shown in Fig. P3.28.

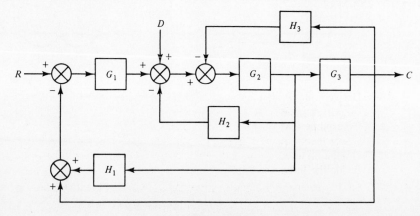

Figure P3.28

3.29. For the two-input, two-output system shown in Fig. P3.29, determine the transfer functions C_1/R_1 and C_1/R_2 relating C_1 to R_1 and R_2, respectively, and express C_1 when both inputs are present.

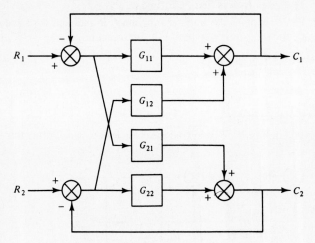

Figure P3.29

3.30. Draw the signal flow graph corresponding to the block diagram in Fig. P3.27(a) and use Mason's gain formula to find the closed-loop transfer function.

3.31. Repeat Problem 3.30 for the block diagram in Fig. P3.27(b).

3.32. Draw the signal flow graph corresponding to the block diagram in Fig. P3.28 and determine:
(a) The closed-loop transfer function.
(b) The transfer function for the response of the output to the disturbance input.

3.33. Draw the signal flow graph corresponding to the block diagram in Fig. P3.29 and determine:
(a) The transfer function relating the first output to the first input.
(b) The transfer function relating the first output to the second input.

3.34. For the signal flow graphs shown in Fig. P3.34:
(a) Determine the overall transfer functions.
(b) Draw the equivalent block diagrams.
Note that the presence or absence of the unity gain branch is important.

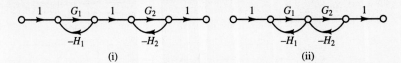

Figure P3.34

4

The Performance of Feedback Systems

4.1 INTRODUCTION

The first part of this chapter is concerned with the motivations for the use of feedback and its effect on performance. The prime reasons for using feedback are the following:

1. Reducing the sensitivity of the performance to parameter variations of the plant and imperfections of the plant model used for design
2. Reducing the sensitivity to disturbance inputs and noise

This includes the sensitivity to plant parameter variations, but also that to known and unknown imperfections and uncertainties of the model. An example of this is the unmodeled high-frequency dynamics such as the higher modes of vibration mentioned in Example 3.4.2.

It is apparent that sensitivity reduction is central to control system design. In the design of feedback to achieve this, two additional considerations enter which in fact will turn out to be the focus of most design techniques:

3. Improving transient response
4. Reducing steady-state errors

These will be discussed in the second part of this chapter. The techniques in Chapter 1 permit transient responses to be calculated, but do not provide much insight into this behavior. Such insight is necessary for design and also leads to the formulation of performance criteria for the transient response.

4.2 EFFECT OF FEEDBACK ON SENSITIVITY AND DISTURBANCE RESPONSE

Sensitivity to Parameter Variations and Model Uncertainty

Consider the standard feedback loop in Fig. 4.1. G is the transfer function of the plant or process to be controlled, G_c is that of a controller, and H may represent the feedback sensor. The plant model G is usually an approximation to the actual dynamic behavior, with various simplifying assumptions and neglecting dynamics at frequencies thought to be sufficiently above the range of interest in the closed-loop control system. Even then the parameter values in the model are often not precisely known and may also vary widely with operating conditions. An aircraft at low level responds differently to control surface deflections than at high altitude. A power plant model linearized about the 30% of full power operating point has different parameter values than that linearized about the 75% point. For very wide parameter variations, adaptive control schemes, which adjust the controller parameters, may be necessary, but a prime advantage of feedback is that it can provide a strong reduction of the sensitivity without such changes of G_c.

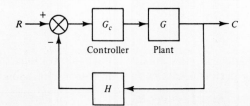

Figure 4.1   Standard loop.

The sensitivity is determined primarily by the *loop gain $G_c GH$*, defined earlier in (3.31):

$G_c GH$ = loop gain function   (the product of the transfer functions around the loop)

The closed-loop transfer function is

$$T = \frac{C}{R} = \frac{G_c G}{1 + G_c GH} \tag{4.1}$$

and shows that if $G_c GH \gg 1$,

$$T = \frac{C}{R} \approx \frac{G_c G}{G_c GH} = \frac{1}{H} \tag{4.2}$$

because the term 1 in the denominator is then relatively negligible. This approximation becomes more and more accurate as the loop gain increases above 1. Hence, as examples will later confirm:

- If the loop gain $G_c GH \gg 1$, C/R depends almost entirely on the feedback H alone and is virtually independent of the plant and other elements in the forward path and of the variations of their parameters.
- The sensitivity of the closed-loop performance to the elements in the forward path reduces as the loop gain is increased.

This is a major reason for the use of feedback. With open-loop control ($H = 0$) the input–output relation is $C/R = G_c G$. Choice of G_c on the basis of an approximate plant model or a model of which the parameters are incorrect will cause errors in C proportional to those in G_c, and any changes in G due to parameter variations will cause proportional changes in C. With feedback, these effects can be strongly attenuated so that approximations and parameter variations in G are much less objectionable. But it is also apparent from (4.2) that the feedback sensor H must be chosen for small parameter variations. However, unlike G, it is usually under the control of the designer.

Example 4.2.1

To illustrate how feedback can reduce the effects of parameter variations, consider Fig. 4.2, where the blocks are simple gains. Without feedback, if the value of K_2 is halved, then C/R is halved also. With the feedback loop present, (4.1) shows that the effect on C/R of halving K_2 depends on the value of $K_1 K_2$. Let

$$\frac{C}{R} = \frac{K_1 K_2}{1 + K_1 K_2} = A$$

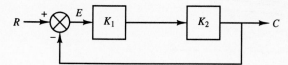

Figure 4.2   Example 4.2.1.

For $K_1 K_2 = 1$ this gives $A = 1/2$, and halving K_2 will reduce this to $A = 1/3$, or to 67% of the original value. But if $K_1 K_2 = 9$, then $A = 0.9$, which reduces to $A = 0.818$ if K_2 is halved. This is 91% of the original value. The larger loop gain has reduced the sensitivity.

Example 4.2.2 Level Control

In Fig. 4.3(a) the closed-loop transfer function is

$$T = \frac{KA}{\tau s + 1 + KAh}$$

Frequently, the variations of A and τ will be interrelated. Suppose that Fig. 4.3(a) models a single-tank level control. Then, from Example 2.6.1, $A = K_v R_1$ and $\tau = A_1 R_1$, where K_v is the control valve gain and A_1 and R_1 are the tank area and outlet valve resistance. R_1 could vary due to obstructions or inadvertent adjustment, and level sensor gain h could change due to a malfunction. To illustrate the effects, unit step responses when these parameters are halved and doubled from their nominal values will be compared, assuming that $K = 9$, $K_v = 1$, $A_1 = 1$, R_1 (nom.) $= 1$, and h (nom.) $= 1$. Leaving R_1 and h as free parameters, the closed-loop transfer function yields the unit step response

$$c(t) = \frac{9}{9h + 1/R_1}\left\{1 - \exp\left[-\left(9h + \frac{1}{R_1}\right)t\right]\right\}$$

This is plotted in Fig. 4.3(b) for the nominal parameters and when one of R_1 and h is 2 or 0.5 times the nominal value. As expected, the response is far more sensitive to h than to R_1. In view of the large parameter changes, the differences between the curves for $R_1 = 1$, 2, and 0.5 appear small. If K were reduced, reducing loop gain, these differences would be larger. This trend is evident in the curves in that the loop gain is smaller for $R_1 = 0.5$ than for $R_1 = 2$. Accordingly, the response for $R_1 = 0.5$ differs more from the nominal plot than that for $R_1 = 2$. With open-loop control ($h = 0$), halving or doubling of R_1, and hence of A, would cause proportional changes in the steady-state values of the response.

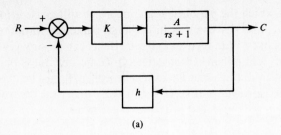

(a)

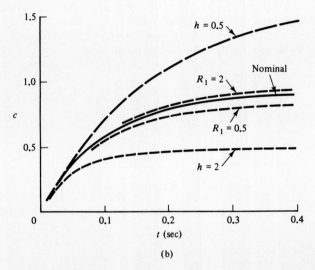

(b)   **Figure 4.3** Examples 4.2.2 and 4.2.3.

Formally, the sensitivity properties can be studied by use of the *sensitivity function S.* For example, the sensitivity of the closed-loop transfer function T to changes in the forward path transfer function $G_f = G_c G$ is the percentage of change $\partial T/T$ divided by the percentage of change $\partial G_f/G_f$ of G_f that causes it:

$$S = \frac{\partial T/T}{\partial G_f/G_f} = \frac{G_f}{T}\frac{\partial T}{\partial G_f} = \frac{G_f}{T}\frac{\partial}{\partial G_f}\frac{G_f}{1 + G_f H}$$

$$= \frac{G_f}{T}\frac{1}{(1 + G_f H)^2} = \frac{1}{1 + G_f H} \tag{4.3}$$

The *static sensitivity* is the value of S for $s \to 0$. *Dynamic sensitivities* are usually calculated by replacing s by $j\omega$ and plotting S as a function of frequency ω. Such results indicate, as the discussion of frequency response methods will show, how sensitivity changes with the frequency of a sinusoidal input R. These methods will also provide a clear picture of the sensitivity to unmodeled high-frequency dynamics. Furthermore, they will show that the condition loop gain $\gg 1$ is generally satisfied only over a limited range of frequencies.

Example 4.2.3   Sensitivity Functions

Application of the sensitivity function S of (4.3) to the closed-loop transfer function T in Example 4.2.2 yields the following sensitivities S_a, S_h, and S_τ for small changes of A, h, and τ:

$$S_a = \frac{\partial T/T}{\partial A/A} = \frac{A}{T}\frac{\partial T}{\partial A} = \frac{A}{T}\frac{K(\tau s + 1)}{(\tau s + 1 + KAh)^2} = \frac{\tau s + 1}{\tau s + 1 + KAh} \tag{4.4a}$$

$$S_h = \frac{\partial T/T}{\partial h/h} = \frac{h}{T}\frac{\partial T}{\partial h} = \frac{h}{T}\frac{-(KA)^2}{(\tau s + 1 + KAh)^2} = \frac{-KAh}{\tau s + 1 + KAh} \tag{4.4b}$$

$$S_\tau = \frac{\partial T/T}{\partial \tau/\tau} = \frac{\tau}{T}\frac{\partial T}{\partial \tau} = \frac{\tau}{T}\frac{-KAs}{(\tau s + 1 + KAh)^2} = \frac{-\tau s}{\tau s + 1 + KAh} \tag{4.4c}$$

The static sensitivity S_{as} is seen to reduce as loop gain KAh increases, but the magnitude of S_{hs} will approach 1. It is seen also that the static sensitivity $S_{\tau s}$ is zero.

Effect of External Disturbances

Several examples in Chapter 3 included the modeling of disturbance inputs. These fit the model shown in Fig. 4.4. The effect of disturbance D on output C is given by the transfer function

$$\frac{C}{D} = \frac{G_2 L}{1 + G_c G_1 G_2 H} = -\frac{E}{D} \qquad \text{for } H = 1 \tag{4.5}$$

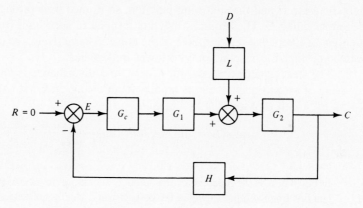

Figure 4.4   Disturbance inputs.

and the effect of input R on C by the closed-loop transfer function

$$\frac{C}{R} = \frac{G_c G_1 G_2}{1 + G_c G_1 G_2 H} \tag{4.6}$$

For loop gains $G_c G_1 G_2 H \gg 1$,

$$\frac{C}{D} \approx \frac{L}{G_c G_1 H} \qquad \frac{C}{R} \approx \frac{1}{H} \tag{4.7}$$

To minimize the response to the disturbance input D, C/D should be as small as possible. From (4.7), this can be achieved by high gain $G_c G_1 H$ in the feedback loop between C and the point where D enters the loop.

But it is also necessary that the system respond well to input R, and the question is whether both requirements can be met simultaneously. From (4.6), with $H = 1$, loop gains $G_c G_1 G_2 \gg 1$ will ensure that $C/R = G_c G_1 G_2/(1 + G_c G_1 G_2)$ is near $C/R = 1$, as desired. So here the high gain should be in the forward path between R and C. Thus both requirements can indeed be met by locating the high gains in $G_c G_1$, between the points where R and D enter the loop. Increasing the loop gain improves these features. This shows the following:

- If loop gain $G_c G_1 G_2 H \gg 1$, then feedback strongly reduces the effect of disturbance D on C if $G_c G_1 H \gg 1$, so if the high gain is in the feedback path between C and D.
- To ensure a good response to input R as well, the location of the high gain should be further restricted to $G_c G_1$, between the points where R and D enter the loop.
- The sensitivity to disturbances reduces as this gain increases.

Disturbances may also enter the feedback path, due to the sensor measuring C. The importance of measures to avoid this is clear, since normally there will not be a high gain in the feedback between C and such disturbances to attenuate their effect. This sensor noise is often at high frequencies, and frequency response methods will again provide the clearest picture of when these frequencies are high enough that the noise is not objectionable. The sensor noise may be modeled to enter the feedback loop between H and the summing junction, so it must pass through G_c, G_1, and G_2 before reaching C. These blocks may not respond to such high frequencies and may thus act as filters that remove most of the noise. Alternatively, the closed-loop system can be designed to respond only to lower frequencies.

The steady-state response, for $t \to \infty$, to a disturbance is clearly an important measure of system quality and is readily found using the final value theorem. From (4.5), for a unit step input $D = 1/s$,

$$\lim_{t \to \infty} c(t) = \lim_{s \to 0} sC(s) = \lim_{s \to 0} \frac{G_2 L}{1 + G_c G_1 G_2 H} \qquad (4.8)$$

Example 4.2.4

In Fig. 4.5, R is the input which the output C should follow as closely as possible, and D is a disturbance input to which the output should ideally not respond at all. The blocks are simple gains. It is desired to find K_1 and K_2 for two sets of specifications:

$$\text{(i)} \ \frac{C}{D} = 0.1, \ \frac{C}{R} = 0.9 \qquad \text{(ii)} \ \frac{C}{D} = 0.01, \ \frac{C}{R} = 0.99$$

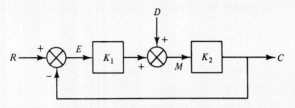

Figure 4.5 Example 4.2.4.

The transfer functions are

$$\frac{C}{R} = \frac{K_1 K_2}{\Delta} \qquad \frac{C}{D} = \frac{K_2}{\Delta} \qquad \Delta = 1 + K_1 K_2$$

and the solutions are readily found by substituting the specifications into these equations:

$$\text{(i)} \ K_1 = 9, K_2 = 1 \qquad \text{(ii)} \ K_1 = 99, K_2 = 1$$

Note that the more severe specifications (ii) require a large increase of gain between the points where R and D enter the loop, that is, in the forward path for R and in the feedback path for D.

Example 4.2.5   Motor Position Control (Fig. 3.11a)

For this system, if $G_c = K_c$, an amplifier with gain K_c, then

$$G_c G_1 = \frac{K_c K_a K_t / R_f}{T_f s + 1} \qquad L = H = 1 \qquad G_2 = \frac{1/B}{s(T_m s + 1)}$$

Hence, with the negative sign at the load disturbance torque $T_l = 1/s$, use of (4.5) and the final value theorem yields

$$
\begin{aligned}
\lim_{t \to \infty} c(t) &= \lim_{s \to 0} sC(s) \\
&= \lim_{s \to 0} \frac{-(T_f s + 1)/B}{s(T_m s + 1)(T_f s + 1) + K_c K_a K_t / (R_f B)} \qquad (4.9) \\
&= -\frac{R_f}{K_c K_a K_t}
\end{aligned}
$$

High controller gain, therefore, makes the system less sensitive to load disturbances, as well as to parameter variations. Again, the high gain is needed between the points where the reference input and the disturbance input enter the loop. It is useful for the sake of insight to verify the result (4.9) directly from Fig. 3.11(a). In the steady state, the net torque $T - T_l$ must be zero, so if $T_l = 1$, T must be 1 as well. From the diagram, this requires that $e = R_f / (K_c K_a K_t)$. Since input r is zero while considering the response to T_l, $e = r - c = -c$, so indeed $c = -R_f / (K_c K_a K_t)$.

4.3 STEADY-STATE ERRORS IN FEEDBACK SYSTEMS

High loop gains were shown to be advantageous to reduce the sensitivity to modeling accuracy, parameter variations, and disturbance inputs. They will now prove equally desirable from the point of view of the reduction of steady-state errors in feedback systems, as intuitive reasoning in Section 1.4 and some examples already suggested.

Consider the *unity feedback system* in Fig. 4.6. From the input-to-error transfer function (3.32), the transform of the error is $E = R/(1 + G)$. The steady-state error e_{ss} can be found directly, without the need for inverse transformation, by the final value theorem:

$$e_{ss} = \lim_{t \to \infty} e(t) = \lim_{s \to 0} sE(s) = \lim_{s \to 0} \frac{sR(s)}{1 + G(s)} \qquad (4.10)$$

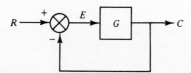

Figure 4.6   Unity feedback system.

For $G(s)$, the following general form is assumed:

$$G(s) = \frac{K}{s^n} \frac{a_k s^k + \cdots + a_1 s + 1}{b_l s^l + \cdots + b_1 s + 1} \qquad (4.11)$$

Definitions

- *Gain:* K as given, with the constant terms in numerator and denominator polynomials made unity, is formally the gain of the transfer function G. It should be distinguished from the *root locus gain,* defined earlier as that for which the highest power coefficients are unity, and equal to Ka_k/b_l in (4.11). Note that the gain can be expressed as

$$\text{gain } K = \lim_{s \to 0} s^n G(s)$$

- *Type number:* The type number of G is the value of the integer n. As discussed in Section 1.6, a factor s in the denominator represents an integration, so the type number is the number of integrators in G.
- *Position error constant K_p:* The value of gain K for $n = 0$.
- *Velocity error constant K_v:* The value of gain K for $n = 1$.
- *Acceleration error constant K_a:* The value of gain K for $n = 2$.

These three special names for the gain K are often used.

Equation (4.11) shows that $\lim_{s \to 0} G(s) = \lim_{s \to 0} (K/s^n)$, so (4.10) can be written as

$$e_{ss} = \lim_{s \to 0} \frac{sR(s)}{1 + (K/s^n)} \tag{4.12}$$

This readily yields Table 4.3.1 for the steady-state errors corresponding to different type numbers and inputs, of which the transforms are given in Table 1.6.1. For example, for a type 2 system with a unit ramp input,

$$e_{ss} = \lim_{s \to 0} \frac{s(1/s^2)}{1 + (K/s^2)} = \lim_{s \to 0} \frac{s}{s^2 + K} = 0$$

TABLE 4.3.1   STEADY-STATE ERRORS

Type Number:	$n = 0$	$n = 1$	$n = 2$
Step $u(t)$; $R = 1/s$	$\dfrac{1}{1 + K_p}$	0	0
Ramp t; $R = 1/s^2$	∞	$\dfrac{1}{K_v}$	0
Acceleration $t^2/2$; $R = 1/s^3$	∞	∞	$\dfrac{1}{K_a}$

For insight into these results it is useful to consider Fig. 4.7, which shows type 0 and type 1 systems for $s \to 0$. For $n = 0$, $C = K_p E$, so there cannot be a nonzero output without a proportional, nonzero error. If the output must increase along a ramp, the error must increase according to a ramp as well. So a type 0 system has a steady-state error for a step input and cannot follow a ramp.

For $n = 1$,

$$C = \frac{K_v}{s} E \qquad \text{so } c(t) = K_v \int e(t)\, dt$$

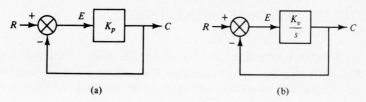

Figure 4.7   Types (a) 0 and (b) 1 systems.

Therefore, the output cannot level off to a constant value unless the error levels off at zero. A steady state with nonzero steady-state error cannot exist for a step input because the integrator would cause the output to change. This means that if, as is often the case, the performance specifications require zero steady-state error after a step input, the designer must ensure that the system is at least of type 1. A type 2 system would be necessary if zero steady-state errors following both steps and ramps were specified. The hydraulic servo and electric motor control examples in Chapter 3 are type 1 systems, but the level control is type 0, unless it is made into type 1 via the controller.

The nonzero finite errors in Table 4.3.1 decrease as gain is increased, as do parameter sensitivity and disturbance response. As suggested in Section 1.4, for larger K a smaller error can achieve the same effect on the output.

Example 4.3.1

In Fig. 4.6:

(a) $G(s) = \dfrac{K_1(as + b)}{(cs^2 + ds + e)(fs + g)}$:   type 0   gain $K_p = \dfrac{K_1 b}{eg}$

   Unit step: $e_{ss} = \dfrac{1}{1 + K_p}$ Unit ramp: $e_{ss} \rightarrow \infty$

(b) $G(s) = \dfrac{K_1(as + b)}{s(cs^2 + ds + e)(fs + g)}$:   type 1   gain $K_v = \dfrac{K_1 b}{eg}$

   Unit step: $e_{ss} = 0$ Unit ramp: $e_{ss} = \dfrac{1}{K_v}$

Example 4.3.2   Steady-State Errors for Unit Steps in R and D (Fig. 4.8)

The system is type 1, so $e_{ss} = 0$ for a step input R. For input D, the steady-state value m_{ss} of M is also zero (since otherwise c_{ss} could not be constant), but the feedback from C to D is not unity but K. For a unit value of D the condition for $m_{ss} = 0$ is $1 + Ke_{ss} = 0$, so $e_{ss} = -1/K$. Direct application of the final value theorem to $C(s)$ will verify these results.

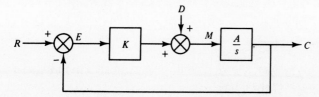

Figure 4.8   Example 4.3.2.

The final value theorem can also be applied to the output in Fig. 4.6. But note that for a ramp input this will give the correct but useless result that the final value of the output is infinite.

4.4 TRANSIENT RESPONSE CHARACTERISTICS AND SYSTEM STABILITY

For a system with transfer function $G(s)$, whether open loop or closed loop, and input $R(s)$, the output is $C(s) = G(s)R(s)$. For distinct poles, whether real or complex, the partial fraction expansion of $C(s)$ and the corresponding solution $c(t)$ are, from (1.50),

$$C(s) = \frac{K_1}{s + p_1} + \frac{K_2}{s + p_2} + \cdots + \frac{K_n}{s + p_n} \tag{4.13}$$

$$c(t) = K_1 \exp(-p_1 t) + K_2 \exp(-p_2 t) + \cdots + K_n \exp(-p_n t)$$

The denominator of $C(s) = G(s)R(s)$ and its partial fraction expansion contain terms due to the poles of input $R(s)$ and those of the system $G(s)$. The terms due to $R(s)$ yield the forced solution, corresponding to the particular integral solution of a differential equation, such as the first terms in the step response solutions (1.38) and (1.39).

The system poles give the *transient solution,* and this is the part of the response into which more insight is needed. To help develop this, it is noted that since the system poles are real or occur in complex pairs, according to (4.13) the transient solution is the sum of the responses for these two types. These two basic types are called simple lag and quadratic lag and will be considered in turn, with emphasis on the correlations between the nature of the response and the pole positions in the s-plane. The purpose is to establish correlations that will permit requirements to be specified that ensure satisfactory performance.

System Stability

This is the most important characteristic of the transient response. For a system to be useful, the transient solution must decay to zero. This leads to the following definition, which will serve for most of this book.

> **Definition.**   A system is stable if the transient solution decays to zero and is unstable if this solution grows.

The fundamental stability theorem can be formulated by examination of (4.13). If any system pole $-p_i$ is positive or has a positive real part, then the corresponding exponential grows, so the system is unstable. A positive real part means that the pole lies in the right half of the s-plane. Hence:

> **Stability Theorem 4.4.1.**   A system is stable if and only if all system poles lie in the left half of the s-plane.

Simple Lag: First-Order Systems

Chapter 2 has shown that this system, shown with its pole–zero pattern in Fig. 4.9, is very common. For a step input $R(s) = 1/s$,

$$C(s) = \frac{1/T}{s(s + 1/T)} = \frac{K_1}{s} + \frac{K_2}{s + 1/T}$$

$$K_1 = \frac{1/T}{s + 1/T}\bigg|_{s=0} = 1 \qquad K_2 = \frac{1/T}{s}\bigg|_{s=-1/T} = -1$$

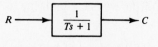

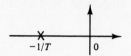

Figure 4.9   Simple lag.

Hence the transient response is

$$c(t) = 1 - e^{-t/T} \qquad (4.14)$$

The first term is the forced solution, due to the input, and the second the transient solution, due to the system pole. Figure 4.10 shows this transient as well as $c(t)$. The transient is seen to be a decaying exponential. If it takes long to decay, the system response is slow, so the speed of decay is of key importance. The commonly used measure of this speed of decay is the time constant.

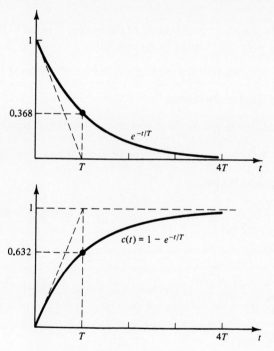

Figure 4.10   Step response of simple lag.

Time constant: This is the time in seconds for the decaying exponential transient to be reduced to $e^{-1} = 0.368$ of its initial value.

Since $e^{-t/T} = e^{-1}$ when $t = T$, it is seen that:

- The time constant for a simple lag $1/(Ts + 1)$ is T seconds.
- This is, in fact, the reason why a simple lag transfer function is often written in this form. The coefficient of s then immediately indicates the speed of decay.

- It takes $4T$ seconds for the transient to decay to 1.8% of its initial value.
- At $t = T$, $c(T) = 1 - 0.368 = 0.632$.

The values at $t = T$ provide one point for sketching the curves in Fig. 4.10. Also, the curves are initially tangent to the dashed lines, since

$$\frac{d}{dt}\left(e^{-t/T}\right)\Big|_{t=0} = -\frac{1}{T}e^{-t/T}\Big|_{t=0} = -\frac{1}{T}$$

These two facts provide a good sketch of the response.

Now consider the correlation between this response and the pole position at $s = -1/T$ in Fig. 4.9. The purpose of developing such insight is that it will permit the nature of the transient response of a system to be judged by inspection of its pole–zero pattern.

For the simple lag, two features are important:

1. *Stability:* As discussed, for stability, the system pole $-1/T$ must lie in the left half of the s-plane, since otherwise the transient $e^{-t/T}$ grows instead of decays as t increases.

2. *Speed of response:* To speed up the response of the system (that is, to reduce its time constant T), the pole $-1/T$ must be moved left.

How such movement is to be achieved is a problem of design, considered later.

Quadratic Lag: Second-Order Systems

This very common transfer function can always be reduced to the standard form

$$G(s) = \frac{\omega_n^2}{s^2 + 2\zeta\omega_n s + \omega_n^2} \tag{4.15}$$

where ω_n = undamped natural frequency
ζ = damping ratio

The significance of these parameters will be discussed. For a unit step input $R(s) = 1/s$, the transform of the output is

$$C(s) = \frac{\omega_n^2}{s(s^2 + 2\zeta\omega_n s + \omega_n^2)} \tag{4.16}$$

For $\zeta < 1$, this is an entry in Table 1.6.1, verified in Example 1.9.3. However, there are three possibilities, depending on the roots of the system characteristic equation

$$s^2 + 2\zeta\omega_n s + \omega_n^2 = 0 \tag{4.17}$$

These system poles depend on ζ:

$$\begin{aligned} \zeta > 1: & \quad \text{overdamped: } s_{1,2} = -\zeta\omega_n \pm \omega_n\sqrt{\zeta^2 - 1} \\ \zeta = 1: & \quad \text{critically damped: } s_{1,2} = -\omega_n \\ \zeta < 1: & \quad \text{underdamped: } s_{1,2} = -\zeta\omega_n \pm j\omega_n\sqrt{1 - \zeta^2} \end{aligned} \tag{4.18}$$

Figure 4.11 shows the s-plane for plotting the pole positions. For $\zeta > 1$, these are on the negative real axis, on both sides of $-\omega_n$. For $\zeta = 1$, both poles coincide at $-\omega_n$. For $\zeta < 1$, the poles move along a circle of radius ω_n centered at the origin, as may be seen from the following expression for the distance of the poles to the origin:

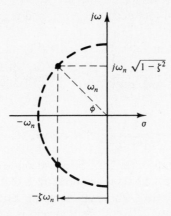

Figure 4.11   System poles quadratic lag.

$$|s_{1,2}| = [(\zeta\omega_n)^2 + (\omega_n\sqrt{1 - \zeta^2})^2]^{1/2} = \omega_n$$

From the geometry in Fig. 4.11, it is seen also that $\cos\phi = \zeta\omega_n/\omega_n = \zeta$. Hence

The damping ratio $\zeta = \cos\phi$, where ϕ is the position angle of the poles with the negative real axis.

An angle of $\phi = 45°$ corresponds to $\zeta = 0.707$ and angle $\phi = 60°$ to a damping ratio $\zeta = 0.5$. This allows the damping ratio to be estimated by inspection of the pole positions, provided that, as was already emphasized in Chapter 1, the same scales are used on both axes.

★ For $\zeta > 1$, when the poles are real and distinct, the transient is a sum of two decaying exponentials, each with its own time constant. The exponential corresponding to the pole closest to the origin has the largest time constant and takes longest to decay. This is called the *dominating pole,* and to increase the speed of response it would have to be moved to the left.

Example 4.4.1

$$G(s) = \frac{2}{s^2 + 3s + 2} = \frac{2}{(s + 1)(s + 2)}$$

For a unit step,

$$C(s) = \frac{2}{s(s + 1)(s + 2)} = \frac{K_1}{s} + \frac{K_2}{s + 1} + \frac{K_3}{s + 2}$$

and it is found that

$$c(t) = 1 - 2e^{-t} + e^{-2t}$$

The nature of this result, that is, that the transient consists of exponentials with time constants $T_1 = 1$ and $T_2 = 0.5$, could have been predicted from inspection of the system pole-zero pattern in Fig. 4.12. The system is a series connection of two simple lags. The dominating pole is the one at -1.

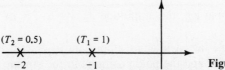

Figure 4.12   Example 4.4.1.

ʌ ʌ   For $\zeta = 1$, a repeated root occurs at $-\omega_n$, and responses can be calculated as in Examples 1.9.1 and 1.9.2.

ʌ ʌ ʌ   For $\zeta < 1$, the result in Table 1.6.1, verified by (1.49), applies,

$$c(t) = 1 - \frac{1}{\sqrt{1 - \zeta^2}} e^{-\zeta\omega_n t} \sin(\omega_n\sqrt{1 - \zeta^2}\, t + \phi) \qquad (4.19a)$$

or its alternative form in (1.55),

$$c(t) = 1 + \frac{1}{\sqrt{1 - \zeta^2}} e^{-\zeta\omega_n t} \cos(\omega_n\sqrt{1 - \zeta^2}\, t + \theta) \qquad (4.19b)$$

where the phase angle $\theta = \phi + (\pi/2)$ is the phase angle of the residue at the pole in the upper half-plane. Figure 4.13 shows a normalized plot of this response for different values of the damping ratio ζ. For small damping ratios the percentage overshoot over the steady-state response is large. The transient is a decaying oscillation of frequency $\omega_n\sqrt{1 - \zeta^2}$, of which the amplitude decays according to $e^{-\zeta\omega_n t}$. For a complex pair of poles:

Time constant T: This is the time in seconds for the amplitude of oscillation to decay to e^{-1} of its initial value: $e^{-\zeta\omega_n t} = e^{-1}$. Hence

$$T = \frac{1}{\zeta\omega_n} \qquad (4.20)$$

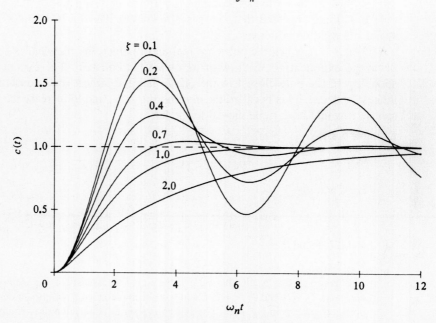

Figure 4.13   Unit step responses of a quadratic lag.

Analogous to the simple lag, the amplitude decays to 2% of its initial value in $4T$ seconds.

As for the simple lag, it is again important to determine the correlations between dynamic behavior and the pole positions in the s-plane in Fig. 4.11:

1. *Absolute stability:* The real part $-\zeta\omega_n$ of the poles must be negative for the transient to decay; that is, the poles must lie in the left half of the *s*-plane, as dictated by stability theorem 4.4.1.

2. *Relative stability:* To avoid excessive overshoot and unduly oscillatory behavior, damping ratio ζ must be adequate. Since $\zeta = \cos\phi$, the angle ϕ may not be close to 90°.

3. *Time constant:* The time constant is reduced (that is, the speed of decay of the transient is increased) by increasing the negative real part of the pole positions.

4. *Speed of response:* Noting that the horizontal axis in Fig. 4.13 is in terms of $\omega_n t$, it is seen that the speed of response of the system is increased by increasing the distance ω_n of the poles to the origin.

5. *Undamped natural frequency ω_n:* This equals the distance of the poles to the origin. Moving the poles out radially (that is, with ζ constant) increases the speed of response while the percentage overshoot remains constant.

6. *Frequency of transient oscillations $\omega_n\sqrt{1 - \zeta^2}$:* This frequency, also called the *resonant frequency* or *damped natural frequency,* equals the imaginary part of the pole positions.

Higher-Order Systems and Dominating Poles

According to (4.13) or (1.53), the total transient for higher-order systems is the sum of the transients corresponding to all system poles. As discussed in Section 3.9, higher-order systems may easily arise. Whatever the order of the transfer function, however, since each real pole causes a decaying exponential transient and each complex pair a decaying oscillation:

The *total transient response* is a superposition of exponential decays and decaying oscillations.

Repeated roots do not change this in an essential way. If all parameters are known, the response can be calculated, but its nature can also be judged without this by inspection of the pole positions. The dominating-poles concept simplifies both analysis and design greatly. The response of many systems is dominated by one pair of complex poles relatively close to the imaginary axis, so they behave approximately as second-order systems. The number of additional poles in near-dominant locations that can cause deviations from this second-order behavior is usually very limited.

4.5 TRANSIENT RESPONSE VERSUS STEADY-STATE ERRORS

Preceding sections have shown that parameter sensitivity, disturbance response, and steady-state errors are all improved by increased gain. Transient response was the fourth consideration in feedback design mentioned in Section 4.1. Here examples will be given to show how transient response and accuracy are related. They will show that dynamic response considerations usually limit the permissible gain. As suggested in Section 1.4, a higher gain means a larger signal to the actuator for a certain change of error. This suggests smaller error, but also increasing danger of overcorrection and oscillations. In fact, feedback design is generally concerned with achieving a satisfactory compromise between relative stability, that is, adequate damping ratio ζ of the dominating closed-loop poles, and accuracy, that is, small errors. The second example is unusual in that both improve with increasing gain. It is useful, however, in its striking demonstration of the effects of feedback.

Example 4.5.1 Step Responses of a DC Motor Position Servo

From (2.11), let the transfer function from field voltage to shaft position of a field-controlled dc motor be

$$G(s) = \frac{0.5}{s(0.25s + 1)}$$

The output of the sensor measuring shaft position is compared with a signal in the same form to obtain the error signal E in the block diagram in Fig. 4.14. Assuming a fast response power amplifier and a constant-gain controller, the loop can be considered to be closed, as shown, by an amplifier block of gain K. The closed-loop transfer function is

$$\frac{C(s)}{R(s)} = \frac{2K}{s^2 + 4s + 2K}$$

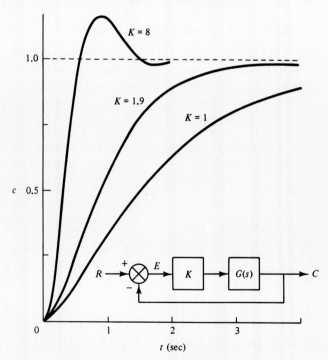

Figure 4.14 Position servo, Example 4.5.1.

and for a unit step input $[R(s) = 1/s]$, the transform of the shaft position is

$$C(s) = \frac{2K}{s(s^2 + 4s + 2K)}$$

Partial fraction expansion and (4.19) yield the following solutions, plotted in Fig. 4.14:

$$K = 1 \qquad c(t) = 1 - 1.207e^{-0.586t} + 0.207e^{-3.414t}$$
$$K = 1.9 \qquad c(t) = 1 - 2.737e^{-1.553t} + 1.737e^{-2.447t}$$
$$K = 8 \qquad c(t) = 1 + 1.1547e^{-2t} \cos\left(3.4641t + \frac{5\pi}{6}\right)$$

The physical discussion in Section 1.4 suggested a faster response for larger K, because the motor input changes more for a given change of error, and a danger of overcorrection and oscillations if K is raised further. This is confirmed by the plots. The time constants of the dominating poles are, in order, $1/0.586$, $1/1.553$, and $1/2$, and for the highest gain the poles are complex and the step response overshoot is considerable.

The steady-state or final value of the error is seen to be zero for all three values of K, as expected for this type 1 system.

For servos, the steady-state error in following a unit ramp input $[R(s) = 1/s^2]$ is also an important measure of performance. For a machine tool slide moving at constant velocity, it can cause errors in part dimensions. Here application of the final value theorem to $C(s)$ yields the expected but useless result that $c_{ss} \to \infty$. The theorem should be applied to the error $E(s)$. Easier yet, however, is to note that the gain of this type 1 system is $0.5K$, so by Table 4.3.1 the steady-state error after a unit ramp is

$$e_{ss} = \frac{1}{0.5K}$$

As expected, this is reduced by increasing K.

Example 4.5.2   Proportional Control of a Simple Lag

Figure 4.15(a) shows a simple lag plant with *proportional control*, or *P control*, a pure gain controller. The closed-loop transfer function is

$$\frac{C}{R} = \frac{KG}{1 + KG} = \frac{K}{Ts + 1 + K} = \frac{K}{1 + K} \frac{1}{[T/(1 + K)]s + 1} \qquad (4.21)$$

Hence the closed-loop system is also a simple lag, but with time constant $T/(1 + K)$ instead of T. So the speed of response increases with K. The steady-state error e_{ss} for a unit step input is $1/(1 + K)$, since the system is type 0 with gain K. This reduces with increasing K. The unit step response is

$$c(t) = \frac{K}{1 + K} (1 - e^{-(1 + K)t/T}) \qquad (4.22)$$

and is plotted in Fig. 4.15(b) for $T = 1$ and several values of K. It is apparent that for larger K the response not only comes closer to the desired value of unity, but also approaches the steady state sooner.

The steady-state value of the output is

$$c_{ss} = r - e_{ss} = 1 - \frac{1}{1 + K} = \frac{K}{1 + K}$$

This can also be derived from (4.21), which gives

$$c_{ss} = \lim_{s \to 0} sC(s) = \lim_{s \to 0} \frac{K}{1 + K} \frac{s(1/s)}{[T/(1 + K)]s + 1} = \frac{K}{1 + K}$$

Evidently, increasing gain K in this example does not reduce relative stability. To allow correlation of the change of transient response with change of system pole position in the s-plane, Fig. 4.15(c) shows the root locus of the system. It indicates, for $T = 1$, how the system pole $-(1 + K)/T$ moves as K is increased from zero. This plot is in fact enough to show that there is no relative stability problem and that, because of increasing distance of the pole to the imaginary axis, the system responds faster with increasing K.

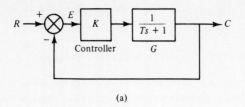

(a)

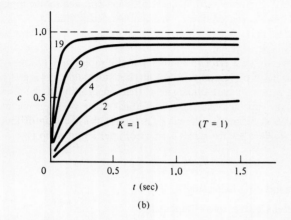

(b)

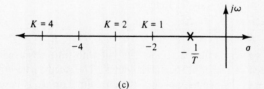

(c)   **Figure 4.15**   Example 4.5.2: P control.

Apparently, improving the transient response can be a major motivation for the use of feedback. In this example a possibly very slow process could be made to respond at any desired speed!

However, a practical restriction should be emphasized. Changing a water level or temperature in a tank fast requires large-capacity and costly water or heat supply systems. The results for the example imply the assumption of infinite-capacity supplies. In practice, the use of excessive gain will cause the supplies to saturate at their maximum output levels. The system then operates in a nonlinear regime, and its response changes from the predictions based on linear analysis.

Furthermore, it is likely that the simple lag model is no longer adequate when the system is "pushed" to high performance. Certain poles or zeros that could be neglected probably have to be included.

Example 4.5.3 P Control with Two Simple Lags

In Fig. 4.16(a) a second simple lag has been included in the system of Fig. 4.15(a). Among other possibilities, it could represent a two-tank level control as in Fig. 2.19, a temperature control with two heat capacitances in series, or a motor speed control. As in Example 4.5.2, the system is type 0 with gain K, so the steady-state error after a unit step is $1/(1 + K)$.

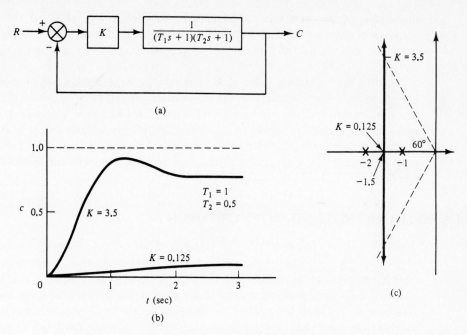

Figure 4.16   Example 4.5.3: P control.

Consider now stability, using $T_1 = 1$, $T_2 = 0.5$ for a numerical example. The closed-loop transfer function and the system characteristic equation are, respectively,

$$\frac{C}{R} = \frac{2K}{s^2 + 3s + 2 + 2K}$$
$$s^2 + 3s + 2 + 2K = 0 \tag{4.23}$$

The closed-loop poles are

$$s_{1,2} = -1.5 \pm \sqrt{0.25 - 2K} \tag{4.24}$$

and a root locus showing how these pole positions in the s-plane change with K can easily be plotted in this case. The plot is shown in Fig. 4.16(c). For $K = 0$ the poles coincide with those of the loop gain function, -1 and -2. As K is increased, the poles initially are still real and move together until for $K = 0.125$, both are located at -1.5. For larger K the real part is constant at -1.5 and the imaginary part increases with K. The cosine of the angle of the poles with the negative real axis is the damping ratio ζ and decreases with increasing K.

Thus the increase of K that is desirable to increase the speed of response and to reduce steady-state errors, parameter sensitivity, and disturbance response is limited by considerations of relative stability. The unit step response can be calculated from (4.23) and is shown in Fig. 4.16(b) for $K = 0.125$ and $K = 3.5$. The improvement in accuracy and speed of response for the higher gain is evident, but the steady-state error is still large, although the overshoot over the steady-state value is already significant.

This example is more typical than the preceding one, in that relative stability considerations usually limit the permissible gain. It is still not representative, however, of most practical systems since at least absolute stability is maintained for all K. Usually, systems actually become unstable beyond a certain value of K. As discussion of the root locus technique in Chapter 6 will show, if the vertical branches of the root locus in Fig. 4.16(c) would instead bend off to the right, then beyond a certain value of K the closed-loop poles would be located in the right-half s-plane.

Often the maximum value of K permitted by these considerations is not large enough to meet specifications on the steady-state error $1/(1 + K)$. Indeed, in many cases this error is required to be zero. This motivates the use of *dynamic compensation*. In proportional control, or P control, only gain adjustment is available to improve performance. If performance specifications cannot be met by P control, it must be replaced by a dynamic controller or dynamic compensator to provide more flexibility.

4.6 ROUTH–HURWITZ STABILITY CRITERION

Absolute stability requires only that all roots of the system characteristic equation
$$a_n s^n + a_{n-1} s^{n-1} + \cdots + a_1 s + a_0 = 0 \qquad (4.25)$$
(that is, the poles of its transfer function) lie in the left-half s-plane. It is known that if any of the coefficients are zero or if not all coefficients have the same sign, there will be roots on or to the right of the imaginary axis. If all coefficients are present and have the same sign, which can be taken to be positive without loss of generality, the Routh–Hurwitz criterion provides a quick method for determining absolute, but not relative, stability from the coefficients, without calculating the roots. It shows how many system poles are in the right-half s-plane or on the imaginary axis.

For the following *Routh array,* the first two rows are produced by arranging the coefficients $a_n, \ldots, a_0$ in the order indicated by the arrows.

$$
\begin{array}{llll}
s^n: & a_n & a_{n-2} & a_{n-4} & \cdots & 0 \\
s^{n-1}: & a_{n-1} & a_{n-3} & a_{n-5} & \cdots & 0 \\
s^{n-2}: & b_1 & b_2 & b_3 & \cdots & 0 \\
s^{n-3}: & c_1 & c_2 & c_3 & \cdots & 0 \\
\vdots & & & & \\
s^1: & g_1 & 0 \\
s^0: & h_1 & 0
\end{array} \qquad (4.26)
$$

Each of the remaining entries b_i, $c_i, \ldots$ is found from the two rows preceding it according to a pattern that can be recognized from the following equations:

$$b_1 = \frac{-1}{a_{n-1}} \begin{vmatrix} a_n & a_{n-2} \\ a_{n-1} & a_{n-3} \end{vmatrix} = \frac{1}{a_{n-1}}(a_{n-1}a_{n-2} - a_n a_{n-3})$$

$$b_2 = \frac{-1}{a_{n-1}} \begin{vmatrix} a_n & a_{n-4} \\ a_{n-1} & a_{n-5} \end{vmatrix} = \frac{1}{a_{n-1}}(a_{n-1}a_{n-4} - a_n a_{n-5})$$

$$b_3 = \frac{-1}{a_{n-1}} \begin{vmatrix} a_n & a_{n-6} \\ a_{n-1} & a_{n-7} \end{vmatrix} = \frac{1}{a_{n-1}}(a_{n-1}a_{n-6} - a_n a_{n-7}) \qquad (4.27)$$

$$c_1 = \frac{-1}{b_1} \begin{vmatrix} a_{n-1} & a_{n-3} \\ b_1 & b_2 \end{vmatrix} = \frac{1}{b_1}(b_1 a_{n-3} - b_2 a_{n-1})$$

$$c_2 = \frac{-1}{b_1} \begin{vmatrix} a_{n-1} & a_{n-5} \\ b_1 & b_3 \end{vmatrix} = \frac{1}{b_1}(b_1 a_{n-5} - b_3 a_{n-1})$$

Calculations in each row are continued until only zero elements remain. In each of the last two rows the second and following elements are zero. It can be shown that the elements in any row can be multiplied by an arbitrary positive constant without affecting the following results. This can be useful to simplify the arithmetic. For large n, computer algorithms can be written based on (4.27).

The *Routh–Hurwitz criterion* states:

1. A necessary and sufficient condition for stability is that there be no changes of sign in the elements of the first column of the array (4.26).
2. The number of these sign changes is equal to the number of roots in the right-half s-plane.
3. If the first element in a row is zero, it is replaced by a very small positive number ε, and the sign changes when $\varepsilon \to 0$ are counted after completing the array.
4. If all elements in a row are zero, the system has poles in the right half-plane or on the imaginary axis.

Example 4.6.1

$$s^5 + s^4 + 6s^3 + 5s^2 + 12s + 20 = 0$$

Using (4.26) and (4.27), the Routh array is

$$
\begin{array}{llll}
s^5: & 1 & 6 & 12 \quad 0 \\
s^4: & 1 & 5 & 20 \quad 0 \\
s^3: & 1 & -8 & 0 \\
s^2: & 13 & 20 & 0 \\
s^1: & \dfrac{-124}{13} & 0 & \\
s^0: & 20 & 0 &
\end{array}
$$

There are two sign changes in the first column, first from plus to minus and then from minus to plus. Hence this characteristic equation represents an unstable system, with two poles in the right-half s-plane.

Example 4.6.2

$$s^5 + s^4 + 5s^3 + 5s^2 + 12s + 10 = 0$$

The zero element in the first column for s^3 in the following Routh array is replaced by a small positive number ε. Then

$$c_1 = \frac{5\varepsilon - 2}{\varepsilon} \approx \frac{-2}{\varepsilon} \qquad d_1 = \frac{2c_1 - 10\varepsilon}{c_1} \approx 2$$

$$
\begin{array}{llll}
s^5: & 1 & 5 & 12 & 0 \\
s^4: & 1 & 5 & 10 & 0 \\
s^3: & \varepsilon & 2 & 0 \\
s^2: & c_1 & 10 & 0 \\
s^1: & d_1 & 0 \\
s^0: & 10
\end{array}
$$

Thus c_1 is a large negative number, implying two unstable poles, as in Example 4.6.1.

Example 4.6.3

For the electromechanical servo modeled in Fig. 4.17, find the limits on K for stability. The characteristic equation is $s^3 + 3s^2 + 2s + K = 0$ and the Routh array is

$$
\begin{array}{llll}
s^3: & 1 & 2 & 0 \\
s^2: & 3 & K & 0 \\
s^1: & \dfrac{6 - K}{3} & 0 \\
s^0: & K & 0
\end{array}
$$

Evidently, $0 < K < 6$ is the range of K for a stable system.

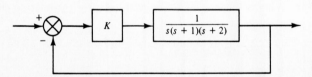

Figure 4.17   Example 4.6.3.

At both limits in Example 4.6.3, one row of the Routh array consists of only zero elements. It can be shown that this means that root pairs are present that are located symmetrically about the origin, usually on the imaginary axis. It can also be shown that such root pairs are the roots of an *auxiliary equation* formulated from the elements in the row of the array that precedes the row with the zeros.

For Example 4.6.3, at the upper limit $K = 6$ the auxiliary equation is $3s^2 + 6 = 0$. The elements for row s^1 are zero, so it is formed from those for s^2. The highest power in the auxiliary equation is generally that of the row, and the powers of successive terms reduce by 2.

The roots of the auxiliary equation are $s = \pm j\sqrt{2} = \pm 1.414j$. So at the limit $K = 6$, system poles occur at these points on the imaginary axis. When several parameters vary, these techniques can be used to determine the relations to be satisfied to ensure system stability.

4.7 CONCLUSION. COMPUTER-AIDED ANALYSIS AND DESIGN

In this chapter the motivations for the use of feedback and its effect on performance were examined, including the characteristics of the transient response. Considerations of relative stability usually limit the gain increase, which is desirable to increase the speed of response and to reduce the sensitivity to plant parameter variations and model approximations, as well as to reduce errors and the response to disturbances. Examples were given that motivate the use of dynamic compensation to overcome this limitation. This is the subject introduced in Chapter 5.

The dynamic response characteristics have been interpreted in terms of the locations of system poles in the s-plane. These poles must lie in the left half-plane to ensure stability. The Routh–Hurwitz criterion permits this to be determined without actual calculation of the poles.

For high-order systems and for routine work on those of lower order, computational aids are indispensable for analysis and design. For example, the techniques for transient response calculation clearly become laborious for high-order systems. Computer aids can range from batch-type programs for specific purposes, such as finding the roots of a characteristic equation or calculating the response for a given transfer function, to interactive analysis and design packages that include computer graphics.

Because computer aids are not essential to the development of concepts and techniques, it is desirable to allow for freedom of choice as to the degree to which such aids are exploited in the study of the subject. Computational aids, with examples of their use, have therefore been collected in Appendix B, and reference to particular programs will be made at appropriate points in the text.

In the context of the present chapter, Appendix B gives a program for transient response computation based on partial fraction expansion, assuming distinct poles. For closed-loop systems, the poles needed for this must usually be found from the system characteristic equation (4.25).

PROBLEMS

4.1. To show the use of feedback in reducing the effect of disturbance inputs:
    (a) Express C/D in Fig. 4.5, with $R = 0$, both with and without the feedback.
    (b) Give the condition for reducing C/D to 10% of its value without feedback.

4.2. It is desired to examine how adding a feedback loop and a gain K as shown in Fig. P4.2(ii) to the block in part (i) can be used to reduce the effect of variations of the system parameter a.
    (a) Calculate the unit step responses in Fig. P4.2(i) and (ii).
    (b) Use the final value theorem to verify the values of $\lim_{t\to\infty} c(t)$ found in part (a).
    (c) If a is nominally 1, so that the desired steady-state output is 1, compare the open-loop and closed-loop systems for $K = 10$ and $K = 100$ if the parameter a doubles in value.

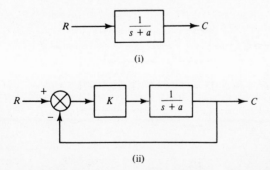

(i)

(ii)   **Figure P4.2**

4.3. In Fig. P4.3, R is the input that output C should follow as closely as possible, and D is a disturbance input to which ideally C should not respond at all.

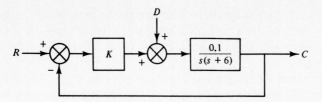

Figure P4.3

(a) Express the transfer functions C/R and C/D and hence determine the ratio of the outputs due to unit step inputs of D and R, and the choice of K that will reduce this ratio.

(b) Calculate the response to a unit step R, both with and without a unit step D occurring simultaneously, if $K = 80$. Note the effect of D relative to that of R.

4.4. A slightly generalized version of the water-level control block diagram of Fig. 3.2 is shown in Fig. P4.4(i). Figure P4.4(ii) shows a simplified model. Here the controller is a gain, and supply pressure disturbances P_s' and the time constant τ_f of the actuator are assumed to be negligibly small. The valve flow gain K_x, incorporated into K, and the tank outflow valve resistance R are linearized gains that change with operating point and are generally not precisely known. K may also change due to gain variations of the valve actuator, and the level sensor gain H may vary from its nominal value of 1. In this model:

(a) Express the sensitivity functions to changes in K, R, and H.

(b) Determine the static sensitivities and hence comment on the effect of K, R, and H under static conditions.

(c) Compare the sensitivity to R with that for open-loop control and hence discuss the effect of feedback.

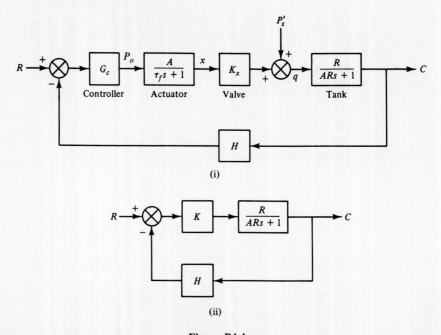

(i)

(ii)

Figure P4.4

4.5. In Fig. P4.5, the actuator time constant τ_f in Fig. P4.4(i) has not been neglected as it was in Fig. P4.4(ii).

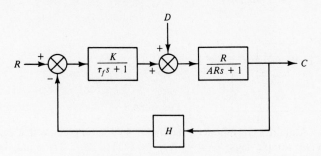

Figure P4.5

(a) Determine the sensitivity function of the closed-loop transfer function C/R with respect to the parameter τ_f, and find the static value of this function.

(b) In frequency response methods, $s = j\omega$, and the magnitude of the sensitivity function as a function of radian frequency ω is of interest. If $AR = 1$ and $\tau_f = 0.2$, find the magnitude of the sensitivity function at $\omega = 1$ if KRH is such that the system damping ratio is 0.5. What is the value of S_τ at very high frequencies?

4.6. In Fig. P4.5, determine the sensitivities of the transfer function C/D for the disturbance input D to variations of the parameters R and K. What are the static values of these sensitivities, and what is the significance of these results for large loop gains KRH?

4.7. For the systems in Fig. P4.7:

(a) Determine the transfer functions E/R.

(b) Use the final value theorem to find the steady-state values of the system error $e = r - c$ in response to unit step inputs.

(c) Observe how this evidently important aspect of system behavior depends on the value of K and the type of system.

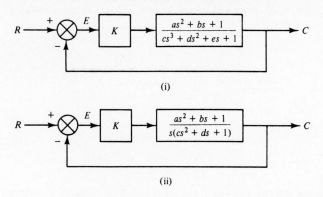

Figure P4.7

4.8. In the motor position servo of Fig. P4.8, let $G(s) = 1/[s(s + 1)]$ represent the motor and load and $G_c = K$ the controller–amplifier. Calculate the steady-state errors of the system for unit step and unit ramp signals applied in turn to reference input R and disturbance input D. Explain the results obtained physically.

4.9. In Fig. P4.8, let $G(s)$ be a simple lag plant $G(s) = K_p/(Ts + 1)$ and G_c an integral controller $G_c = K/s$. Calculate the steady-state errors for unit step and unit ramp inputs applied in turn to R and D. Compare the results and discuss the desired gain distribution over K and K_p. Explain the differences with Problem 4.8.

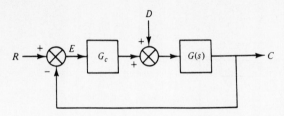

Figure P4.8

4.10. In Fig. P4.8, let $G_c = K$ and let $G(s)$ model a thermal or hydraulic process that can be approximated by two simple lags: $G(s) = K_p/[(s + 1)(0.2s + 1)]$. Find the steady-state errors for unit step inputs applied in turn to R and D and compare the results. Discuss the desired gain distribution over K and K_p.

4.11. In Fig. P4.11, with $G(s) = A/(Ts + 1)$, the parameters A and T are nominally 1, but each may vary by a factor of 2 in either direction with operating conditions. Find K so that despite these variations the steady-state errors for step inputs will not exceed 10% and the system time constant will stay below 0.2 sec.

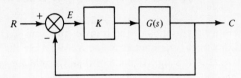

Figure P4.11

4.12. For the system in Fig. P4.11, with $G(s) = (s + 1)(s + 3)/[s(s + 2)(s + 4)]$:
  (a) What is the system type number?
  (b) What is the gain of the loop gain function?
  (c) What are the steady-state errors following unit step and unit ramp inputs?

4.13. The open-loop system in Fig. P4.13(a) responds the same to reference inputs R and disturbance inputs D. In Fig. P4.13(b), calculate K so that the steady-state response to R is the same as in part (a), and compare parts (a) and (b) on the basis of dynamic response to R and steady-state and dynamic performance in response to D.

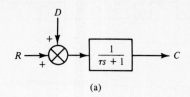

(a)

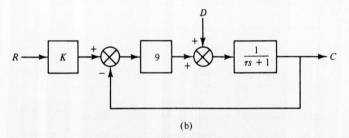

(b)

Figure P4.13

4.14. A system is given by its transfer function

$$G(s) = \frac{5(1 - 0.4s)}{(s + 1)(0.2s + 1)}$$

    (a) What are the time constants of the components of its transient response, and how long does it take for the transient to decay almost completely?

    (b) The unit step response is readily found to be

$$c(t) = 5 - 8.75e^{-t} + 3.75e^{-5t}$$

Use the alternative definitions of a time constant to obtain reasonable sketches of the components of the response, and use these to sketch $c(t)$.

4.15. What should be the time constant of a simple lag system if it is specified that in 1 sec the transient must have been reduced to half its initial value?

4.16. Calculate the unit step response of the system $G = 4/(s^2 + 5s + 4)$ and use the definitions of a time constant to sketch the transient components and to obtain a reasonable sketch of the overall response.

4.17. For the system in Fig. P4.11, with $G(s) = 1/[(s + 1)(s + 4)]$:

    (a) What is the dominating time constant of the plant?

    (b) For what value of K will feedback make the dominating system time constant half of that in part (a)?

    (c) Calculate the unit step response for K in part (b) and find the steady-state error.

    (d) What limits the increase of K desirable to reduce the steady-state error in part (c)? Find K and the corresponding steady-state error for a system damping ratio of about 0.7.

4.18. In Fig. P4.11, with $G(s) = 1/[(s + 2)(s + 10)]$, to examine the effect of feedback on performance:

    (a) Calculate the unit step responses for $K = 7$ and $K = 20$.

    (b) Verify the steady-state error values of these responses directly.

    (c) Compare the responses on the basis of time constant and nature of the response.

4.19. The system in Fig. P4.11, with the simple lag plant $G(s) = 1/(s + 5)$ could model, among other possibilities, a temperature, pressure, level, or speed control system.

    (a) Calculate and plot unit step responses for $K = 10$ and $K = 45$.

    (b) Compare the responses for speed of response and steady-state error (that is, the difference between r and c for $t \to \infty$).

4.20. In Fig. P4.11, the plant transfer function $G(s) = 20/[(s + 5)(s + 20)]$ could be a more precise model for the same systems as in Problem 4.19.

    (a) Calculate and plot unit step responses for the same values of K as in Problem 4.19(a).

    (b) Note the effect of gain on steady-state error, and note the difference with Problem 4.19 with respect to the possibility of making this error quite small.

4.21. Use the Routh–Hurwitz stability criterion to determine the stability of systems with the following characteristic equations.

    (a) $s^4 + 10s^3 + 33s^2 + 46s + 30 = 0$ **(b)** $s^4 + s^3 + 3s^2 + 2s + 5 = 0$

    (c) $s^3 + 2s^2 + 3s + 6 = 0$ **(d)** $s^4 + 3s^3 + s^2 + 3s + 5 = 0$

For part (c), express the auxiliary equation and find one pair of roots.

4.22. For the system of Fig. P4.11, with $G(s) = 1/[s(s + 1)(s + 4)]$, use the Routh–Hurwitz criterion to find the limit on K for stability. At this limit, determine the position of the system poles on the imaginary axis of the s-plane.

4.23. In Problem 4.22, determine the limit on the value of K for stability if the amplifier K is replaced by a dynamic compensator

$$\frac{K(0.5s + 1)}{0.1s + 1}$$

Where on the imaginary axis is a root pair located at this limit?

5

Introduction to Feedback System Design

5.1 INTRODUCTION

In proportional control, or P control, already employed in Examples 4.5.2 and 4.5.3 and on many other occasions in Chapter 4, only gain adjustment is available to improve performance. Its limitations for achieving both satisfactory accuracy and acceptable relative stability, that is, limited transient overshoot and oscillations, were discussed. If the performance specifications cannot be met by P control, it must be replaced by a dynamic controller or dynamic compensator to provide more flexibility. This is a usually quite simple transfer function, and analytical tools still to be provided are not required to gain insight into the essential concepts of how such controllers can change system behavior.

First, however, the performance criteria that a design must satisfy are expressed. Also, the effect of system zeros needs to be discussed, to help understand how a controller zero can affect closed-loop system behavior.

System configurations and methods of dynamic compensation are introduced as well as controllers. PID (proportional plus integral plus derivative) control is discussed not only because it is the most commonly used standard form of controller in practice, but also because the basic proportional, integral, and derivative actions are fundamental to all compensators.

5.2 TRANSIENT RESPONSE PERFORMANCE CRITERIA

Relative Stability and Dominating Poles

As discussed in the last part of Section 4.4, the response of many systems is dominated by one pair of complex poles. This dominating-poles concept simplifies design greatly, because design can concentrate on locating this dominating pair satisfactorily.

The requirements that this pair must satisfy are much more stringent than those of absolute stability, which imposes only the condition that the poles be in the left-half plane. Satisfactory relative stability means the absence of severe overshoot and unduly oscillatory behavior. This requires a large enough damping ratio so that the angle of the poles with the negative real axis may not be too large. Also, to achieve a small enough time constant, the poles may not be too close to the imaginary axis.

The fact that one pair of complex poles usually dominates is also the reason why the performance features of a quadratic lag, discussed in Section 4.4, are the basis for criteria that apply to higher-order systems.

Transient Response Performance Measures

Figure 5.1 identifies important performance criteria for the transient response, that is, features of the response that may have to meet certain performance specifications. Expressions for these criteria are given that are based on dominance of one pair of poles. Then, from (4.19) or Table 1.6.1, the unit step response is approximated by

$$c(t) = 1 - \frac{1}{\sqrt{1 - \zeta^2}} e^{-\zeta \omega_n t} \sin\left(\omega_n \sqrt{1 - \zeta^2}\, t + \tan^{-1}\left[\frac{\sqrt{1 - \zeta^2}}{\zeta}\right]\right) \qquad (5.1)$$

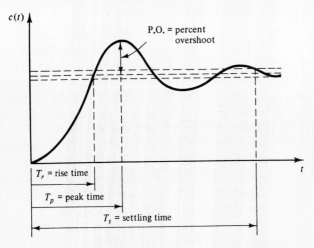

Figure 5.1   Performance criteria.

- *Settling time* T_s: The time required for the response to come permanently within a 2% band around the steady-state value. From below (4.20), it equals four time constants for a quadratic lag:

$$T_s = 4T = \frac{4}{\zeta \omega_n} \qquad (5.2)$$

- *Percentage overshoot* (P.O.): The maximum percentage overshoot over the steady-state response.
- *Peak time* T_p: The time to the maximum peak of the response.
- *Rise time* T_r: The time at which the response first reaches the steady-state level.

The overshoot is a critical measure of performance. Equating the derivative of $c(t)$ in (5.1) to zero, to determine the extrema of the response, easily yields the equation

$$\tan\left(\omega_n\sqrt{1-\zeta^2}\,t + \tan^{-1}\frac{\sqrt{1-\zeta^2}}{\zeta}\right) = \frac{\sqrt{1-\zeta^2}}{\zeta} \tag{5.3}$$

This implies that at the peaks

$$\omega_n\sqrt{1-\zeta^2}\,t = i\pi \qquad i = 1, 3, \ldots$$

since then left and right sides are equal. Hence the time at the maximum peak ($i = 1$), the peak time, is

$$T_p = \frac{\pi}{\omega_n\sqrt{1-\zeta^2}} \tag{5.4}$$

If the tan of the angle in (5.3) is $\sqrt{1-\zeta^2}/\zeta$, its sin is $\pm\sqrt{1-\zeta^2}$, and substituting (5.4) for t in (5.1) yields

$$\text{P.O.} = 100\,\exp\left(\frac{-\pi\zeta}{\sqrt{1-\zeta^2}}\right) \tag{5.5}$$

The rise time T_r is closely related to peak time T_p.

It is noted that, while T_s, T_p, and T_r depend on both ω_n and ζ, P.O. depends only on the damping ratio ζ. Figure 5.2 shows a graph of P.O. versus damping ratio ζ. Permissible overshoot and hence minimum acceptable ζ depend on the application. For a machine tool slide, overshoot may cause the tool to gouge into the material being machined, so $\zeta \geqslant 1$ is required. But in most cases a limited overshoot is quite acceptable, and then $\zeta < 1$ is preferable, because it reduces peak time T_p and rise time T_r. This is discussed later but may already be seen in the step response curves in Fig. 4.13. For $\zeta = 0.7$ the overshoot is only 5% and the response approaches steady state much sooner.

The correlations in Section 4.4 between response and pole positions for a quadratic lag can now be extended to correlate the performance criteria with the location of the dominating pair of system poles. From Fig. 4.11, these have real parts $-\zeta\omega_n$ and imaginary parts $\pm j\omega_n\sqrt{1-\zeta^2}$, with $\zeta = \cos\phi$.

1. *Percentage overshoot* (P.O.): For adequate relative stability, to avoid excessive overshoot, damping ratio ζ must be adequate. Since $\zeta = \cos\phi$, the angle ϕ may not be close to 90°.

2. *Settling time:* Time constant and settling time are reduced by increasing the real part of the pole positions.

3. *Peak time and rise time:* These are reduced by increasing the imaginary part $\omega_n\sqrt{1-\zeta^2}$ of the pole positions. From (5.1), this is the frequency of the transient oscillations, also called the resonant frequency or damped natural frequency.

4. *Undamped natural frequency ω_n:* This equals the distance of the poles to the origin. Moving the poles out radially (that is, with ζ constant) reduces settling time, peak time, and rise time while the percentage overshoot remains constant.

5. *Speed of response:* For a constant real part, the settling time and speed of decay of the transient are constant as well. But the speed of response is improved by increasing the imaginary part until ζ is at the minimum permissible level. This follows from point 3 above and is illustrated next.

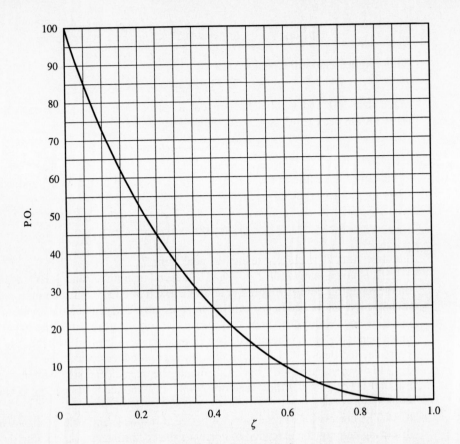

Figure 5.2 P.O. versus ζ.

If the damping ratio could be held constant while ω_n is increased, the poles would move radially outward and both settling time and rise time would decrease. The next example demonstrates what constraints may exist on how the poles of a closed-loop system can be adjusted. It also illustrates point 5.

Example 5.2.1   DC Motor Position Servo

Figure 5.3(a) shows the block diagram of the servo considered in Example 4.5.1. The closed-loop transfer function is

$$\frac{C(s)}{R(s)} = \frac{2K}{s^2 + 4s + 2K}$$

and the system poles are $s_{1,2} = -2 \pm \sqrt{4 - 2K}$. For $K < 2$ these poles lie along the negative real axis in the s-plane in Fig. 5.3(b). This corresponds to $\zeta > 1$, and the transient is a superposition of two decaying exponentials. (Figure 4.14 shows step responses for $K = 1$ and 1.9.)

For $K = 2$ both poles coincide at -2. By the technique used in Example 1.9.2, the unit step response, the inverse of $C(s) = 4/[s(s + 2)^2]$, is

$$c(t) = 1 - 2te^{-2t} - e^{-2t}$$

This is plotted in Fig. 5.3(c), where also the plot for $K = 8$ from Fig. 4.14 is repeated. Now, for $K = 8$ the poles are the roots $s_{1,2} = -2 \pm 3.464j$ of $s^2 + 4s + 16 = 0$,

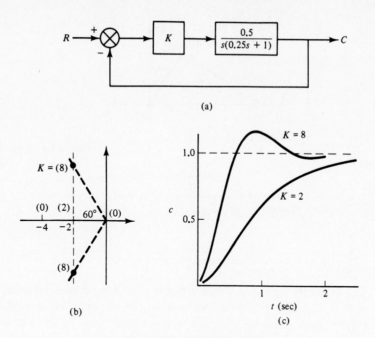

Figure 5.3   Example 5.2.1.

which corresponds to $\omega_n = 4$, $\zeta = 0.5$. The real part of the system poles is -2 for all $K \geqslant 2$, so the time constant and settling time are the same for $K = 2$ and $K = 8$. Yet, clearly, the plot for $K = 8$ reflects a higher speed of response. The difference lies in the smaller peak time and rise time associated with the larger values of ω_n and the larger imaginary part $\omega_n \sqrt{1 - \zeta^2}$ of the poles.

Note that with a controller constrained to a gain the pole positions are limited to the real axis between 0 and -4 and the vertical at -2. This is, in fact, the root locus for the system, studied in Chapter 6.

5.3 EFFECT OF SYSTEM ZEROS. POLE–ZERO CANCELLATION

Much attention has been given to system poles and the correlations between their positions in the s-plane and the nature of the transient response. A natural question is therefore what correlations exist between the response and the positions of the system zeros. This question is especially important in connection with design, because controller zeros play an essential role in improving system dynamics.

First the unit step response of a system with transfer function

$$G(s) = \frac{\omega_n^2}{z_1} \frac{s + z_1}{s^2 + 2\zeta\omega_n s + \omega_n^2} \tag{5.6}$$

will be studied to examine the effect of adding a zero to an underdamped quadratic lag.

The pole–zero pattern of $C(s) = G(s)R(s)$ is shown in Fig. 5.4. The residues K_1 and $(\mathbf{K}_2 \underline{/\phi_2})$ in the solution $c(t)$ given by (1.53) are, by the graphical rule,

$$K_1 = \frac{\omega_n^2}{z_1} \frac{z_1}{(\omega_n \underline{/-\phi})(\omega_n \underline{/\phi})} = 1$$

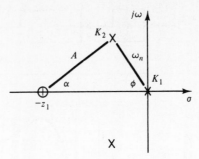

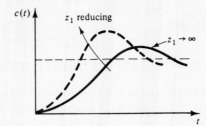

Figure 5.4   Pole–zero pattern of $C(s) = G(s)/s$ for (5.6).

$$K_2 = \frac{\omega_n^2}{z_1} \frac{A \underline{/\alpha}}{(\omega_n \underline{/\pi - \phi})(2\omega_n \sqrt{1 - \zeta^2} \underline{/\pi/2})} = \frac{A}{2z_1 \sqrt{1 - \zeta^2}} \underline{/\alpha + \phi + \pi/2}$$

so, by (1.53),

$$c(t) = 1 + \frac{A}{z_1 \sqrt{1 - \zeta^2}} e^{-\zeta\omega_n t} \cos\left(\omega_n \sqrt{1 - \zeta^2} t + \phi + \alpha + \frac{\pi}{2}\right) \qquad (5.7)$$

The geometry of Fig. 5.4 shows that if z_1 is large compared to ω_n then $A \approx z_1$ and $\alpha \approx 0$. Equation (5.7) then reduces to the response for a quadratic lag given by (4.19). Thus a zero far along the axis has little effect on the transient response of the quadratic lag. But if the zero is moved to the right, it will cause a gradually increasing percentage overshoot (P.O.), and the peak time T_p will reduce, as indicated in Fig. 5.5. The proof is analogous to that of (5.3) to (5.5) for the quadratic lag: With $\phi = \tan^{-1}(\sqrt{1 - \zeta^2}/\zeta)$, equating the derivative of $c(t)$ in (5.7) to zero yields

$$\tan[\omega_n \sqrt{1 - \zeta^2} t + \alpha + \tan^{-1}(\sqrt{1 - \zeta^2}/\zeta)] = \frac{\sqrt{1 - \zeta^2}}{\zeta} \qquad (5.8)$$

Figure 5.5   Effect of a zero.

At the extrema, $\omega_n \sqrt{1 - \zeta^2} t + \alpha = i\pi$, since then left and right sides are equal. The time T_p for the maximum peak ($i = 1$) is

$$T_p = \frac{\pi - \alpha}{\omega_n \sqrt{1 - \zeta^2}} \qquad (5.9)$$

Since the tan of the angle in (5.8) is $\sqrt{1 - \zeta^2}/\zeta$, its sin is $\pm\sqrt{1 - \zeta^2}$, and using this and (5.9) in (5.7) yields the maximum overshoot:

$$\text{P.O.} = 100 \frac{A}{z_1} \exp\left[-\frac{(\pi - \alpha)\zeta}{\sqrt{1 - \zeta^2}}\right] \qquad (5.10)$$

Here A and α can be measured or calculated from Fig. 5.4. Two examples follow:

1. If the zero is moved in from far left to the position where $\alpha = \pi/2$, (5.9) shows that the peak time is reduced by half.
2. For $\alpha = \pi/2$ and $\phi = 60$ $(\zeta = 0.5)$, $A/z_1 = 1.732$, and (5.10) yields an overshoot of almost 70%, a large increase over the 16% when the zero is far away.

The effect of zeros can be quite significant even if the ratio A/z_1 does not exceed 1. This is demonstrated in the next example, in which a number of feedback systems of the same structure are designed to have the same (closed-loop) poles and different zeros.

Example 5.3.1   PI Control of a Simple Lag

Figure 5.6(a) shows a control system with a simple lag process. The controller is a *PI controller,* an extremely common form of control. This system could model a level, temperature, or pressure control. Another possibility is motor speed control. In this case the system output transform is $s\theta(s)$, the transform of speed $\dot{\theta}(t)$, and the motor transfer function in (2.11) reduces to a simple lag. The feedback sensor could be a small generator, giving a feedback voltage proportional to shaft speed $\dot{\theta}(t)$. The closed-loop transfer function in Fig. 5.6(a) is

$$\frac{C(s)}{R(s)} = \frac{Ka(s + z)}{s^2 + a(1 + K)s + Kaz}$$

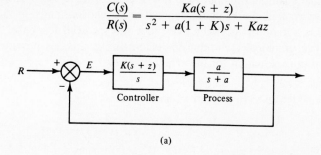

(a)

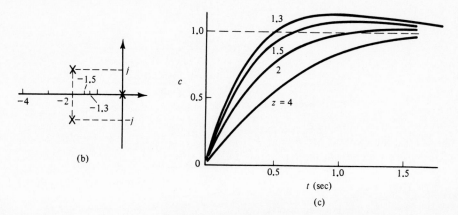

(b)

(c)

Figure 5.6   Example 5.3.1.

It is desired to make the closed-loop system poles equal to $-2 \pm j$; that is, the denominator of C/R should be $s^2 + 4s + 5$, or

$$a(1 + K) = 4 \qquad Kaz = 5$$

Systems for which $z = 4$, 2, 1.5, and 1.3, respectively, will then result for $Ka = 1.25$, 2.5, 3.3333, and 3.8462, with $a = 2.75$, 1.5, 0.6667, and 0.1538. For a unit step input $R(s) = 1/s$, the output transform is

$$C(s) = \frac{Ka(s + z)}{s(s^2 + 4s + 5)} = \frac{5(1 + s/z)}{s(s^2 + 4s + 5)}$$

The pole–zero pattern is shown in Fig. 5.6(b), and the graphical residue rule readily yields

$$c(t) = 1 + Ae^{-2t} \cos(t + \theta)$$

with the sets of values for (z, A, θ) equal to $(4, 1.25, -216.9°)$, $(2, 1.118, -153.4°)$, $(1.5, 1.667, -126.9°)$, and $(1.3, 2.1, -118.5°)$.

These responses are plotted in Fig. 5.6(c). The system poles at $(-2 \pm j)$ correspond to a damping ratio 0.894, for which the overshoot is negligible and the peak time, from (5.4), is 3.13 sec. Moving the zero in from the left is seen to reduce the peak time and increase the overshoot. For $z = 1.3$ the maximum overshoot is about 14% and occurs at about 1 sec. Thus care must be exercised in using the correlations between response and pole positions for a quadratic lag if zeros are present in relatively dominant locations.

The example illustrates the importance of the effect of zeros in design. In this case the zero of the PI controller had to be chosen, and Fig. 5.6(c) shows that the choice is a trade-off between overshoot and speed of response.

The graphical technique for determining residues provides a particularly enlightening explanation for the effect of zeros.

Example 5.3.2

Find the unit step response of the system

$$G(s) = \frac{4}{a} \frac{s + a}{(s + 1)(s + 4)} \tag{5.11}$$

The pole–zero pattern of $C(s) = G(s)R(s)$ is shown in Fig. 5.7. The residues are

$$K_1 = \frac{(4/a)a}{(1)(4)} = 1$$

$$K_2 = \frac{(4/a)(a - 1)}{(-1)(3)} = \frac{-4(a - 1)}{3a}$$

$$K_3 = \frac{(4/a)(-4 + a)}{(-3)(-4)} = \frac{a - 4}{3a}$$

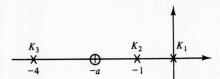

Figure 5.7   Example 5.3.2.

So the unit step response is

$$c(t) = 1 - \frac{4(a - 1)}{3a} e^{-t} + \frac{a - 4}{3a} e^{-4t} \tag{5.12}$$

It is seen that if $a = 1$, so if the zero coincides with, or cancels, the pole at -1, the transient corresponding to that pole is zero. Similarly, if $a = 4$, the zero cancels the pole at -4, making the corresponding transient zero.

The zero in this example can again be thought of as being part of a controller. In fact, pole–zero cancellation, to minimize the effect of certain poles, is an important approach to control system design. In the example, the dominating pole would be expected to be the one at -1. But if the zero is close enough to it, the residue corresponding to this pole becomes small enough that the other pole will dominate the transient.

The following remarks generalize the implications of the example for the significance of the zeros in the system pole–zero pattern:

1. A residue K_i at pole $-p_i$ corresponds to a transient term $K_i \exp(-p_i t)$, so the significance of the residue is that its magnitude is the initial size of the transient corresponding to the pole.
2. If a zero is close to a pole, the residue at the pole tends to be small, because of a short vector in the numerator, so the corresponding transient is probably small.
3. If the zero coincides with the pole, it cancels it, and the transient term is zero.
4. Thus the significance of zeros is that they affect the residues at the poles and hence the sizes of the corresponding transients.
5. A "slow" pole (close to the imaginary axis, so with a large time constant) or a highly oscillatory pair (small damping ratio) may be acceptable if nearby zeros make the corresponding transients small.

Because pole–zero cancellation is so commonly used in design, it is important to observe that the response to initial conditions is not affected. The preceding results were based on the system transfer function, with the implied assumption of zero initial conditions. In Example 5.3.2, cross multiplication in $G(s) = C(s)/R(s)$ yields $(s^2 + 5s + 4)C = (4 + 4s/a)R$. For $r = 0$, the corresponding differential equation is $\ddot{c} + 5\dot{c} + 4c = 0$. Analogous to Example 1.8.2, transformation for nonzero initial conditions gives

$$C(s) = \frac{(s + 5)c(0) + \dot{c}(0)}{s^2 + 5s + 4}$$

This initial condition response is unaffected by the numerator of $G(s)$, and so by any cancellations that may have been achieved in the input–output response.

It is also useful to clarify why the addition of a faraway zero, and so of a long vector in their numerators, does not greatly increase the sizes of residues and transients. By the final value theorem, the steady-state response of a system $G(s) = K/[(s + a)(s + b)]$ to a unit step is

$$c_{ss} = \lim_{s \to 0} sG(s)R(s) = \frac{K}{ab} \tag{5.13}$$

If a zero factor $(s + z)$ or pole factor $(s + p)$ is included to change the dynamic behavior, K must be multiplied by p/z if the steady state is not to be affected also; that is, $G(s)$ should be changed to

$$G_1(s) = K \frac{p}{z} \frac{s + z}{(s + p)(s + a)(s + b)} \tag{5.14}$$

A large vector $(s + z)$, say, now in effect becomes $(1 + s/z)$ and does not cause large residues. In fact, for large z this factor is close to 1, so the effect of faraway zeros on the transient response is small, as was noted earlier in this section.

5.4 DESIGN PRELIMINARIES: CONFIGURATIONS, CONTROLLERS, AND CONSTRAINTS

System Configurations

One common system configuration for dynamic compensation is minor loop feedback compensation, or simply feedback compensation, shown in Fig. 5.8(b). This was encountered earlier in Examples 3.3.2 and 3.7.1 and is discussed in the next section. The most common configuration, however, is *series compensation,* indicated in Fig. 5.8(a). The controllers $G_c(s)$ and $H_c(s)$ to be designed are usually very simple transfer functions.

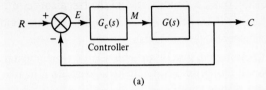

(a)

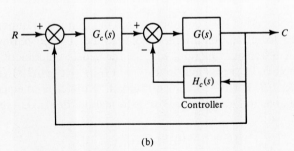

(b)

Figure 5.8 (a) Series and (b) feedback compensation.

Model Simplification

The plant transfer function $G(s)$ in these block diagrams is subject to the various approximations outlined earlier. However, this model if often also simplified intentionally to facilitate design. For example, as discussed in the last section, the effect of faraway zeros on the transient response is small. The same is true for faraway poles, at which the residues will tend to be small. Such poles and zeros, except for their steady-state effects, are therefore often neglected to simplify the model. This yields a further advantage, because for many systems the values of such faraway poles and zeros, and even their presence and the precise form of the model, constitute the most uncertain part of the transfer function. However, the designer must verify that such "unmodeled high-frequency dynamics" do not cause instability.

Physical Realizability

Consider a transfer function, for example, of a controller that must be realized as a physical system of the general form

$$G_c(s) = K \frac{(s - a_1)(s - a_2)\cdots(s - a_m)}{(s - b_1)(s - b_2)\cdots(s - b_n)} \tag{5.15}$$

The following condition may be stated.

> Any transfer function, to be physically realizable, must have at least as many poles as zeros (that is, $n \geq m$).

For very large s the constant terms in (5.15) are negligible relative to s, so $G_c \approx Ks^{m-n}$. Later work will show that, to determine the response to sinusoidal inputs of frequency ω, s can be replaced by $j\omega$ and that $|G_c| \approx |K(j\omega)^{m-n}| = K\omega^{m-n}$ then gives the amplification that the function applies to a sinusoidal input of a high-frequency ω. If $m > n$, this would mean infinite amplification of an input of infinite frequency. This is not possible for physical systems. If the frequency of the input is increased, sooner or later the system will in effect give up, and its output amplitude will stop growing. In fact, for most plants $n > m$, and the response tends to zero at high frequencies. For controllers, however, the $n = m$ case is quite common.

Controllers

PID (proportional plus integral plus derivative) control is the most commonly used standard form of dynamic compensation in practice. Electronic, pneumatic, and digital PID controllers are available off the shelf in a wide variety of makes and types. Pneumatic and electronic realizations were modeled and discussed in Sections 3.5 and 3.6. Alternatively, their action may be programmed in the form of a *control algorithm* on digital computers or microprocessors. If the series compensator $G_c(s)$ in Fig. 5.8(a) is a proportional plus integral plus derivative controller, its output m and input e are related by the equation

$$m = K_c e + K_i \int e \, dt + K_d \dot{e} \tag{5.16}$$

From Table 1.6.1, this implies the transfer function

$$G_c(s) = K_c + \frac{K_i}{s} + K_d s = \frac{K_d s^2 + K_c s + K_i}{s}$$

PI control: $G_c(s) = K_c + \dfrac{K_i}{s} = \dfrac{K_c s + K_i}{s}$ (5.17)

PD control: $G_c(s) = K_c + K_d s$

PD control finds application, for example, in robot control, and PI control is extremely common in *process control,* a very large and important area of application of control systems. Process control is usually concerned with regulator systems, to maintain controlled variables such as temperatures, pressures, and levels at constant values despite disturbances and parameter variations. The water-level control system in Fig. 3.1 is an example where the controller is most likely to be of the PI type. Furthermore, the basic proportional, integral, and derivative actions are also fundamental to other dynamic compensators, so a good understanding of the P, I, and D actions, both in analytical and physical terms, is important for an appreciation of how dynamic compensation can change system behavior. PID controllers are discussed later in this chapter.

The most common alternatives to PID controllers are *phase-lag compensators* and *phase-lead compensators*. Passive electrical networks to realize these very important forms of compensation were given in Fig. 2.8., and active realizations based on operational amplifiers in Section 3.6. The transfer functions can be written as

$$G_c(s) = \frac{K_c(s + z)}{s + p} \tag{5.18}$$

For a lead, $z < p$, and for a lag, $z > p$. Figure 5.9 shows the pole–zero patterns. Phase-lead compensation can be considered as an approximation to PD control, with a pole added to the zero of the PD controller in (5.17). If this pole is chosen far enough away, its effect will be small. Similarly, phase-lag compensation approximates the pole at the origin of the PI controller in (5.17) by a pole close to the origin.

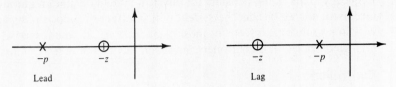

Figure 5.9   Phase lead and phase lag.

A *lag–lead compensator* can be thought of as a series connection of a lag and a lead and is frequently suitable if these cannot meet the specifications individually. The pole and zero of the lag are generally closer to the origin than those of the lead. It can also be considered as an approximation to a PID controller.

Derivative Control and Noise

The PD controller $(K_c + K_d s)$ in (5.17) may be called an "idealized controller" because it ignores the physical realizability constraint. It is actually realized as $(K_c + K_d s)/(Ts + 1)$, but T is taken to be so small that the pole is far enough away to be negligible. This pole changes the controller to a phase-lead compensator. In fact, a phase-lead is often preferable to PD control to reduce the effect of signal noise. In PD control, compensator output and input are related by $m = K_c e + K_d \dot{e}$. Here $\dot{e}$ is the slope of a plot of e versus time, and this term will amplify the effect of even small irregularities in e due to noise. To reduce this effect, the ratio p/z should not be made larger than necessary to meet the performance specifications.

It is often possible to avoid taking the derivative of a signal by measuring the rate directly. For example, in the position servo with velocity feedback modeled in Example 3.3.2 the velocity is measured directly instead of being obtained by taking the derivative of the position signal. If taking the derivative is unavoidable, a simple lag filter with time constant T can be used to filter out the noise. This is again equivalent to adding a pole to PD control.

Pole–Zero Cancellation Control

There are systems in which the plant G has poles in undesirable locations that cannot easily be changed by feedback. In this case the preceding controllers may not be sufficiently effective. Pole–zero cancellation control, introduced in the last section, may then provide a solution. As was noted, it is only necessary to place the zeros of $G_c(s)$ close enough to the undesirable poles to make the corresponding transients acceptably small.

In the typical case it may be desired to approximately cancel a complex conjugate pair of plant poles. The series compensator should provide two zeros at or near these locations, and to ensure physical realizability it also has two poles, selected in more desirable positions. A possible realization of such a controller is the bridged-T network in Fig. 2.8(f), of which the transfer function is given by (2.8f).

Two points should be emphasized. The first, already raised in the last section, is that the system response to initial conditions is not affected by any cancellations that may have been achieved in the input–output response. The second is that so-called open-loop unstable poles of G, that is, poles of G in the right-half s-plane, may never be canceled in this way by zeros of G_c. The root locus technique in Chapter 6 will be used to show that this leads inevitably to an unstable closed-loop system.

5.5 DYNAMIC COMPENSATION: VELOCITY FEEDBACK

The most common example of feedback compensation is velocity feedback. Figure 5.10 repeats the block diagram in Fig. 3.10, which was derived in Example 3.3.2 for a motor position servo with velocity feedback. T_m has been replaced by its definition J/B in terms of inertia J and damping constant B, and K_m is equivalent to K_mB in Fig. 3.10. The physical motivation for velocity feedback was discussed in Example 3.3.2. If damping B is small, velocity feedback can provide the equivalent of a damping torque, proportional to velocity and in the opposite direction. As mentioned in the preceding section, to avoid problems due to noise, the velocity signal may be obtained by direct measurement using a tachogenerator instead of by differentiation of the position signal. The corresponding device in aircraft control systems is the rate gyro.

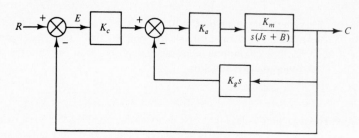

Figure 5.10   Velocity feedback.

To verify the improved servo damping, the minor loop in Fig. 5.10 is reduced to obtain the transfer function

$$\frac{C}{E} = \frac{K_c K_a K_m}{s(Js + B + K_a K_m K_g)} \tag{5.19}$$

This is the loop gain function in the unity feedback major loop and yields the system characteristic equation

$$s^2 + \frac{B + K_a K_m K_g}{J} s + \frac{K_c K_a K_m}{J} = 0 \qquad (5.20)$$

on dividing through by J. To see the significance of such quadratic equations, it is always useful to compare them with the normalized form $s^2 + 2\zeta\omega_n s + \omega_n^2 = 0$. The constant term shows that ω_n is independent of the velocity feedback gain K_g. Comparison of the damping term coefficients now shows immediately that K_g has indeed increased the effective damping constant of the system and improved the damping ratio.

In systems where the inherent damping B is quite small, the characteristic equation without velocity feedback would be

$$s^2 + \frac{K_c K_a K_m}{J} = 0$$

so the system poles would lie on the imaginary axis, where the system damping ratio is zero.

The other aspect of performance considered is the effect of velocity feedback on steady-state errors. The plant in Fig. 5.10 is type 1, so without velocity feedback the system has zero steady-state errors for step inputs. In general, when a minor feedback loop is closed around such a plant, the loop gain function of the major loop will no longer have a factor s in its denominator, so the system will no longer be of type 1. For this particular minor loop feedback, however, C/E in (5.19) shows that the system type number is still 1. So velocity feedback does not affect the steady-state error for step inputs.

Example 5.5.1   Velocity Feedback

The loop gain function in Fig. 5.11(a) is

$$\frac{C}{E} = \frac{25}{s(s + 2 + 25K_g)}$$

and the characteristic equation

$$s^2 + (2 + 25K_g)s + 25 = 0$$

Let a system damping ratio $\zeta = 0.7$ be desired. Comparison with $s^2 + 2\zeta\omega_n s + \omega_n^2 = 0$ shows that $\omega_n = 5$ and $2\zeta\omega_n = 10\zeta = 2 + 25K_g$. Thus the required velocity feedback gain $K_g = 0.2$. The unit step response is shown in Fig. 5.11(b), determined from the closed-loop transfer function

$$\frac{C}{R} = \frac{25}{s^2 + (2 + 25K_g)s + 25}$$

The overshoot is less than 5%, as expected from Fig. 5.2 for a damping ratio of 0.7.

Since the characteristic equation is of second order, a root locus showing how the system poles and damping ratio change with K_g can easily be plotted. It is shown in Fig. 5.11(c). Except for the large enough K_g, the poles lie on a circle of radius $\omega_n = 5$. The angle ϕ of the poles with the negative real axis depends on K_g according to $\zeta = \cos\phi = 0.2 + 2.5K_g$. The improvement due to velocity feedback is evident. The angle ϕ for $K_g = 0$ is not far from 90° and corresponds to a damping ratio of only 0.2.

However, the effect of this compensation on steady-state errors must also be considered. The loop gain function C/E shows that the system is still of type 1, so steady-state errors after step inputs remain zero. But the gain is $25/(2 + 25K_g)$, so the steady-state error following a unit ramp is $(0.08 + K_g)$ and increases with increasing

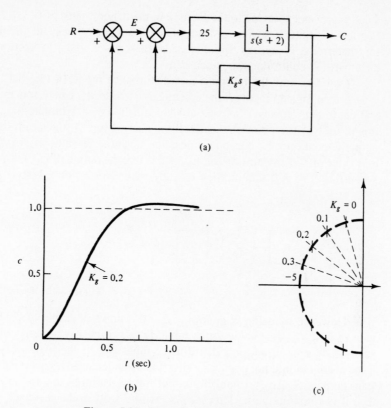

(a)

(b)   (c)

Figure 5.11   Example 5.5.1: velocity feedback.

velocity feedback gain K_g. Between $K_g = 0$ and $K_g = 0.2$, it changes from 0.08 to 0.28. Thus K_g should not be made larger than necessary to raise ζ to a satisfactory level.

5.6 SERIES COMPENSATION USING PID CONTROLLERS

Proportional plus Integral Control

From (5.17), with PI control the series compensator $G_c(s)$ in Fig. 5.8(a) is given by

$$m = K_c e + K_i \int e \, dt$$

$$G_c(s) = K_c + \frac{K_i}{s} = K_c \frac{s + z}{s} \qquad z = \frac{K_i}{K_c} \tag{5.21}$$

Let the type number of the plant, that is, the number of integrators in G, be zero. Then with P control the system is type 0 and has a steady-state error following step inputs. The essential improvement due to the addition of I control now follows immediately from Table 4.3.1 on steady-state errors:

By adding I control, the system has been changed from type 0 to type 1 and hence now has zero steady-state error following a step input.

This improvement is due to the factor s in the denominator of $G_c(s)$. As explained in Section 4.3, for a step input a steady state cannot exist unless the input to the integrator, the error e, is zero. A direct physical reason was given in Section 3.5, when the pneumatic PI controller was discussed.

Consider now the response to a disturbance input. In Fig. 5.12, if the plant is type 1, then for step inputs R the system has zero steady-state errors even if G_c is only a P controller. However, as was already illustrated in Example 4.3.2 and Fig. 4.8, this is not true for a disturbance D. If the plant is type 1, the steady-state value c_{ss} of output C could not be constant unless the steady-state value m_{ss} of M is zero. For a unit value of D the condition for $m_{ss} = 0$ is that the output of G_c be equal to -1. With P control $G_c = K_c$, this implies a steady-state error $e_{ss} = -1/K_c$.

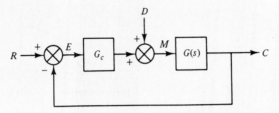

Figure 5.12   Disturbance inputs.

However, by using PI control (5.21) this nonzero output of G_c can be produced with zero steady-state value for its input. In fact, it is easy to verify that this provides zero steady-state error for a disturbance input also if the plant is type 0. Actually, these results should not come as a surprise. If under steady-state conditions with constant outputs G_c gives a nonzero output for a zero input due to the integral control, then it has infinite gain. This means infinite gain in the feedback path between C and the point where D enters the loop. From Section 4.2, this implies zero steady-state error.

Thus there can be reasons for using integral control even if the plant is type 1. Note that this changes the system to type 2, so the steady-state errors following ramp inputs R will then also be zero. However, the disadvantage is that integrations anywhere in the loop complicate the design for adequate relative stability.

The need for PI control to reduce steady-state errors is already evident in Examples 4.5.2 and 4.5.3 for plants consisting of one and two simple lags. With one simple lag, the high gains needed for small errors imply large actuating signals to the plant, and so a high cost of control. With two simple lags, Fig. 4.16 shows excessive steady-state error even when a system damping ratio of 0.5 is allowed. To illustrate the basic control actions, the P control in these examples will be replaced with PI control here and by PD control later in this section.

Example 5.6.1   PI Control of a Simple Lag Plant

In Fig. 5.8(a), let

$$G(s) = \frac{1}{s + 1} \qquad G_c(s) = \frac{K_c(s + z)}{s} \tag{5.22}$$

The integral of the error as well as the error itself is used for control. The loop gain function is

$$G_c G = \frac{K_c(s + z)}{s(s + 1)}$$

and the closed-loop transfer function is

$$\frac{C}{R} = \frac{K_c(s + z)}{s^2 + (1 + K_c)s + K_c z}$$

Design involves the choice of K_c and z. In the present case the effect of different choices of z will be compared for designs of which the closed-loop poles all have real part -2. So all have time constant 0.5 and identical settling times. If the imaginary part is a, the system characteristic polynomial is then

$$(s + 2 - ja)(s + 2 + ja) = s^2 + 4s + 4 + a^2$$

and equating this with the denominator of C/R yields

$$K_c = 3 \qquad a^2 = 3z - 4$$

The pole–zero patterns of C/R corresponding to the choices $z = 2, 3,$ and 4 are shown in Fig. 5.13(b). Also shown is the pattern for the special case $z = 1$, when the zero cancels the plant pole at -1. Then the closed-loop transfer function with the closed-loop pole at -2 is $C/R = 2/(s + 2)$.

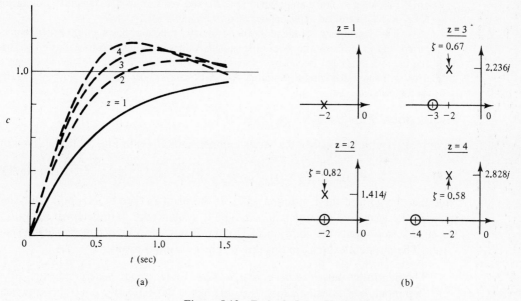

(a)   (b)

Figure 5.13   Example 5.6.1: PI control.

The unit step response for $z = 1$ is $c(t) = 1 - e^{-2t}$ and is shown in Fig. 5.13(a) together with those for $z = 2, 3,$ and 4. The latter may be found using graphical determination of the residues or by means of the computer programs in Appendix B. The responses are of the form $c(t) = 1 + Ae^{-2t}\cos(at + \theta)$, where for $z = 2, 3,$ and 4 the values of (A, a, θ) are, respectively, $(1.22, 1.414, -144.7°)$, $(1.096, 2.236, -155.9°)$, and $(1.060, 2.828, -160.6°)$.

As expected, all designs have zero steady-state error and the same settling time. The smaller rise time and greater speed of response associated with larger imaginary parts of the poles, illustrated by Example 5.2.1, are also again in evidence.

Comparison of the responses with the pole–zero patterns again demonstrates that, due to the zeros in relatively dominant positions, overshoot cannot be predicted on the basis of the damping ratios associated with the pole positions. These ratios are indicated on Fig. 5.13(b) and would for $z = 2, 3,$ and 4 predict smaller overshoots than the 7.5%, 13.5%, and 18.5% actually present.

Example 5.6.2   PI Control of Plant with Two Simple Lags

With the PI controller $G_c(s)$ of (5.21) and the plant

$$G(s) = \frac{1}{(s + 1)(0.5s + 1)}$$

as in Example 4.5.3, the loop gain function in Fig. 5.8(a) is

$$G_c G = K_i \frac{1 + s/z}{s(s + 1)(0.5s + 1)} = 2K_c \frac{s + z}{s(s + 1)(s + 2)}$$

and the closed-loop transfer function is

$$\frac{C}{R} = \frac{K_c(s + z)}{0.5s^3 + 1.5s^2 + (1 + K_c)s + K_c z}$$

The loop gain function shows that the system has again been made to be of type 1, with zero steady-state error following step inputs. The gain as defined in Section 4.3, with all constant terms in the loop gain function made unity, is seen to be K_i. So the steady-state error following a unit ramp input is $1/K_i$ and reduces as the integral gain factor $K_i = K_c z$ is increased. This applies also in Example 5.6.1.

The denominator of the closed-loop transfer function shows a third-order characteristic equation. Design to determine suitable values of K_c and z can be carried out using the root locus and frequency response methods discussed in the following chapters. This will show that careful design is necessary to avoid oscillatory or slow system dynamic behavior.

Proportional Plus Derivative Control

From (5.17), with PD control the series compensator $G_c(s)$ in Fig. 5.8(a) is given by

$$\begin{aligned} m &= K_c e + K_d \dot{e} \\ G_c(s) &= K_c + K_d s = K_d(s + z) \qquad z = K_c/K_d \end{aligned} \tag{5.23}$$

Note that the gain of the controller is K_c. If the plant is type 0, then both with and without the derivative control this is a type 0 system with gain determined by K_c. Hence, for step inputs the addition of D control will have no effect on steady-state errors. The physical explanation of this result is straightfoward:

> The controller output is $m = K_c e + K_d \dot{e}$. Under steady-state conditions for a step input, e = constant, so $\dot{e} = 0$, and hence the derivative control component has no effect on the steady-state error.

However, the steady-state error can be reduced indirectly. This is because the addition of D control will be found to improve relative stability and therefore allows K_c to be larger than with P control alone.

A physical appreciation of why derivative control improves damping may be gained from the following:

> The derivative control component responds to the rate of change of error and hence gives a stronger control signal if the error changes faster. Thus it anticipates large errors and attempts corrective action before they occur.

The following example illustrates this and also shows how the closed-loop system pole positions in the s-plane depend on the proportional and derivative gains.

Example 5.6.3   PD Control of Plant with Two Simple Lags

Here both the error and its derivative are used for control. In Fig. 5.8(a), with the same plant as in Example 5.6.2,

$$G(s) = \frac{1}{(s + 1)(0.5s + 1)} \qquad G_c(s) = K_c + K_d s \qquad (5.24)$$

The loop gain function is, with $z = K_c/K_d$,

$$G_c G = K_c \frac{1 + s/z}{(s + 1)(0.5s + 1)} = 2K_d \frac{s + z}{(s + 1)(s + 2)} \qquad (5.25)$$

This is a type 0 system with gain K_c, so if K_c is the same as for P control, then the steady-state error for a unit step input is also the same and equals $1/(1 + K_c)$.

The closed-loop transfer function is found to be

$$\frac{C}{R} = \frac{2K_d(s + K_c/K_d)}{s^2 + (3 + 2K_d)s + 2 + 2K_c}$$

To see the effect of D control, equate the denominator of C/R with the standard form $(s^2 + 2\zeta\omega_n s + \omega_n^2)$. This shows that the distance ω_n of the poles to the origin depends only on K_c, and the distance $-\zeta\omega_n$ to the imaginary axis only on K_d:

$$\omega_n^2 = 2 + 2K_c \qquad 2\zeta\omega_n = 3 + 2K_d \qquad (5.26)$$

Thus, varying K_c for constant K_d moves the poles vertically, and varying K_d for constant K_c moves them along a circle.

Let a damping ratio 0.5 be required. For $K_d = 0$, the characteristic equation is $s^2 + 3s + 2 + 2K_c = s^2 + 2\zeta\omega_n s + \omega_n^2 = 0$. To achieve $\zeta = 0.5$ requires $\omega_n = 3$, or $\omega_n^2 = 9 = 2 + 2K_c$, so $K_c = 3.5$. The steady-state error $1/(1 + K_c)$ then equals 22%, and the system poles are located at A in Fig. 5.14. If the specifications permit an error of no more than 10%, $K_c = 9$ is required. For constant $K_d = 0$, increasing K_c moves the poles vertically to B, at a distance $\omega_n = \sqrt{2 + 2K_c} = \sqrt{20} = 4.472$ to the origin. However, as expected with this increase of gain to reduce steady-state errors, the damping ratio associated with B is an inadequate 0.335 ($= 1.5/4.472$).

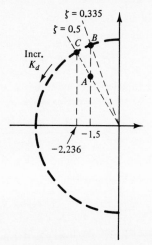

Figure 5.14   Example 5.6.3: PD control.

But by keeping $K_c = 9$ constant, the poles can be moved along a circle through B by increasing K_d from $K_d = 0$. The pole location C in Fig. 5.14 will realize both the desired damping and the desired steady-state accuracy and is obtained, with $K_c = 9$, if

$\zeta\omega_n = 1.5 + K_d = 2.236$, so if $K_d = 0.736$. The zero of the closed-loop transfer function is $-K_c/K_d = -9/0.736 = -12.23$. This is about 5.5 times as far from the imaginary axis as the poles at C, so the unit step responses should closely resemble those for a quadratic lag in Fig. 4.13. It is apparent that the addition of D control improves the damping for a given value of K_c. From a different point of view, PD control may also be used to improve accuracy, because it permits an increase of K_c without loss of system damping. The reduction of settling time from the systems A or B to C should be noted as well.

PID Control and I-PD Control

The transfer function (5.17) shows a pole at the origin, due to the integral control, and two zeros. These may be real or form a complex conjugate pair. A lag–lead controller, introduced in the last section, typically is an approximation for the case of real zeros.

The classical system configuration for PID control is series compensation, in Fig. 5.8(a). However, when derivative control is present, then the implementation is often based on the minor loop feedback configuration in Fig. 5.8(b). With series compensation, for a step input the error also changes by a step, since the output takes time to change. The derivative of a step is an impulse, so the derivative control component causes an impulse in the actuating signal, in addition to a step due to the proportional control, and is likely to produce nonlinear saturation effects. For this reason, the D action may be "moved to" the feedback controller $H_c(s)$ and changes from derivative-of-error to derivative-of-output control, or rate feedback control. Since the output changes more slowly, this avoids impulsive signals. The effects of noise should again be considered, as discussed in the last section. To avoid even step changes of the actuating signal, thus restricting it to only relatively slow changes, the P action term can also be included in $H_c(s)$, leaving only the I controller for $G_c(s)$. This is sometimes called I-PD control.

Example 5.6.4   PID Control and I-PD Control

In Fig. 5.8(a) with a series compensator

$$G_c(s) = K_c + \frac{K_i}{s} + K_d s$$

the closed-loop transfer function is found to be

$$\frac{C}{R} = \frac{\left(K_c + \dfrac{K_i}{s} + K_d s\right)G}{1 + \left(K_c + \dfrac{K_i}{s} + K_d s\right)G}$$

If the minor loop feedback configuration in Fig. 5.8(b) represents I-PD control, then

$$G_c(s) = \frac{K_i}{s} \qquad H_c(s) = K_c + K_d s$$

and the closed-loop transfer function is given by

$$\frac{C}{R} = \frac{\dfrac{K_i}{s}G}{1 + \left(K_c + \dfrac{K_i}{s} + K_d s\right)G}$$

The preceding example shows that the denominators of the closed-loop transfer functions are the same for both forms of PID control, so both have the same closed-loop poles. However, the zeros differ, so the transient responses will be different.

Pneumatic controllers were discussed in Section 3.5, with physical explanations of why PI control gives zero steady-state error and why PD control improves system damping. Block diagram models were derived, and are repeated in Fig. 5.15, but it was never verified that these actually represent PI control and PD control.

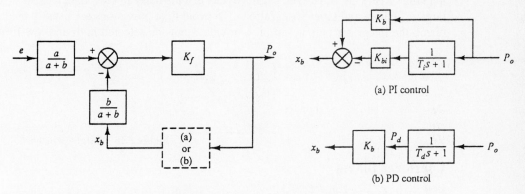

Figure 5.15   Pneumatic PID controllers.

For Fig. 5.15(a), combining the two parallel feedback loops gives

$$\frac{X_b}{P_o} = K_b - \frac{K_{bi}}{T_i s + 1} = \frac{K_b T_i s}{T_i s + 1} \qquad \text{if } K_b = K_{bi} \qquad (5.27)$$

Then the closed-loop transfer function is found to be

$$\frac{P_o(s)}{E(s)} = \frac{aK_f}{a + b} \bigg/ \left(1 + \frac{b}{a + b} \frac{K_b T_i s}{T_i s + 1} K_f\right)$$

$$= \frac{aK_f(T_i s + 1)}{(a + b)(T_i s + 1) + bK_b T_i s K_f}$$

Since K_f is quite large, as was emphasized in Section 3.5, this approximates

$$\frac{P_o(s)}{E(s)} \approx \frac{a}{bK_b} \frac{T_i s + 1}{T_i s} = \frac{a}{bK_b} + \frac{a}{bK_b T_i} \frac{1}{s} = K_c + \frac{K_i}{s}$$

$$K_c = \frac{a}{bK_b} \qquad K_i = \frac{a}{bK_b T_i} \qquad (5.28)$$

This is PI control. Note that the proportional and integral gain constants K_c and K_i can only be adjusted independently via T_i.

In a similar manner, the block diagram in Fig. 5.15(b) can be shown to represent PD control, with the approximate transfer function

$$\frac{P_o(s)}{E(s)} \approx K_c + K_d s \qquad K_c = \frac{a}{bK_b} \qquad K_d = \frac{aT_d}{bK_b} \qquad (5.29)$$

Integral Windup

Reset windup or *integral windup* is a problem that can arise in all physical implementations of integral control, including those by digital computers and microprocessors.

For large changes of input or large disturbances, it could lead to severe transient oscillations. Referring to Fig. 5.16, if a large step input r is applied, the error $e = r - c$ will be of the initial sign for a considerable period, until the output c passes the level of r. During this period the integrator accumulates a large output. It winds up to a much larger value than that which makes the actuator go to its physical limit, such as the fully open position of a control valve. After c passes r, it takes long for the integrator to wind down to a value where the controller output is back inside the range where the valve even begins to close. This delay in closing can cause a very large overshoot and severe oscillations. To avoid this, provisions are usually made to clamp the integrator output and prevent values outside selected high and low limits.

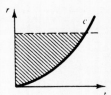

Figure 5.16   Integral windup.

Controller Tuning

It is important to point out that the use of controllers in practice does not depend on finding values of K_c, K_d, and K_i by analytical means. Instead, these are set by following established procedures for tuning a controller after installation. This is why experienced personnel with knowledge of the process can apply PID controllers without any formal knowledge of mathematical modeling or control theory. This important reason for the success of PID controllers should be recognized as a major advantage, even in a book on these theories.

Two tuning methods of Ziegler and Nichols are quite popular. They provide initial settings that are refined by simulation or on-site tuning. The methods are based on experiments and analysis and are especially convenient if mathematical models for the plant are not available. Both tuning rules are designed for a step response overshoot of about 25%. The controller transfer function is taken to be in the usual form:

$$G_c(s) = K_c + \frac{K_i}{s} + K_d s \tag{5.30}$$

First, consider the Ultimate-Cycle Method. Here, with only proportional control present, the gain K_c is increased to the crtical or ultimate gain K_u, where the output shows sustained oscillations of measured period T_u. The following controller settings are then suggested:

$$\text{P:} \quad K_c = 0.5K_u$$

$$\text{PI:} \quad K_c = 0.45K_u \qquad K_i = 0.54\frac{K_u}{T_u} \tag{5.31}$$

$$\text{PID:} \quad K_c = 0.6K_u \qquad K_i = 1.2\frac{K_u}{T_u} \qquad K_d = 4.8\frac{K_u}{T_u}$$

This method does not apply if on increasing K_c sustained oscillations do not develop at some point (as an indication that the system is at the limit of stability).

The second method is based on measuring the unit step response of the plant experimentally. Most plants in process control satisfy the restrictions of this method, which are that the plant must be of type 0 and have no dominating complex poles. The step response tends to have the S-shaped character shown in Fig. 5.17. By drawing a tangent to this curve at its inflection point, a delay time L and time constant T can be identified as shown. Identifying T as a time constant may be verified to be consistent with Fig. 4.10. And calling L a delay time is consistent with an approximation in which the response over this period is taken to be zero. Using the delay theorem in Table 1.6.1, the plant may then be modeled approximately as a series connection of a delay and a simple lag, with transfer function

$$G(s) = \frac{Ke^{-Ls}}{Ts + 1} \tag{5.32}$$

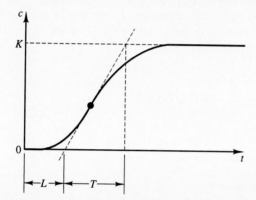

Figure 5.17   Controller tuning.

With this model, the suggested controller settings are

$$\text{P:} \quad K_c = \frac{T}{L}$$

$$\text{PI:} \quad K_c = 0.9\,\frac{T}{L} \qquad K_i = 0.27\,\frac{T}{L^2} \tag{5.33}$$

$$\text{PID:} \quad K_c = 1.2\,\frac{T}{L} \qquad K_i = 0.6\,\frac{T}{L^2} \qquad K_d = 0.6T$$

In both methods, if the overshoot is considered too large, further tuning can be carried out to improve relative stability.

5.7 CONCLUSION

In this chapter, feedback system design has been introduced. Performance criteria were formulated and related to pole positions. The importance of the effect of zeros was discussed and related to pole–zero cancellation design. The most common types of controllers were listed. Feedback compensation was treated, but the emphasis was on proportional plus integral plus derivative control (PID), which is extremely common in practice. Also, the essential effects of the basic integral and derivative control actions were emphasized, because the very widely used phase-lag and phase-lead compensators rely on approximations to these basic actions.

The key questions raised in Section 1.4, that is, how systems behave and how this behavior may be changed, have now been considered. Indeed, profound changes of performance result by changing only the gain of the loop gain function. However, the inadequacy of the tools made available so far for system analysis and design has become apparent. For example, how can the effect of gain or compensator parameters on relative stability be determined if the system characteristic equation is of order 3 or higher? Furthermore, while the final value theorem easily yields steady-state errors, what about accuracy under dynamic conditions? For example, if the frequency spectrum of the system input is known to be in a certain range, what is the largest error that may occur, or how does one design systems for which this error satisfies given specifications?

The root locus method and frequency response techniques, which are discussed next, are the classical tools for both system analysis and design. The concept of a root locus was introduced in several examples. Frequency response considerations will also provide new and important insights into the behavior of feedback systems, including the question of sensitivity.

PROBLEMS

5.1. A system with transfer function

$$G(s) = \frac{10}{s^2 + s + 5}$$

is subjected to a step input.
  (a) Plot the system pole–zero pattern and that of the system output.
  (b) From the system pattern, find the undamped natural frequency of the transient, the damping ratio ζ, the frequency of oscillations, and the time constant of response.
  (c) Determine the percentage peak overshoot and the time at which it occurs.
  (d) Find the unit step response.

5.2. In Problem 5.1, what is the effect of halving the imaginary part of the system poles on:
  (a) Settling time and time constant?
  (b) ω_n and ζ?
  (c) The number of oscillations during the decay (that is, on how oscillatory the response is)?
  (d) Percentage overshoot, peak time, and rise time?
  (e) Discuss the differences with Problem 5.1.

5.3. Let Fig. P5.3 represent a simple motor position control system, with motor input being the amplified error between desired and actual shaft positions.
  (a) Find C/R, E/R, and the characteristic equation.
  (b) Use the final value theorem to express the steady-state shaft position error e_{ss} following unit step and unit ramp inputs. How does the value of K affect these errors?
  (c) Find the value of K for which the system poles, the roots of the characteristic equation, will have a damping ratio $\zeta = 0.7$.
  (d) How does the percentage overshoot for a step input change if K is raised above the value in part (c)?

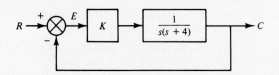

Figure P5.3

5.4. In Problem 5.3:
   (a) Plot on one s-plane the system pole positions for $K = 0, 1, 4, 8$, and 13.
   (b) For the last four, find the percentage overshoot in response to a step input.
   (c) Calculate the unit step responses for $K = 1, 4$, and 8.

5.5. **(a)** Plot the required locations of the dominating pair of poles of a system if the damping ratio is to be about 0.5 and the time constant about 0.1.
   (b) Show permissible pole locations if the system time constant may not exceed 1 sec.
   (c) Similarly to part (b), if the damping ratio may not be less than about 0.7.
   (d) Similarly to part (b), if conditions (b) and (c) must both be met.

5.6. Fig. P5.6 shows the pole–zero pattern of the system

$$G(s) = \frac{K}{(s + a)(s^2 + 2s + 2)}$$

   for $a = 0.5, 1$, and 2.
   (a) Calculate the unit step responses for $K = 2a$.
   (b) Compare these responses to gain an idea of the effect of the position of the real pole relative to the complex pair on the relative dominance of the transient terms, considering both size and speed of decay.

5.7. The pole–zero pattern of a system with root locus gain $3/a$ is shown in Fig. P5.7. The zero $-a$ lies between -1 and -3. Using graphical residue calculation:
   (a) Calculate the unit step response.
   (b) What happens if the zero is moved very close to either system pole? Which zero position is preferable, and why?

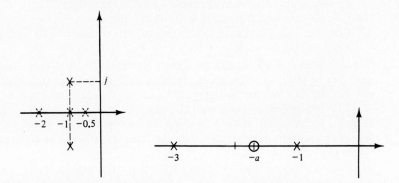

Figure P5.6   **Figure P5.7**

5.8. A system has the transfer function

$$G(s) = 4.8 \frac{s + 0.25}{(s + 0.3)(s^2 + 2s + 5)}$$

   (a) Plot its pole–zero pattern.
   (b) What are the time constants or undamped natural frequency and damping ratio, as appropriate, of the components of the transient response to a step input? How

long does the slowest component take to decay almost completely?

(c) Which component would you expect to dominate, and why?

5.9. For a system with root locus gain $\frac{5}{3}$ and pole–zero pattern as shown in Fig. 5.9:

(a) Express the transform of the system output for a unit step input.

(b) Express the residues for this input by the graphical rule and find their values.

(c) Express the unit step response.

(d) Which transient component dominates, and why?

(e) To verify part (d), calculate the value of time at which the exponential decay term and the amplitude of the oscillatory term are equal, and find this amplitude.

5.10. For the system in Fig. P5.10 with $G(s) = (s + 1)/(s + 2)$:

(a) Find the value of K required for a system time constant $T = 0.667$ sec.

(b) Plot the system pole–zero pattern for this K.

(c) Calculate the corresponding unit step response.

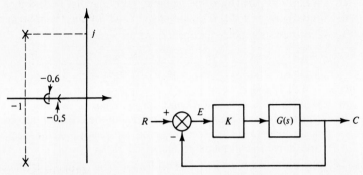

Figure P5.9   Figure P5.10

5.11. In Fig. P5.10 with $G(s) = (s + 1)/[(s + 2)(s + 20)]$:

(a) Find K so that the dominant system time constant will be $T = 0.667$ sec, and for this K also determine the second pole of the system.

(b) Calculate the unit step response for K of part (a).

(c) Compare this problem and the results with Problem 5.10 and discuss the observations.

5.12. The system in Fig. P5.10 with the simple lag plant $G(s) = 1/(s + 5)$ could represent, among others, a temperature, pressure, level, or speed control system.

(a) Plot the root loci for varying gain K, that is, the loci showing how the closed-loop system pole changes with changing K.

(b) Find the positions of the pole along the loci for $K = 10$ and $K = 45$.

(c) From these positions, predict how the transient response is affected by the increase of K.

(d) Compare this prediction with the step response results calculated in Problem 4.19.

5.13. Repeat Problem 5.12 for the same values of K if the plant model is refined to $G(s) = 20/[(s + 5)(s + 20)]$, for which step responses were calculated and plotted in Problem 4.20. For what value of K does the nature of the response change, and how does it change?

5.14. In Fig. P5.10 with $G(s) = 1/[(s + 1)(s + 7)]$:

(a) How long would transients take to decay almost completely if there were no feedback?

(b) What is the characteristic equation of the closed-loop system, and where must the dominating system pole be located if the time in part (a) is to be halved?

   (c) What value of K will achieve this?
   (d) What is the corresponding steady-state error for a unit step input?
5.15. In Problem 5.14 let an increase of gain be desirable to reduce the steady-state error.
   (a) Express the system poles as functions of K.
   (b) Find the lowest value of K that will minimize the settling time.
   (c) Find K and the corresponding steady-state error for a unit step to obtain a system damping ratio of about 0.7.
   (d) Compare the settling times of parts (b) and (c). Which is the best to minimize rise time, and why?
5.16. In Fig. P5.10 with $G(s) = (s + 1)/[s(s + 3)]$:
   (a) Find K so that the dominating system time constant will be 2 sec, and find for this K the second system pole.
   (b) What are the resulting steady-state errors following unit step and unit ramp inputs?
   (c) Calculate the unit step response from the results in part (a).
5.17. A plant $G(s)$ has the transfer function $1/(s - 2)$.
   (a) Determine stability from its pole–zero pattern.
   (b) Show that the dynamic behavior may be changed by adding a feedback loop with an amplifier K around G.
   (c) Find K to obtain a stable system with a time constant $T = 0.1$ sec.
   (d) What is the corresponding steady-state error for a unit step input?
5.18. The system in Fig. P5.10 with $G(s) = (s + 2)/[(s - 2)(s + 4)]$ would be unstable without feedback. Why? Show that it can be stabilized by feedback as shown, and find K so that the dominating system time constant is 1 sec.
5.19. **(a)** For the motor position servo in Fig. P5.19 without the rate or velocity feedback ($K_g = 0$), find K for a system damping ratio 0.5 and the corresponding steady-state error following unit ramp inputs.
   (b) Still with $K_g = 0$, what value of K will give a steady-state unit ramp following error of 0.1, and what is the corresponding damping ratio?
   (c) With K as in part (b), what value of K_g will give a system damping ratio 0.5? How does the steady-state error compare with that in part (b)?

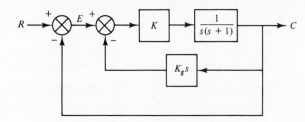

Figure P5.19

5.20. In Fig. P5.20, the system in Fig. P5.19 for $K = 10$ has been extended by including the amplifier K_a. Determine whether it is now possible to achieve both the steady-state error 0.1 in Problem 5.19(b) and the damping ratio 0.5 in Problem 5.19(c). If so, find the values of K_a and K_g required.
5.21. The transient velocity feedback in the motor position servo in Fig. P5.21 is a modification of the velocity feedback in Fig. P5.20. It may be realized by following the tachogenerator by a simple lag RC circuit and may be used to reduce the effect of high-frequency noise, as Chapter 8 will show.
   (a) Determine the steady-state errors for unit step and unit ramp inputs, and compare with those for pure velocity feedback for the same values of K_c and K_g.

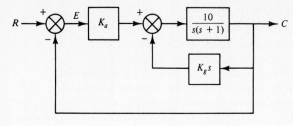

Figure P5.20

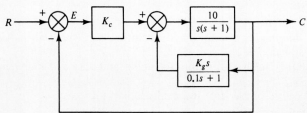

Figure P5.21

(b) Use the Routh–Hurwitz criterion to investigate stability for all values of K_c.

5.22. The block diagram of a roll stabilizer for a ship such as considered in Problem 3.4 is shown in Fig. P5.22. Minor loop rate feedback is included because of the low damping associated with the ship dynamics.

    (a) Express the transfer function for the effect of wave disturbance torque T_d on ship roll angle C.

    (b) Find the equations that must be satisfied by K_a, K_1, and K_g to ensure both a steady-state value of no more than 0.1 for C in response to a unit step T_d and a system damping ratio 0.5.

    (c) Which of K_1 and K_a must be adjustable to enable both specifications in part (b) to be met?

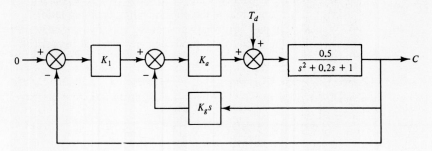

Figure P5.22

5.23. In Fig. P5.23 P control and PI control are to be compared. For $K_c = 4$, $K_i = 5$:

    (a) Find the transfer functions E/R and E/D.

    (b) Find the steady-state errors $e = r - c$ for unit step inputs of R and D in turn.

    (c) Compare the results for parts (i) and (ii) and comment on the effects on the steady-state performance of adding the integral control.

5.24. In Fig. P5.23, determine the steady-state values of the system error e if the input r is a unit ramp signal, for $K_c = 4$, $K_i = 5$. Which of parts (i) and (ii) is in effect not capable of following a unit ramp input?

5.25. In Fig. P5.25 the plant is a simple lag $G = 1/(s + 1)$. To demonstrate why stability considerations usually dictate the choice of PI control $G_c = K_c + K_i/s$ over pure I

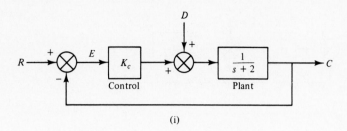

(i)

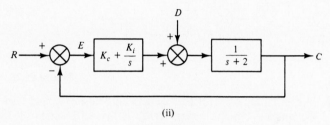

(ii)

Figure P5.23

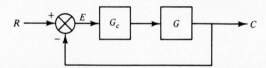

Figure P5.25

control $G_c = K_i/s$ if zero steady-state errors for constant inputs are desired, try to design these controllers for a system damping ratio 0.5 and either of the following conditions:
1. A steady-state error of 0.25 following unit ramp inputs.
2. A settling time of about 4 sec.

5.26. In Problem 5.25, compare P control and PI control on the basis of steady-state errors and the nature of the response if both are designed for a time constant $T = 0.5$ sec. Note the design freedom still left with PI control, and use it to minimize rise time subject to a constraint of about 0.7 on damping ratio, a desirable solution.

5.27. In Problem 5.25:
   (a) For I control $G_c = K_i/s$, find K_i for a steady-state error of 0.25 for unit ramp inputs, and find the corresponding damping ratio.
   (b) Suppose that D control is added in part (a) to improve damping; that is, $G_c = (K_i/s) + K_d s$ is used. Investigate the stability and accuracy properties with this ID control, and conclude whether this is a desirable addition.
   (c) Evaluate whether D control by itself, $G_c = K_d s$, realizes desirable properties for steady-state error and settling time. Explain the steady-state error results physically.

5.28. Compare the effect of the location of the integrator in the loop gain functions in Fig. P5.28 on stability and steady-state errors for step and ramp inputs of both reference input R and disturbance input D. Express and compare all steady-state errors:
   (a) How do the errors for R and D compare in part (i)?
   (b) How do these errors compare for part (ii)?
   (c) Explain these differences in behavior.

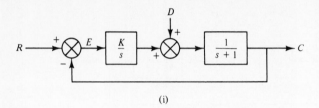

(i)

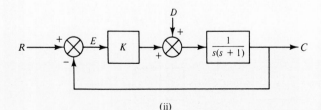

(ii)   **Figure P5.28**

5.29. In Fig. P5.25, let $G(s)$ be the open-loop unstable plant $G(s) = 1/(s - 1)$. Design the simplest possible controller $G_c(s)$ that will satisfy all the following specifications:

1. The steady-state error for constant inputs must be zero.
2. The system settling time must be about 4 sec.
3. The system damping ratio should be 0.5.

5.30. Let Fig. P5.25 model a temperature control system with plant transfer function $G(s) = 1/[(s + 1)(s + 5)]$.
   (a) With P control $G_c = K_c$, what is the system type number, and what is the gain?
   (b) For $G_c = K_c$, find K_c for a damping ratio 0.5 and the corresponding steady-state error for a unit step input.
   (c) Choose the form of controller that will make this steady-state error zero, and write the characteristic equation to note why this choice complicates stability analysis.
   (d) Use the Routh–Hurwitz criterion to determine for what relations among the parameters, if any, either of these systems may be unstable.

5.31. **(a)** In Problem 5.30, compare P control $G_c = K_c$ and PD control $G_c = K_c + K_d s$. A system damping ratio 0.5 is required and the steady-state error for step inputs should not exceed 5%.
   (b) If a 15% steady-state error is acceptable, compare the solutions on the basis of settling time and rise time.

5.32. Similar to the motor position servo with a load disturbance torque T_l in Fig. 3.11(a), Fig. P5.32 has been extended to include velocity feedback.
   (a) If $G_c = K_c$, a gain, find K_g and K_c to obtain a system damping ratio 0.5 and 5% steady-state error for step inputs T_l.
   (b) Does K_g affect this steady-state error directly? If not, why not?
   (c) How does the velocity feedback affect steady-state errors?

5.33. In Fig. P5.32:
   (a) Why is there a difference in steady-state error behavior to step inputs of R and of T_l for this type 1 system?
   (b) PI control for G_c, $G_c = K_c + K_i/s$, would be a natural choice if the steady-state error for step changes of T_l must be zero. Verify that this is indeed achieved.

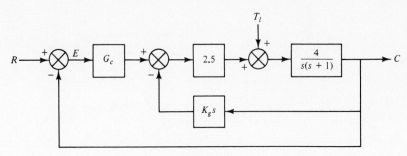

Figure P5.32

(c) Write the system characteristic equation for the choices of K_c and K_g made in Problem 5.32(a), and use the Routh–Hurwitz criterion to determine the limiting value of K_i for stability.

5.34. A system for the control of water level in the steam drum of a power station boiler is shown schematically in Fig. P5.34(i), and Fig. P5.34(ii) gives its block diagram model. W_f and W_s are the mass flow rates of feedwater to the drum and steam from the drum. G_c is the controller transfer function, and the time constant of the feedwater control valve is neglected. The transfer function of the drum, of which the net inflow is $W_f - W_s$ in the model, is $1/(As)$, as for a hydraulic cylinder. Variations of steam flow W_s due to changes of steam turbine control valve opening are the main disturbances affecting the system. If $K_v = 10$, $A = 5$, examine stability and determine the steady-state errors for step changes of L_r and W_s for P control $G_c = K_c$ and PI control $G_c = K_c + K_i/s$. For PI control, find the relation between K_c and K_i for a system damping ratio of about 0.7. Ignore the dashed links in Fig. P5.34.

5.35. In the drum-level control of Fig. P5.34, the drum model $1/(As)$ is often inadequate, because it does not reflect the swell and shrink of water level that can occur in dynamic operation. *Swell* occurs when an increase of W_s causes a temporary drop of drum pressure, which in turn causes steam bubbles in the water to grow. *Shrink* causes the level to fall temporarily when relatively cool feedwater, to balance the increased steam flow, enters and shrinks the bubbles. These level changes can introduce severe transients and often make PI control inadequate.

  Feedforward control is very important in practice to reduce the effect of measurable disturbances. In the present case it is indicated by the dashed link K_s and in effect introduces a second, parallel, path from W_s to L.
  (a) Ignoring link K_f, find the gain K_s in terms of the parameters of a P or PD controller G_c such that the steady-state effect of step changes of W_s on L will be zero.
  (b) What would happen in the steady state if in part (a) G_c included integral control?

5.36. In Fig. P5.34, to allow for disturbances in the feedwater system, the feedwater flow W_f is often also measured and fed back via K_f. The system can be viewed as a flow control system, to make W_f equal to W_s, with an outer correction loop to ensure that W_f and W_s will balance at the desired level.
  (a) Find the conditions on the parameters needed to ensure zero steady-state error for step changes of W_s, both with P or PD control and with PI control.
  (b) In view of parameter uncertainties, are zero errors actually likely in either case? Where would you locate the PI controller?

5.37. In Fig. P5.25, let $G(s) = 1/[(s + 1)(s + 4)]$ and let the compensator G_c have the

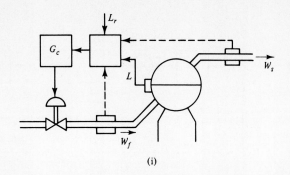

(i)

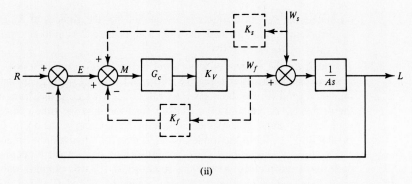

(ii)

Figure P5.34

form $G_c = K(T_1s + 1)/(T_2s + 1)$. This is phase-lead compensation if $T_1 > T_2$ and phase-lag compensation if $T_1 < T_2$. Pole–zero cancellation design is considered, where the zero of G_c is chosen to cancel one of the plant poles. If the system is to be designed for a damping ratio 0.5 using

1. Proportional control $G_c = K$
2. Phase-lag compensation $K(s + 1)/(5s + 1)$
3. Phase-lead compensation $K(0.25s + 1)/(0.05s + 1)$

then:

(a) Determine the values of K required.
(b) Find and compare the steady-state errors following unit step inputs.
(c) Determine the closed-loop system time constants and compare the speeds of response.

5.38. In Problem 5.37, evaluate the effect of phase-lag compensation by calculating and plotting, on the same graph, the unit step responses for compensators 1 and 2.

5.39. Compare the phase-lead compensation 3 in Problem 5.37 by adding the unit step response curve for this compensator to the graph in Problem 5.38.

6

The Root Locus Method

6.1 INTRODUCTION

Figure 6.1 shows a system with loop gain function $G_c GH$. The closed-loop transfer function is $C/R = G_c G/(1 + G_c GH)$, and the closed-loop poles are the roots of the characteristic equation $1 + G_c GH = 0$. The root loci show how these poles move in the s-plane when a parameter of $G_c GH$ is varied. Calculation is easy when the characteristic equation is of first or second order, and loci for such cases were already constructed and used in the examples in preceding chapters. Figures 4.15, 4.16, 5.11, and 5.14 show, respectively, loci for the following loop gain functions:

$$\frac{K}{s + 1} \qquad \frac{K}{(s + 1)(0.5s + 1)} \qquad \frac{25(K_g s + 1)}{s(s + 2)} \qquad \frac{K_c + K_d s}{(s + 1)(0.5s + 1)}$$

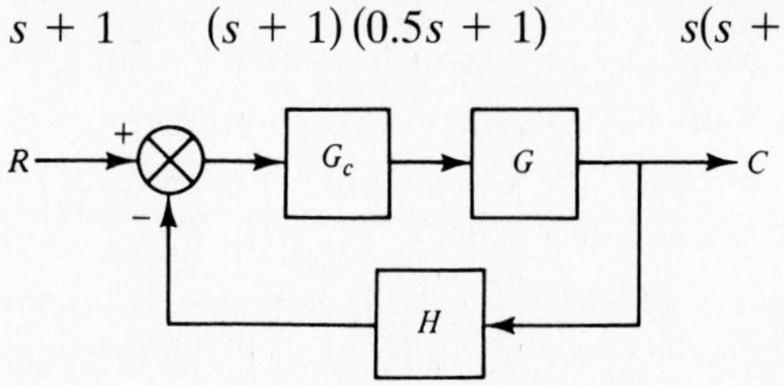

Figure 6.1   System configuration.

For the first two, and this is the most common case, the root loci show how the closed-loop poles change when gain K is changed. For the third, they show the effect of velocity feedback gain K_g on pole position, and for the fourth, both loci for varying K_c with K_d constant and for varying K_d with K_c constant were constructed.

These examples already illustrate the power of the root locus method in analysis and design, because the loci give a graphic picture of the effect of selected parameters on the system poles and suggest what values should be chosen to meet specifications on time constant and damping ratio and to improve the speed of response. The closed-loop poles are also needed to determine system stability and to calculate transient responses by the partial fraction expansion technique.

While in the examples mentioned the systems are of first or second order, most practical systems are unfortunately of at least third order. In Example 5.6.2 on PI control of a plant consisting of two simple lags, for instance, $H = 1$, $G_c = K(s + z)/s$, and $G = 1/[(s + p_1)(s + p_2)]$, so the closed-loop transfer function is

$$\frac{C}{R} = \frac{K(s + z)}{s(s + p_1)(s + p_2) + K(s + z)} \tag{6.1}$$

and the closed-loop poles are the roots of a cubic characteristic equation.

The root locus method originated as a graphical technique for determining how the system poles move when a parameter, say K in (6.1), is changed and to find these poles for particular values of the parameter. This method and the graphical construction rules are developed in this chapter and applied to analysis and design. The graphical construction technique provides the insight needed to enable the general shape of root loci to be sketched rapidly. The ability to do this remains quite important for analysis and design even though computer methods are much faster and provide accurate plots and are commonly used for root locus plotting in practice.

An interactive computer program using graphics is given in Appendix B, with an example. It is noted that, given a program to calculate the roots of polynomials, these roots can be generated for a range of values of the varying parameter and the results plotted.

In the development, the loop gain function in Fig. 6.1 will be assumed to be of the general form

$$G_c(s)G(s)H(s) = K \frac{(s - a_1)(s - a_2) \cdots (s - a_m)}{(s - b_1)(s - b_2) \cdots (s - b_n)} \tag{6.2}$$

where $n \geq m$ for physical realizability, as discussed in Section 5.4. The following definitions are used:

- *Open-loop zeros:* the roots $a_1, \ldots, a_m$ of the numerator of $G_c GH$
- *Open-loop poles:* the roots $b_1, \ldots, b_n$ of the denominator of $G_c GH$
- *Open-loop pole–zero pattern:* the s-plane plot of open-loop poles and zeros
- *Root locus gain:* defined below (1.38) as the gain factor K that results if the coefficients of the highest powers of s in the numerator and denominator polynomials of $G_c GH$ are made unity, as in (6.2). It is again emphasized that this should be distinguished from the gain used in Section 4.3 to calculate steady-state errors.

As indicated in Fig. 6.2, in (6.2) the typical factor $(s - a_i)$ is a vector from a_i to s, and $(s - b_k)$ from b_k to s. This was discussed in connection with Fig. 1.13, and these vectors may be expressed alternatively as follows:

$$\begin{aligned} s - a_i &= A_i e^{j\alpha_i} \\ s - b_k &= B_k e^{j\beta_k} \end{aligned} \tag{6.3}$$

Here A_i and B_k are vector lengths and α_i and β_k vector angles, measured positive counterclockwise from the direction of the positive real axis.

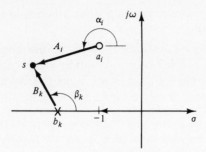

O : open-loop zeros

X : open-loop poles   **Figure 6.2**   Vectors in the s-plane.

6.2 ROOT LOCI

The closed-loop poles are the roots of the system characteristic equation $G_c GH + 1 = 0$, and so of the equation

$$G_c GH = -1 \qquad (6.4)$$

Interpreting both sides as vectors in the s-plane, the vector -1 indicated on Fig. 6.2 is a vector from the origin to the point -1 on the negative real axis. This vector has a length, or magnitude, of unity and a phase angle that is an odd multiple of $\pm 180°$, or $\pm(2n + 1)180°$, where n is any integer. Therefore, the closed-loop poles are the values of s for which the vector $G_c GH$ has a length of unity and a phase angle of $\pm(2n + 1)180°$. Substituting (6.3) into (6.2) gives, assuming that K is positive,

$$\text{magnitude } (G_c GH) = K \frac{A_1 A_2 \cdots A_m}{B_1 B_2 \cdots B_n} \qquad (6.5a)$$

$$\text{phase } (G_c GH) = \alpha_1 + \cdots + \alpha_m - \beta_1 \cdots - \beta_n \qquad (6.5b)$$

Hence

The closed-loop poles are the values of s that satisfy both of the following conditions:

1. *Angle condition:*

$$\text{phase } (G_c GH) = \alpha_1 + \cdots + \alpha_m - \beta_1 - \cdots - \beta_n = \pm(2n + 1)180° \qquad (6.6a)$$

2. *Magnitude condition:*

$$\text{magnitude } (G_c GH) = K \frac{A_1 \cdots A_m}{B_1 \cdots B_n} = 1 \quad \text{or} \quad K = \frac{B_1 \cdots B_n}{A_1 \cdots A_m} \qquad (6.6b)$$

By inspection of (6.2), the A_i may be taken to be 1 if there are no open-loop zeros. Equations (6.6) lead to a two-stage process, of which the first is the construction of the root loci, that is, the loci of the closed-loop pole positions in the s-plane for variations of the root locus gain K (or of another parameter):

1. The root loci are constructed from the angle condition alone, as the loci of all points s for which the sum of the vector angles α_i from all open-loop zeros to

s minus the sum of the vector angles β_i from all open-loop poles to s equals an odd multiple of $\pm 180°$.

2. After the loci have been constructed, the magnitude condition shows that the value of K for which a closed-loop pole will be located at a given point s along a locus equals the product of the vector lengths B_i from all open-loop poles to s divided by the product of the vector lengths A_i from all open-loop zeros to s.

Note that the reverse problem, that of finding where along the loci the closed-loop poles are located for a given value of K, generally involves trial and error.

Root locus work always starts with the construction of the open-loop pole–zero pattern from the loop gain function. The following simple example is given to clarify the ideas. It also introduces the very useful concept of a *trial point*. This is an, in principle, arbitrary point s. Vectors are drawn from all open-loop poles and zeros to s. The trial point lies on the locus if the sum of the vector angles from all open-loop zeros to s minus the sum of the vector angles from all open-loop poles to s equals an odd multiple of $\pm 180°$.

Example 6.2.1   Design Problem

For a system with loop gain function $G_c GH = K/(s + a)$, use root loci to find K for which the (closed-loop) system time constant will be T seconds. The open-loop pole–zero pattern, plotted first, consists of just a pole at $-a$, shown in Fig. 6.3. Following the procedure, a trial point s is chosen, and the vector is drawn from the open-loop pole to s. In this example, the sum of vector angles from the open-loop zeros minus that from the open-loop poles is $-\beta$. Here it is clear that only trial points on the real axis to the left of $-a$ satisfy the angle condition that the net sum be an odd multiple of $\pm 180°$. All such points satisfy the angle condition. Hence, and this completes the first stage, the root locus is the real axis to the left of the open-loop pole at $-a$.

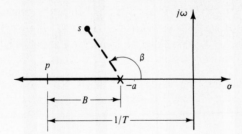

Figure 6.3   Loci for $K/(s + a)$.

To solve the design problem, for a time constant T, the system pole is required to be at p, at a distance $1/T$ from the imaginary axis. The magnitude condition immediately gives the value of K needed for the pole to be at this point along the locus:

$$K = \frac{B_1 B_2 \cdots}{A_1 A_2 \cdots} = B$$

Here B is the distance from $-a$ to p in Fig. 6.3. (Without zeros, the A_i factors are not present; that is, the denominator is in effect unity.)

Although construction of the loci and solution of the design problem were easy for this example, it is clear that something better than arbitrarily choosing trial points is required to make the technique feasible for less simple systems.

The *180° locus,* which is emphasized in this chapter, is based on the assumption that the varying gain of the loop gain function for which the loci are plotted is positive. When the effect of a parameter other than the root locus gain is studied, allowance must also be made for negative gains. This will be encountered in Section 6.9 and on occasion elsewhere. In such cases the *0° locus* is desired, because a trial point *s* will lie on the locus if the specified sum of vector angles is an even multiple of ±180°.

6.3 RULES FOR ROOT LOCUS PLOTTING

The following guides are provided to facilitate plotting the loci of the roots of the characteristic equation $G_c GH + 1 = 0$, or of

$$(s - b_1)(s - b_2)\cdots(s - b_n) + K(s - a_1)(s - a_2)\cdots(s - a_m) = 0:$$

1. For $K = 0$, the closed-loop poles coincide with the open-loop poles, since the b_k then satisfy the characteristic equation.
2. For $K \to \infty$, closed-loop poles approach the open-loop zeros, since then the a_i satisfy the characteristic equation.
3. There are as many locus branches as there are open-loop poles. A branch starts, for $K = 0$, at each open-loop pole. As K is increased, the closed-loop pole positions trace out loci, which end, for $K \to \infty$, at the open-loop zeros.
4. If there are fewer open-loop zeros than poles ($m < n$), those branches for which there are no open-loop zeros left to go to tend to infinity along asymptotes. The number of asymptotes is equal to the number of open-loop poles minus the number of open-loop zeros, $n - m$.
5. The directions of the asymptotes are found from the angle condition. In Fig. 6.4, choose a trial point s at infinity at angle α. The vectors from all m open-loop zeros and n open-loop poles to s then have angle α, so the net sum of the vector angles is $(m - n)\alpha$. For s to lie on the locus, this must equal an odd multiple of ±180°. Hence the asymptote angles α must satisfy

$$\alpha = \frac{\pm(2i + 1)180}{n - m} \qquad i = \text{any integer} \qquad (6.7)$$

If $n - m = 1$, α is 180°; if $n - m = 2$, α is +90° and −90°; if $n - m = 3$, α is +60°, −60°, and 180°; and so on. The angles are uniformly distributed over 360°.

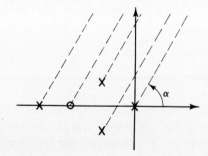

Figure 6.4   Asymptote angles.

6. All asymptotes intersect the real axis at a single point, at a distance ρ_0 to the origin:

$$\rho_0 = \frac{\text{(sum of o.l. poles)} - \text{(sum of o.l. zeros)}}{\text{(number } n \text{ of o.l. poles)} - \text{(number } m \text{ of o.l. zeros)}} \qquad (6.8)$$

The proof will be omitted. If ρ_0 is positive, the intersection occurs on the positive real axis. Note that, say, the sum of a complex conjugate pair of open-loop (o.l.) poles is real:

$$(a + bj) + (a - bj) = 2a$$

7. Loci are symmetrical about the real axis since complex open-loop poles and zeros occur in conjugate pairs.

8. Sections of the real axis to the left of an odd total number of open-loop poles and zeros on this axis form part of the loci. This is because any trial point on such sections satisfies the angle condition. For example, at a trial point s in Fig. 6.5(a), the vectors from the poles left of point s contribute zero angles, and the contributions from the complex poles cancel each other. The pole and zero on the axis to the right of s contribute ($\pm180 \mp180$), which is $0°$ or $\pm360°$, not an odd multiple of $\pm180°$, so s is not part of the loci.

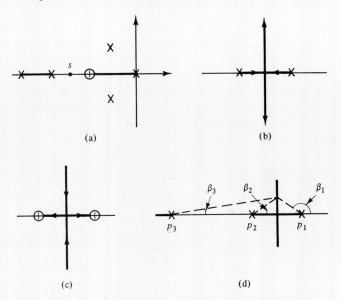

(a)   (b)

(c)   (d)

Figure 6.5   (a) Root locus rule 8; (b)–(d) root locus rule 9.

9. Points of breakaway from or arrival at the real axis may also exist. If, as indicated in Fig. 6.5(b) and (c), the part of the real axis between two o.l. poles (o.l. zeros) belongs to the loci, there must be a point between them where the loci break away from (arrive at) the axis.

   The loci that start at each open-loop pole as K is increased from zero cannot disappear into thin air, nor can the locus branches that must approach the zeros as $K \to \infty$ appear out of it. Algebraic rules for determining the locations of these points are available in numerous books. However, usually, as will be discussed later, these locations are not of much interest and a rough approximation is satisfactory.

If no other poles and zeros are close by, the breakaway point will be halfway. In Fig. 6.5(d), if pole p_3 is not present, only points on a vertical halfway between p_1 and p_2 satisfy the angle condition $\beta_1 + \beta_2 = 180°$. From the geometry in Fig. 6.5(d), adding the pole p_3 pushes the breakaway point away. But even if the distance $p_2 p_3 = p_1 p_2$, the point only moves from $0.5(p_1 p_2)$ to $0.42(p_1 p_2)$ distance to p_1. A zero at the position of p_3 would similarly attract the breakaway point. If better accuracy is desired, sections of the breakaway branches constructed elsewhere in the s-plane can be extrapolated by applying the angle condition to trial points at decreasing distance to the real axis.

10. The angle of departure of loci from complex o.l. poles (or of arrival at complex o.l. zeros) is a final significant feature. Apply the angle condition to a trial point very close to p_1 in Fig. 6.6. Then the vector angles from the other poles and the zero are the same as those to p_1 shown in the plot. If the point is offset from p_1 in the direction of $\gamma°$, it will lie on the locus if $60° - 90° - 135° - \gamma° = \pm(2i + 1)180°$, so $\gamma = 15°$. Thus the loci will depart from pole p_1 at an angle of $15°$.

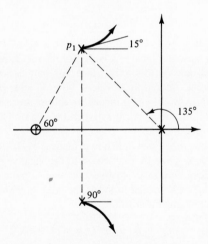

Figure 6.6   Angle of departure.

6.4 ROOT LOCUS EXAMPLES: PLOTTING AND SKETCHING

Both plotting and sketching are considered, but for routine work a computer-aided construction method such as that in Appendix B should be used.

The ability to sketch the general shape of the loci, without regard for accuracy, is very useful in analysis and design. In analysis, it can often provide a quick explanation of why, say, the step response changes in a certain direction with changes of gain, even if the numerical values of the coefficients in the transfer function are not known. In design, it may show quickly whether a contemplated form of compensation could be successful.

Example 6.4.1 Simple Motor Position Servo (Fig. 6.7)

(a) Find the loci of the (closed-loop) system poles for varying K.
(b) *Design:* Find the value of K to obtain a system damping ratio $\zeta \approx 0.7$.

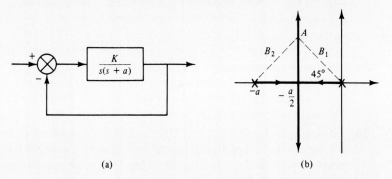

(a) (b)

Figure 6.7 Example 6.4.1.

First the open-loop pole–zero pattern is plotted, consisting of poles at the origin and $-a$, as shown in Fig. 6.7(b). There are two poles, so two locus branches, starting at 0 and $-a$ for $K = 0$. Since there are no open-loop zeros, there must be two asymptotes. From rule 5, these will be at $+90°$ and $-90°$, and from (6.8) of rule 6, they will intersect the real axis at

$$\rho_0 = \frac{0 - a - 0}{2 - 0} = -\frac{a}{2}$$

By rule 8, the real axis between 0 and $-a$ is part of the locus, because it lies to the left of one (an odd number) pole on this axis. Since this part is between two poles, there must be a breakaway point. According to rule 9, with no other poles and zeros present, breakaway will occur halfway, at $-a/2$. Indeed, in this example asymptotes and loci coincide because any point on this vertical satisfies the angle condition. This completes part (a), the construction of the loci.

For part (b), Chapter 4 has shown that to achieve $\zeta = 0.7$ the (closed-loop) system poles must be at an angle ϕ given by $\zeta = \cos \phi$ to the negative real axis, that is, at $45°$. So the poles must lie where lines at $45°$ intersect the loci, at point A and its complex conjugate position. The value of K needed to locate the poles at these points is found immediately from the magnitude condition (6.6b):

$$K = \frac{B_1 B_2 \cdots}{A_1 A_2 \cdots} = B_1 B_2 = (\tfrac{1}{2} a \sqrt{2})(\tfrac{1}{2} a \sqrt{2}) = \frac{a^2}{2}$$

It may be recalled that Fig. 5.3 shows loci for a numerical example of this type and also compares step responses for $K = a^2/4$ and the value of K corresponding to a damping ratio 0.5. Both have the same settling time and time constant, but the last has greater speed of response due to a smaller rise time.

Example 6.4.2(a) Root Locus Sketch for a Position Servo

For the system in Fig. 6.8(a), sketch the loci of the system poles for varying K.

The open-loop pole–zero pattern consists of poles at 0, -1, and -2. There are three o.l. poles ($n = 3$) and no o.l. zeros ($m = 0$), so there are three asymptotes. Rule 5 gives their directions as $+60°$, $-60°$, and $180°$ and rule 6 their intersection with the real axis at $\rho_0 = (0 - 1 - 2 - 0)/(3 - 0) = -1$. Figure 6.8(b) shows these asymptotes. The real axis between 0 and -1 and left of -2 belongs to the loci, by rule 8.

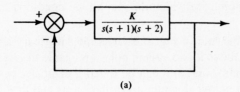

(a)

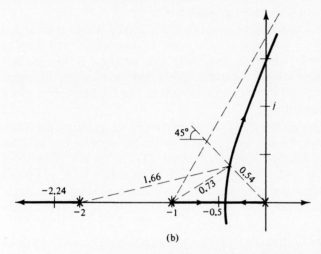

(b)

Figure 6.8 Example 6.4.2.

There must be a breakaway point between 0 and -1. But for the pole at -2, it would be halfway, at -0.5. This third pole pushes it away to the right, so it is taken to be somewhat to the right of -0.5. (In rule 9, the true value for this case was given as -0.42.) It is logical that the breakaway branches should move toward the $+60°$ and $-60°$ asymptotes and that the branch from the pole at -2 forms the third asymptote.

A root locus sketch such as that shown in Fig. 6.8(b) can now be completed without any use of trial points and angle measurements.

Before proceeding with this example, certain points should be noted.

1. The loci need only be constructed with satisfactory accuracy where they are needed to solve the particular analysis or design problem.

2. If, analogous to Example 6.4.1, the dominating poles on the complex branches in Fig. 6.8 are required to have a damping ratio of 0.7, then it is only necessary to know with reasonable accuracy where the locus intersects the line at $45°$ to the negative real axis.

   This can be done by applying the angle condition to some trial points along the $45°$ line and interpolating or extrapolating.

   With this point found, the value of K needed so that the closed-loop poles will be at this location along the locus can be obtained from the magnitude condition.

3. Similarly, if the design problem would be to find K to achieve a specified time constant T for the dominating poles, then these poles are required to be on a vertical at a distance of $1/T$ to the imaginary axis. Trial points along this vertical

can provide the crossing point, for which the magnitude condition then gives the gain K.

4. The value of K at which the loci in Fig. 6.8 cross the imaginary axis is the limit for stability because above it two system poles are inside the right half of the s-plane. Similar to points 2 and 3, it can be found graphically by applying the angle condition to trial points along the imaginary axis and then using the magnitude condition. Alternatively, the Routh–Hurwitz stability criterion can be used, as was done for this case in Example 4.6.3.

    If the system characteristic equation does not exceed fourth order in s, a third technique deserves note and permits both the intersection and K to be found analytically. Points along the imaginary axis satisfy $s = j\omega$. As illustrated in the following example, substituting this into the characteristic equation and separating real and imaginary parts leads to two equations from which K and ω can be calculated.

5. This approach also provides an analytical alternative to the graphical technique in points 2 and 3 above. For example, a line at 45° to the negative real axis can be described by the equation $s = \omega(-1 + j)$. Then $s^2 = \omega^2(-1 + j)^2 = -2j\omega^2$ and $s^3 = -2j\omega^3(-1 + j) = 2(1 + j)\omega^3$. Since the s-value of interest must lie on the line, these are substituted into the characteristic equation. Separation into real and imaginary parts then yields the solution, as illustrated in the next example.

6. A final point concerns finding the locations of the remaining closed-loop poles after the dominating pair has been determined. One way is to find K for points along these branches, using the magnitude condition, and thus by trial and error to find the positions for the known value of gain by interpolation. A simpler way may be available, as illustrated in the next example. If the system is third order and the closed-loop poles are λ_1, λ_2, and λ_3, then the characteristic equation is

$$(s - \lambda_1)(s - \lambda_2)(s - \lambda_3) = s^3 - (\lambda_1 + \lambda_2 + \lambda_3)s^2 + \cdots = 0 \qquad (6.9)$$

This illustrates a general result:

> The sum of the system poles is equal to the negative of the coefficient of the next-to-highest power term of the characteristic equation.

With the sum of two of these poles in (6.9) available from the dominating pair, the third pole can be calculated.

Example 6.4.2(b)

For the system in Fig. 6.8(a):

(a) *Design:* Find K to realize a damping ratio $\zeta = 0.7$ for the dominating pair of closed-loop poles, as defined in Section 4.4.

(b) Find the closed-loop poles for K of part (a).

(c) Determine the limiting value of K for stability.

The root locus sketch obtained in Example 6.4.2(a) is refined as suggested in point 2 by using trial points along the 45° line. This yields the plot with the measured vector lengths indicated on Fig. 6.8(b). The magnitude condition then gives the gain needed for the closed-loop poles to be at these points along the loci:

$$K = \frac{0.54 \times 0.73 \times 1.66}{1} = 0.65$$

It is useful to emphasize that the magnitude condition gives the values of the root locus gain. From Fig. 6.8(a), in this example K is indeed the root locus gain.

As an alternative to this graphical approach, the analytical method of point 5 may be used. The characteristic equation, the denominator of C/R, is

$$s(s + 1)(s + 2) + K = s^3 + 3s^2 + 2s + K = 0 \qquad (6.10)$$

Substituting for the powers of s as indicated in point 5 and separating the real and imaginary parts yields

$$(2\omega^3 - 2\omega + K) + 2j\omega(\omega^2 - 3\omega + 1) = 0 \qquad (6.11)$$

Both the real and imaginary parts must be zero, and solving the quadratic for ω gives $\omega = 0.382$. So the locus must intersect the 45° line at $0.382j$ distance to the real axis. For this ω, the real part in (6.11) gives $K = 0.65$, as before.

The position of the third pole for this value of K must still be found. Using the analytical method suggested in point 6 for this example instead of the graphical one, comparing (6.9) and (6.10) gives $\lambda_1 + \lambda_2 + \lambda_3 = -3$ for the sum of the closed-loop poles. But the sum of two of these, say $\lambda_1 + \lambda_2$, has already been found in Fig. 6.8(b): $\lambda_1 + \lambda_2 = -2 \times 0.54 \cos 45° = -0.76$. Hence $\lambda_3 = -3 + 0.76 = -2.24$ is the position of the third pole.

With the closed-loop poles now known, the closed-loop transfer function is

$$\frac{C}{R} = \frac{0.65}{(s + 2.24)(s + 0.38 + 0.38j)(s + 0.38 - 0.38j)} \qquad (6.12)$$

and the transient response for given inputs could be calculated. By applying the magnitude condition to other points, the loci can also be used to determine pole sensitivity to gain changes.

The value of K at the limit for stability may be found by one of the methods in point 4. Using the analytical method, substituting $s = j\omega$ into the characteristic equation (6.10) and separating real and imaginary parts leads to the equation

$$K - 3\omega^2 + j\omega(2 - \omega^2) = 0 \qquad (6.13)$$

Both real and imaginary parts must be zero, so $\omega = \sqrt{2}$, $K = 6$. Hence the limiting value for stability is $K = 6$, and the loci cross the imaginary axis at $\pm 1.414j$, as was found in Example 4.6.3.

As noted early in this section, the ability to sketch the general shape of the loci is very useful. Some sketch examples follow, with the loop gain functions and the corresponding loci shown in Fig. 6.9.

Example 6.4.3

In Fig. 6.9(a), three asymptotes, at $+60°$, $-60°$, and $180°$, intersect the real axis at the average real part of the poles. The real axis left of $-a$ is part of the loci. Branches from the complex poles approach the other asymptotes. An approximate equation for the angle of departure α is $-90 - 110 - \alpha = -180$, so $\alpha \approx -20°$.

Example 6.4.4

Consider Fig. 6.9(b). Since $n = 3$ and $m = 1$, there are two asymptotes, at $+90°$ and $-90°$, intersecting the real axis at $0.5(-b - c + a)$. Breakaway occurs somewhat to the

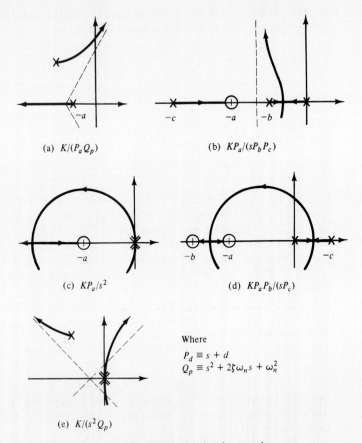

(a) $K/(P_a Q_p)$ (b) $KP_a/(sP_b P_c)$

(c) KP_a/s^2 (d) $KP_a P_b/(sP_c)$

(e) $K/(s^2 Q_p)$

Where
$P_d \equiv s + d$
$Q_p \equiv s^2 + 2\zeta\omega_n s + \omega_n^2$

Figure 6.9   Root locus sketch examples.

left of halfway between 0 and $-b$, because the zero at $-a$ is closer and therefore pulls it more than the pole at $-c$ pushes it. The breakaway branches approach the asymptotes.

Example 6.4.5

In Fig. 6.9(c) there is a double pole at the origin. Verify that only trial points close to the origin above and below the double pole satisfy the angle condition, so the loci must depart from the origin in vertical direction. Since $n = 2$, $m = 1$, only the negative real axis is an asymptote. The real axis left of $-a$ is part of the locus since it lies left of three poles and zeros on this axis. But there must be an arrival point of the loci here, because as $K \to \infty$ one branch must approach $-a$ and the other tend to infinity along the asymptote. For this example it is easily verified, by applying the angle condition to trial points slightly off the axis, that arrival must occur at $-2a$. The locus can also be shown to be a circle.

Example 6.4.6

Figure 6.9(d) illustrates that nothing changes if open-loop poles, or zeros, occur in the right-half s-plane. A system with an open-loop pole in the right-half plane is *open-loop unstable;* that is, it is unstable unless a suitably designed feedback loop is closed around it. A tall rocket is an example that requires feedback control for stability.

Example 6.4.7

In Fig. 6.9(e) there are four poles and no zeros, so four asymptotes, at $+45°$, $-45°$, $+135°$, and $-135°$, intersect the real axis at the average real part. Departure from the double pole at the origin is in the vertical direction.

6.5 ROOT LOCI AND SYSTEM DESIGN

In the preceding sections, root loci were used for design to the extent of choosing the gain to obtain a specified damping ratio or time constant. Such P control design does not change the shape of the loci. But if dynamic compensation is used, such as a series compensator $G_c(s)$ in Fig. 5.8, then G_c will add poles and zeros to the open-loop pole–zero pattern, in order to change the shape of the loci in a desirable direction.

The design of PID controllers and of phase-lead, phase-lag, and lag–lead compensators is considered subsequently. In this section some other aspects of design are discussed via examples. These are the general effects of adding a pole or a zero, pole–zero cancellation, and feedback compensation.

Example 6.5.1   Effect of Adding a Pole or Zero

Figure 6.10(a) shows loci that could represent P control of a process consisting of two simple lags, as in Fig. 4.16 for Example 4.5.3. These loci are equivalent to those for a simple motor position servo in Fig. 6.7. The loci in Fig. 6.10(b) are equivalent to those in Fig. 6.8 and show the effect of adding a pole. The loci in Fig. 6.10(c) are much like those in Fig. 6.9(c) and (d), and show the effect of adding a zero. The following general effect is evident:

Adding a pole pushes the loci away from that pole, and adding a zero pulls the loci toward that zero.

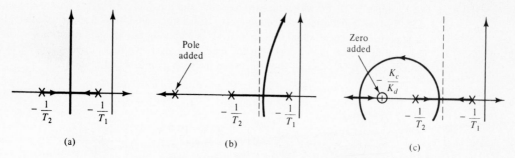

Figure 6.10   Effect of adding a pole or a zero.

These effects increase in strength with decreasing distance. A zero can improve relative stability because it can pull the loci, or parts thereof, away from the imaginary axis, deeper into the left-half plane.

Example 6.5.2   Use of Pole–Zero Cancellations

A common practice if G has poles in undesirable locations that cannot easily be changed by feedback alone is to choose zeros of G_c at the same or nearby locations. This was discussed in Section 5.4. In many cases when this method is used there is a root locus branch, of ideally zero length, between such pole–zero pairs. The zero of G_c is also a closed-loop zero and will be close to the closed-loop pole along this branch. By Section 5.3, this implies a small, ideally zero, residue at the pole and therefore a small transient. This situation is illustrated in Fig. 6.11(a). Figure 6.11(b) shows the alternative. A zero has been added near the pole at −1, far from the closed-loop pole, which might typically be at A. Now the effect of the pole at −1 on the transient corresponding to A is effectively eliminated because the vector to A from −1 is canceled by that from the zero. A zero at −1 cancels the pole at −1 and the loci are those for a loop gain function $K/[s(s + 2)]$.

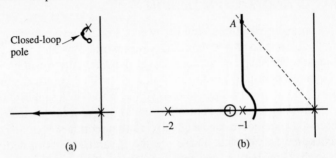

Figure 6.11 Example 6.5.2: modes of pole–zero cancellation.

Two points already raised earlier should be emphasized. The first is that the system response to initial conditions is not affected by any cancellations that may have been achieved in the input–output response. The second is that open-loop unstable poles of G, that is, poles of G in the right-half s-plane, may never be canceled in this way by zeros of G_c. However short the locus branch between such a pole–zero pair, there would be a closed-loop pole along it, in the right-half plane. Here feedback must be used to pull the pole into the left-half plane, as in Fig. 6.9(d).

Example 6.5.3 Feedback Compensation

Figure 6.12 shows a more complex motor position servo with velocity feedback than that of Example 5.5.1, for which the step response and root loci for varying K_g are shown in Fig. 5.11. G is the transfer function (2.10) of the motor and its load if the load damping is assumed to be negligible. If error analysis is required, the loop gain function C/E must be used to keep E in evidence, but for stability analysis the two feedback loops can be combined into $(K_g s + 1)$. By inspection, the loop gain function is then

$$\frac{KK_m(K_g s + 1)}{s^2(T_f s + 1)} = \frac{KK_m K_g}{T_f} \frac{s + 1/K_g}{s^2(s + 1/T_f)} \tag{6.14}$$

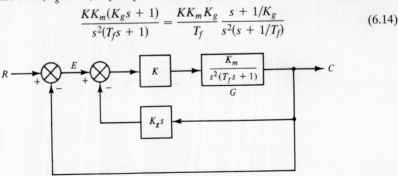

Figure 6.12 Example 6.5.3: velocity feedback.

The second form gives the root locus gain, of which the magnitude condition provides numerical values at points along the loci.

Figure 6.13 shows root locus sketches for a range of choices of the zero $-1/K_g$. For $K_g = 0$ the system is apparently unstable for any value of gain K, because two locus branches are entirely in the right-half plane. With the zero to the left of $-1/T_f$, in (b), the system is still unstable for all K, and for $K_g = T_f$, in (c), it is marginally stable. The loci (d) and (e) cover a suitable range of values of K_g.

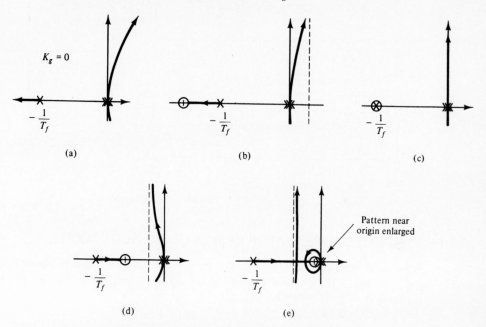

Figure 6.13   Effect of zero due to velocity feedback.

In design (e) the zero is so close to the double pole at the origin that it pulls the loci back to the real axis, with one branch then going to the zero and the other to the left. The change of the nature of the loci from (d) to (e) as the zero moves close enough to the origin illustrates a situation where sketching is no longer adequate and construction using trial points or, better, the computer aids of Appendix B are required. But the sketches easily showed that P control is inadequate and quickly homed in on the suitable range of parameters of the compensator, without any numerical values.

Example 6.5.4   Satellite Attitude Control

Figure 6.14(a) shows a block diagram as in Fig. 3.13 for a satellite attitude control with rate feedback. It is a simpler version of the preceding example, with $T_f = 0$, and design could again proceed by adding the two parallel feedback loops. Note that from the point of view of stability this is equivalent to the design of PD control $(K_c + K_r s)$. As an alternative to adding the feedback loops, Fig. 6.14(b) shows the minor loop reduced to a single block. It is seen that the rate feedback has in effect moved one of the plant poles from the origin to $-K_a K_r/J$. Figure 6.14(c) shows the root loci. K_r can be chosen to obtain a desired settling time or time constant, since on the vertical branches the real part is $-0.5 K_a K_r/J$. The root locus gain is $K_a K_c/J$, so K_c can be determined to obtain a desired damping ratio. It may be verified that the value of K_c that will place the poles at $45°$, for a damping ratio 0.7, is $K_c = K_a K_r^2/(2J)$.

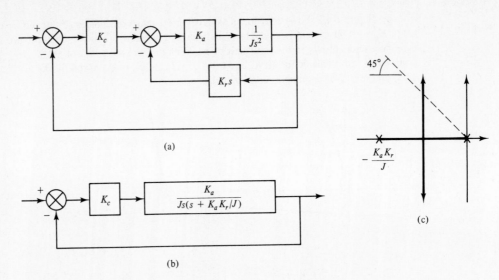

(a)

(b)

(c)

Figure 6.14   Satellite attitude control.

6.6 PHASE-LEAD COMPENSATOR DESIGN USING ROOT LOCI

Phase-lead compensation, introduced in Section 5.4, can be considered as an approximation to PD control and has a stabilizing effect. From (5.18), the transfer function is

$$G_c(s) = \frac{K_c(s + z)}{s + p} \qquad (6.15)$$

with $z < p$; that is, the zero is closer to the origin than the pole, as shown in Fig. 5.9. Figure 6.15 helps to illustrate the basic stabilizing effect. Suppose that A_1 is a point on the loci and must be shifted left, for greater relative stability, to A_2. The algebraic sum of the vector angles from plant zeros and poles at A_1 is $-180°$, by definition. At A_2 it will be more negative than $-180°$, because usually the plant has more poles than zeros, so the negative angles increase more than the positive angles. If the algebraic sum of the vector angles at A_2 due to the plant poles and zeros is $-180 - \phi_d°$, then for A_2 to lie on the loci of the compensated system a "phase lead" of $\phi_d°$ is needed, as indicated in Fig. 6.15(b).

   In the following design procedure, the desired dominant closed-loop poles are determined first, for example, from specifications on overshoot or damping ratio and settling time. This possibility of specifying the closed-loop poles is an advantage of root locus design. But a disadvantage is that specifications on steady-state accuracy cannot be applied directly, but must be checked after design. Design is then iterative in character, and frequency response methods, discussed later, are more convenient.

DESIGN PROCEDURE

1. Determine the desired dominant closed-loop poles from the specifications.
2. At these positions, determine the angle deficiency ϕ_d that the phase-lead must contribute, as discussed above.

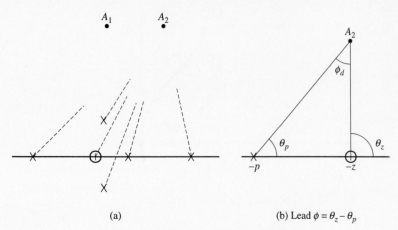

(a)   (b) Lead $\phi = \theta_z - \theta_p$

Figure 6.15   Stabilizing effect of a lead.

3. Choose the compensator zero on the real axis straight below the desired pole or to the left of this position.

4. With the zero chosen, determine the compensator pole by drawing a line at angle ϕ_d as indicated in Fig. 6.15(b).

5. Determine the root locus gain at the desired pole position from the magnitude condition.

6. Determine the actual loop gain and hence the steady-state errors.

7. Repeat the design with a new choice of desired poles if these errors are too large.

The choice of zero in step 3 must be such that the desired poles remain dominant. For example, if the plant has poles at 0, −1, and −2, then the zero should not be to the right of the second pole, at −1. Otherwise, there would be a locus branch between the origin and the zero, and the closed-loop pole along this branch would dominate the transient.

Furthermore, as discussed in Section 5.4, to reduce the effect of noise, the ratio p/z should not be made larger than necessary to achieve satisfactory relative stability.

Example 6.6.1   Phase-Lead Compensation of a Satellite Attitude Control

From Example 3.4.1, let the transfer function for the satellite be

$$G(s) = \frac{1}{s^2}$$

and let the specifications require a damping ratio 0.5 and a settling time of 4 sec. Since, from (5.2), the settling time $T_s = 4/(\zeta\omega_n)$, it follows that

$$\zeta = 0.5 \qquad \omega_n = 2 \qquad \zeta\omega_n = 1.0 \qquad \omega_n\sqrt{1 - \zeta^2} = \sqrt{3}$$

Hence the desired dominant closed-loop poles are $(-1 \pm j\sqrt{3})$. These are shown in Fig. 6.16(a). At these poles the two plant poles at the origin contribute $2 \times -120 = -240°$, so for the desired poles to lie on the locus the phase-lead must make up for an angle deficiency $\phi_d = 240 - 180 = 60°$. If the compensator zero is chosen at −1, straight below the desired closed-loop pole, then this 60° angle fixes the pole at −4, as indicated in Fig. 6.16(a). The loop gain function is

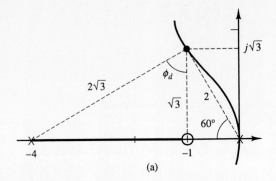

(a)

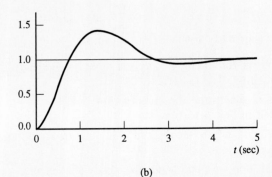

(b)

Figure 6.16   Example 6.6.1: satellite attitude control.

$$G_c G = \frac{K_c(s + 1)}{s^2(s + 4)}$$

The root locus gain is K_c, and its desired value is found by applying the magnitude condition at the desired pole position:

$$K_c = \frac{2 \times 2 \times 2\sqrt{3}}{\sqrt{3}} = 8$$

The controller transfer function is

$$G_c = 8 \frac{s + 1}{s + 4}$$

The system step response is shown in Fig. 6.16(b) and indicates a much larger overshoot than the approximately 16% that is expected for a quadratic lag with damping ratio 0.5. This is due to the third closed-loop pole. This pole may be found as suggested by note 6 in Section 6.4. The closed-loop characteristic equation is $s^3 + 4s^2 + 8s + 8 = 0$, so the sum of the three poles is -4. Since the sum of the dominant pair is -2, the third closed-loop pole is at -2. This is only twice as far from the imaginary axis as the desired pair. This can be improved by choosing the compensator zero farther left. However, it cannot be moved far with the present choice of desired poles. From the geometry in Fig. 6.16(a), it may be seen that a choice at -2 would imply that the compensator pole should move to $-\infty$ to realize the desired 60° lead.

Example 6.6.2   Phase-Lead Compensation of a Motor Position Servo

The motor plus load transfer function for a position servo is approximated by

$$G(s) = \frac{1}{s(s + 2)}$$

The dominating closed-loop time constant should be 0.25 sec, and the step response overshoot is permitted to be about 16%. From (4.20) and Fig. 5.2, these specifications imply

$$\zeta = 0.5 \qquad \zeta\omega_n = 4 \qquad \omega_n = 8 \qquad \omega_n\sqrt{1 - \zeta^2} = 4\sqrt{3} = 6.928$$

Hence the desired dominant closed-loop poles are $(-4 \pm j6.93)$. These are shown in Fig. 6.17(a). At these poles the plant poles at the origin and -2 contribute $-120 - 106.1 = -226.1°$, so for the desired poles to lie on the locus the phase-lead must make up for an angle deficiency of $\phi_d = 226.1 - 180 = 46.1°$. If the compensator zero is chosen straight below the desired pole, at -4, then this places the compensator pole at -11.2, as indicated on Fig. 6.17(a). The loop gain function is

$$G_c G = \frac{K_c(s + 4)}{s(s + 2)(s + 11.2)}$$

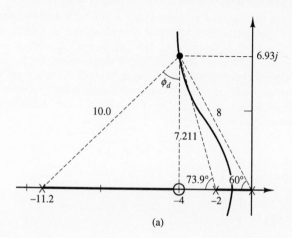

(a)

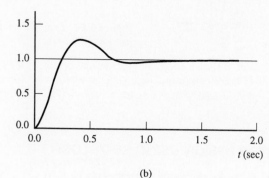

(b)

Figure 6.17 Example 6.6.2: motor position servo.

Application of the magnitude condition at the desired pole position gives the root locus gain from the measured or calculated vector lengths shown:

$$K_c = \frac{8 \times 7.211 \times 10.0}{6.93} = 83.3$$

The compensator is

$$G_c = \frac{83.3(s + 4)}{(s + 11.2)}$$

The loop gain is $83.3 \times 4/(2 \times 11.2) = 14.87$, so the steady-state error after a unit ramp of this type 1 system is $1/14.87 = 0.067$. If this is too large, the design must be repeated with a new set of desired poles at greater distance to the origin.

The step response is shown in Fig. 6.17(b). The overshoot is larger than the 16% expected for a damping ratio 0.5. This is due to the third closed-loop pole. The closed-loop characteristic equation is $s^3 + 13.2s^2 + 105.665s + 333.06 = 0$, so the sum of the poles is -13.2. Since the sum of the dominant pair is -8, the third pole must be -5.2, not far left of the dominating pair. The compensator zero and pole can both be moved left to improve this.

In these examples the design technique was applied somewhat formally. A more informal approach is illustrated by the next example and is frequently useful, for example, when the performance requirements are not stated precisely.

Example 6.6.3   Phase-Lead Compensation for a Third-Order Plant

In the system of Fig. 6.8, the P control of Example 6.4.2 is to be replaced by phase-lead. It is intended to "pull" the locus branches for P control, shown as dashed curves in Fig. 6.18, to the left. The zero is chosen at -1.3, a little to the left of the plant pole at -1. As discussed earlier, a choice to the right of -1 would imply the presence of a strongly dominating closed-loop pole on a locus branch between the origin and the zero. With the choice made there will be a closed-loop pole on the branch between -2 and -1.3, but this is more to the left, and the zero is close enough to help attenuate its effect on the transient response.

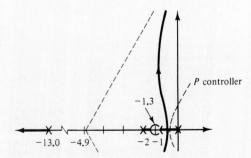

Figure 6.18   Example 6.6.3: phase-lead compensation.

The added pole is chosen at -13, ten times as far from the origin as the zero, on the assumption that considerations of noise permit this ratio. The asymptotes intersect the axis at $(-1 - 2 - 13 + 1.3)/3 = -4.9$, instead of at -1 as for P control. The resulting loci show the desired stabilizing effect.

6.7 PHASE-LAG COMPENSATOR DESIGN USING ROOT LOCI

Phase-lag compensation is given by the same transfer function (6.15) as phase-lead compensation, but now $z > p$, so the pole is closer to the origin than the zero. It is commonly used, like PI control, to improve steady-state accuracy.

The loci for P control are drawn first, and the desired dominant closed-loop poles are chosen on these loci. Now a pole–zero pair is added close to the origin, far from the desired closed-loop poles. The vectors $(s + z)$ and $(s + p)$ from the pair to the desired poles then almost cancel each other so that for such a pair the net contribution to the vector angle is small. Therefore, the main branches of the loci change only little in the region of interest, and the effect of this compensation on the location of the desired dominant closed-loop poles is quite small. A design guide is that the "phase lag," that is, the negative phase angle contribution, of the pole–zero pair should not exceed 5° and should preferably stay below 2°.

To see how, then, phase-lag compensation can improve steady-state accuracy as claimed above, consider again its transfer function

$$G_c(s) = \frac{K_c(s + z)}{s + p}$$

The root locus gain is K_c, but the actual gain that determines steady-state error is $K_c z/p$, and so has been increased by a factor z/p. Thus, by placing the compensator pole ten times as close to the origin as the zero, steady-state accuracy can be improved by a factor of 10 with small effect on the transient response if the pole and zero are close enough to the origin.

As illustrated later, the pole–zero pair will introduce a closed-loop pole near the origin. But the compensator zero is close to this pole and, from $C/R = G_c G/(1 + G_c G)$, this zero is also a closed-loop zero. Usually, it is close enough to the pole to ensure an acceptably small residue. The design procedure can be summarized as follows.

DESIGN PROCEDURE

1. Draw the root loci for a proportional gain controller.
2. Determine the desired position of the dominating pair of closed-loop poles on these loci from the specifications.
3. Determine the root locus gain at this position from the magnitude condition, and hence the value of K_c for P control.
4. For this value of K_c, determine as discussed above the value of the factor z/p needed to satisfy the specifications on steady-state accuracy.
5. Choose p and z with this ratio and close enough to the origin that the vector angles to the dominant poles differ only a few degrees.
6. Draw the loci of the compensated system and find the dominating poles. Reduce K_c if needed to counter any reduction of the relative stability.

Example 6.7.1   Phase-Lag Compensation of a Motor Position Servo

As in Example 6.6.2, let the plant be

$$G(s) = \frac{1}{s(s + 2)}$$

and the desired damping ratio of the closed-loop poles be 0.5. As noted in step 2 above, these poles are chosen on the loci for P control, shown in Fig. 6.19. The desired poles are $-1 \pm j\sqrt{3}$, and applying the magnitude condition to these locations yields the root locus gain $K_c = 2 \times 2 = 4$. The loop gain function with P control is then $4/[s(s + 2)]$, so the gain is $4/2 = 2$. Suppose that this must be increased by a factor of 10 for satisfactory steady-state accuracy. Then according to step 4 the ratio z/p of the phase-lag must be 10. If $z = 0.1$ is chosen, much closer to the origin than the desired poles, then $p = 0.01$, and the phase-lag becomes

$$G_c = 4 \frac{s + 0.1}{s + 0.01}$$

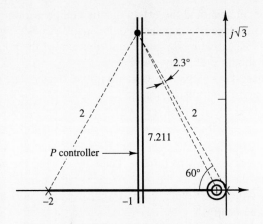

Figure 6.19   Example 6.7.1: phase-lag compensation.

The loci with this compensator are shown. As expected, the change is small, because the difference in vector angles from the compensator pole and zero is only 2.3°.

The picture near the origin is of the type shown in Fig 6.10(c). The zero pulls the loci back to the axis, where one branch moves right to the zero and the other left to the breakaway point of the main branches.

Example 6.7.2   Phase-Lag Compensation of a Third-Order Plant

In the system of Fig. 6.8, the P control of Example 6.4.2 is to be replaced by phase-lag compensation to improve steady-state accuracy. A gain of 1.5 is specified. The damping ratio of the dominant poles is to be 0.707 as in Example 6.4.2, so the root locus gain K_c and the desired poles are also the same:

$$K_c = 0.65 \qquad s_{1,2} = -0.382 \pm j0.382$$

The loop gain function with P control is $0.65/[s(s + 1)(s + 2)]$, so the gain is $0.65/2 = 0.325$. To realize the desired gain of 1.5, this must be increased by a factor of 5. Thus the ratio z/p of the phase-lag must be 5. For illustration, Fig. 6.20 shows loci for two (z, p) combinations:

$$\text{(a)} \quad (0.1, 0.02) \qquad \text{(b)} \quad (0.05, 0.01)$$

It is evident that for (a) the compensator zero and pole are not close enough to the origin compared to the 0.54 distance to the desired dominant poles. Observe that the zero does not manage to pull the loci back on the real axis. The closed-loop poles for (a) will differ considerably from those for P control.

Lag–Lead Compensation

Not infrequently, the specifications on speed of response and relative stability impose a choice of dominant closed-loop poles for which steady-state accuracy specifications are difficult to meet by phase-lead compensation alone.

The phase-lead design may then be followed by design of a phase-lag, considering the series connection of the lead and the original plant as the new plant. Thus the design procedures for a lead and a lag are applied in sequence.

6.8 PID CONTROL DESIGN USING ROOT LOCI

The design procedures given in the preceding section for lag, lead, and lag–lead design apply also to, respectively, PI, PD, and PID controller design. Therefore, the following examples focus on performance rather than on design procedure.

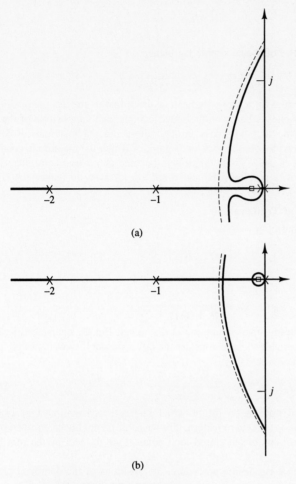

(a)

(b)

Figure 6.20   Example 6.7.2: phase-lag compensation.

Example 6.8.1   PD Controllers

A PD control

$$G_c(s) = K_c + K_d s$$

adds a zero to the open-loop pole–zero pattern. Figures 6.21(a) and (b) are effectively the same as Figures 6.9(c) and 6.10(c) and illustrate the stabilizing effect for the following:

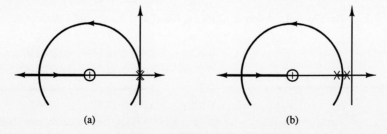

(a)   (b)

Figure 6.21   Example 6.8.1: PD control.

(a) Attitude control of a rigid satellite:

$$G(s) = \frac{1}{s^2}$$

(b) Temperature control of a dual simple lag plant:

$$G(s) = \frac{2}{(s + 1)(s + 2)}$$

The design procedure for a phase-lead can be used; that is, the desired position of the closed-loop poles is chosen based on the specifications, and the position of the zero is determined so that the loci will pass through these desired locations. Note that $G_c = K_d(s + K_c/K_d)$, so for (b) the root locus gain, of which the magnitude condition gives numerical values, is $2K_d$. These controllers become phase-lead controllers if, as discussed in Section 5.4, a pole must be added not too far away to reduce the effects of noise.

Example 6.8.2   PI Control of Temperature

In Example 5.6.2, PI control

$$G_c(s) = K_c + \frac{K_i}{s}$$

was applied to the plant

$$G = \frac{2}{(s + 1)(s + 2)}$$

and the advantage for steady-state accuracy was discussed. But the characteristic equation is third order, and suitable tools to consider the effect on stability were not yet available. The loop gain function is

$$G_c G = 2K_c \frac{s + z}{s(s + 1)(s + 2)} \qquad z = \frac{K_i}{K_c} \qquad (6.16)$$

The controller has added a pole at the origin, which makes the system into type 1, and a zero at $-z$ to the open-loop pole–zero pattern. The shape of the loci is determined by the choice of z, and the positions of the closed-loop poles along the loci by the root locus gain $2K_c$.

Following the phase-lag design procedure, the positions of the dominant closed-loop poles are chosen on the loci for P control to obtain a damping ratio 0.5. The zero is then chosen close to the pole at the origin so that the difference of their vector angles to the desired position is only a few degrees. Figure 6.22(a) shows the root locus for P control. The magnitude condition gives the root locus gain $2K_c = 7$, so $K_c = 3.5$. The closed-loop transfer function is

$$\frac{C}{R} = \frac{7(s + z)}{s^3 + 3s^2 + 9s + 7z}$$

The roots of the denominator are listed next for several values of z.

$z = 0$	:	-1	-2
0.02:		$-1.49 \pm j2.59$	-0.0156
0.1 :		$-1.46 \pm j2.58$	-0.0798
0.2 :		$-1.42 \pm j2.55$	-0.164
0.5 :		$-1.28 \pm j2.50$	-0.445

These can be used with the graphical residue technique to calculate the unit step responses shown in Fig. 6.22(b). Examination of these responses is instructive. For $z = 0$ this is a type 0 system and has a steady-state error. For all other values of z this error will

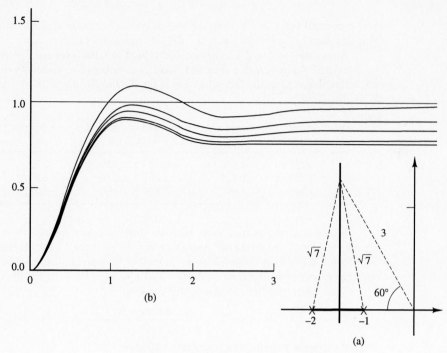

Figure 6.22   Example 6.8.2: PI control.

ultimately become zero, but convergence will take longer if the zero is chosen closer to the origin. This applies also to phase-lag design.

The real poles in the above table are in dominant locations, but their effect on the transient is small because of nearby open-loop zeros, which are also closed-loop zeros. For $z = 0.5$ the complex poles have moved appreciably and the zero is no longer relatively very close to the origin, but this response is attractive and avoids the slow-convergence penalty associated with a choice of zero close to the origin. This aspect of slow convergence should be taken into consideration in both PI and phase-lag control design.

An interesting alternative for PI controller design is to view the pole at the origin as part of the plant. Then choice of the zero is equivalent to a PD control design for a type 1 plant. This alternative can avoid the slow error convergence and lead to a design that will often be preferred. It is discussed in the next example.

Example 6.8.3   Example 6.8.2 Continued

Consider again the plant with PI control and desired damping ratio 0.5 as in the preceding example. Instead of choosing the desired poles on the loci for P control, results for three different choices of z will be compared: 4, 0.5, and 1.2. In each case the system is designed for a damping ratio 0.5 of the complex pair of closed-loop poles. For the design, only the intersections with the 60° lines are needed. These may be found graphically, using trial points along the line, or by calculation analogous to that in note 5 of Section 6.4. Points along the 60° line are described by $s = -a + 1.732ja$ ($s^2 = -2a^2 - 3.464ja^2$; $s^3 = 8a^3$). Substituting into the characteristic equation

$$s^3 + 3s^2 + (2 + 2K_c)s + 2K_i = 0$$

and equating the real and imaginary parts to zero yield

$$K_c = 3a - 1 \qquad K_i = -6a^2 - 4a^3$$

The sum of all poles is -3, from the second coefficient in the characteristic equation. Since the sum of the complex pair is $-2a$, the third pole is $(-3 + 2a)$.

If the time constant $1/a$ were specified instead of z, the poles and $z = K_i/K_c$ could be calculated directly. With z specified, some trial values of a, guided by the sketches, and interpolation yield the following results for K_c and K_i, the poles $p_{1,2}$ and p_3, and $\omega_n = |p_{1,2}|$:

z	K_c	K_i	$p_{1,2}$	p_3	ω_n
4	0.17	0.676	$-0.39 \pm 0.676j$	-2.22	0.78
0.5	2.852	1.425	$-1.284 \pm 2.224j$	-0.432	2.568
1.2	1.577	1.892	$-0.859 \pm 1.488j$	-1.282	1.718

Figure 6.23 shows root locus sketches that are drawn through these calculated complex pole locations on the 60° lines. Note that the values of K_c and the complex poles for $z = 0.5$ differ from those calculated in the last example. In that case the pole locations for P control were chosen on the 60° line, instead of those for the PI control.

The output transform for a unit step input is

$$C(s) = \frac{2K_c(s + z)}{s(s - p_3)(s^2 + \omega_n s + \omega_n^2)}$$

and the graphical residue technique yields responses

$$c(t) = 1 + A_1 e^{p_3 t} + A_2 e^{-at} \cos(bt + \theta)$$

where $a = \text{Re}\,(p_{1,2})$, $b = \text{Im}\,(p_{1,2})$, and corresponding sets of values $(z, A_1, A_2, \theta°)$ are (4, -0.072, 1.215, $-220°$), (0.5, -0.158, 0.989, $-212°$), and (1.2, $+0.084$, 1.218, $-207°$). These responses are plotted in Fig. 6.23(d).

Figure 6.23(a): For $z = 4$ the zero is too far left. The asymptotes are in the right-half plane ($\rho_0 > 0$), and the branches to these asymptotes cross into the right-half plane at a relatively low gain. These branches leave the real axis to the right of -0.5, suggesting a slow response, as confirmed in Fig. 6.23(d).

Figure 6.23(b): For $z = 0.5$, the response is close to that in Fig. 6.22(b), as expected. While in this example the effect of the real closed-loop pole is small due to the nearby zero, in general this effect tends to become more significant as the zero is moved farther from the origin and beyond the range where it might be chosen based on the PI design procedure. This closed-loop pole then begins to dominate the response, and the overall transient may be visualized as consisting of an oscillatory component due to the complex poles being superimposed on a slowly decaying exponential due to this real pole. In general, therefore, a choice of zero this far from the origin will be avoided unless it satisfies the condition that the difference of the vector angles from the zero and the origin to the desired complex poles is only a few degrees.

Figure 6.23(c): The zero for $z = 1.2$ is somewhat to the left of the dominating plant pole at -1. The value of ρ_0 is to the right of that for design (b), suggesting the somewhat slower response verified in Fig. 6.23(d). The effect of the real closed-loop pole on the locus branch between -2 and -1.2 is small; the overshoot is about 18%, only slightly above the 16% associated with $\zeta = 0.5$.

This is the alternative design referred to at the end of the last example. The pole of the PI control is considered as part of the plant. Then choice of the zero is equivalent to

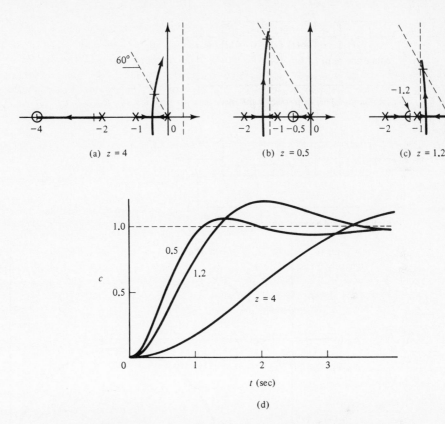

Figure 6.23   Example 6.3.8: choices of the zero in PI control.

a PD control design for a type 1 plant. The phase-lead design procedure then leads to a choice of zero in this range, a little left of the dominating plant pole, as indicated following the stepwise design procedure in Section 6.6. In fact, a choice of the zero location in this range is usually the preferred design.

Example 6.8.4   Satellites with Structural Resonance

Figure 3.14 shows block diagrams derived in Example 3.4.2 for the attitude control of a satellite with structural resonance. A numerical case of these diagrams, with $J_m = 1$, $J_l = 0.2$, $b = 0.02$, and $k = 1.5$, is shown in Fig. 6.24. The resonance has an undamped natural frequency of 3 rad/sec and a damping ratio 0.02. It is recalled that Fig. 6.24(a), where G has one zero, applies when actuators and sensors are taken to be located on the two different masses of the two-mass model. In Fig. 6.24(b), where G has a complex pair of zeros, both are modeled to be located on the same mass.

As in Example 3.4.1, a reasonable choice for the controller is PD control

$$G_c = K(Ts + 1) \qquad (6.17)$$

implemented by means of an attitude sensor and a rate gyro. Figures 6.25(a) and 6.25(b) show root loci with this control for the systems of Fig. 6.24(a) and (b), respectively. The time constants T used are indicated, and the values shown along the curves are those of K.

The importance of sensor location is evident. In design (a), closed-loop poles on the branches from the resonance poles will enter the right-half plane for very small values

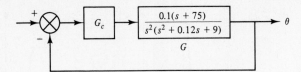

(a) Resonance between actuator and sensor

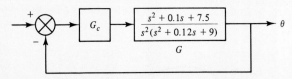

(b) No resonance between actuator and sensor

Figure 6.24 Satellites with structural resonance.

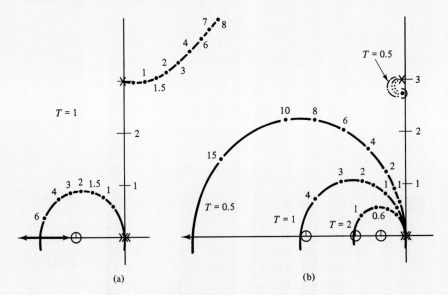

Figure 6.25 Example 6.8.4: satellites with resonance.

of K. In design (b), all branches stay in the left-half plane and K can be chosen to realize a desired damping ratio for the main system poles. The residues at the poles due to the resonance are limited by the nearby pair of zeros, so the resonance response will not be large. Figure 6.26 shows unit step responses for some selected cases of design (b). This design also suggests a possible solution if case (a) is unavoidable. The pair of complex open-loop zeros that facilitated design for case (b) can be introduced into (a) by means of the compensator. Electrical networks that provide such zeros are variations of the bridged-T network in Fig. 2.8(f). Of course, the zeros need not be the same as those in (b), but they should be chosen closer to the origin than the resonance pole pair for all plant parameter variations. In Fig. 6.25(b), if the zero is above the resonance pole, the angle of departure from this pole reverses, and the locus branch curves toward and into the right-half plane. The poles that must be added to make the compensator realizable are chosen farther left in the s-plane. In effect, in this method of compensation for (a), the pair of zeros is used to attenuate the effect of the resonance pole pair before the PD control is applied.

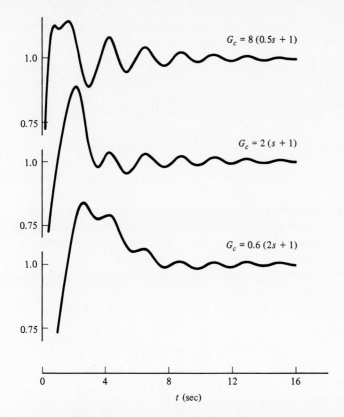

Figure 6.26   Example 6.8.4: unit step responses.

6.9 OTHER USES OF ROOT LOCI AND CONCLUSION

The root locus technique has been discussed and its application to analysis and design. Both series and feedback compensation were considered. The loci are used to select parameters that realize a specified performance and to find the corresponding closed-loop poles. The denominator of the closed-loop transfer function C/R is then available in factored form, and the step response can be calculated, for example, by graphical evaluation of the residues.

The emphasis was on the construction and use of loci for varying root locus gain. However, loci are used also in several other types of applications, discussed briefly next.

(a) Loci for parameters other than gain. This was introduced in Chapter 5, and Figures 5.11 and 5.14 show loci for the loop gain functions

$$\frac{25(K_g s + 1)}{s(s + 2)} \qquad \frac{K_c + K_d s}{(s + 1)(0.5s + 1)}$$

For the first, the root loci show how the closed-loop poles change when the velocity feedback gain K_g is changed. For the second, both loci for varying K_c with K_d constant and for varying K_d with K_c constant were constructed. These examples also demonstrate how differently shaped loci may describe the same system, depending on whether they picture the effect of a controller parameter or the controller gain on the closed-loop poles.

(b) Finding the roots of polynomials. The root locus technique applies also to find the roots of polynomials generally. For example, the equation

$$a_5s^5 + a_4s^4 + a_3s^3 + a_2s^2 + a_1s + a_0 = 0 \qquad (6.18)$$

can be written as follows:

$$(a_5s^5 + a_4s^4 + a_3s^3)\left[1 + \frac{a_2}{a_5}\frac{s^2 + (a_1/a_2)s + a_0/a_2}{s^3(s^2 + (a_4/a_5)s + a_3/a_5)}\right] \qquad (6.19)$$

The last term can be considered as a loop gain function, with root locus gain a_2/a_5, for which loci can be constructed. Other forms are possible in (6.19), but the one shown has the advantage that the open-loop poles and zeros can be found from the solution of quadratics.

(c) Determining the sensitivity to plant parameter variations. Root loci are also used to examine the sensitivity of the closed-loop pole positions to plant parameter variations. As an example, in the system of Fig. 6.8, with open-loop poles at $0, -1, -2$ and with gain $K = 0.65$, let the position of the pole at -1 be uncertain. Loci can be used to show the effect of variations δ of this pole on the closed-loop system poles. The system characteristic equation

$$s(s + 1 + \delta)(s + 2) + 0.65 = s^3 + (3 + \delta)s^2 + (2 + 2\delta)s + 0.65 = 0 \qquad (6.20)$$

is rearranged to isolate δ and written in the form of (6.19):

$$(s^3 + 3s^2 + 2s + 0.65) + \delta s(s + 2)$$

$$= (s^3 + 3s^2 + 2s + 0.65)\left[1 + \delta\frac{s(s + 2)}{s^3 + 3s^2 + 2s + 0.65}\right] \qquad (6.21)$$

The last term is the loop gain function of interest, with δ as the root locus gain. Figure 6.27(a) shows the loci for $\delta > 0$. The open-loop zeros are 0 and -2, and the open-loop poles are the roots of (6.10) for $K = 0.65$, given by the denominator of (6.12). For $\delta < 0$, the root locus gain is negative, and the *0° locus* introduced at the end of Section 6.2 is needed. Because the minus sign represents 180°, the angle condition now requires the net sum of the vector angles to be an even multiple of $\pm180°$. This implies that directions of asymptotes change by 180°, and angles of departure or arrival at complex poles and zeros also. Furthermore, parts of the real axis to the left of an even number of poles and zeros on this axis, and to the right of all of these, belong to the loci.

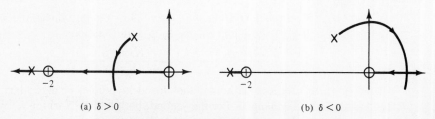

(a) $\delta > 0$	(b) $\delta < 0$

Figure 6.27   Loci for parameter sensitivity.

Figure 6.27(b) shows the locus sketch for the example. Evidently, negative variations δ are destabilizing, as expected because the pole -1 is then moving closer to the origin. The magnitude condition or the Routh–Hurwitz criterion can be used to determine the value of δ at which stability is lost.

PROBLEMS

6.1. Plot the loci of the closed-loop system poles for varying K of systems with the following loop gain functions.

(a) $\dfrac{K}{s}$ (b) $\dfrac{K}{s+1}$ (c) $\dfrac{K}{s-1}$

(d) $\dfrac{K}{s^2}$ (e) $\dfrac{K}{s^2+4}$ (f) $\dfrac{K}{s^2-4}$

(g) $\dfrac{K}{s^2+2s+2}$ (h) $\dfrac{K(s+2)}{s(s+3)}$

6.2. Sketch the general shape of the loci of the system poles for systems with the following loop gain functions.

(a) $\dfrac{K(s+1)}{s^2}$. Use the angle condition to show that the locus contains a circle centered at -1 with radius 1.

(b) $\dfrac{K(s+2)}{s(s+1)}$ (c) $\dfrac{K}{(s^2+2s+2)(s^2+6s+10)}$

(d) $\dfrac{K(s+2)}{(s+1)(s^2+6s+11.25)}$

6.3. For the system in Fig. P6.3 with $H=1$, $G=[K(s+1)]/[s(s+2)]$:
(a) Draw the root loci.
(b) Use them to find K for a 2-sec time constant of the dominating closed-loop pole.
(c) Find the other system pole, by root locus methods (not analytically, although for this problem root loci are not at all required).
(d) Use the roots of the system characteristic equation found in parts (b) and (c) to write the closed-loop transfer function $C(s)/R(s)$ and plot its pole–zero pattern.
(e) Add the pole–zero pattern of a step input $R(s)=1/s$ to obtain that of the corresponding $C(s)$.
(f) Calculate $c(t)$, evaluating the residues graphically, and also find the steady-state errors for unit step and unit ramp inputs.

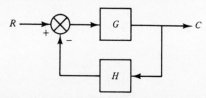

Figure P6.3

6.4. In Fig. P6.4 with $G(s)=1/[s(s+1)]$, to evaluate the effect of adding a pole or a zero to the open-loop pole–zero pattern, sketch and compare the loci for

(a) $G_c=K$ (b) $G_c=K/(s+2)$ (c) $G_c=K(s+2)$

How is relative stability affected?

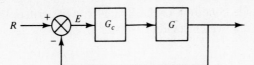

Figure P6.4

6.5. For a unity feedback hydraulic position servo with loop gain function

$$G = \frac{K}{s(s^2 + 2s + 5)}$$

  (a) Sketch the loci of the closed-loop poles for varying K.
  (b) Find reasonably accurately the value of K for which the time constant of the dominating pair of closed-loop poles is 2 sec.
  (c) Determine the position of the third pole.
  (d) Calculate the unit step response.

6.6. Figure P6.4 models a temperature control system with a simple lag plant $G(s) = 1/(0.25s + 1)$. PI control $G_c(s) = K(s + 2)/s$ is used to reduce steady-state errors.
  (a) Plot the loci and use them to find K for a dominating system time constant of 1 sec.
  (b) For this K, find the second pole on the loci, and determine the unit step response from the pole–zero pattern of the output $C(s)$.

6.7. In Fig. P6.3 with

$$G = \frac{4K}{(0.5s + 3)(2s^2 + 8s + 12)} \qquad H(s) = \frac{0.1}{0.05s + 0.4}$$

  (a) Sketch the loci of the closed-loop poles for varying K.
  (b) Find the approximate value of K so that the damping ratio of the dominating closed-loop poles will be 0.5.

6.8. For a system with loop gain function

$$G = \frac{K(s + 3)}{s^2 - 2s + 10}$$

  (a) Plot the open-loop pole–zero pattern and determine whether the system is open-loop stable.
  (b) Sketch the root loci for varying K.
  (c) Find the range of K for which the system is stable.
  (d) Find K for a damping ratio of about 0.7.

6.9. Figure P6.4 models a pressure control system with a plant $G = 1/[(s + 1)(s + 3)]$ consisting of two simple lags, and PI control $G_c = K(s + 2)/s$:
  (a) Sketch the loci of the closed-loop system poles for varying K.
  (b) Find, reasonably accurately, the value of K for a damping ratio 0.5 for the dominating pair of poles.

6.10. For a fluid power position servo with loop gain function

$$G = \frac{K}{s(s^2 + 6s + 13)}$$

  (a) Sketch the loci of the system poles for varying K.
  (b) Find K for a damping ratio 0.707 of the dominating poles.
  (c) Where is the third pole for this K?

6.11. Figure P6.11 shows root loci for a robotic manipulator with a flexible joint, modeled by

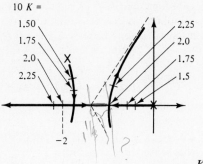

Figure P6.11

$$GH = \frac{K}{s(0.1s^2 + 0.4s + 0.5)}$$

Note that in contrast with Problems 6.5 and 6.10 the loci from the complex pair of open-loop poles dip down to the real axis. This change of configuration may occur if the pair lies relatively far to the left.

(a) How does the nature of the transient response, in terms of time constants and damping ratios of its components, change as K is increased?

(b) What is the limiting value of K for stability?

6.12. For a system with loop gain function

$$G = \frac{K(s^2 + 4s + 8)}{s^2(s - 1)}$$

sketch the loci and find the range of values of K for which the system is stable.

6.13. For an open-loop unstable system with loop gain function

$$GH = \frac{K(s + 1)}{s(s - 1)}$$

sketch the loci of the system poles and find K for a system time constant of 1 sec from the loci.

6.14. Plot the root loci for a servo subjected to load resonance with the loop gain function

$$\frac{K}{s(s + 6)(s^2 + 6s + 13)}$$

6.15. Plot the root loci for a system with the loop gain function

$$\frac{K}{(s + 2)(s + 6)(s^2 + 8s + 20)}$$

and find the limiting value of K for stability.

6.16. Sketch the loci of the closed-loop system poles for varying K of a flexible manipulator system with loop gain function

$$\frac{K}{s(0.5s^2 + s + 1)}$$

and find the limiting value of K for stability.

6.17. From Problem 6.2(a), the loci of $K(s + 2)/s^2$ contain a circle of radius 2 centered at -2. Use these loci to find K for a damping ratio 0.707.

6.18. If Fig. P6.4 models a motor position servo with $G(s) = 1/[s(s + 1)]$ and $G_c = K$:

(a) Plot the loci and use them to find K for a damping ratio of about 0.7.

(b) Calculate the unit step response, using the system poles as found from the loci.

6.19. Plot the loci for a loop gain function $GH = K/[s(s + 4)]$ and use them to find the lowest value of K that will minimize system settling time. Also, is there a change of nature of the transient response as K is increased, and where does it occur?

6.20. Sketch the loci of

$$GH = \frac{K}{s(s + 1)(s + 3)(s + 4)}$$

and determine the limiting value of K for stability by use of the characteristic equation and the condition $s = j\omega$.

6.21. If Fig. P6.3 with $G = K/[s(0.1s + 1)]$ and $H = 1/(0.02s + 1)$ represents a position servo with appreciable sensor time constant:

   (a) Plot loci and find K for a damping ratio 0.5 for unity feedback (that is, assuming an ideal feedback sensor).

   (b) If the actual sensor has in fact a significant time constant, what will be the actual damping ratio if the value of K found in part (a) is used? [Use the angle condition for some points along a horizontal through the "design point" of part (a) to sketch the section of the actual loci that is of interest more accurately.] An approximate answer will suffice. Compare it with the design value in part (a).

6.22. For a unity feedback hydraulic position servo with loop gain function

$$G = \frac{K}{s(s^2 + 2s + 2)}$$

   (a) Sketch the loci of the closed-loop system poles for varying K, and find K for a damping ratio 0.5 for the complex poles.

   (b) Calculate the unit step response and the unit ramp steady-state following error for the design of part (a).

6.23. For a motor position servo with loop gain function

$$G = \frac{K}{s(0.25s + 1)(0.1s + 1)}$$

sketch the loci of the closed-loop system poles for varying K, and find K for a damping ratio 0.5 of the dominating pair.

6.24. If Fig. P6.4 models a two-tank level control system with $G(s) = 1/[(s + 1)(s + 4)]$, design a physically realizable controller $G_c(s)$ such that:

   1. The steady-state error for step inputs is zero.

   2. The dominating time constant is 0.5 sec.

   3. The damping ratio is about 0.7.

6.25. In Fig. P6.4 with $G(s) = 1/[(0.1s + 1)(0.02s + 1)]$, a system damping ratio 0.707 is required.

   (a) For $G_c(s) = K_c$, find K_c and the corresponding steady-state error after a unit step.

   (b) Design PD control $G_c = K + K_d s$ to halve the steady-state error in part (a). Solve this analytically after verifying that root locus design would involve considerable trial and error.

6.26. **(a)** If Fig. P6.4 models a pressure control system with $G(s) = 1/[(s + 1)(0.5s + 1)]$, sketch root loci for PI control $G_c = K_p + K_i/s$ if (i) K_i/K_p is large, (ii) $K_i/K_p = 2$, (iii) $K_i/K_p = 1$, and (iv) $K_i/K_p = 0.1$.

   (b) Which of conditions (i) to (iii) is preferred, and why?

   (c) For condition (iv), will the locus branches at considerable distance to the origin differ much from those for P control $G_c = K_p$? If not, why not?

6.27. In Problem 6.26 it is desired to compare condition (iv) with P control $G_c = K_p$.
  (a) For $G_c = K_p$, find K_p and the system poles for a damping ratio 0.707.
  (b) Since condition (iv) with the same K_p will have little effect on these poles (why not?), use K_p as in part (a) and assume that the poles in part (a) are also system poles for condition (iv). Use the magnitude condition to determine the system pole near the origin for the value of K_p, from the open-loop pole–zero pattern.
  (c) Calculate the initial magnitude of the slowly decaying exponential transient due to this pole. Why is it rather small?
  (d) Which of conditions (i) to (iv) would you choose if fast convergence to a value near steady state is important and slow convergence to zero steady-state error is acceptable?

6.28. Figure P6.4 with $G(s) = 3/[s^2(s + 3)]$ can represent the pitch control system of a missile.
  (a) Sketch loci to determine whether $G_c(s)$ could be chosen to be a simple gain.
  (b) Choose an idealized controller that would stabilize the system and for which the system time constant at high gains will be 1 sec.
  (c) What are the steady-state errors for unit step and unit ramp inputs?

6.29. In Fig. P6.4 a plant transfer function $G(s) = 1/[s(s - 2)]$ may represent a tall rocket. Such a plant is open-loop unstable (that is, unstable without feedback control), like a pencil standing on its end.
  (a) Plot the loci to determine whether the system can be stabilized by P control $G_c = K_c$.
  (b) If not, could the unstable pole of G be canceled by a zero of G_c to stabilize the system, and if not, why not?
  (c) Choose an idealized controller that can stabilize the system, and find the corresponding range of gains for stability.

6.30. To appreciate that the solution in Problem 6.29(c) is not unique, sketch root loci for the following controllers:
  (a) Phase-lead compensation $G_c = K(s + 1)/(s + 8)$
  (b) Idealized quadratic lead $G_c = K(s^2 + 4s + 5)$
  Note how rough locus sketches can give a quick idea of the potential of contemplated forms of compensators.

6.31. Using root locus sketches:
  (a) Investigate the stability of the minor loop in Fig. P6.31.
  (b) Determine whether the overall system can be stable if G_c is just a gain.
  (c) What form of idealized controller G_c would you propose, and how would you find its parameters?

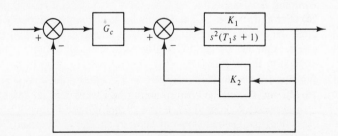

Figure P6.31

6.32. The feedback $H = (1 + K_t s)$ in Fig. P6.3 can represent a parallel combination of direct feedback and minor loop rate feedback. If $G(s) = K/(s^2 + 2s + 3.25)$ represents a spring–mass–damper system with a position output and a force input:

(a) Find the constraints on K and/or K_t for a steady-state error of 10% following step inputs.

(b) Write the characteristic polynomial with the constraints of part (a), and rearrange it to obtain the loop gain function for the construction of loci of the closed-loop poles for varying K_t.

(c) Calculate the value of K_t for a damping ratio 0.707 from the quadratic characteristic equation, and use the corresponding roots in sketching the loci for varying K_t.

6.33. In the motor position servo with rate feedback shown in Fig. P6.33:

(a) Sketch the loci and find K for a system damping ratio 0.5 for the dominating poles.

(b) Find the steady-state errors for step and ramp inputs for K of part (a).

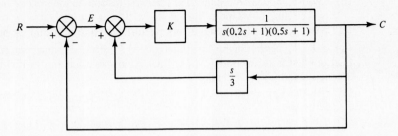

Figure P6.33

6.34. If Fig. P6.4 models a position servo with $G(s) = 1/[s(s + 3)]$, choose p and K in the controller $G_c = K(s + 1)/(s + p)$ such that:

1. The steady-state unit ramp following error is zero.

2. The damped natural frequency of oscillatory components in the transient response is 4 rad/sec.

6.35. In Fig. P6.4, if the open-loop unstable plant $G(s) = 1/[s(s - 4)]$ represents a tall rocket, use root loci to determine the range of K for which the system will be stable if G_c is a phase-lead network:

$$G_c(s) = \frac{K(s + 2)}{s + 20}$$

6.36. In Fig. P6.4, let $G(s)$ be a motor plus load transfer function $1/[s(Js + B)]$ in which damping B is negligible so that, for $J = 1$, G can be approximated by $G = 1/s^2$. Sketch loci for $G_c = K_p$ and for phase-lead compensation $G_c = K(s + 2)/(s + 8)$. Note the stabilizing effect of the lead and give the system time constant for high gains. Also, sketch the different general shape of the loci if the zero is moved close enough to the origin.

6.37. For Fig. P6.4 with $G(s) = 1/[s(s + 1)]$, choose the pole(s) and zero(s) of a controller G_c, and sketch the corresponding loci, such that no closed-loop poles will have a time constant above 0.5 sec or a damping ratio below 0.5 for some value of gain. How does the general shape of the loci change if the pole of G_c is moved relatively far away?

6.38. In Fig. P6.4, let G be a hydraulic servo with transfer function $G(s) = \omega_n^2/(s^2 + \omega_n^2)$.

(a) Can a compensator $G_c = K$ be used?

(b) Use root loci to investigate the use of a compensator $G_c = K(s + a)$, $a > 0$.
(c) Similarly, examine the use of $G_c = K(s^2 + a^2)$ with a both larger and smaller than ω_n.

6.39. In Fig. P6.4, $G(s) = K_h/(ms^2 + bs + k)$ represents a lightly damped spring–mass–damper system controlled by a hydraulic servo for which the transfer function can be approximated by a constant K_h over the frequency range of interest. It is specified that the system must follow ramp inputs with zero steady-state error.
(a) Sketch loci to determine whether the simplest possible controller that will meet the error specification can be used.
(b) Sketch the general shapes of the loci for $G_c = K(s + a)/s^2$ for some positions of the zero $-a$ from far left to close to the origin.

6.40. In Fig. P6.4 with $G = 1/[(s + 1)(s + 2)(s + 3)]$:
(a) Sketch the loci for $G_c = K$ and determine the limiting value of K for stability.
(b) Repeat for the lag compensator $G_c = (K/5)(s + 2.5)/(s + 0.5)$ and compare the steady-state errors for step inputs with part (a).

6.41. In Fig. P6.4 with $G = K/[s(s + 1)]$:
(a) With $G_c = 1$ find K for a damping ratio 0.707 and the steady-state error for a unit ramp input.
(b) Sketch the loci and repeat the above for $G_c = (10s + 1)/(40s + 1)$. How does this phase-lag compensator affect behavior?

6.42. In Fig. P6.4 with $G = 1/(s^2 + s + 5/4)$, sketch the loci for a phase-lead compensator $G_c(s) = K(s + 0.5)/(s + 5)$.

6.43. If Fig. P6.4 with $G(s) = 1/[s(s + 1)]$ models a position servo, design phase-lag compensation for a dominant damping ratio 0.5 and a factor of 10 improvement of steady-state accuracy over that for P control.

6.44. If Fig. P6.4 with $G(s) = 1/[s(s + 1)]$ models an axis of an antenna positioning system, design phase-lead compensation for a dominant damping ratio 0.5 and a settling time of 4 sec. Sketch the corresponding loci and find the steady-state errors following unit step and unit ramp inputs.

6.45. If Fig. P6.4 with $G(s) = 1/[s(s + 1)(s + 4)]$ models an aircraft roll angle control, design phase-lead compensation for a dominating damping ratio 0.5 and time constant of 1 sec. Sketch the corresponding loci and find the steady-state errors following unit step and unit ramp inputs.

6.46. If Fig. P6.4 with $G(s) = 1/[(s + 1)(s + 4)]$ models a motor speed control, design PD control for a damping ratio 0.707 and a settling time of 1 sec. Sketch the corresponding loci and find the steady-state errors for unit step and unit ramp inputs.

6.47. If Fig. P6.4 with $G(s) = 1/[s(s + 1)(s + 4)]$ models a robot joint control, design PD control for a dominant damping ratio 0.5 and time constant of 1 sec. Sketch the corresponding loci and find the steady-state errors for unit step and unit ramp inputs.

6.48. If Fig. P6.4 with $G = 1/(s + 1)$ models a temperature control system, design a PI controller for a damping ratio 0.5 and a time constant of 1 sec. Sketch the corresponding loci and find the steady-state errors for unit step and unit ramp inputs.

6.49. If Fig. P6.4 with $G = 1/[(s + 1)(s + 4)]$ models a level control system, design PI control for a dominant damping ratio 0.5 and a settling time of 4 sec. Sketch the corresponding loci and find the steady-state errors for unit step and unit ramp inputs.

6.50. Use root locus sketches to determine the stability of systems with the following characteristic equations.
(a) $s^3 + 3s^2 + 2s + 1 = 0$ **(b)** $s^3 + 3s^2 + 2s + 8 = 0$

6.51. Repeat Problem 6.50 for:

(a) $s^4 + 5s^3 + 6s^2 + 2s + 1 = 0$ (b) $s^5 + 5s^4 + 6s^3 + 2s^2 + 4s + 4 = 0$

6.52. For a system with loop gain function

$$\frac{K(s + 5)}{s(s + 2)(s + 3)}$$

(a) Sketch the loci of the closed-loop poles for varying K. It may be shown that for $K = 8$ the closed-loop poles are located at -4, $-0.5 \pm j3.12$.

(b) Sketch loci to show the effect of variations δ of the open-loop pole at -2 on the closed-loop poles for $K = 8$. Which direction of variation is dangerous?

7

Frequency Response Analysis

7.1 INTRODUCTION

In frequency response methods, system behavior is evaluated from the steady-forced response to a sinusoidal input

$$r(t) = A \sin \omega t \qquad (7.1)$$

These methods for analysis and design are very widely used, and frequency response testing, over a range of frequencies ω, is often the most convenient method for measurement of systems dynamics. The techniques of the preceding chapters are not well adapted to certain frequency response concepts of great practical importance. For example, if high-frequency "noise" is superimposed on the input, how can the system be designed to respond well to the input, but "filter out" the noise? Or, for audio systems, instrumentation, control systems, and so on, how does one design filters that pass only input signal components in a selected range of frequencies? This relates to the problem of designing a system for a specified *bandwidth,* that is, range of frequencies over which it responds well. Also, what if a maximum percentage error has been specified for input signals in a normal operating range of frequencies, thus extending accuracy specifications from static to dynamic conditions? The frequency domain also permits the clearest picture of the sensitivity properties in Chapter 4, such as the sensitivity to unmodeled high-frequency dynamics.

Compensator design is often more convenient in the frequency domain and can be carried out, for the general configuration in Fig. 7.1, by the use of frequency response plots of the loop gain function $G_c GH$.

In this chapter, several common types of frequency response plots are introduced, and absolute and relative stability are discussed on the basis of the famous Nyquist stability criterion. This criterion is central to the great practical importance

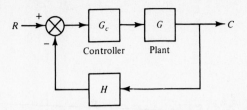

Figure 7.1 Feedback system.

of frequency response methods in control engineering. Design by frequency response methods is discussed in Chapter 8, together with frequency domain performance criteria. This design will strongly emphasize the use of Bode plots, which are more convenient than other plots for most practical analysis and design.

7.2 FREQUENCY RESPONSE FUNCTIONS AND PLOTS

Figure 7.2 shows a system with transfer function $G(s)$ and a sinusoidal input

$$r(t) = A \sin \omega t \tag{7.2}$$

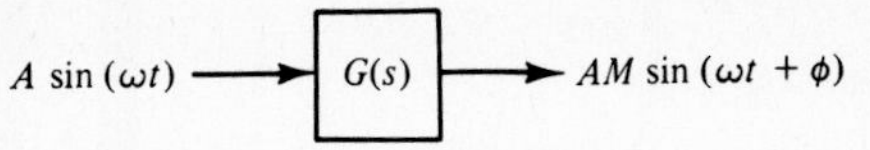

$$A \sin (\omega t) \longrightarrow \boxed{G(s)} \longrightarrow AM \sin (\omega t + \phi)$$

Figure 7.2 Frequency response.

Using the transform of $r(t)$ given in Table 1.6.1, the transform $C(s)$ of the system output is

$$C(s) = \frac{A\omega G(s)}{s^2 + \omega^2} = \frac{K_1}{s + j\omega} + \frac{K_2}{s - j\omega} + \cdots \tag{7.3}$$

Of the partial fraction expansion, only the terms due to the roots of the denominator of $R(s)$ are shown. As discussed following (4.13), these give the steady-forced part of the solution needed in frequency response methods. As in Section 1.9, the residues are

$$K_1 = \frac{A\omega G(s)}{s - j\omega}\bigg|_{s=-j\omega} = \frac{AG(-j\omega)}{-2j} \qquad K_2 = \frac{AG(j\omega)}{2j}$$

and the forced response is

$$c_s(t) = \frac{A[-G(-j\omega)e^{-j\omega t} + G(j\omega)e^{j\omega t}]}{2j} \tag{7.4}$$

$G(j\omega)$ is a complex variable, which can be represented in the alternative ways indicated in Fig. 7.3, either as a sum of a real part $a(\omega)$ and an imaginary part $jb(\omega)$ or as a vector of length $M(\omega)$ and phase angle $\phi(\omega)$:

$$G(j\omega) = a(\omega) + jb(\omega) = M(\omega)e^{j\phi(\omega)}$$

$$M(\omega) = |G(j\omega)| = \sqrt{a^2(\omega) + b^2(\omega)} \tag{7.5}$$

$$\phi(\omega) = \tan^{-1}\frac{b(\omega)}{a(\omega)}$$

For $G(-j\omega)$, the imaginary part and hence the phase angle ϕ change sign:

$$G(-j\omega) = M(\omega)e^{-j\phi(\omega)} \tag{7.6}$$

Substituting (7.5) and (7.6) into (7.4) yields

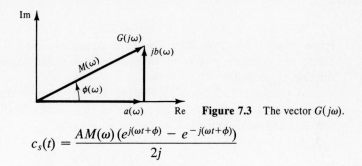

Figure 7.3   The vector $G(j\omega)$.

$$c_s(t) = \frac{AM(\omega)\,(e^{j(\omega t + \phi)} - e^{-j(\omega t + \phi)})}{2j}$$

or

$$c_s(t) = AM(\omega)\,\sin[\omega t + \phi(\omega)] \tag{7.7}$$

This result is indicated in Fig. 7.2, and shows the following:

1. For a sinusoidal input, the forced response is also sinusoidal and of the same frequency.
2. The magnitude M of the *frequency response function* $G(j\omega)$, obtained by replacing s by $j\omega$ in the transfer function $G(s)$, equals the ratio of output amplitude to input amplitude.
3. The phase angle ϕ of $G(j\omega)$ is the phase angle of the output relative to that of the input.

The Laplace operator $s = \sigma + j\omega$ in $G(s)$ is replaced by $s = j\omega$, so for $G(j\omega)$ only values of s along the imaginary axis ($\sigma = 0$) are considered.

$G(j\omega)$ can be plotted on a complex plane or polar plot as a vector of length $M(\omega)$ and phase angle $\phi(\omega)$, positive counterclockwise from the positive real axis. As ω varies, the end point of the vector describes the polar plot.

For a simple lag

$$G(j\omega) = \frac{K}{j\omega T + 1}$$

$$M(\omega) = \frac{K}{\sqrt{1 + (\omega T)^2}} \qquad \phi(\omega) = -\tan^{-1}\omega T \tag{7.8}$$

M and ϕ are verified by inspection of Fig. 7.4(a) and yield the polar plot in Fig. 7.4(b).

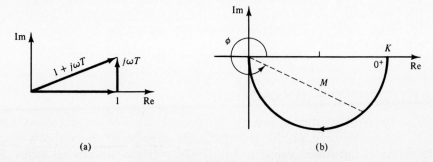

(a)   (b)

Figure 7.4   Polar plot of simple lag.

Examples of polar plots, one of several ways of representing frequency response results, are given later. The most commonly used plots are Bode diagrams. These are separate graphs of magnitude and phase angle versus frequency, on semilog graph paper.

A graphical technique for finding the frequency response function $G(j\omega)$ by measurement from the pole–zero pattern of $G(s)$ shows the direct link between these domains and is of considerable conceptual interest.

Let

$$G(s) = K \frac{(s + z_1)(s + z_2)\cdots}{(s + p_1)(s + p_2)\cdots} \qquad (7.9)$$

with the pole–zero pattern shown in Fig. 7.5(a). Note that on the imaginary axis $s = j\omega$. Thus the plot of $G(s)$ for s traveling up the imaginary axis from $\omega = 0^+$ to $\omega \rightarrow +\infty$ is in effect just the polar plot of the frequency response function $G(j\omega)$. As discussed on several occasions, the factors $(s + z_i)$ and $(s + p_i)$ are vectors from $-z_i$ and $-p_i$ to s. Hence, for any value of $s = j\omega$ on the imaginary axis, the magnitude and phase of $G(j\omega)$ can be determined graphically by measuring the vector lengths and angles from the pole–zero pattern. Figure 7.5(b) shows the resulting magnitude M of $G(j\omega)$ plotted versus frequency on a very common type of diagram. It is significant to observe the correlation between the nature of this plot and the pole–zero pattern. If the pole pair $-p_1$, $-p_2$ has a small damping ratio, then the vector $(s + p_2)$ will be short at the values of ω where the point s passes $-p_2$. This means that M will be large and leads to the general result that a small damping ratio corresponds to a high *resonance peak* on the frequency response plot.

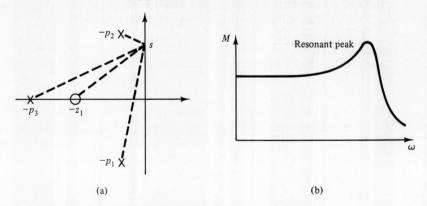

(a)   (b)

Figure 7.5   Frequency response from pole–zero pattern.

7.3 NYQUIST STABILITY CRITERION

The Nyquist criterion is central to the importance of frequency response techniques of analysis and design. In Fig. 7.1 the closed-loop system poles are the roots of the characteristic equation $1 + G_c GH = 0$ and so can be called the zeros of $(1 + G_c GH)$. The poles of $(1 + G_c GH)$ (that is, the roots of its denominator) equal those of the loop gain function $G_c GH$, the open-loop poles. Let

$$1 + G_c GH = K \frac{(s + z_1)(s + z_2) \cdots}{(s + p_1)(s + p_2) \cdots} \qquad (7.10)$$

The poles $-p_1$, $-p_2$, ... of $(1 + G_c GH)$ are usually known, but the zeros $-z_1$, $-z_2$, ... are not. If they were, stability analysis would be unnecessary.

Let the pole–zero pattern be as shown in Fig. 7.6, where $-z_1$ and $-z_2$ are not known. To prove stability, it is necessary and sufficient to show that no zeros $-z_i$ are inside the *Nyquist contour D* which encloses the entire right half of the s-plane. D consists of the imaginary axis from $-j\infty$ to $+j\infty$ and a semicircle of radius $R \to \infty$. In principle, stability analysis is based on plotting $(1 + G_c GH)$ in a complex plane as s travels once clockwise around the closed contour D.

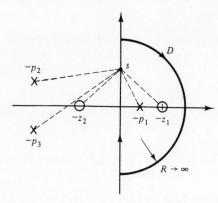

Figure 7.6   Nyquist contour D.

As discussed in the preceding section, the plot of $(1 + G_c GH)$ for s traveling up the imaginary axis from $\omega = 0^+$ to $\omega \to +\infty$ is in effect just the polar plot of the frequency response function $(1 + G_c GH)$. By (7.6), its mirror image relative to the real axis is the plot between 0^- and $-\infty$ on the imaginary axis.

For the part of the Nyquist contour along the semicircle of radius $R \to \infty$, the value of $(1 + G_c GH)$ is constant, usually 1. This is because, as discussed in Section 5.4, physical realizability prevents the order of the numerator of $G_c GH$ from being larger than that of the denominator, and in practice it is usually smaller. Hence, for $|s| \to \infty$, $G_c GH$ generally tends to zero.

Figure 7.6 shows that if s moves once clockwise around D, vectors $(s + z_i)$ and $(s + p_i)$ rotate 360° clockwise for each pole and zero inside D, and undergo no net rotation for poles and zeros outside D.

From (7.10), if the vector $(s + z_i)$ in the numerator rotates 360° clockwise, this will contribute a 360° clockwise rotation of the vector $(1 + G_c GH)$ in the complex plane in which it is plotted. If vector $(s + p_i)$ in the denominator rotates 360° clockwise, this will contribute a 360° counterclockwise revolution of $(1 + G_c GH)$. Poles and zeros outside D do not contribute any net rotation. The result can be expressed as follows:

Principle of the Argument.   If $(1 + G_c GH)$ has Z zeros and P poles inside the Nyquist contour D, a plot of $(1 + G_c GH)$ as s travels once clockwise around D will encircle the origin of the complex plane in which it is plotted $N = Z - P$ times in the clockwise direction.

In principle, this completes the attributes needed for stability analysis. For stability the condition $Z = 0$, so $N = -P$, is necessary and sufficient. Apparently, then, if $(1 + G_c GH)$ is plotted from (7.10), but with its numerator given in polynomial rather than factored form, as s travels once around D, the system will be stable if and only if the plot encircles the origin P times in the counterclockwise direction.

The following observation allows this result to be stated in a more convenient form:

The encirclements of a plot of $(1 + G_c GH)$ around the origin equal the encirclements of a plot of $G_c GH$ around the -1 point on the negative real axis.

The plot of $G_c GH$ as s travels once around D is called a *Nyquist diagram*. With this, the following has been proved.

Nyquist Stability Criterion.   A feedback system is stable if and only if the number of counterclockwise encirclements of the Nyquist diagram about the -1 point is equal to the number of poles of $G_c GH$ inside the right-half plane, called open-loop unstable poles.

Usually, systems are open-loop stable, that is, $P = 0$. In this case, the criterion becomes:

An open-loop stable feedback system is stable if and only if the Nyquist diagram does not encircle the -1 point.

For open-loop stable systems, it is in fact not necessary to plot the complete Nyquist diagram; the polar plot, for ω increasing from 0^+ to $+\infty$, is sufficient. The Nyquist diagram examples given later will help to verify that the following is equivalent to zero encirclements.

Simplified Nyquist Criterion.   If $G_c GH$ does not have poles in the right-half s-plane, the closed-loop system is stable if and only if the -1 point lies to the left of the polar plot when moving along this plot in the direction of increasing ω, that is, the polar plot passes on the right side of -1.

For example, if Fig. 7.7(a) shows the polar plot of the loop gain function of an open-loop stable system, then the system is also closed-loop stable. In the marginal case where $G_c GH$ has poles on the imaginary axis, these will be excluded from the Nyquist contour by semicircular indentations of infinitesimal radius around them. This is shown in Fig. 7.7(b) for the common case of a pole at the origin. The following examples will illustrate how this affects the Nyquist diagrams.

7.4 POLAR PLOTS AND NYQUIST DIAGRAMS

Polar Plots and Stability

An ability to sketch the general form of polar plots is frequently useful. To this end, $\omega \to 0$ is considered first. For example, the diagram in Fig. 7.4(a) shows that a simple lag $K/(j\omega T + 1)$ then approaches K on the positive real axis. Next, for

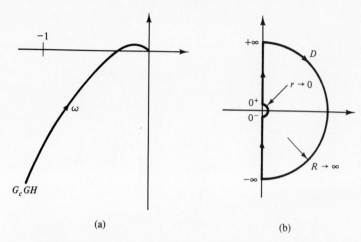

(a)   (b)

Figure 7.7   (a) Polar plot and (b) indented Nyquist contour D.

$\omega \to \infty$ the magnitude tends to zero, and Fig. 7.4(a) shows that the phase of the denominator approaches $+90°$, so that of the simple lag itself $-90°$. Thus for $\omega \to \infty$ the plot approaches the origin from the bottom, as shown in Fig. 7.4(b). Alternatively, for $\omega \to \infty$ the simple lag approaches $K/(j\omega T)$, and each factor j in a denominator represents $-90°$ since

$$\frac{1}{j} = \frac{1}{j}\frac{j}{j} = \frac{j}{j^2} = -j \qquad \text{(i.e., } -90°\text{)}$$

The diagram of Fig. 7.4(a) is also useful to show the behavior of the simple lag between 0^+ and $+\infty$. Both the magnitude and phase of the denominator are seen to increase continuously with ω, because the imaginary part $j\omega T$ increases with ω. Therefore, the phase of the simple lag must increase continuously in the negative direction, from $0°$ to $-90°$, while the distance to the origin decreases continuously.

Figure 7.8 shows similarly derived sketches for a number of common functions. In Fig. 7.8(a), for $\omega \to 0^+$ the function values are positive and real, and for $\omega \to +\infty$ all tend to the origin. For (a1) and (a3) the functions approach a constant divided by $(j\omega)^2$ as $\omega \to \infty$, and for (a2) a constant divided by $(j\omega)^3$. Each denominator factor $j\omega$ represents an angle $-90°$, so (a1) and (a3) approach the origin from $180°$, and (a2) from $-270°$ (or $+90°$). The diagram in Fig. 7.4(a) helps to show that for (a1) and (a2) the vector from the origin must rotate clockwise and shorten as ω increases. This is also true for (b1), (b2), and (c1) in Fig. 7.8(b) and (c). Note that for $\omega \to \infty$, (b3) approaches a constant divided by $(j\omega)^2$, and (c2) a constant divided by $(j\omega)^3$. So (b3) approaches the origin from the left, and (c2) from the top. For $\omega \to 0^+$, the functions (b) approach $K/(j\omega)$, and (c) approach $K/(j\omega)^2$. Therefore, the plots (b) start far out along the negative imaginary axis, and (c) far out along the negative real axis. In (b3) and (c2) the numerator factors contribute positive phase angles (lead), so counterclockwise rotation. The equations in Fig. 7.9 can be used to calculate points, if needed. For example, for case (b3),

$$M = \frac{KM_3}{\omega M_1 M_2} \qquad \phi = -90° + \phi_3 - \phi_1 - \phi_2 \qquad (7.11)$$

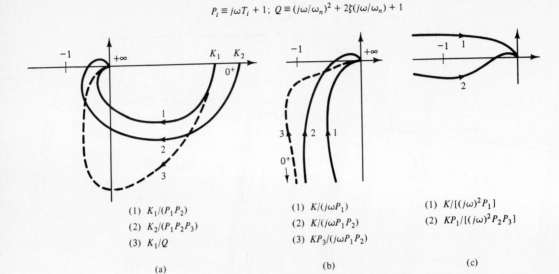

$$P_i \equiv j\omega T_i + 1; \quad Q \equiv (j\omega/\omega_n)^2 + 2\zeta(j\omega/\omega_n) + 1$$

(1) $K_1/(P_1 P_2)$

(2) $K_2/(P_1 P_2 P_3)$

(3) K_1/Q

(a)

(1) $K/(j\omega P_1)$

(2) $K/(j\omega P_1 P_2)$

(3) $KP_3/(j\omega P_1 P_2)$

(b)

(1) $K/[(j\omega)^2 P_1]$

(2) $KP_1/[(j\omega)^2 P_2 P_3]$

(c)

Figure 7.8 Polar plot examples.

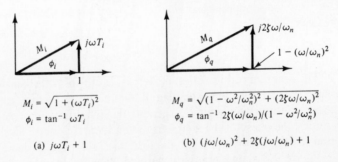

$M_i = \sqrt{1 + (\omega T_i)^2}$

$\phi_i = \tan^{-1} \omega T_i$

(a) $j\omega T_i + 1$

$M_q = \sqrt{(1 - \omega^2/\omega_n^2)^2 + (2\zeta\omega/\omega_n)^2}$

$\phi_q = \tan^{-1} 2\zeta(\omega/\omega_n)/(1 - \omega^2/\omega_n^2)$

(b) $(j\omega/\omega_n)^2 + 2\zeta(j\omega/\omega_n) + 1$

Figure 7.9 Simple and quadratic lags.

Appendix B gives a design-oriented program for computer construction.

As to stability, assume that Fig. 7.8 shows polar plots of loop gain functions $G_c GH$ of open-loop stable systems and that the -1 points are located where shown. Then, by the simplified Nyquist criterion, all systems except (c1) will be stable.

Nyquist Diagrams and Stability

The polar plots are also part of Nyquist diagrams, for values of s on the imaginary axis from 0^+ to $+\infty$. As discussed in Section 7.3, the mirror image of the polar plot relative to the real axis is the part of the Nyquist diagram corresponding to the section of the Nyquist contour between 0^- and $-\infty$ on the imaginary axis, and for values of s on the large semicircle $G_c GH$ generally tends to zero.

It remains to consider the case of poles on the imaginary axis. The form of a loop gain function with n poles at the origin is

$$G_c GH = \frac{K(s + z_1)(s + z_2)\cdots}{s^n(s + p_1)(s + p_2)\cdots} \tag{7.12}$$

Just as $(s + p_1)$ is a vector from the pole $-p_1$ to s, so each factor s in the denominator is a vector from a pole at the origin to points s. In Fig. 7.7(b), as s travels around the small semicircle of radius $r \to 0$, each of these vectors rotates 180° counterclockwise, at a constant $|s| \to 0$, between the points for frequencies 0^- and 0^+ on the imaginary axis. This means that $|G_c GH| \to \infty$ and $G_c GH$ rotates 180° clockwise between frequencies 0^- and 0^+ for each pole of $G_c GH$ at the origin.

Figure 7.10 shows some of the polar plots in Fig. 7.8 completed into Nyquist diagrams. In Fig. 7.10(a), only the mirror image need be added to the plot (a2) in Fig. 7.8. If the system is open-loop stable, and the numerical values are such that the -1 point is located where shown, the system is stable, because the -1 point is not encircled. However, the effect of an increase of gain K is that each point on the plot is moved radially outward proportionally, since the phase angle of $G_c GH$ is unaffected, but the magnitude is proportional to K. Beyond a certain gain, the -1 point will be encircled twice in the clockwise direction as ω increases from $-\infty$ to $+\infty$, and the system will be unstable. Note that for (a1) and (a3) in Fig. 7.8 the -1 point will be approached for high gains, but cannot be encircled, so these systems, if open-loop stable, cannot lose stability. Root locus sketches readily confirm these results.

$$P_i \equiv j\omega T_i + 1$$

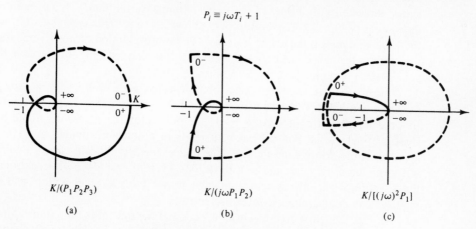

$$K/(P_1 P_2 P_3) \qquad\qquad K/(j\omega P_1 P_2) \qquad\qquad K/[(j\omega)^2 P_1]$$

(a)   (b)   (c)

Figure 7.10   Nyquist diagram examples.

In Fig. 7.10(b), after completing the mirror image of the polar plot, the ends 0^- and 0^+ are connected by a clockwise 180° rotation at large radius due to the pole at the origin. The system is stable if it is open-loop stable, but would become unstable if radially expanded by an increase of gain such that -1 is encircled.

In Fig. 7.10(c) there must be a 360° clockwise rotation at large radius between 0^- and 0^+, due to the double pole at the origin. Following the curve from $-\infty$ to $+\infty$ shows that the -1 point is encircled twice in clockwise direction, so the system is unstable for any value of gain K.

The Nyquist diagram in Fig. 7.11(a), which incorporates the polar plot (c2) in Fig. 7.8, is useful to show that care is necessary when deciding how many encirclements are present. It is convenient to visualize an elastic with one end fixed at -1 and the other moved along the closed contour from $-\infty$ to $+\infty$. The appearance to the contrary, this will show that there is no net encirclement of -1. Therefore,

$$P_i \equiv j\omega T_i + 1$$

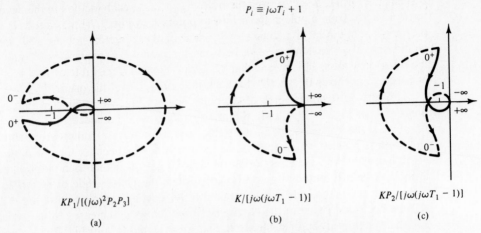

$$KP_1/[(j\omega)^2 P_2 P_3] \qquad\qquad K/[j\omega(j\omega T_1 - 1)] \qquad\qquad KP_2/[j\omega(j\omega T_1 - 1)]$$

(a)   (b)   (c)

Figure 7.11 Additional examples of Nyquist diagrams.

assuming that it is open-loop stable, the system shown is stable. This, incidentally, shows the stabilizing effect of phase-lead compensation $(j\omega T_1 + 1)/(j\omega T_3 + 1)$ on the system in Fig. 7.10(c). Figure 6.18 showed a root locus interpretation, while on the Nyquist diagram the effect is seen to be an introduction of positive phase angles that deform the curve in a counterclockwise direction, causing it to pass on the other side of the -1 point.

Figure 7.11(b) shows an open-loop unstable case. For $\omega \to 0^+$ the function approaches $-K/(j\omega)$. With the minus sign equivalent to $\pm 180°$ and with the factor $j\omega$ in the denominator contributing $-90°$, the polar plot "starts" far out along the positive imaginary axis. For $\omega \to +\infty$, it tends to a constant divided by $(j\omega)^2$ and so approaches the origin from the left. The pole at the origin again causes a $180°$ clockwise rotation at large radius from 0^- to 0^+. There is one clockwise encirclement of the -1 point. Hence the system is unstable for all gains K, since with $P = 1$ stability requires one counterclockwise encirclement.

In Fig. 7.11(c) a zero has been added, and the system is now stable, since there is one counterclockwise encirclement of the -1 point. In this example, instability would result from a reduction of gain below the value at which the locus will still cross the negative real axis to the left of -1.

7.5 BODE PLOTS

Bode diagrams are an alternative to polar plots and are very widely used. These plots are much easier to make than polar plots and can readily be interpreted in terms of different aspects of system performance. Consider the general frequency response function

$$G(j\omega) = \frac{K}{(j\omega)^n} \frac{S_1 S_2 \cdots Q_1 Q_2 \cdots}{S_{k+1} S_{k+2} \cdots Q_{l+1} Q_{l+2} \cdots} \qquad (7.13)$$

where

$$S_i \equiv j\omega T_i + 1 \qquad Q_i \equiv \left(\frac{j\omega}{\omega_{ni}}\right)^2 + 2\zeta_i \frac{j\omega}{\omega_{ni}} + 1$$

K is the gain defined in Section 4.3, since the constant terms are unity. T_i, ω_{ni}, and ζ_i are all positive, so no poles and, for now, no zeros lie in the right-half plane. $G(j\omega)$ is the product of only four types of *elementary factors*:

1. Gain K
2. Integrators $1/(j\omega)^n$ or differentiators $(j\omega)^n$
3. Simple lag $1/S_i$ or simple lead S_i
4. Quadratic lag $1/Q_i$ or quadratic lead Q_i

For a product $G(j\omega) = M_1 e^{j\phi_1} M_2 e^{j\phi_2} \cdots = M e^{j\phi}$, $M = M_1 M_2 \cdots$ and $\phi = \phi_1 + \phi_2 + \cdots$. The phase angle ϕ is expressed as a sum. The magnitude M will also be expressed as a sum by using decibels (dB) as units:

$$M \text{ in dB} = M_{\text{dB}} = 20 \log_{10} M$$
$$20 \log M = 20 \log M_1 + 20 \log M_2 + \cdots \tag{7.14}$$

A decibel conversion table is given in Table 7.5.1. For a value outside the table, say $M^1 (10)^n$, where M^1 is a value in the table, the conversion is $20 \log M^1 (10)^n = 20 \log M^1 + 20 \log (10)^n = 20 \log M^1 + 20n$.

TABLE 7.5.1 DECIBEL CONVERSION: $m = 20 \log_{10} M$

M	0	1	2	3	4	5	6	7	8	9
0.0	m =	−40.00	−33.98	−30.46	−27.96	−26.02	−24.44	−23.10	−21.94	−20.92
0.1	−20.00	−19.17	−18.42	−17.72	−17.08	−16.48	−15.92	−15.39	−14.89	−14.42
0.2	−13.98	−13.56	−13.15	−12.77	−12.40	−12.04	−11.70	−11.37	−11.06	−10.75
0.3	−10.46	−10.17	−9.90	−9.63	−9.37	−9.12	−8.87	−8.64	−8.40	−8.18
0.4	−7.96	−7.74	−7.54	−7.33	−7.13	−6.94	−6.74	−6.56	−6.38	−6.20
0.5	−6.02	−5.85	−5.68	−5.51	−5.35	−5.19	−5.04	−4.88	−4.73	−4.58
0.6	−4.44	−4.29	−4.15	−4.01	−3.88	−3.74	−3.61	−3.48	−3.35	−3.22
0.7	−3.10	−2.97	−2.85	−2.73	−2.62	−2.50	−2.38	−2.27	−2.16	−2.05
0.8	−1.94	−1.83	−1.72	−1.62	−1.51	−1.41	−1.31	−1.21	−1.11	−1.01
0.9	−0.92	−0.82	−0.72	−0.63	−0.54	−0.45	−0.35	−0.26	−0.18	−0.09
1.0	0.00	0.09	0.17	0.26	0.34	0.42	0.51	0.59	0.67	0.75
1.1	0.83	0.91	0.98	1.06	1.14	1.21	1.29	1.36	1.44	1.51
1.2	1.58	1.66	1.73	1.80	1.87	1.94	2.01	2.08	2.14	2.21
1.3	2.28	2.35	2.41	2.48	2.54	2.61	2.67	2.73	2.80	2.86
1.4	2.92	2.98	3.05	3.11	3.17	3.23	3.29	3.35	3.41	3.46
1.5	3.52	3.58	3.64	3.69	3.75	3.81	3.86	3.92	3.97	4.03
1.6	4.08	4.14	4.19	4.24	4.30	4.35	4.40	4.45	4.51	4.56
1.7	4.61	4.66	4.71	4.76	4.81	4.86	4.91	4.96	5.01	5.06
1.8	5.11	5.15	5.20	5.25	5.30	5.34	5.39	5.44	5.48	5.53
1.9	5.58	5.62	5.67	5.71	5.76	5.80	5.85	5.89	5.93	5.98
2.	6.02	6.44	6.85	7.23	7.60	7.96	8.30	8.63	8.94	9.25
3.	9.54	9.83	10.10	10.37	10.63	10.88	11.13	11.36	11.60	11.82
4.	12.04	12.26	12.46	12.67	12.87	13.06	13.26	13.44	13.62	13.80
5.	13.98	14.15	14.32	14.49	14.65	14.81	14.96	15.12	15.27	15.42
6.	15.56	15.71	15.85	15.99	16.12	16.26	16.39	16.52	16.65	16.78
7.	16.90	17.03	17.15	17.27	17.38	17.50	17.62	17.73	17.84	17.95
8.	18.06	18.17	18.28	18.38	18.49	18.59	18.69	18.79	18.89	18.99
9.	19.08	19.18	19.28	19.37	19.46	19.55	19.65	19.74	19.82	19.91

In Bode plots, the magnitude M in decibels and the phase angle ϕ in degrees are plotted against ω on semilog paper. The development has shown the following:

Bode magnitude and phase-angle plots of $G(j\omega)$ are obtained by summing those of its elementary factors.

Bode plots of the four factors will be obtained first.

1. *Gain K > 0* [Fig. 7.12(a)]:
$$M_{dB} = 20 \log K, \qquad \phi = 0, \qquad \text{both independent of } \omega$$

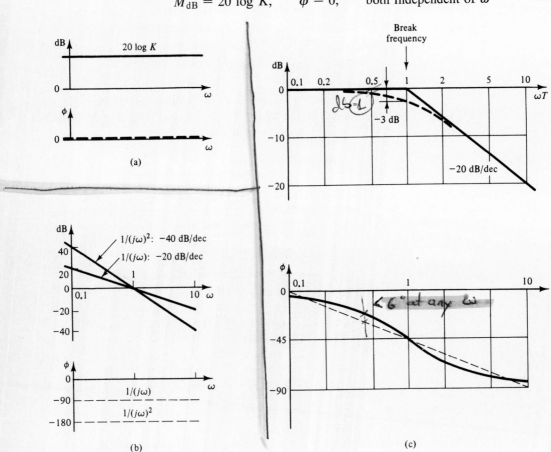

Figure 7.12 Bode plots of (a) gain, (b) integrators, and (c) simple lag.

2. *Integrators $1/(j\omega)^n$* [Fig. 7.12(b)]:
$$M_{dB} = 20 \log |j\omega|^{-n} = 20 \log \omega^{-n} = -20n \log \omega \qquad (7.15)$$
At $\omega = 1$, $M_{dB} = 0$, and at $\omega = 10$, 1 decade (dec) away from $\omega = 1$, $M_{dB} = -20n$. Hence on a log scale the magnitude plot is a straight line crossing the 0-dB axis at $\omega = 1$ at a slope of $-20n$ dB/dec. Phase angle $\phi = -n90°$, since each factor $j\omega$ in the denominator contributes $-90°$, and is independent of frequency.

For *differentiators* $(j\omega)^n$, the plots are the mirror images relative to the 0-dB and 0° axes. This is also true for the *leads* corresponding to the simple and quadratic lag below.

3. *Simple lag* $1/(j\omega T + 1)$ [Fig. 7.12(c) or Table 7.5.2]: By inspection or using Fig. 7.9(a),

$$M_{\text{dB}} = 20 \log \frac{1}{\sqrt{1 + (\omega T)^2}} \qquad \phi = -\tan^{-1} \omega T \qquad (7.16)$$

$$\omega T \ll 1: \quad M_{\text{dB}} \to 20 \log 1 = 0 \text{ dB} \qquad \phi \to 0°$$

TABLE 7.5.2   SIMPLE LAG $1/(j\omega T + 1)$

ωT	0.1	0.2	0.5	1	2	5	10
Deviation from asymptotes (dB)	−0.04	−0.2	−1.	−3.	−1.	−0.2	−0.04
Phase angle ϕ (deg)	−5.7	−11.3	−26.6	−45	−63.4	−78.7	−84.3

So the 0-dB axis is the *low-frequency asymptote*.

$$\omega T \gg 1: \quad M_{\text{dB}} \to 20 \log (\omega T)^{-1} = -20 \log \omega T \qquad \phi \to -90°$$

Analogous to the integrator, this *high-frequency asymptote* crosses the 0-dB axis at $\omega T = 1$, at a slope of −20 dB/dec. This is the *asymptotic approximation*. The asymptotes meet at the break frequency or corner frequency given by $\omega T = 1$ on the normalized plot. On a log ω scale, it is at $\omega = 1/T$.

Closer to $\omega T = 1$, the actual values can be calculated from (7.16). The deviations from the asymptotes and the phase angles ϕ at several frequencies are shown in Table 7.5.2. At $\omega T = 1$, the deviation is −3 dB and the phase −45°. At $\omega T = 0.5$ and $\omega T = 2$ the deviation is −1 dB. Note that at 0.1 of its break frequency the simple lag contributes −5.7° phase angle, and at 10 times the break frequency, −84.3°, 5.7° away from −90°. Figure 7.12(c) shows by a dashed line between $(\omega T = 0.1, \phi = 0)$ and $(\omega T = 10, \phi = -90°)$ an asymptotic approximation to the phase-angle curve. Its error is below 6° at all frequencies in this range.

4. *Quadratic lag* (Fig. 7.13):

$$\frac{1}{(j\omega/\omega_n)^2 + 2\zeta(j\omega/\omega_n) + 1}$$

From Fig. 7.9(b),

$$M_{\text{dB}} = 20 \log \left[\left(1 - \frac{\omega^2}{\omega_n^2}\right)^2 + \left(\frac{2\zeta\omega}{\omega_n}\right)^2 \right]^{-1/2}$$

$$(7.17)$$

$$\phi = -\tan^{-1} \frac{2\zeta\omega/\omega_n}{1 - \omega^2/\omega_n^2}$$

$\omega/\omega_n \ll 1$: Then the quadratic lag approaches 1, so $M_{\text{dB}} \to 20 \log 1 = 0$ dB and $\phi \to 0°$. So the 0-dB axis is again the low-frequency asymptote.

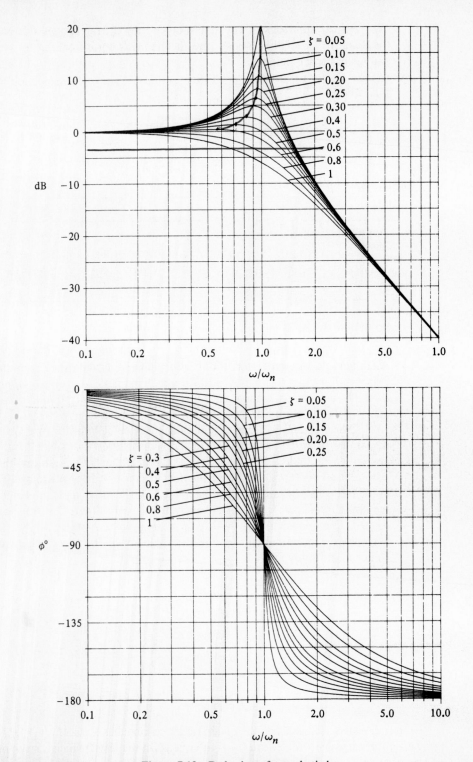

Figure 7.13 Bode plots of a quadratic lag.

$\omega/\omega_n \gg 1$: Now the quadratic lag approaches $1/(j\omega/\omega_n)^2$ because the other terms in the denominator become relatively negligible. Since each factor $j\omega$ in the denominator contributes $-90°$, the magnitude and phase become

$$M_{dB} \to 20 \log\left(\frac{\omega}{\omega_n}\right)^{-2} = -40 \log\left(\frac{\omega}{\omega_n}\right) \qquad \phi \to -180°$$

Analogous to integrators, this high-frequency asymptote crosses the 0-dB axis at $\omega/\omega_n = 1$, at a slope of -40 dB/dec. Closer to $\omega/\omega_n = 1$, (7.17) gives the actual curves. At $\omega/\omega_n = 1$, $M_{dB} = 20 \log 0.5/\zeta$; $\phi = -90°$. Smaller damping ratios ζ cause more severe peaking of M_{dB} and more abrupt change of ϕ near $\omega/\omega_n = 1$. The Bode plots for a quadratic lead are again mirror images relative to the 0-dB and 0° axes.

The asymptotes meet at the *break frequency* or *corner frequency* given by $\omega/\omega_n = 1$ on the normalized plot. On a log ω scale, it is at $\omega = \omega_n$. Note that for a quadratic lag the actual magnitude plot can differ greatly from the asymptotic approximation formed by its low- and high-frequency asymptotes.

Before showing examples of the construction of Bode plots by use of those of the elementary factors, the translation of the concept of relative stability to frequency response plots is considered.

7.6 RELATIVE STABILITY: GAIN MARGIN AND PHASE MARGIN

Most practical systems are open-loop stable, so the simplified Nyquist criterion applies. The polar plot of the loop gain function shown in Fig. 7.14 then indicates a stable system. If the curve passes through -1, the system is on the verge of instability. For adequate relative stability it is reasonable that the curve should not come too close to the -1 point. Gain margin and phase margin are two common design criteria, which specify the distance of a selected point of the polar plot to -1. Both are defined in Fig. 7.14:

1. *Gain margin* = $1/0C$. $\qquad\qquad\qquad\qquad\qquad\qquad\qquad\qquad$ (7.18a)
2. *Phase margin* $\phi_m = 180°$ plus the phase angle of $G_c GH$ at the *crossover frequency* ω_c at which $|G_c GH| = 1$. It is also the negative phase shift (that is, clockwise rotation) of $G_c GH$ that will make the curve pass through -1.(7.18b)

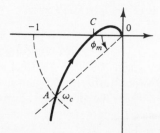

Figure 7.14 Phase margin and gain margin.

Each specifies the distance to -1 of only one point on the curve, so misleading indications are possible. These criteria will be used mostly in the context of Bode plots, but they can also be calculated from polar plots, as illustrated in the following examples. Here the gain margin is calculated using the condition that the imaginary part of $G_c GH$ is zero along the negative real axis.

Example 7.6.1

Consider Fig. 7.11(a):

$$G_c GH = \frac{K(1 + j\omega T_1)}{(j\omega)^2(1 + j\omega T_2)(1 + j\omega T_3)} \tag{7.19}$$

The imaginary part can be identified by multiplying the numerator and denominator by $(1 - j\omega T_2)(1 - j\omega T_3)$. Then

$$G_c GH = K \frac{1 - \omega^2(T_2 T_3 - T_1 T_2 - T_1 T_3) + j\omega(T_1 - T_2 - T_3 - \omega^2 T_1 T_2 T_3)}{-\omega^2(1 + \omega^2 T_2^2)(1 + \omega^2 T_3^2)}$$

The imaginary part is zero if $\omega^2 = (T_1 - T_2 - T_3)/(T_1 T_2 T_3)$, and substituting this yields $|G_c GH| = 0C$. Note that only one intersection can occur here. If

$$G_c GH = \frac{K(j\omega T_1 + 1)(j\omega T_4 + 1)}{(j\omega)^2(j\omega T_2 + 1)(j\omega T_3 + 1)(j\omega T_5 + 1)} \tag{7.20}$$

then two intersections could occur, as in Fig. 7.15, which represents a stable system since the -1 point lies left of the curve.

Figure 7.15   Polar plot for (7.20).

Example 7.6.2

Consider Fig. 7.10(b):

$$G_c GH = \frac{K}{j\omega(j\omega T_1 + 1)(j\omega T_2 + 1)} \tag{7.21}$$

This simpler example can be treated as the preceding one, but it may also be recognized that since the factor $j\omega$ provides $-90°$ of the $-180°$ required along the negative real axis, the product

$$(j\omega T_1 + 1)(j\omega T_2 + 1) = 1 - \omega^2 T_1 T_2 + j\omega(T_1 + T_2)$$

must yield the remaining $-90°$. This requires that $\omega^2 = 1/(T_1 T_2)$, for which

$$G_c GH = \frac{K}{(j\omega)^2(T_1 + T_2)} = \frac{-K}{\omega^2(T_1 + T_2)} = \frac{-KT_1 T_2}{T_1 + T_2}$$

Hence the gain margin is $(T_1 + T_2)/(KT_1 T_2)$.

Now consider the translation of the concepts of phase margin and gain margin to the Bode plot. From Fig. 7.14, the phase margin ϕ_m is the sum of $180°$ and the phase angle at the frequency where $|G_c GH| = 1$ (that is, 0 dB), so where the magnitude plot crosses the 0-dB axis. Similarly, the gain margin equals 1 divided by the magnitude at the frequency where the Bode plot shows a phase angle of $-180°$. In decibels, the division becomes a subtraction of the magnitude from 0 dB. Hence, as shown by the partial plots in Fig. 7.16:

1. The *phase margin* ϕ_m is the distance of the phase-angle curve above $-180°$ at the *crossover frequency* ω_c, where the magnitude plot crosses the 0-dB axis.

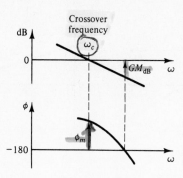

Figure 7.16 Phase margin and gain margin.

2. The *gain margin, GM*$_{dB}$ in decibels, is the distance of the magnitude plot below the 0-dB axis at the frequency where the phase-angle curve shows an angle of $-180°$.

The following should be noted:

> Bode plot stability analysis is limited to open-loop stable systems. This is because these plots do not show the encirclements specified by the Nyquist criterion.

Root loci or Nyquist diagrams should be used for open-loop unstable systems.

7.7 EXAMPLES OF BODE PLOTS

As discussed in Section 7.5, Bode plots are obtained by summing those of the elementary factors present.

Example 7.7.1   Asymptotic Magnitude Plots

In each case in Fig. 7.17 the Bode plots of the elementary factors, obtained in Section 7.5, are shown by dashed lines. Summation then gives the desired plots. On the normalized simple lag plot the break frequency occurs at $\omega T = 1$, but on an ω scale this break occurs at $\omega = 1/T$.

In Fig. 7.17(a), up to $1/T_1$ both simple lags coincide with the 0-dB axis, so the sum coincides with the gain. Then, up to $1/T_2$ the corresponding simple lag is still 0 dB, but the other one has a slope of -20 dB/dec. Therefore, up to $1/T_2$ the sum will also have a slope of -20 dB/dec. Above $1/T_2$ this simple lag also contributes a change of slope of -20 dB/dec, so here the slope of the sum will change from -20 to -40 dB/dec. Note that all this amounts to a direct algebraic addition of the ordinates. The other plots can be interpreted similarly. Certain other features of these plots are discussed later.

In these *asymptotic approximations* the *low-frequency asymptote* is the one in the lowest range, below $1/T_1$ in Figs. 7.17(a) and (b), and the *high-frequency asymptote* is the one above $1/T_2$ in these figures.

In practice, the elementary factors are not plotted, but the final plot is made immediately:

> Asymptotic Bode plots are constructed by adding those of the elementary factors sequentially rather than simultaneously. To do this, the low-frequency asymptote is plotted first, and then the other elementary factors are added in the

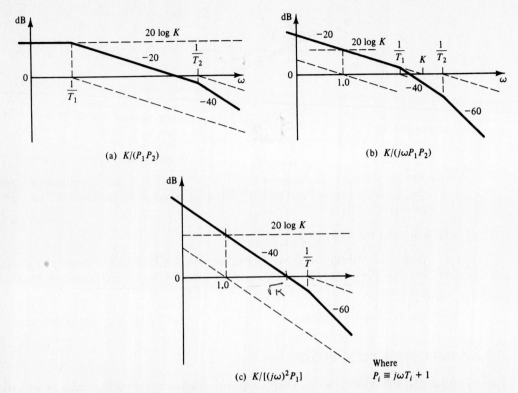

(a) $K/(P_1P_2)$ (b) $K/(j\omega P_1 P_2)$

(c) $K/[(j\omega)^2 P_1]$ Where
$P_i \equiv j\omega T_i + 1$

Figure 7.17   Asymptotic magnitude plots.

order of increasing break frequencies. This works because each factor has a 0-dB asymptote below its break frequency, and so does not affect the asymptotic plot up to this frequency.

At its break frequency a simple lag contributes a change of slope of -20 dB/dec and a quadratic lag contributes a change of -40 dB/dec.

The low-frequency asymptote is found from (7.13). From this equation, as $\omega \to 0$,

$$\text{low-frequency asymptote} = \frac{K}{(j\omega)^n} \qquad (7.22)$$

This consists of a gain and integrators and crosses the 0-dB axis at ω given by

$$\left| \frac{K}{(j\omega)^n} \right| = \frac{K}{\omega^n} = 1 \quad n = 1 : \omega = K \qquad (7.23)$$

$$n = 2 : \omega = \sqrt{K} \qquad (7.24)$$

For $n = 0$ (Fig. 7.17a) the low-frequency asymptote has zero slope, for $n = 1$ (Fig. 7.17b) the slope is -20 dB/dec, and for $n = 2$ (Fig. 7.17c), -40 dB/dec.

Note that, instead of from the intersection with the 0-dB axis, the low-frequency asymptote can also be plotted by calculating the magnitude of $K/(j\omega)^n$ at any convenient value of ω. For, say, $K = 10$ and $n = 1$ the dB magnitude is

20 log $(10/\omega)$. At, say, $\omega = 0.1$, this is 40 dB, so the -20 dB/dec asymptote must pass $\omega = 0.1$ at a level of 40 dB. This will be used if the intersection of the asymptote with the 0-dB axis does not fall on the graph.

With the low-frequency asymptote known, the other factors are considered in the order of increasing break frequencies. For example, in Fig. 7.17(b), by (7.23) the low-frequency asymptote (in this case its extension) intersects the 0-dB axis at $\omega = K$. Moving to the right along this asymptote, the first break frequency encountered, at $1/T_1$, is due to a simple lag. It contributes a change of slope of -20 dB/dec, so at $1/T_1$ the slope changes from -20 to -40 dB/dec. The next break, at $1/T_2$, is again due to a simple lag, so here the slope changes from -40 to -60 dB/dec.

If $G(j\omega)$ does not have right-half-plane poles or zeros, there is a unique relation between the magnitude plot and the phase-angle curve. Recalling that a factor $j\omega$ in the denominator means $-90°$, it is evident from (7.22) that a zero slope of the low-frequency asymptote means $0°$ phase angle at low frequencies, -20 dB/dec means $-90°$, and -40 dB/dec means $-180°$. Similarly, if the slope of the high-frequency asymptote is $-20\ m$ dB/dec, the phase angle at high frequencies is $-m \cdot 90°$.

Appendix B gives a design-oriented interactive program for asymptotic and actual magnitude plots and for phase-angle curves. But manual plotting is much less objectionable than for polar plots. The asymptotic magnitude plot is drawn first. Very often the actual magnitude curve and the phase-angle curve are not required over the entire frequency range, and it is sufficient to obtain these values only where needed to solve the particular problem, say to determine phase margin or gain margin. An approach that is often convenient is, therefore, to calculate phase angles and deviations from asymptotic magnitude plots at selected frequencies. This is done by adding the contributions of the elementary factors, as in the following example.

Example 7.7.2

Consider a system with loop gain function

$$G_c GH = \frac{0.325}{s(s + 1)(0.5s + 1)} \tag{7.25}$$

Figure 6.8 shows a root locus design for this example. Determine:

- The asymptotic magnitude plot
- The crossover frequency
- The phase margin
- The frequency where the phase angle is $-180°$
- The gain margin

Verify that the constant terms of the polynomials in (7.25) are unity, so the gain is 0.325. From (7.23), the low-frequency asymptote crosses the 0-dB axis at $\omega_c = 0.325$ at a slope of -20 dB/dec, as shown in Fig. 7.18. Moving to the right along this asymptote, the first break encountered is at $\omega = 1$ and is due to a simple lag. It contributes here a change of slope of -20 dB/dec, so here the slope changes from -20 to -40 dB/dec. Moving along this new asymptote, the next break occurs at $\omega = 1/0.5 = 2$ and is again due to a simple lag. So here the slope changes from -40 to -60 dB/dec. This completes the asymptotic magnitude plot.

To obtain an accurate value for the phase margin, the actual crossover frequency is needed. Therefore, it is always important to determine the deviation of the actual

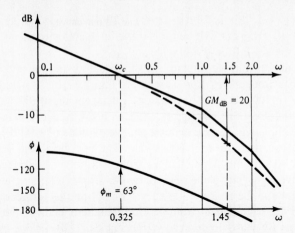

Figure 7.18   Example 7.7.2.

magnitude curve from the asymptotes in this range of frequencies. This can be done by calculating the deviations at one or two frequencies and sketching this part of the true magnitude plot. The deviation at 0.325 consists of the sum of the contributions at this frequency from all simple and quadratic lags that may be present. To obtain the contribution at 0.325 due to the simple lag with break frequency 1, Fig. 7.12(c) or Table 7.5.2 is entered at the normalized frequency $0.325/1 = 0.325$. It is seen that this lag contributes at 0.325 a deviation of less than 0.5 dB between asymptote and true curve. The contribution at 0.325 due to the simple lag with break frequency 2 is even less, as may be seen by entering Fig. 7.12(c) or Table 7.5.2 at the normalized frequency $0.325/2 = 0.1625$. So $\omega_c = 0.325$ can be taken to be the crossover frequency for phase-margin calculation. Actually, this result was clear by inspection, because the closest break frequency is a factor of 3 away from the frequency of interest.

The phase margin can now be calculated, again by adding the contributions to the phase angle at the crossover frequency from all elementary factors. The integrator contributes $-90°$ at all frequencies. The simple lag break at 1 contributes at frequency 0.325 an angle of $-\tan^{-1}(0.325/1) = -18°$, as may be seen from (7.16). Similarly, the simple lag break at 2 contributes $-\tan^{-1}(0.325/2) = -9°$. Hence the phase angle at 0.325 and the phase margin are

$$\phi(0.325) = -90 - 18 - 9 = -117° \qquad \phi_m = 180 - 117 = +63° \quad (7.26)$$

2. Determination of the gain margin requires the frequency at which the phase is $-180°$. From the slopes in Fig. 7.18, the angle is $-90°$ at low and $-270°$ at high frequencies. Inspection suggests that $-180°$ will occur between 1 and 2 rad/sec, so the phase angle is calculated for some trial frequencies in this range:

$$\phi(1) = -90 - 45 - 26.6 = -161.7$$
$$\phi(1.5) = -90 - 56.3 - 36.9 = -183.2 \qquad (7.27)$$

Interpolating by drawing a curve through the points of (7.26) and (7.27) yields

$$\phi(1.45) \approx -180 \qquad \text{gain margin frequency} = 1.45 \qquad (7.28)$$

Figure 7.18 suggests considerable deviation of the actual magnitude curve from the asymptotes at this frequency. To obtain an accurate value of the gain margin, this deviation can be calculated. Or a local section of the actual curve can be constructed. At $\omega = 1$, the simple lag break at this frequency contributes a deviation of -3 dB, and the break at 2 rad/sec contributes at $\omega = 1$ a deviation of -1 dB [from Fig. 7.12(c) or Table 7.5.2 at the normalized frequency of 1/2]. Hence the total deviation at $\omega = 1$ is

-4 dB. This is also the total deviation at $\omega = 2$. The actual curve in Fig. 7.18 is sketched through these two points and yields at $\omega = 1.45$

$$GM_{dB} = 20 \text{ dB} \qquad GM = 10 \tag{7.29}$$

The next example is concerned with design, because the gain K of a proportional controller is chosen to obtain a specified phase margin. Changing K will not affect the phase-angle curve and will just raise or lower the magnitude plot. On polar plots, as discussed earlier, it moves points radially.

Example 7.7.3   Proportional Control: Choice of Gain for $G_c = K$

For a unity feedback system ($H = 1$), a partial Bode plot of the loop gain function

$$G_c G = \frac{K}{s(s + 1)(0.5s + 1)} \tag{7.30}$$

is shown in Fig. 7.19 for $K = 0.325$. This is part of the plot in Fig. 7.18. For $K = 0.325$, $\phi_m = 63°$, and the steady-state error for a unit ramp input of this type one system is $1/K$. Suppose that to reduce this steady-state error the lower phase margin $\phi_m = 53°$ will be accepted. This means that the magnitude plot should cross the 0-dB axis at a frequency where the phase angle is $-180 + 53 = -127°$. This desired crossover frequency can be found by calculating phase angles at several frequencies and interpolating. As indicated in Fig. 7.19, $\phi_m = 53°$ will result if the actual curve crosses 0 dB at 0.43 rad/sec. To realize this, the plot must be raised by 3 dB; that is, the gain 0.325 must be raised by a factor α given by $20 \log \alpha = 3$. Hence $\alpha = 1.41$, and the new gain is $K = 1.41 \times 0.325 = 0.45$. This gain value is also evident by construction, since according to (7.23) it is the frequency at which the low-frequency asymptote, or its extension, intersects the 0-dB axis.

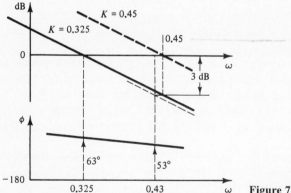

Figure 7.19   Choice of K.

7.8 CLOSED-LOOP FREQUENCY RESPONSE AND *M* CIRCLES

By means of the Nyquist stability criterion, it has been possible to determine the stability of the closed-loop system from a frequency response plot of its loop gain function. It is now shown how, for systems with unity feedback, the closed-loop frequency response may also be found from these plots. With input R, output C, and loop gain function $G(j\omega)$ in the forward loop, the closed-loop frequency response is given by

$$\frac{C(j\omega)}{R(j\omega)} = \frac{G(j\omega)}{1 + G(j\omega)} \tag{7.31}$$

Consider now the polar plot of $G(j\omega)$ in Fig. 7.20(a). By definition, $G(j\omega)$ is the vector from 0 to points A, with length $0A$ and phase angle ϕ_a. But, also, $1 + G(j\omega)$ is the vector from B to A, because it is the vector sum of $G(j\omega)$ and the vector 1, which is the vector from B to 0. Therefore, the magnitude $M(\omega)$ and phase angle $\phi(\omega)$ of the closed-loop frequency response function C/R are given by

$$M(\omega) = \left|\frac{C}{R}\right| = \frac{0A}{BA}$$

$$\phi(\omega) = \text{phase angle } \frac{C}{R} = \phi_a - \phi_b$$

(7.32)

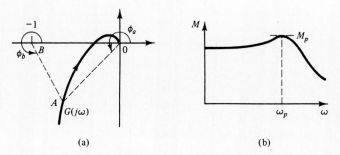

(a)   (b)

Figure 7.20   Closed-loop frequency response.

By measuring the vector lengths as A moves along the curve, the magnitude M can be plotted versus ω, as shown in Fig. 7.20(b) for a typical case. Low relative stability means that the curve comes close to -1, so BA will be small over a range of frequencies. Thus a low phase margin will be reflected in severe resonance peaking in $M(\omega)$.

For any given point A, $0A$, BA, ϕ_a, and ϕ_b have given values, and therefore M and ϕ in (7.32) also. This permits loci to be constructed for constant values of M and ϕ. For example, along a vertical midway between 0 and -1, $0A = BA$, so this is the locus for $M = 1$. The real axis to the left of -1 and to the right of 0 is the locus for $\phi = 0$, since $\phi_a = \phi_b$, and this axis between -1 and 0 is the locus for $\phi = \pm 180°$.

It can be shown that, with the exceptions noted above, the loci for constant M and for constant ϕ are circles, called *M circles* and *N circles*, respectively. Some M circles are shown in Fig. 7.21. The centers are on the real axis, and their locations x and the circle radii r are given by

$$x = \frac{M^2}{1 - M^2} \qquad r = \left|\frac{M}{1 - M^2}\right|$$

(7.33)

In Fig. 7.21, a polar plot of a function $G(j\omega)$ has been superimposed on a graph on which the M circles are printed. Then M for the values of frequency ω along $G(j\omega)$ can be read off from the intersections with these loci. Thus M can be plotted versus ω as in Fig. 7.20(b).

In particular, in Fig. 7.21 the maximum value of M occurs at the point of tangency of $G(j\omega)$ with the locus for $M = 2$. Thus in this case the resonant peak $M_p = 2$ and occurs at the resonant frequency ω_p given by $G(j\omega)$ at the point of tangency.

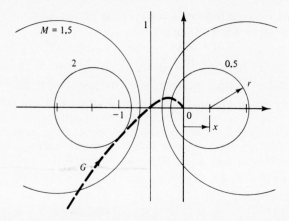

Figure 7.21 *M* circles.

These loci help to clarify the relation between open-loop and closed-loop response, because of the idea represented in Fig. 7.20 on which they are based. However, they are not used nearly as often for system analysis and design as is their translation to another type of plot, the Nichols chart, discussed in Chapter 8.

7.9 CONCLUSION

Frequency response methods and frequency response plots have been discussed, as well as frequency domain criteria for absolute and relative stability. For most systems, Bode plot techniques are most convenient for analysis and design and allow the clearest visualization of the effects of compensation on performance. Their application to design is discussed in Chapter 8, where the Nichols chart is also introduced and on occasion will serve to confirm predictions made on the basis of Bode plots.

Construction of accurate polar plots is laborious, and computer aids are essential for their application in detailed analysis and design. Design-oriented interactive programs with graphics are given in Appendix B for polar plots and Bode plots.

PROBLEMS

7.1. Determine the steady-forced response of a system with transfer function

$$T(s) = \frac{1}{(s + 1)(0.1s + 1)}$$

to the following sinusoidal inputs.

(a) $r(t) = 2 \sin 0.5t$ **(b)** $r(t) = 2 \sin 5t$

Note the effect of frequency on output amplitude and phase shift.

7.2. In Problem 7.1, determine the frequency at which the output lags the input by 90°, and find the ratio of output to input amplitude at this frequency. [*Hint:* $T(s) = 1/(0.1s^2 + 1.1s + 1)$, and the phase shift is $-90°$ when $T(j\omega)$ plots along the negative imaginary axis.]

7.3. For $T(s) = 1/[s(s + 1)(0.1s + 1)]$, find the intersection of $T(j\omega)$ with the real axis of the complex plane.

7.4. To relate the frequency response function and the s-plane, determine the magnitude and phase angle of the frequency response function

$$G(j\omega) = \frac{10(j\omega + 1)}{j\omega + 5}$$

at $\omega = 1$, $\omega = 5$, and $\omega = 25$ from measurement on the pole–zero pattern of $G(s)$.

7.5. Figure P7.5 shows the poles of a quadratic lag transfer function with root locus gain 17 ($1 + 16 = \omega_n^2 = $). To show that low damping (that is, poles at an angle relatively close to the imaginary axis) is equivalent to high resonance peaking in a plot of the magnitude of the frequency response function versus frequency, obtain an adequate number of points for such a plot by measurements on the pole–zero pattern. Note that the high peak is due to a short vector in the denominator.

Figure P7.5

7.6. In Problem 7.5, also use measurement on the pole–zero pattern to determine and plot the phase angle of the frequency response function versus frequency at $\omega = 1$, 3, 4.123 ($= \sqrt{17}$), and 5.

7.7. Sketch polar plots of (a) $2/(s + 1)$ and (b) $2/(0.1s + 1)$ based on calculation of at least three points. How do the plots differ? Prove that the shape is circular.

7.8. Construct polar plots for

(a) $G(s) = \dfrac{10}{(s + 1)(0.1s + 1)}$ (b) $G(s) = \dfrac{10}{(s + 1)(0.5s + 1)}$

from calculations at frequencies of 0.5, 1, 2, and 6. Comment on the nature of the plot if the second time constant is quite small relative to the dominating one of 1 sec.

7.9. Construct and compare polar plots of $G(s) = \omega_n^2/(s^2 + 2\zeta\omega_n s + \omega_n^2)$ for $\omega_n = 1$ and (a) $\zeta = 0.75$, (b) $\zeta = 0.25$ from calculated points for $\omega = 0.5$, 1, and 2.

7.10. Sketch polar plots of

(a) $G(s) = \dfrac{1}{s(s + 1)}$

(b) $G(s) = \dfrac{1}{s(s + 1)(0.5s + 1)}$

(c) $G(s) = \dfrac{1}{s^2(s + 1)}$

based on the behavior at low and high frequencies and points calculated for $\omega = 0.5$, 1, and 2.

7.11. Extend the polar plots of Problem 7.10 into Nyquist diagrams and use these to determine stability and the number of unstable poles of systems of which these are the loop gain functions.

7.12. Sketch Nyquist diagrams for the loop gain functions

(a) $\dfrac{K}{s-1}$ (b) $\dfrac{1}{s(s-1)}$

and in part (a) give the range of K for which the system is stable. Verify the results by root locus sketches.

7.13. Figure P7.13 shows the polar plot of an open-loop stable system for gain $K = 1$.
(a) Determine the limits on K for stability.
(b) What are the system phase and gain margins?
(c) Sketch the Nyquist diagram corresponding to Fig. P7.13 and verify the conclusion on stability by considering encirclements.

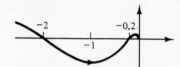

Figure P7.13

7.14. Sketch the polar plot for a system with loop gain function

$$G(s) = \frac{10}{s(1 + s/4)(1 + s/16)}$$

and determine the gain and phase margins.

7.15. Gain margin and phase margin should be measures of the distance of the polar plot to -1. Accordingly, determine in each of Fig. P7.15(a) and (b) which measure would and which would not provide a misleading idea of the degree of relative stability.

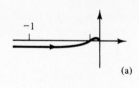

(a)

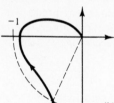

(b)   **Figure P7.15**

7.16. Sketch the polar plot for a loop gain function

$$G(s) = \frac{7}{s(0.1s + 1)(0.02s + 1)}$$

based on calculated points at $\omega = 5$, 10, and 15 and the intersection with the real axis, and determine the gain margin and the approximate phase margin.

7.17. In Fig. P7.17, let

$$G(s) = \frac{5}{s(s + 1)(0.5s + 1)}$$

Based on calculations at $\omega = 0.5$, 1, and 1.5:
(a) Construct the polar plot of $G_c G$ for $G_c = 1$ and determine stability.

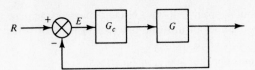

(b) Construct a polar plot and find phase and gain margins to determine the effect of a phase-lag network $G_c = (10s + 1)/(50s + 1)$.

(c) Has the compensation affected the steady-state errors for step and ramp inputs?

7.18. In Problem 7.17, replace the phase-lag by phase-lead compensation

$$G_c = \frac{2s + 1}{0.2s + 1}$$

Sketch the polar plot of $G_c G$ based on the frequencies $\omega = 2$ and 4 to estimate gain margin and phase margin. What is the effect on steady-state errors? Note that both phase lag and phase lead bend the plot around the other side of -1. Compare how each achieves this (that is, how each moves a point on G to change the plot in the desired direction).

7.19. Construct asymptotic Bode magnitude plots for the following transfer functions.

(a) $\dfrac{4}{s + 2}$ **(b)** $\dfrac{4}{(0.4s + 1)(s + 1)}$ **(c)** $\dfrac{8}{s(1.25s + 1)(s + 2)}$

7.20. Construct asymptotic Bode magnitude plots for the following transfer functions. In the case of underdamped quadratics, always sketch the true magnitude curve locally based on the peak value alone.

(a) $\dfrac{5(s + 0.6)}{s(2.5s + 1)(s + 2)(0.25s + 1)}$ **(b)** $\dfrac{3.125}{s(s^2 + 0.625s + 1.5625)}$

(c) $\dfrac{1.6}{(s + 0.4)(s + 0.8)(s + 1)}$

7.21. Construct asymptotic Bode magnitude plots for the following loop gain functions, and determine by inspection, noting the contributions of the elementary factors, whether or not the corresponding systems are stable.

(a) $\dfrac{72}{s(s^2 + 2s + 9)}$ **(b)** $\dfrac{45}{s^2(s + 5)}$

7.22. For a system with loop gain function

$$G(s) = \frac{20}{(s + 5)(0.1s + 1)(0.025s + 1)}$$

(a) Plot the asymptotic Bode magnitude plot.

(b) At two convenient frequencies, 10 and 20, find the deviation of the true magnitude curve from the asymptotes, and sketch a section of this curve for use in part (c).

(c) Determine the phase margin and gain margin.

7.23. In Fig. P7.17, $G_c = K$ and $G = 2000/[s(s + 20)(0.01s + 1)]$.

(a) Construct the asymptotic Bode magnitude plot and find the approximate deviation with the true curve for a frequency near crossover to get a better approximation of the crossover frequency for $K = 1$.

(b) Determine the phase margin for $K = 1$.

(c) How does a change of gain K affect Bode magnitude and phase-angle plots? What is the effect of reducing K by a factor m?

(d) Find K to get a phase margin of 50°.

7.24. In Fig. P7.17 with $G = 4/[s(s + 4)(0.0625s + 1)]$ and $G_c = K$, find K to obtain a phase margin of $\phi_m = 54°$.

7.25. In Fig. P7.17, G is the transfer function of a field-controlled motor in a position control system:

$$G = \frac{1}{s(0.1s + 1)(0.02s + 1)}$$

If $G_c = K$, find K for a phase margin of 45° and determine the corresponding gain margin.

7.26. Plot the closed-loop frequency response function magnitude characteristic of a system versus frequency if the system has unity feedback and the polar plot of its loop gain function is as shown in Fig. P7.26. Note how nearness of this plot to -1 translates into a high-resonance peak.

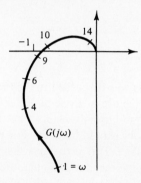

Figure P7.26

7.27. A system with unity feedback has the loop gain function

$$G(s) = \frac{K}{(s + 1)(s^2 + 2s + 2)}$$

(a) Find the limiting value of K for stability.

(b) Sketch the polar plot for $K = 2.5$ based on points calculated at $\omega = 1, 1.5, 1.8,$ and 2.

(c) How does doubling K to $K = 5$ affect this plot?

(d) For $K = 5$, estimate the maximum peaking of the closed-loop frequency response, and plot the corresponding M circle to verify the estimate.

7.28. In Problem 7.27, use M circles to find the value of K for which the closed-loop frequency response peaking is about 20%.

8

Frequency Response Design

8.1 INTRODUCTION

Frequency response analysis was discussed in the preceding chapter. This chapter is concerned with design and the use of Bode plots for open-loop stable and minimum-phase systems. For open-loop-unstable plants (that is, with right-half-plane poles), Nyquist diagrams or root loci should be used. For nonminimum-phase plants there are zeros in the right half-plane. This is discussed separately in this chapter. Systems with transport lag, due to pipelines for example, are also considered, as are certain particular control configurations, such as feedforward control for reducing the effect of disturbances. The Nichols chart is introduced and will be used on occasion to confirm predictions made on the basis of Bode plots. Similarly, sometimes the effects of dynamic compensation will be interpreted on polar plots.

The frequency-domain performance criteria on which design is based are developed and are also compared with the transient performance criteria discussed earlier.

8.2 DYNAMIC COMPENSATION

For most of this chapter the system configuration will be that shown in Fig. 8.1, requiring a series compensator to be designed for a system with unity feedback. The form of the compensator must be chosen and then the values of its parameters. The

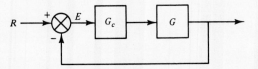

Figure 8.1 Series compensation.

common dynamic compensators, introduced in Section 5.4 and designed using root loci in Chapter 6, are shown in Fig. 8.2 together with their Bode plots. PID-type controllers are on the left and phase-lag, phase-lead, and lag–lead compensators on the

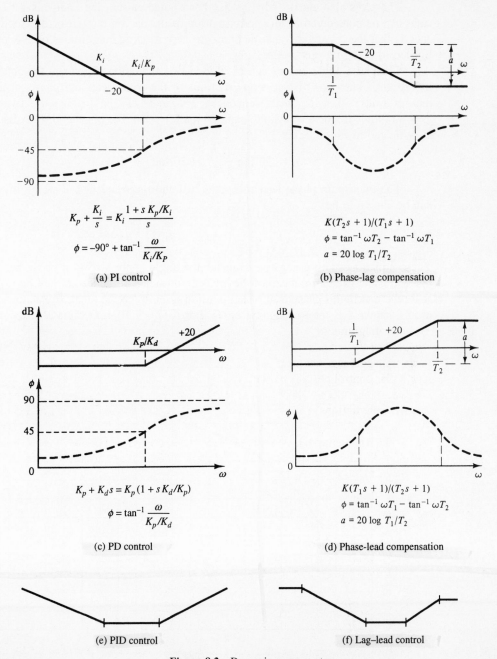

(a) PI control

$$K_p + \frac{K_i}{s} = K_i \frac{1 + s\,K_p/K_i}{s}$$

$$\phi = -90° + \tan^{-1} \frac{\omega}{K_i/K_p}$$

(b) Phase-lag compensation

$$K(T_2 s + 1)/(T_1 s + 1)$$
$$\phi = \tan^{-1} \omega T_2 - \tan^{-1} \omega T_1$$
$$a = 20 \log T_1/T_2$$

(c) PD control

$$K_p + K_d s = K_p(1 + s\,K_d/K_p)$$

$$\phi = \tan^{-1} \frac{\omega}{K_p/K_d}$$

(d) Phase-lead compensation

$$K(T_1 s + 1)/(T_2 s + 1)$$
$$\phi = \tan^{-1} \omega T_1 - \tan^{-1} \omega T_2$$
$$a = 20 \log T_1/T_2$$

(e) PID control

(f) Lag–lead control

Figure 8.2   Dynamic compensators.

right. It is seen that a phase-lag compensator contributes negative phase angles to a loop gain function at all frequencies, hence its name, and a phase-lead contributes positive phase angles.

The plots also clarify why, as has been noted earlier, the compensator on the right can be considered as an approximation to the one on the left of it. The difference is that the asymptotes that theoretically rise to large magnitudes on the left are limited to certain levels on the right.

The phase-angle curves can be plotted by calculation of ϕ from the equations in Fig. 8.2. The ± 20 dB/dec slope sections of the phase-lead and phase-lag imply that for a ratio 10 of the break frequencies $1/T_2$ and $1/T_1$ the distance a between the low- and high-frequency asymptotes is $20 \log 10 = 20$ dB. In general, this distance, as indicated in Fig. 8.2, is

$$a = 20 \log \frac{1/T_2}{1/T_1} = 20 \log \frac{T_1}{T_2} \quad \text{dB} \qquad (8.1)$$

The maximum phase lead in Fig. 8.2(d) and the frequency at which it occurs can be shown to be

$$\omega_m = \sqrt{\frac{1}{T_1}\frac{1}{T_2}} \qquad \phi_{\max} = \sin^{-1}\frac{(T_1/T_2) - 1}{(T_1/T_2) + 1} \qquad (8.2)$$

These equations with a negative sign for the angle apply also to a phase-lag. Note that ω_m is the geometric mean of the break frequencies of the lead or lag. So on a log scale the maximum phase lead or lag is midway between the break frequencies. Figure 8.3 shows $\phi_{\max}$ plotted against T_1/T_2. For $T_1/T_2 = 10$, for example, $\phi_{\max} = 55°$.

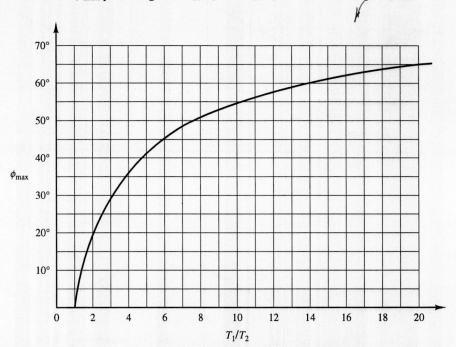

Figure 8.3   Maximum phase angles of phase-lead compensators.

From a frequency response point of view, phase-lead compensation is stabilizing because it contributes a positive angle to the phase margin. Note the leveling off of the curve in Fig. 8.3 for larger ratios T_1/T_2. Two phase-leads in series with ratios 3 contribute more lead than one with ratio 9.

The shape of the magnitude characteristics of these compensators will also prove to be important. Since G_c and G are in series, their Bode plots add. Thus G_c can be used to change the Bode plot of loop gain function $G_c G$ in a desired direction. For example, a phase-lag can be used to raise the Bode plot of $G_c G$ at lower frequencies and/or lower it at higher frequencies. With a phase-lead, the $+20$ dB/dec slope section can be exploited to reduce the slope of $G_c G$ in the region of its crossover frequency. Both of these effects will prove to be important to improve specific features of the performance.

8.3 FREQUENCY RESPONSE PERFORMANCE CRITERIA

In Sections 5.2 and 4.4 transient performance was discussed by examination of the quadratic lag (closed-loop) system transfer function (4.15):

$$\frac{C(s)}{R(s)} = \frac{\omega_n^2}{s^2 + 2\zeta\omega_n s + \omega_n^2} \tag{8.3}$$

This was based on the fact that the performance of many systems is dominated by one complex conjugate pair of poles. It is appropriate, therefore, to consider the frequency response of the system (8.3):

$$\frac{C(j\omega)}{R(j\omega)} = \frac{1}{(j\omega/\omega_n)^2 + 2\zeta(j\omega/\omega_n) + 1} \tag{8.4}$$

From Fig. 7.9

$$M = \left|\frac{C}{R}\right| = \frac{1}{\sqrt{(1 - \omega^2/\omega_n^2)^2 + (2\zeta\omega/\omega_n)^2}} \tag{8.5}$$

The key frequency response performance measures are indicated on the typical frequency response plot shown in Fig. 8.4. These criteria are:

1. *Resonant peak M_p at the resonant frequency ω_p:* a measure of relative stability.
2. *Bandwidth ω_b:* a measure of speed of response, defined as the range of frequencies over which M equals at least $0.707(=\sqrt{2}/2)$ of its value at $\omega = 0$:

$$M(\omega_b) = 0.707M(0) \tag{8.6}$$

If the magnitude is plotted in decibels, it is the frequency at which M falls below -3 dB ($\equiv 0.707$). The bandwidth is the range of frequencies over which the response is considered to be satisfactory.

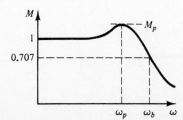

Figure 8.4   Frequency response performance criteria.

As discussed in connection with Fig. 7.5, a small damping ratio corresponds to a high *resonance peak* on the frequency response plot.

In transient response, ζ and ω_n corresponding to the dominating poles determine speed of response and overshoot. The real part of the poles determines settling time, and the imaginary part determines the frequency of transient oscillations and hence the rise time. Increasing ω_n for given ζ reduces both settling time and rise time, improving the speed of response. The relations between ω_n and ζ and the frequency response performance criteria may be expressed as follows. The resonant peak is found by setting the derivative of M to zero. This yields the following results for ω_p/ω_n and M_p, where M_p is obtained by substituting ω_p/ω_n into (8.5):

$$\frac{\omega_p}{\omega_n} = \sqrt{1 - 2\zeta^2} \qquad M_p = \frac{1}{2\zeta\sqrt{1 - \zeta^2}} \qquad \zeta \leq 0.707 \qquad (8.7)$$

These equations are plotted in Fig. 8.5. The resonant peaking is linked here directly to the damping ratio ζ, the relative stability measure for transient response. Low damping ratio ζ means both severe step response overshoot and severe resonance peaking. The behavior of ω_p/ω_n is as expected from the magnitude plots for a quadratic lag in Fig. 7.13. For low damping the ratio is about 1 because the frequency at the resonant peak and the undamped natural frequency, which is also the break frequency of the asymptotes, about coincide. But as the damping ratio increases the peak moves to lower frequencies.

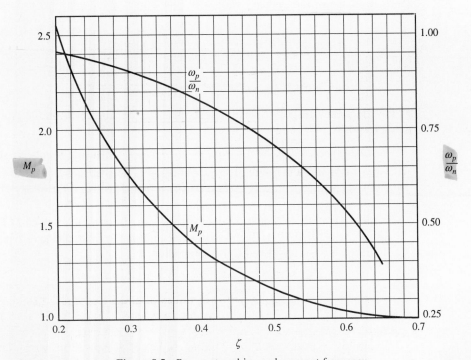

Figure 8.5 Resonant peaking and resonant frequency.

To express the bandwidth, substituting 0.707 for M in (8.5), squaring both sides, and solving the resulting quadratic equation for ω_b/ω_n yield

$$\frac{\omega_b}{\omega_n} = [1 - 2\zeta^2 + \sqrt{2 - 4\zeta^2(1 - \zeta^2)}]^{1/2} \qquad (8.8)$$

This is shown plotted in Fig. 8.6. When $\zeta = 0.707$, the bandwidth ω_b equals the undamped natural frequency ω_n. The nature of this curve can be verified by inspection of the quadratic lag magnitude plot in Fig. 7.13. The bandwidth is the frequency at which this magnitude drops below -3 dB. The intersection with this level is seen to move left as the damping ratio increases, and ω_b/ω_n reduces from a value above 1 to a value below 1.

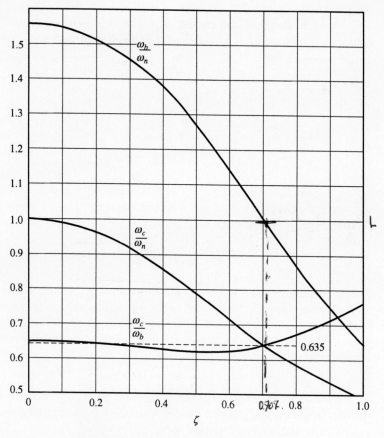

Figure 8.6   Bandwidth and crossover frequency.

Beyond ω_b, in particular for lower values of ζ, the response drops off rapidly. For given ζ, ω_b is proportional to ω_n and is a measure of speed of response. Raising ω_b reduces the settling time and rise time of the step response.

In the next section these closed-loop performance measures will be translated to specifications on the loop gain function, in particular on its Bode plot. The system sensitivity properties introduced in Chapter 4 depend on the loop gain function and are therefore also reflected by this plot.

8.4 PERFORMANCE SPECIFICATIONS ON THE BODE PLOT

Open-Loop Frequency Response Performance Measures

Resonance peaking and bandwidth are performance specifications for the closed-loop system. However, the polar plots and Bode plots in Chapter 7 are plots of the loop gain function. The question is, therefore: What requirements should be imposed on these plots to satisfy specifications on closed-loop bandwidth ω_b and relative stability ζ or M_p?

To correlate open- and closed-loop criteria, it is noted that C/R of (8.3) can be considered to be realized by a unity feedback system with loop gain function

$$G_c G = \frac{\omega_n^2}{s(s + 2\zeta\omega_n)} \tag{8.9}$$

The frequency response function is

$$G_c G(j\omega) = \frac{\omega_n^2}{-\omega^2 + 2j\zeta\omega_n\omega} \tag{8.10}$$

$$\underline{/G_c G} = -\tan^{-1}\frac{2\zeta\omega_n}{-\omega} \qquad |G_c G| = \frac{\omega_n^2}{\sqrt{\omega^4 + (2\zeta\omega_n\omega)^2}}$$

Of particular importance for analysis and design is the *crossover frequency* ω_c, where $G_c G$ crosses the 0-dB axis and where the phase margin is defined. From (8.10), $|G_c G| = 1$ when $\omega_c^4 + (2\zeta\omega_n\omega_c)^2 - \omega_n^4 = 0$, of which the positive root is given by

$$\frac{\omega_c}{\omega_n} = (\sqrt{4\zeta^4 + 1} - 2\zeta^2)^{1/2} \tag{8.11}$$

This is plotted in Fig. 8.6. Note that with increasing ζ the crossover frequency of the loop gain function lies farther below the undamped natural frequency of the closed-loop system. Also shown is the ratio ω_c/ω_b of crossover frequency to bandwidth frequency, calculated from (8.8) and (8.11). It is quite interesting to note that the ratio is rather constant and can be approximated by

$$\omega_c \approx 0.635\omega_b \tag{8.12}$$

This shows how the crossover frequency of the loop gain function should be chosen to meet a given specification of the closed-loop system bandwidth. Raising the crossover frequency increases the speed of response.

The phase margin ϕ_m equals 180° plus the phase angle of $G_c G$ at ω_c, so $\phi_m = 180° - \tan^{-1} 2\zeta\omega_n/(-\omega_c) = 180° - (180° - \tan^{-1} 2\zeta\omega_n/\omega_c)$, or

$$\phi_m = \tan^{-1}\frac{2\zeta}{(\sqrt{4\zeta^4 + 1} - 2\zeta^2)^{1/2}} \tag{8.13}$$

This is plotted in Fig. 8.7. The ϕ_m–ζ plot represents an important and useful correlation between frequency response and transient response measures of relative stability. A good approximation to the curve is given by the dashed line:

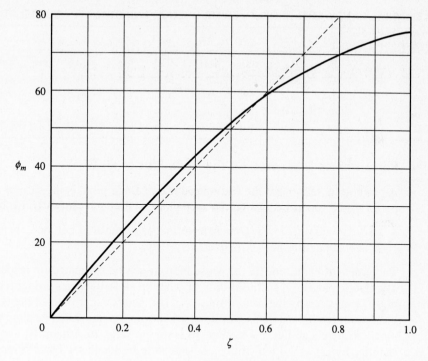

Figure 8.7 Phase margin versus damping ratio.

$$\phi_m \approx 100\zeta \tag{8.14}$$

With this relation, the graphs in Figs. 8.5 and 8.6 can also be thought of as plots against phase margin.

Table 8.4.1 summarizes some significant correlations from the graphs presented and is useful for design. For ease of reference, the percentage overshoot P.O. in step responses has been included from Fig. 5.2. The actual values are shown instead of the approximations given by (8.12) and (8.14). If relative stability is specified in terms of damping ratio or percentage overshoot, the corresponding phase margin can be found in the table. Some corresponding pairs of values of damping ratio of the dominating closed-loop poles and phase margin of the loop gain function are the following:

$$\zeta = 0.7: \qquad \phi_m = 65$$
$$\zeta = 0.6: \qquad \phi_m = 59$$
$$\zeta = 0.5: \qquad \phi_m = 52$$

For specified relative stability, the crossover frequency ω_c corresponding to a specified bandwidth ω_b can then be found from the ratios ω_c/ω_b in the table. Similarly, if the settling time $T_s = 4/(\zeta\omega_n)$ is specified, ω_n can be calculated for given ζ and the corresponding desired crossover frequency found from the ratios ω_c/ω_n in the table.

Consider now, again for a unity feedback system, how these and other performance specifications are reflected in the Bode plot of the loop gain function.

TABLE 8.4.1 BANDWIDTH-CROSSOVER FREQUENCY AND STABILITY CORRELATIONS

ζ	0	0.1	0.2	0.3	0.4	0.5	0.6	0.7	0.8	0.9	1.0
ϕ_m	0	11.4	22.6	33.3	43.1	51.8	59.2	65.2	69.9	73.5	76.3
P.O.	100	73	53	37	25	16	9.5	4.6	1.5		
ω_c/ω_b	0.64	0.64	0.64	0.63	0.62	0.62	0.62	0.64	0.67	0.71	0.75
ω_c/ω_n	1	0.99	0.96	0.91	0.85	0.79	0.72	0.65	0.59	0.53	0.49
ω_b/ω_n	1.55	1.54	1.51	1.45	1.37	1.27	1.15	1.01	0.87	0.75	0.64

Performance Measures on the Bode Plot of $G_c G$

An important reason for the widespread use of Bode plots is the ease of interpreting performance specifications on the asymptotic magnitude plot of $G_c G(j\omega)$.

1. *Relative stability:* $G_c G$ must have adequate length of not more than a -20 dB/dec slope at or near crossover frequency ω_c.

The order of magnitude of the phase margin can be estimated by inspection of the asymptotic plot of $G_c G$. In Fig. 8.8, if other break frequencies are far removed from ω_c, as suggested by the dashed lines, (a), (b), and (c) approximate the functions and phase margins indicated, as may be verified from Section 7.7. The plot in (a) reflects ample relative stability, the system (b) is on the verge of instability, and the relative stability for (c) will often still be considered to be adequate. Thus, if inspection of a Bode plot reveals little or no length of not more than a -20 dB/dec slope at or near ω_c, it is immediately evident that ϕ_m will be inadequate.

2. *Steady-state accuracy:* To improve steady-state accuracy, the low-frequency asymptote must be raised or its slope changed.

From (7.22), the low-frequency asymptote of $G_c G$ is $K/(j\omega)^n$, where K is the gain and n the system type number. For $n = 0$ (type 0) the slope of the low-frequency asymptote is zero and, from Table 4.3.1, the steady-state error after a unit step is $1/(1 + K)$. So the steady-state error reduces if the low-frequency asymptote $20 \log K$ is raised. If zero steady-state error after a step is required, the system must be at least type 1 ($n = 1$), so the slope of the low-frequency asymptote must be at least -20 dB/dec. If a Bode plot shows a -40 dB/dec low-frequency asymptote, the system is type 2 and so has zero steady-state errors for both steps and ramps. Steady-state errors for $n = 1$ and 2 also reduce when the low-frequency asymptotes are raised. This is because these errors reduce as K is increased and, according to (7.23) and (7.24), the low-frequency asymptotes intersect the 0-dB axis at $\omega = K$ and $\omega = \sqrt{K}$, respectively.

3. *Accuracy in the operating range:* To ensure specified accuracy over a normal range of frequencies, the plot may not fall below a given level over this range. To improve accuracy, this level must be raised.

In Fig. 8.1, $E/R = 1/(1 + G_c G)$. If up to 10% error is permitted up to a certain frequency ω_a, then $|G_c G| \geqslant 10 = 20$ dB is approximately the minimum allowed level

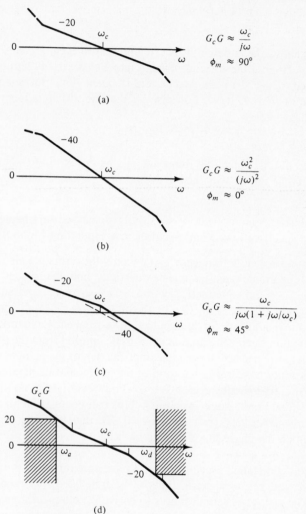

$$G_c G \approx \frac{\omega_c}{j\omega}$$

$$\phi_m \approx 90°$$

(a)

$$G_c G \approx \frac{\omega_c^2}{(j\omega)^2}$$

$$\phi_m \approx 0°$$

(b)

$$G_c G \approx \frac{\omega_c}{j\omega(1 + j\omega/\omega_c)}$$

$$\phi_m \approx 45°$$

(c)

(d)

Figure 8.8   Phase margin and sensitivity estimates from Bode plots.

up to this frequency, as in Fig. 8.8(d). Thus the notion of accuracy is extended to dynamic operation. Raising low-frequency asymptotes not only improves steady-state accuracy but also the accuracy at other frequencies where this has raised the plot.

4. *Crossover frequency and bandwidth:* Crossover frequency ω_c is a measure of bandwidth ω_b and so of speed of response. From (8.12), $\omega_c \approx 0.635\omega_b.$

See Table 8.4.1 Roughly, if $|G_c G| \gg 1$, then $C/R = G_c G/(1 + G_c G) \approx 1$, since the 1 in the denominator is relatively negligible, so output follows input almost perfectly. For $|G_c G| \ll 1$, $C/R \approx G_c G \ll 1$, since now the term $G_c G$ in the denominator is negligible relative to 1. So in this range, beyond its bandwidth, the system's response is quite poor. The crossover $|G_c G| = 1 = 0$ dB separates these two ranges.

5. *Noise rejection:* To ensure a specified attenuation (reduction) of noise compo- nents in the input above a certain frequency, the plot above that frequency should be below a certain level.

Noise at high frequencies, above the bandwidth, is often present in input signals. To cause the system to act as a filter at these frequencies, it may be specified that noise amplitudes shall be reduced to, say, 10% above a certain frequency ω_d. Sufficiently above the bandwidth, where $C/R \approx G_c G \ll 1$ as discussed above, this requires that

$$\left|\frac{C}{R}\right| \approx |G_c G| \leq 0.1 = -20 \text{ dB}$$

so the plot must be below -20 dB above the frequency ω_d, as in Fig. 8.8(d). Noise rejection considerations are one reason why the bandwidth of a system should not be made larger than necessary.

6. *System sensitivity:* The various aspects of sensitivity discussed in Chapter 4 are closely related to specifications 3 and 5 and the corresponding Fig. 8.8(d).

Where the loop gain is large, say below frequency ω_a, not only the errors but also the sensitivity to model approximations, plant parameter variations, and disturbance in- puts are small. Where the loop gain is small, say above frequency ω_d and suffi- ciently above the system bandwidth $\omega_b \approx \omega_c/0.635$, the output is quite small and of little interest. This is in fact what allows for "unmodeled high-frequency dynamics", because it allows open-loop poles and zeros in this range to be omitted from the plant model. This is fortunate not only because it permits simplification of the plant model but also because the uncertainty concerning the form of the model and its parameter values is often greatest in this region. However, it is important to verify that no high- frequency resonances exist that could cause instability. This can occur if the damp- ing is so small that the resonance peak reaches above 0 dB, thus creating a new crossover frequency, at which the phase margin could be negative.

These criteria show how different aspects of performance are reflected in indi- vidual features of the plot and permit specifications to be translated into require- ments on the Bode plot. The task of system design is to derive the compensators that will meet these requirements. Observe that design to more severe specifications 3 and 5 will also improve the related sensitivity properties.

8.5 CLOSED-LOOP FREQUENCY RESPONSE AND THE NICHOLS CHART

The correlations represented in Figs. 8.5 to 8.7 and Table 8.4.1 allow predictions of closed-loop frequency response peaking M_p, resonant frequency ω_p, and bandwidth ω_b to be made from Bode plot data. The Nichols chart will give the actual values of these performance measures and therefore can be used, if desired, to verify the pre- dictions. The Nichols chart is discussed because it finds considerable use for design, though not as much as in the past. Here it is used only occasionally, to verify predic- tions from Bode plots. This section is therefore not essential for those that follow.

For unity feedback systems, the magnitude and phase angle of the closed-loop frequency response function can be found from the intersections of the polar plot with M and N circle loci, introduced in Section 7.8. However, just as Bode plots are more convenient than polar plots, the Nichols chart is more convenient than the use of M circles and N circles. The Nichols chart is shown in Fig. 8.9. The magnitude of the loop gain function in decibels is plotted against its phase angle, in degrees. This

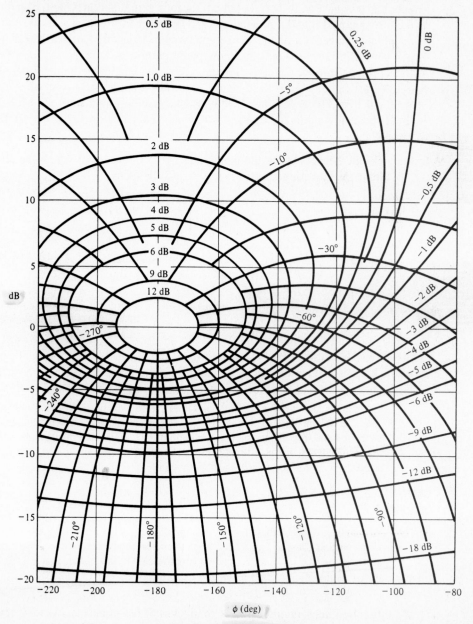

Figure 8.9 Nichols chart.

is a decibel magnitude versus phase-angle plot and is often obtained most easily by transferring these data from Bode plots. The plot becomes a Nichols chart when the M and N circles are translated to the loci for constant magnitude M and phase angle ϕ of the closed-loop frequency response function. The loci for M are identified by the magnitude in decibels. In principle, the M loci could be obtained from the M circles in Fig. 7.21 by plotting the distance of points to the origin vertically, in decibels, and the phase angle horizontally. If the loop gain function is plotted on this graph, the resonant peaking M_p, in decibels, and the resonant frequency ω_p are found from the point of tangency with the locus for maximum M, as for M circles. The detail of the plot shown in Fig. 8.10 indicates how phase margin ϕ_m and gain margin GM_{dB} are found. By definition, GM_{dB} is the distance of the magnitude below 0 dB along the $-180°$ vertical, and ϕ_m is the distance of the phase angle to the right of $-180°$ along the 0-dB line. The system bandwidth ω_b can also be found from the Nichols chart. Since the bandwidth is the frequency at which M falls below $0.707 = -3$ dB, it follows that:

> On the Nichols chart, the bandwidth is the frequency at which the plot of the loop gain function crosses the $M = -3$ dB locus.

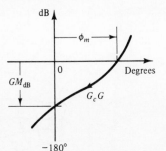

Figure 8.10   Phase and gain margin on a Nichols chart.

So if it is desired to verify predictions of ω_b and M_p made on the basis of Bode plots of the loop gain function, the Bode plot data should be plotted on the Nichols chart.

Example 8.5.1   Resonance Peak and Bandwidth for Example 7.7.2

The numerical values for the Bode plot in Fig. 7.18 were obtained in a root locus design (Fig. 6.8) in which a damping ratio 0.707 was specified for the dominating poles. Based on Table 8.4.1, this suggests a phase margin of 66°. This is not far from the actual value $\phi_m = 63°$ shown in Fig. 7.18. Also, with a crossover frequency $\omega_c = 0.325$, Table 8.4.1 suggests a closed-loop system bandwidth ω_b of

$$\omega_b \approx \frac{0.325}{0.64} = 0.5 \text{ rad/sec} \qquad (8.15)$$

Also, Fig. 8.5 indicates that $M_p \approx 1$, so the closed-loop frequency response should show no peaking. To check these predictions, the Bode plots in Fig. 7.18 have been transferred to the Nichols chart in Fig. 8.11. The plot verifies the absence of a resonance peak; only at low frequencies, where the angle approaches $-90°$, does M approach 0 dB. The plot intersects the $M = -3$ dB locus at the bandwidth $\omega_b \approx 0.54$ rad/sec, not far from the predicted value of 0.5 rad/sec.

In following sections, Nichols charts will be used on occasion to verify closed-loop response predictions. For Nichols chart design, which is not discussed here, if

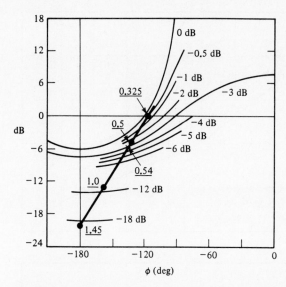

Figure 8.11   Example 8.5.1.

the gain of the loop gain function is changed, the curve is moved parallel to the magnitude axis, since changing K will not affect the phase angle. By Fig. 8.10, to realize a specified phase margin, say $\phi_m = 52°$, the curve should be moved until it intersects the 0-dB axis on the Nichols chart at the corresponding phase angle. Alternatively, the permissible resonance peaking may be specified. For example, from Table 8.4.1, $\phi_m = 52°$ corresponds to $\zeta \approx 0.5$, and Fig. 8.5 shows that $\zeta = 0.5$ is about equivalent to a closed-loop resonance peaking $M_p = 1.15$. Thus in design using the Nichols chart the gain would be adjusted so that $G_c G$ becomes tangent to the corresponding M locus.

8.6 DESIGN OF PHASE-LAG COMPENSATION

Proportional control design on the Bode plot was treated in Section 7.7. In Example 7.7.3 a controller gain was chosen to obtain a specified phase margin. Therefore, only dynamic compensator design is considered in these sections.

The Bode plot of a phase-lag compensator G_c is shown in Fig. 8.2(b). It contributes negative phase angles to $G_c G$ and so tends to reduce the phase margin. As indicated in Section 8.2, the feature of interest is that a phase-lag can be used to raise the magnitude plot of $G_c G$ at lower frequencies and/or lower it at higher frequencies. Raising the plot, as discussed in Section 8.4, improves accuracy for inputs in that range of frequencies. Lowering the plot can bring a -20 dB/dec slope section of $G_c G$ to the 0-dB axis and hence improve relative stability. Generally, however, when improved stability or increased speed of response is desired, phase-lead compensation is used, and phase-lag is used to improve steady-state accuracy.

The effect of phase-lag compensation is illustrated in Fig. 8.12. The plot of KG is that of the plant with a proportional controller of which the gain K has been chosen to satisfy the specifications on steady-state accuracy. The crossover frequency is at ω_1, and even without the phase-angle curve it is clear by inspection that the phase

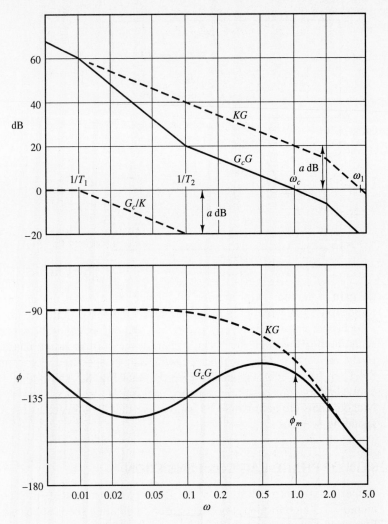

Figure 8.12   Phase-lag compensation.

margin will be small, because crossover does not occur in or near the range of the −20 dB/dec slope.

To realize the specified phase margin, this plot should be lowered by a dB so that it crosses the 0-dB axis at ω_c, well within the −20 dB/dec slope range. This can be done without lowering the low-frequency asymptote, thus avoiding a loss of steady-state accuracy, by using the phase-lag characteristic shown as G_c/K. It is in series with KG, so their Bode plots add to form $G_c G$. The low-frequency asymptote of G_c/K is 0 dB, so at low frequencies $G_c G$ coincides with KG, and at high frequencies it is a dB below KG.

The highest break frequency of G_c is chosen a factor of 10 below ω_c so that G_c will contribute only a small negative phase angle at ω_c, as suggested by the phase-angle curves, where the distance between the plots for KG and $G_c G$ is the angle contributed by the phase lag. With the highest break frequency $1/T_2$ chosen, according to (8.1) the lowest break frequency $1/T_1$ is given by

$$a = 20 \log \frac{T_1}{T_2} \tag{8.16}$$

A design procedure for phase-lag compensation can now be summarized as follows:

1. Determine the controller gain K required to meet the specification on steady-state accuracy and plot the loop gain function KG for this gain.
2. Determine the frequency ω_c at which the phase angle of G is $(-180° + \phi_m + 5°)$, where ϕ_m is the specified phase margin, with a 5° allowance for the angle contributed at ω_c by the compensator. This is the desired crossover frequency.
3. Choose the highest break frequency $1/T_2$ of the phase-lag 1 decade below ω_c.
4. Measure the magnitude a dB of the plot constructed in step 1 at frequency ω_c. From (8.16), this determines the ratio of the break frequencies of the phase lag.
5. Calculate the lowest break frequency $1/T_1$ of the phase-lag from steps 3 and 4. Alternatively, find it by construction: $G_c G$ in Fig. 8.12 is parallel to KG down to $1/T_2$, and an asymptote at -40 dB/dec starting here intersects KG at $1/T_1$.
6. Plot $G_c G$ to verify that all specifications are met, and express the compensator

$$G_c = K \frac{T_2 s + 1}{T_1 s + 1} \tag{8.17}$$

Example 8.6.1   Phase-Lag Compensation in a Position Servo

With $G = 1/[s(0.5s + 1)]$, design phase-lag compensation to meet the following specifications:

1. The steady-state unit ramp following error may not exceed 10%.
2. The phase margin should be at least 55°.
3. The error in response to sinusoidal inputs up to 0.1 rad/sec may not exceed about 10%.

For $G_c = K$ the loop gain function is

$$G_c G = \frac{K}{s(0.5s + 1)}$$

Specification 1 requires at least $K = 10$. From (7.23) this implies that the low-frequency asymptote of $G_c G$ may not intersect the 0-dB axis below 10 rad/sec. The Bode plot of $10G$ in Fig. 8.13 satisfies this requirement. Specification 3 is also satisfied by this plot, since it requires a gain of at least $10 = 20$ dB up to a frequency of 0.1 rad/sec. But, by inspection, the phase margin of $10G$ is quite inadequate. The phase angles at some frequencies are

$$\phi(1) = -117° \qquad \phi(1.5) = -127° \qquad \phi(2) = -135°$$

So, if the crossover frequency of $G_c G$ is chosen to be 1 rad/sec, an allowance of $(180 - 117) - 55 = 8°$ is available to account for the phase lag introduced by G_c at 1 rad/sec. Thus it is desired to lower the higher-frequency portion of $10G$ so that it crosses 0 dB at 1 rad/sec, while leaving the low-frequency part of $10G$ unchanged. This can be achieved by phase-lag compensation as indicated in Fig. 8.13. The highest break frequency of G_c

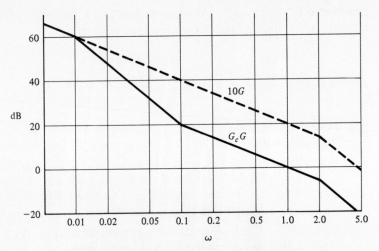

Figure 8.13　Example 8.6.1: phase-lag compensation.

is chosen at 0.1 rad/sec, a factor of 10 below the crossover frequency. This also meets specification 3. By construction, a 40 dB/dec asymptote starting at 0.1 rad/sec meets the plot of $10G$ at 0.01 rad/sec, the lowest break frequency of G_c. Therefore,

$$G_c = 10 \frac{1 + s/0.1}{1 + s/0.01}$$

The gain factor 10 follows because the gain of G is 1 and that of $G_c G$ is 10.

Example 8.6.2　Phase-Lag Compensation in Example 7.7.3

In Example 7.7.3

$$G = \frac{1}{s(s + 1)(0.5s + 1)} \qquad G_c = K \qquad (8.18)$$

so the loop gain function is

$$G_c G = \frac{K}{s(s + 1)(0.5s + 1)}$$

For $G_c = K = 0.325$ the phase margin was found to be $\phi_m = 63°$. Suppose that it is specified that the steady-state error $1/K$ may not exceed 40%. Then $K = 2.5$ would be required. The Bode plot of $2.5G$ in Fig. 8.14, for $G_c = K = 2.5$, shows crossover about midway between the -20 and -60 dB/dec slope sections, so the phase margin is probably near zero.

As has been discussed, phase-lag compensation can raise the plot of $0.325G$ at low frequencies, and leave it unchanged over a wide enough range around its crossover frequency 0.325 that the phase margin changes little. The highest break frequency $1/T_2 = 0.0325$ is chosen a factor of 10 below 0.325. Then the compensator will not contribute more than about $-5°$ phase lag at frequency 0.325 and so will not reduce the phase margin by more than this amount. The value $1/T_1$ of the lowest break frequency is 0.00422 and can be found graphically, where the -40 dB/dec asymptote of $G_c G$ starting at 0.0325 intersects the desired level of the low-frequency asymptote, on the plot of $2.5G$. It can also be calculated from (8.16), where the distance a dB between the plots of $0.325G$ and $2.5G$ can be measured. This distance is due to a gain increase from 0.325 to 2.5, so, as an alternative to measurement, it can be found from $20 \log(2.5/0.325)$. Equating this to (8.16) yields

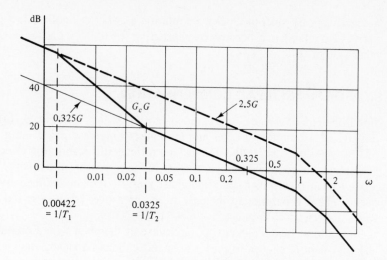

Figure 8.14   Example 8.6.2.

$$\frac{T_1}{T_2} = \frac{2.5}{0.325} \qquad \frac{1}{T_1} = \frac{1}{T_2}\frac{0.325}{2.5} = 0.00422$$

The desired compensator is then

$$G_c = 2.5\,\frac{1 + s/0.0325}{1 + s/0.00422} \tag{8.19}$$

The gain of 2.5 follows because $G_c G$ coincides with $2.5G$ at low frequencies and the gain of G is 1.

Since $|G_c G| > 100$ up to 0.01 rad/sec, the errors for inputs up to this frequency will not exceed 1%. Above 2 rad/sec, $|G_c G| < 0.1$, so less than 10% of noise at higher frequencies will appear in the output. Also, the system is not very sensitive to any plant poles and zeros in this high-frequency range that may have been omitted from the model, either unintentionally because they are not known or intentionally to simplify the model.

The preceding example can be interpreted in two different ways. Relative to $0.325G$, phase-lag compensation has improved low-frequency accuracy without loss of relative stability, by raising the low-frequency asymptote. However, relative to $2.5G$, phase-lag compensation has improved relative stability without loss of low-frequency accuracy: The phase-lag has served to lower the higher-frequency portion of the plot and so to realize an adequate length of -20 dB/dec slope at crossover.

Figure 8.15 shows a polar plot to interpret the stabilizing effect possible with a phase-lag. G_c moves a point A on G to B on $G_c G$. G_c introduces a phase-lag ϕ_c, so

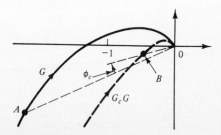

Figure 8.15   Phase-lag compensation on polar plot.

a negative phase shift or clockwise rotation as well as a gain reduction. The benefit derives from the latter, because by shortening $0A$ to $0B$ it causes G_cG to pass on the other side of -1.

8.7 DESIGN OF PHASE-LEAD COMPENSATION

The Bode plot of a phase-lead compensator G_c is shown in Fig. 8.2(d). It contributes positive phase angles to G_cG and so tends to increase phase margin and improve relative stability. The $+20$ dB/dec slope reflects this stabilizing effect, in that it can be exploited to reduce the slope of G_cG in the region of its crossover frequency. The phase lead also permits a specified phase margin to be maintained up to a higher frequency. Thus it permits an increase in the crossover frequency, and hence in bandwidth and speed of response.

In Fig. 8.16 the plots of KG are those of the plant with a proportional controller, with gain K chosen to meet the specifications on steady-state accuracy. By inspection, the phase margin is small, since the crossover frequency ω_a is not close to the range of -20 dB/dec slope.

By phase-lead compensation, shown as G_c/K, stability can be improved without reduction of the crossover frequency and so without a loss of speed of response.

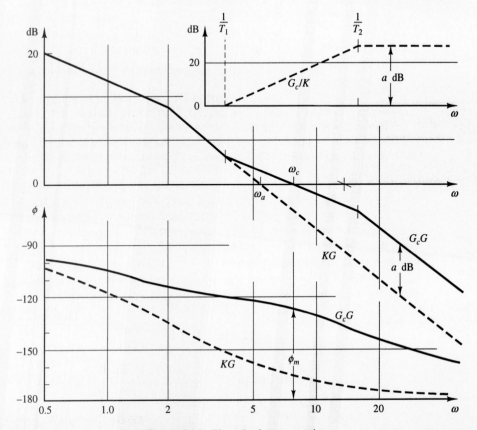

Figure 8.16 Phase-lead compensation.

It is in series with KG, so the sum of their plots is $G_c G$. Below its break frequency $1/T_1$ the asymptote of G_c/K is 0 dB, so here $G_c G$ asymptotically coincides with KG. But the +20 dB/dec slope between the break frequencies contributes a length of −20 dB/dec slope at the crossover frequency of $G_c G$. The phase-angle curves verify the improved stability. The lead is the difference between the curves of KG and $G_c G$, and the latter shows a phase margin ϕ_m.

In design, the phase margin available at the crossover frequency ω_a of KG is compared with the specified phase margin ϕ_m to determine the minimum lead needed. An allowance of, say, 7° is added to this because the as yet unknown crossover frequency ω_c is larger than ω_a. If the plot of KG shows −60 dB/dec slope in the frequency range of interest, an allowance of the order of 20° is a better choice. Figure 8.3 or Eq. (8.2) is then used to determine the ratio T_1/T_2 of the break frequencies needed to make the maximum lead ϕ_{max} provided by the compensator equal to the total lead angle calculated. This is satisfactory provided that the crossover frequency ω_c of $G_c G$ is made to coincide with the frequency $\omega_m = 1/\sqrt{T_1 T_2}$ midway between the break frequencies $1/T_1$ and $1/T_2$, where according to (8.2) the phase lead is maximum.

To make ω_m and ω_c coincide, recall from (8.1) that the distance between the low- and high-frequency asymptotes of the phase-lead is a dB $= 20 \log(T_1/T_2)$. This means that at the midpoint ω_m the lead contributes $10 \log(T_1/T_2)$ dB. Thus, to make ω_m and ω_c coincide, ω_c should occur where KG is $10 \log(T_1/T_2)$ dB below the 0-dB axis, since then its sum with G_c/K crosses at ω_c.

The break frequencies $1/T_1$ and $1/T_2$ can then be calculated from the known ratio T_1/T_2 and the known frequency

$$\omega_c = \omega_m = \frac{1}{\sqrt{T_1 T_2}} \tag{8.20}$$

A design procedure for phase-lead compensation can now be summarized as follows:

1. Determine the controller gain K required to meet the specification on steady-state accuracy.

2. Plot the loop gain function KG for this gain and find the corresponding phase margin.

3. Subtract this phase margin from the desired one, obtained if necessary by translating a specification on closed-loop percentage overshoot or damping ratio using Table 8.4.1.

4. Add to the difference an allowance of 7° or more to account for an increase of the crossover frequency due to the compensator. This gives the maximum lead to be provided by the compensator.

5. From Eq. (8.2) or Fig. 8.3, determine the ratio of the compensator break frequencies needed to provide this maximum lead.

6. Hence determine the desired crossover frequency ω_c where KG is $10 \log T_1/T_2$ below the 0-dB axis.

7. The compensator break frequencies $1/T_1$ and $1/T_2$ can now be found from (8.20) and the ratio in 5.

8. Plot $G_c G$ to verify that all specifications are met, and express the compensator

$$G_c = K \frac{T_1 s + 1}{T_2 s + 1} \qquad (8.21)$$

As discussed in Section 8.4, specifications on bandwidth or settling time can be translated to a specification on crossover frequency. If such a requirement exists and is not met, design may be repeated for a larger phase margin.

Example 8.7.1   Phase-Lead Compensation in a Position Servo

The transfer function of motor plus load is

$$G = \frac{1}{s(s + 2)}$$

A loop gain of 15 is needed for steady-state accuracy. Figure 8.16 therefore shows the magnitude plot of KG for $K = 30$. (Ignore the phase-angle curves; they are not used in the design.) The crossover frequency is 5.4, and the phase margin is $180 - 159.7 = 20.3°$. The specifications require a damping ratio 0.5 of the dominating poles. From Table 8.4.1, this means a phase margin of $52°$. Thus the compensator must provide a lead of $(52 - 20.3)$ plus an allowance of, say, $7°$ to account for an increase of the crossover frequency. The ratio T_1/T_2 of the break frequencies of the phase-lead is chosen to be 4.5. This gives a maximum lead of $39°$. From point 6, the crossover frequency is chosen as $\omega_c = 8$, where KG is $10 \log 4.5 = 6.5$ dB below the 0-dB axis. Their ratio 4.5 and Eq. (8.20) then yield 3.8 and 17.0 as the break frequencies of the phase-lead. The plot of $G_c G$ is shown. The phase margin is $53.4°$. The compensator is

$$G_c = 30 \frac{1 + s/3.8}{1 + s/17.0}$$

Example 8.7.2   Satellite Attitude Control with $G = 1/s^2$

In Example 6.5.4, root loci were used for a design with rate feedback based on the use of a rate gyro. From the point of view of stability, this is equivalent to PD control without the noise problem. Here rate feedback is assumed to be unavailable, and phase-lead compensation is used to approximate the PD control. Figure 8.17 shows the Bode plot of $KG = K/s^2$ for the controller gain $K = 1.75^2 = 3.06$ determined from a steady-state error specification. Recall from Eq. (7.24) that this plot crosses the 0-dB axis at the frequency $\sqrt{K} = 1.75$. Raising or lowering the plot will not produce the -20 dB/dec slope at the 0-dB axis needed for adequate relative stability, nor will phase-lag compensation. But if the phase-lead G_c/K is put in series with KG, their Bode plots add to produce the desired -20 dB/dec at the crossover of $G_c G$. G_c/K has a 0-dB low-frequency asymptote, so up to $1/T_1$ the plot of $G_c G$ coincides with KG. Beyond this the slope of $G_c G$ becomes -20 dB/dec up to frequency $1/T_2$.

Let a damping ratio $\zeta = 0.5$ be specified. From Table 8.4.1, this suggests a phase margin $\phi_m = 52°$. Since the phase of G is $-180°$, the lead compensator must contribute at least $+52°$. From Fig. 8.3, $T_1/T_2 = 10$ will provide a maximum lead of $+55°$ at a frequency $\omega_m = 1/\sqrt{T_1 T_2}$. As discussed, to make ω_m and ω_c coincide, ω_c should occur where KG is $10 \log(T_1/T_2) = 10 \log 10 = 10$ dB below the 0-dB axis. On the plot this is found to be at 3.08 rad/sec, so

$$\omega_c = \omega_m = 3.08 = \frac{1}{\sqrt{T_1 T_2}} = \frac{1}{T_2 \sqrt{10}}$$

The compensator break frequencies and transfer function are then

$$\frac{1}{T_1} = 0.974 \qquad \frac{1}{T_2} = 9.74 \qquad G_c = 1.75^2 \frac{1 + s/0.974}{1 + s/9.74}$$

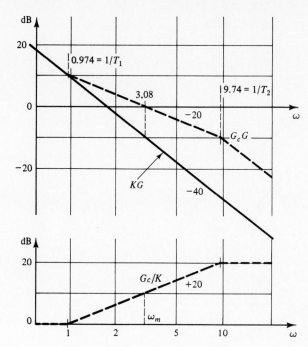

Figure 8.17 Example 8.7.2: satellite attitude control.

If K is not set by a steady-state error specification, but the bandwidth ω_b is specified instead, ω_c is estimated from Table 8.4.1 as $\omega_c = 0.62\omega_b$, and KG must pass at 10 dB below the 0-dB axis at this frequency. The value $\omega_c = 3.08$ corresponds to $\omega_b = 5$.

A third alternative, like the preceding one also mentioned in Section 8.4, is a design derived from a specified settling time $T_s = 2$ sec. Since $T_s = 4/(\zeta\omega_n)$ and $\zeta = 0.5$, this requires that $\omega_n = 4$. From Fig. 8.6, for $\zeta = 0.5$, $\omega_c/\omega_n = 0.77$, so $T_s = 2$ translates to an estimated requirement of $\omega_c = 4 \times 0.77 = 3.08$. Observe, finally, from Fig. 8.17 that both lowering and raising the Bode plot, by changing K, will move the crossover closer to the -40 dB/dec slope sections and hence will reduce the phase margin.

Example 8.7.3   Phase-Lead Compensation with G as in Examples 8.6.2 and 7.7.3

Consider again control of the plant G of Examples 7.7.3 and 8.6.2:

$$G = \frac{1}{s(s+1)(0.5s+1)} \tag{8.22}$$

Figure 8.18 shows the magnitude and phase-angle plots of G, derived in Example 7.7.2. It is evident from the phase-angle characteristic that phase lead must be introduced if a phase margin of 53° is to exist at frequencies above about 0.5 rad/sec. The maximum possible crossover frequency depends on the phase lead added, which in turn depends on T_1/T_2. As discussed earlier, noise considerations suggest that this ratio not be made larger than necessary. Say that $T_1/T_2 = 10$ is acceptable. This provides a maximum phase lead of 55°. Since the phase of G is $-180°$ at 1.45, this makes it possible to achieve a phase margin in excess of 53° at a crossover frequency $\omega_c = 1.45$, provided that ω_c is equal to the frequency ω_m where the phase lead is maximum:

$$\omega_c = \omega_m = 1.45 = \frac{1}{\sqrt{T_1 T_2}} = \frac{1}{T_2\sqrt{10}} \qquad \frac{1}{T_1} = 0.459 \qquad \frac{1}{T_2} = 4.59$$

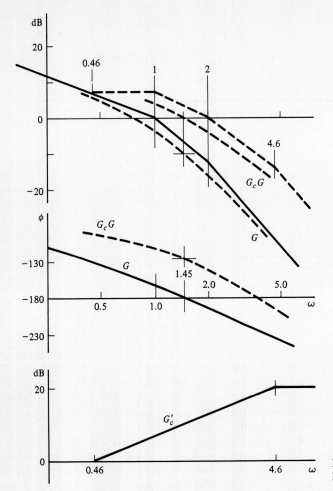

Figure 8.18 Example 8.7.3: phase-lead compensation.

A lead G'_c with these break frequencies and a 0-dB low-frequency asymptote is shown. One way to find the gain K of the compensator $G_c = KG'_c$ is to add the plots of G'_c and G and then raise or lower the resulting plot of $G'_c G$ until it crosses the 0-dB axis at $\omega_c = 1.45$. It turns out that in the present case no raising or lowering is required (that is, $K = 1$). Alternatively, analogous to Example 8.7.2, to make $\omega_c = \omega_m$, KG must be $10 \log 10 = 10$ dB below the 0-dB axis at 1.45. This is satisfied by G itself, so again $K = 1$, and

$$G_c = \frac{1 + s/0.46}{1 + s/4.6} \tag{8.23}$$

Note that sections of the actual magnitude plots are used, because two simple lag breaks close to crossover suggest that the deviations from the asymptotes will be considerable.

The following observations are made:

1. *Closed-loop performance:* From Table 8.4.1, a phase margin of 53° corresponds to a closed-loop damping ratio $\zeta = 0.5$, for which Fig. 8.5 indicates a closed-loop frequency response peaking $M_p = 1.15 = 1.2$ dB, and Table 8.4.1 suggests

a bandwidth $\omega_b = \omega_c/0.62 = 1.45/0.62 = 2.34$. To check these predictions, the Bode plots of $G_c G$ in Fig. 8.18 have been transferred to the Nichols chart in Fig. 8.19. The curve intersects the -3-dB locus at $\omega_b = 2.4$, close to the predicted bandwidth of 2.34 based on Bode plot design and Table 8.4.1, and it is about tangent to the $M = 1$-dB locus, close to the predicted peaking of 1.2 dB.

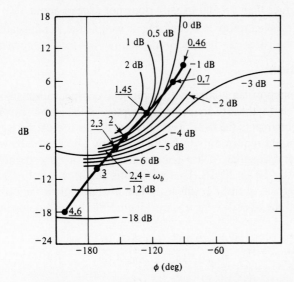

Figure 8.19   Example 8.7.3: Nichols chart.

2. The phase lead extends the -20 dB/dec slope to higher frequencies and hence increases bandwidth without loss of phase margin.

3. Figure 8.20 interprets phase-lead compensation on a polar plot. G_c moves a point A on G to B on $G_c G$. The benefit derives from the positive phase shift (counterclockwise rotation) ϕ_c introduced by G_c. Although $0B > 0A$, this causes the plot to pass on the other side of the -1 point. To obtain maximum phase shift near -1, the break frequencies of the lead must be in the range of those near -1. This contrasts with the phase-lag in Fig. 8.15. Its break frequencies must be in a lower range, to realize the magnitude reduction near the -1 point on which its use relies.

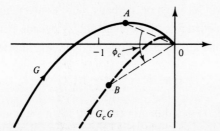

Figure 8.20   Phase-lead compensation on polar plot.

4. It is useful to observe in Example 8.7.3 that, if the break frequencies of the plant were, say, 1 and 10 rad/sec instead of 1 and 2, a smaller ratio of the break frequencies of the lead would have sufficed to achieve $\omega_c = 1.45$ with $\phi_m = 53°$.

5. Phase-lead compensation is not effective if the phase angle decreases rapidly near the crossover frequency, due to two nearby poles or a complex conjugate pair of poles. As may be visualized from Fig. 8.16, adding the phase-angle curve of the compensator to that of the plant then yields little benefit. In such cases pole–zero cancellation design may be preferable.

6. Because of noise considerations, the ratio of the break frequencies of the compensator is rarely larger than 15 and usually does not exceed about 10.

7. This limits the achievable crossover frequency, as well as the steady-state accuracy. If necessary, the latter can then be improved by lag–lead compensation.

Lag–Lead Compensation

The procedures for the design of lag compensation and lead compensation also apply to lag–lead design. In Fig. 8.18, the low-frequency asymptote of the lead-compensated system of Example 8.7.3 intersects the 0-dB axis at frequency 1, indicating a gain of 1. If the specifications should require this to be increased by a factor of 10 without reduction of the crossover frequency, phase-lag can be added to raise the low-frequency asymptote by 20 log 10 = 20 dB. By (8.16), this requires a ratio 10 of the break frequencies. The highest of these is chosen a factor of 10 below the crossover frequency $\omega_c = 1.45$, and the overall compensator, including the lead compensation of (8.23), becomes

$$G_c = 10 \frac{1 + s/0.145}{1 + s/0.0145} \frac{1 + s/0.46}{1 + s/4.6} \tag{8.24}$$

8.8 DESIGN OF PID CONTROLLERS

PI Control

A design procedure may be formulated which is a variation of that for phase-lag compensation:

1. Determine the frequency ω_c at which the phase angle of G is $(-180° + \phi_m + 5°)$, where ϕ_m is the specified phase margin, with a 5° allowance for the angle contributed at ω_c by the compensator. This is the desired crossover frequency.

2. Plot the loop gain function KG that crosses the 0-dB axis at this frequency. K can be determined at this stage or, perhaps more conveniently, later.

3. Choose the break frequency z of the controller 1 decade below ω_c.

4. Plot $G_c G$ to verify that all specifications are met, and express the compensator

$$G_c(s) = K \frac{s + z}{s} = K\left(1 + \frac{z}{s}\right) \tag{8.25}$$

The next example illustrates this procedure for the most common type of application of this controller. This is in industrial process control, with a type 0 plant and with the intention of removing steady-state errors after step inputs.

Example 8.8.1   PI Control of Temperature

The plant is given by

$$G = \frac{2}{(s + 1)(s + 2)} \tag{8.26}$$

The specified damping ratio of the dominating poles is 0.5, suggesting a phase margin $\phi_m = 52°$. Thus, from point 1, the desired crossover frequency is where the phase angle of the plant is $-123°$. By calculating phase angles at some frequencies, $\omega_c = 2.6$ is chosen, for which the phase margin is $53°$.

Figure 8.21 shows the Bode plot of KG with the desired crossover frequency 2.6. It is rather close to the break frequencies, so the deviation of the actual curve from the asymptotes must be accounted for. Recall that for a simple lag the deviation at a frequency ratio 2 to the break frequency is -1 dB. The break at 1 will contribute about -0.5 dB at 2.6, and that at 2, about -2 dB. The total deviation at 2.6 is therefore about -2.5 dB, and the -40 dB/dec asymptote above frequency 2 is drawn to pass 2.6 at a height of 2.5 dB. The asymptotic plot of KG can now be completed and gives a level of 13 dB for the low-frequency asymptote. Since the gain of G is 1, this gives $K = 4.57$.

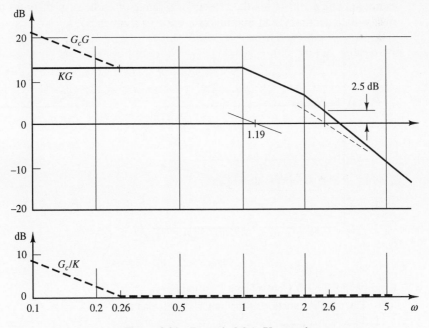

Figure 8.21   Example 8.8.1: PI control.

The plot of G_c/K is also shown, with break frequency 1 decade below the crossover frequency, and a 0-dB high-frequency asymptote. It is in series with KG, so the sum of their Bode plots is G_cG. Using (8.25), the controller becomes

$$G_c = 4.57\left(1 + \frac{0.26}{s}\right)$$

The low-frequency asymptote of G_cG is $(4.57 \times 0.26)/s$, and if K had not yet been determined, it could now be found by construction from the condition that this asymptote intersects the 0-dB axis at 4.57×0.26, from (7.23).

This example is a variation of the phase-lag design procedure in that the high-frequency asymptote instead of the low-frequency asymptote of G_c/K is taken to coincide with the 0-dB axis. This relates to the two viewpoints mentioned at the end of Section 8.6. In the present case the controller is seen as improving steady-state accuracy without loss of relative stability, instead of improving relative stability without loss of steady-state accuracy.

PD Control

Figure 8.22 follows from Figs. 8.2(c) and 7.9 and summarizes the characteristics of a PD controller. In particular, it shows the phase lead ϕ_d provided at an arbitrary frequency ω_c if the break frequency of the controller is z. This lead is small at frequencies significantly below the break frequency, and at frequencies that are more than a factor of 5 above the break frequency the change of lead becomes small. Thus little further improvement of the phase margin at a specified frequency will be obtained by choosing the break frequency more than a factor of 5 lower. This is used in the next example, which illustrates the effects of a range of choices of the controller zero on steady-state accuracy and maximum crossover frequency.

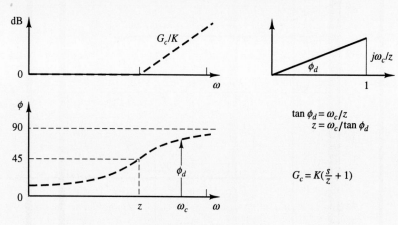

Figure 8.22   PD controller.

Example 8.8.2   PD Control in a Position Servo

The transfer function of the motor and its load is

$$G = \frac{1}{s(s + 1)(0.5s + 1)} \tag{8.27}$$

Figure 8.23 shows a plot of KG for arbitrary K; that is, the 0-dB axis is as yet left undetermined. It may be noted that this is often a useful approach.

The desired phase margin $\phi_m = 52°$. The following table shows the angle deficiency ϕ_d that must be made up to achieve this for a range of desired crossover frequencies ω_c of $G_c G$. Also shown are the phase angles ϕ of G and the largest values of the controller zero z, calculated from Fig. 8.22, that can provide the angle deficiency.

ω_c	1	1.5	2.0	2.5	3.0
ϕ	−161.6	−183.2	−198.4	−209.5	−217.9
ϕ_d	33.6	55.2	70.4	81.5	89.9
z	1.50	1.04	0.71	0.37	

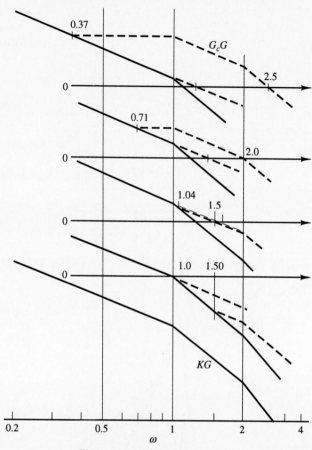

0.37

G_cG

2.5

0

0.71

2.0

0

1.04

1.5

0

1.0   1.50

0

KG

0.2   0.5   1   2   4
ω

Figure 8.23   Example 8.8.2: PD control.

The corresponding Bode plots of $G_c G$ are shown. Below the controller zeros at z they coincide asymptotically with KG since here the asymptote of G_c/K in Fig. 8.22 is 0 dB. Using the asymptotes for convenience, the 0-dB axis levels are drawn where the asymptotes intersect the crossover frequencies. Recall that for this type 1 system the (extension of the) low-frequency asymptote intersects the 0-dB axis at the value of the gain of $G_c G$. This gain equals K, the gain of G_c, since that of G is unity.

Observe that the design with the highest crossover frequency is not the one with the highest gain. Depending on the desired characteristics, a range of choices may be satisfactory.

The next example concerns a system with a mechanical resonance and involves the use of pole–zero cancellation as well as PD control.

Example 8.8.3   Satellites with Structural Resonance

Figure 8.24 repeats the block diagrams in Fig. 6.24 for the attitude control of satellites with structural resonance. System (a) applies when actuators and sensors are on the separate masses of the two-mass model, and (b) when they are on the same mass. Figure 8.25 shows the Bode plots of G. Note that (a) shows more than one crossover frequency. This could also be true for (b) because damping ratios of resonances are usually difficult to estimate. For stability the phase margins at these crossover frequencies must be positive. Design (b) needs phase lead to achieve an adequate phase margin, but the situation is

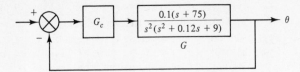

(a) Resonance between actuator and sensor

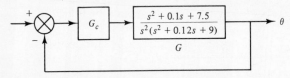

(b) No resonance between actuator and sensor

Figure 8.24　Satellites with structural resonance.

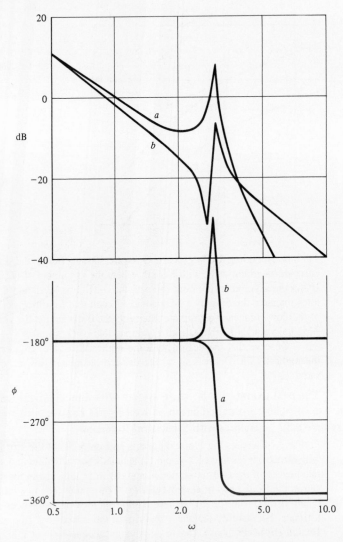

Figure 8.25　Bode plots for plants G in Fig. 8.24.

certainly much better than for (a), as was also seen by root locus design in Example 6.8.4. For (a) the phase angle at the highest crossover frequency is close to $-360°$, and phase-lead compensation or PD control does not provide nearly enough phase lead to realize a positive phase margin. In Example 6.8.4 it was noted that design (a) can be made to look like (b) by using the compensator to introduce a pair of zeros near the resonance poles, and that these zeros should be at lower frequency than the poles for all plant parameter variations. Indeed, on the Bode plot for (b), if the zeros were to the right of the poles, the phase angle peak above $-180°$ would change to a dip below $-180°$.

Figure 8.26 shows Bode plots of (b) with PD control $K(Ts + 1)$ as in (6.17) for $K = 1$ and several values of T. Note that changing the gain K will not affect the phase-angle curves and will only raise or lower the magnitude plots. The system is stable for any combination of K and T. Increasing the gain K moves the crossover frequencies to higher values and increases the phase margins. The PD control provides more phase lead at a given frequency when T is larger, because the zero due to PD control is then at a lower frequency. Figure 6.26 shows unit step responses for $0.6(2s + 1)$, $2(s + 1)$, and $8(0.5s + 1)$. From inspection of Fig. 8.26, these choices of gains provide reasonable phase margins and the cases are in the order of increasing crossover frequencies. Accordingly, they should be in the order of increasing speed of response, as is verified by Fig. 6.26. For the controller $8(0.5s + 1)$ the lower of the crossover frequencies is close to the resonance frequency, so the larger resonance response is not surprising. As one check on the sensitivity, Fig. 8.27 shows Bode plots for system (b) with controller $(s + 1)$

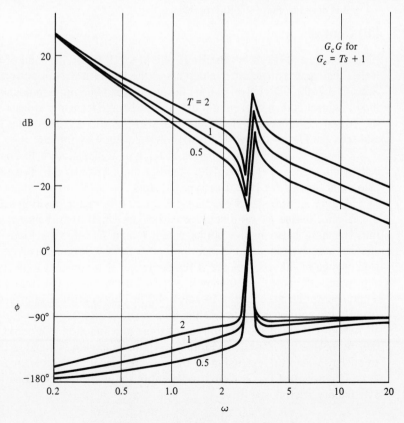

Figure 8.26   Plots for system (b) with $G_c = Ts + 1$.

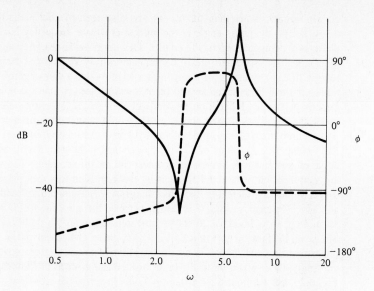

Figure 8.27   Plant parameter variation in Example 8.8.3.

when the undamped natural frequency of the resonance poles is doubled. It is seen that this does not cause stability problems.

PID Control

Figure 8.2(e) and (f) shows the Bode plot of a PID controller, and of a lag–lead controller that approximates it. Analogous to the lag–lead design outlined at the end of Section 8.7, the PD control can be designed first. The high-frequency asymptote of the PI control is then chosen at 0 dB, and its break frequency at least a decade below the crossover frequency. However, it is also possible to design the PI control first, and then the PD control with a 0-dB low-frequency asymptote.

The most common application is for type 0 plants, with PI control for steady-state accuracy and PD control for stability and/or bandwidth. But PID control may also be necessary for type 1 or type 2 plants.

Figure 8.28(a) shows the block diagram of a motor position servo with a load disturbance torque D. As discussed earlier, for small disturbance response the gain must be high between the points where R and D enter the loop. Example 4.2.5 showed that, even though the plant is type 1, there will be a steady-state error after a step change of the disturbance if the controller is a constant gain. But this error is zero with a PI controller, as may be verified by means of the final value theorem in a manner similar to its use in Example 4.2.5. This controller in effect introduces infinite static gain at this point in the loop.

Figure 8.28(b) shows the type of loop gain function that could result from the design for this type 2 system. From (7.24), the loop gain is equal to the square of the frequency where the extension of the low-frequency asymptote intersects the 0-dB axis. Crossover of $G_c G$ occurs well within the -20 dB/dec slope range to ensure satisfactory relative stability.

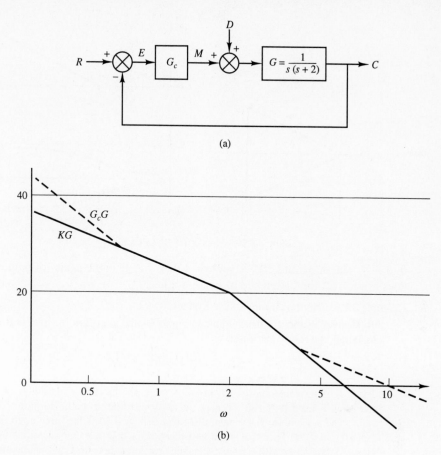

Figure 8.28   PID control for disturbance rejection.

8.9 OPEN-LOOP UNSTABLE OR NONMINIMUM-PHASE PLANTS

If the plant G is open-loop unstable, that is, has poles in the right-half s-plane, then, as mentioned earlier, Nyquist diagrams or root loci should be used to determine system stability. If this shows stability, Bode plots can be applied to find phase margin, as well as loop gain characteristics, accuracy, and crossover frequency, or to choose gains for a specified crossover frequency. Bode plots are always valid for the latter uses.

For nonminimum-phase plants, that is, with right-half-plane zeros, Bode plots are valid for stability analysis, but care is necessary because, as with right-half-plane poles of G, there is no longer a unique relation between the magnitude and phase-angle plots. Thus the very useful design guide that aims for an adequate length of -20 dB/dec slope near crossover is no longer valid. Figure 8.29 shows vector diagrams and Bode diagrams for $(1 + Ts)$ and $(1 - Ts)$. For $(1 + Ts)$ the angle increases from $0°$ to $+90°$, and for $(1 - Ts)$ it decreases from $0°$ to $-90°$ as ω increases from

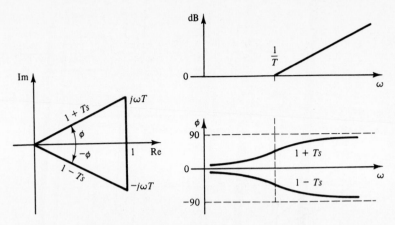

Figure 8.29   Plots of $(1 + j\omega T)$ and $(1 - j\omega T)$.

0 to $+\infty$. The magnitude $|1 + j\omega T| = \sqrt{1 + (\omega T)^2}$ is the same for both, so its plot does not reflect the difference of phase angles.

Example 8.9.1   Nonminimum-Phase System

To sketch root loci for the example in Fig. 8.30, the loop gain function is first rewritten to extract the root locus gain:

$$\frac{(-K T_n/T)(s - 1/T_n)}{s(s + 1/T)}$$

For $K > 0$, the root locus gain is negative, so, as for the example in Fig. 6.27, the asymptote is at $0°$ instead of $180°$ because the angle condition is satisfied in this direction.

The asymptotic Bode magnitude plot in Fig. 8.30 is the same as for a numerator factor $(1 + T_n s)$. Therefore, right-half-plane zeros and poles are identified by small circles. To predict stability from the magnitude plot, note that the phase of a function $G_a = G(1 - T_n s)$ is the same as that of $G_b = G/(1 + T_n s)$. Thus in Fig. 8.30 the angle curve for (a) is the same as for the minimum-phase function shown by (b), and adequate phase margin will exist if (b) has an adequate length of -20 dB/dec slope at the frequency where (a) crosses the 0-dB axis. These ideas are useful to estimate the severity of any stability problems visually from the magnitude plot (a), without actually plotting (b).

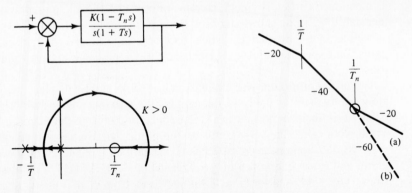

Figure 8.30   Example 8.9.1: nonminimum-phase system.

But for design of nonminimum-phase systems it is usually recommended that the complete phase-angle curve be drawn as well.

With the -60 dB/dec slope of (b) beyond $1/T_n$, phase margins on the order of at least $50°$ can be realized only at crossover frequencies below $1/T$, while for the corresponding minimum-phase system, theoretically any crossover frequency is possible.

As this example illustrates, nonminimum-phase zeros can impose severe restrictions on crossover frequency and bandwidth. However, if their break frequencies occur far above the crossover frequency, they contribute only small negative angles at this frequency and are not objectionable.

8.10 SYSTEMS WITH TRANSPORT LAG

The terms *transport lag, transportation lag, dead time,* or *delay time* have been used to describe elements with the transfer function e^{-Ts}. From Eq. (1.17) or Table 1.6.1, if $F(s)$ is the transform of a function $f(t)$, then $e^{-Ts}F(s)$ is the transform of $f(t - T)$, the function $f(t)$ delayed by T, as illustrated in Fig. 8.31(a). Transport lag is often needed to model the effect of flow through long fluid lines, as indicated in Fig. 8.31(b), and occurs in other contexts as well. If the fluid line has length L and the fluid velocity is v, the dead time during which the process at the outlet is unaware of a change of $f(t)$ at the inlet is $T = L/v$. Long pipelines between subsystems in power plants and other processes can have a profound effect on systems dynamics. In the schematic diagram of a level control system in Fig. 3.1, long pneumatic lines may connect the controller to the actuator, and the pipeline between the control valve and the tank may introduce a considerable delay. Lumping the delays in a loop into a single transport lag T, the loop gain function in the block diagram of Fig. 3.2 can be written in the following form, for P control:

$$G_c G = \frac{Ke^{-Ts}}{(T_1 s + 1)(T_2 s + 1)} \qquad (8.28)$$

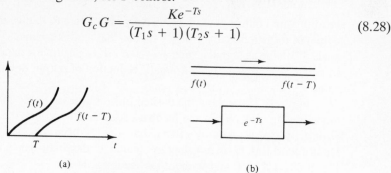

(a)   (b)

Figure 8.31   Transport lag.

The series expansion of e^{-Ts} implies that $G_c G$ theoretically has an infinite number of zeros, complicating root locus analysis. An approximation useful to show the general effect is to use the first two terms of the series: $e^{-Ts} \approx 1 - Ts$. This indicates a nonminimum-phase system, and serves, from Section 8.9, to emphasize the potentially strong effect, depending on the value of T compared to T_1 and T_2. Another approximation often used for initial analysis is $e^{-Ts} \approx 1/(1 + Ts)$. Unless T is small compared to T_1 and T_2, this also shows the bandwidth-limiting effect of transport lag.

Analysis in the frequency domain is both exact and more convenient.

$$|e^{-j\omega T}| = |\cos \omega T - j \sin \omega T| = 1 \qquad \phi(e^{-j\omega T}) = -\omega T \qquad (8.29)$$

Thus transport lag does not affect the magnitude plot of $K/[(T_1 s + 1)(T_2 s + 1)]$ shown in Fig. 8.32, but changes its phase-angle curve by $-\omega T$. The curve identified by ϕ_2 is for larger T than ϕ_1. If the horizontal at angle ϕ_m above $-180°$ indicates the desired phase margin, it is evident that larger transport lag T reduces the crossover frequency possible without phase-lead compensation. This shows that, if a system schematic suggests appreciable transport lag, careful analysis is necessary to ensure that adequate stability is maintained.

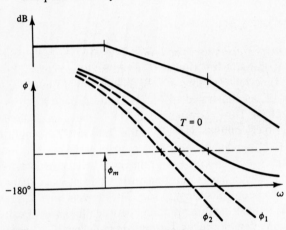

Figure 8.32   Design with transport lag.

8.11 FEEDFORWARD CONTROL

Feedforward control is used extensively in practice to reduce the effect on the system output of measurable disturbance inputs, that is, disturbances that via a sensor can be made available as signals, such as water, oil, or pneumatic supply pressure variations. In many cases it can give a dramatic reduction of output deviations from the desired value.

Figure 8.33, without the dashed link, is similar to the disturbance input system model in Fig. 4.4. While the feedback loop acts to reduce the effect of disturbances D on C, it can do so only after C has already been affected by D. The idea of the feedforward link G_f is in effect to counteract the disturbance before it changes C. In Fig. 8.33, if $R = 0$ and without the feedback loop,

$$M = (L - G_f G_c G_1)D \qquad (8.30)$$

If G_f could be made equal to $L/(G_c G_1)$, the effect of D on M, and on C, would be eliminated.

However, usually this form of G_f has more zeros than poles and hence is not physically realizable. But it is evident, and can be explored by the use of Bode plots, that simple forms of G_f can often greatly reduce the size of the effective disturbance input

$$D^1 = (L - G_f G_c G_1)D \qquad (8.31)$$

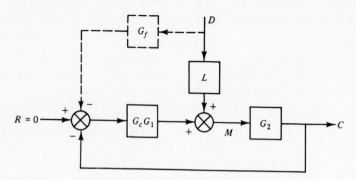

Figure 8.33   Feedforward control.

to the feedback system. This is particularly advantageous if the block G_2 is relatively slow acting compared to the others. Then the steady-state gains of these other blocks are most important, and choosing G_f to be a constant gain can greatly reduce disturbance effects.

8.12 CONCLUSION

Although other forms of plots were discussed, the heavy emphasis in this chapter has been on design by means of Bode plots. With the performance measures of Section 8.4, these combine ease of plotting and ready identification of features of the performance. The particular cases of nonminimum-phase plants and systems with transport lag were considered, as was the use of feedforward control to reduce the effect of measurable disturbances.

For routine analysis and design, Appendix B gives an interactive program for constructing actual and asymptotic magnitude plots and actual phase-angle curves. It is design oriented in that for a given plant the effect of selected dynamic compensators can be shown.

PROBLEMS

8.1. Determine the required positions of the dominating pair of poles of a system if the resonance peak is specified to be 1.15 and the bandwidth is 10 rad/sec.

8.2. Experimental measurements yield a plot of the magnitude of the frequency response function with a resonance peak 1.35 at a frequency of 10 rad/sec.
   (a) Estimate ζ and ω_n of the dominating system poles.
   (b) Estimate the bandwidth.
   (c) Estimate the percentage overshoot in response to step inputs.

8.3. Estimate the phase margin and crossover frequency specifications to which a loop gain function must be designed if the closed-loop system must meet one of the following sets of requirements:
   1. An effective damping ratio $\zeta = 0.6$ and a bandwidth of $\omega_b = 10$ rad/sec.
   2. A resonance peak 1.15 occurring at a frequency of 10 rad/sec.
   3. A step response with an overshoot of 20% and a settling time of 1 sec.

8.4. In Fig. P8.4, G is the plant in a two-tank level control system:

$$G = \frac{1}{(0.1s + 1)(0.02s + 1)}$$

   (a) If $G_c = K$, find K for $\phi_m = 55°$ and determine the corresponding steady-state error for unit step inputs.

   (b) To reduce this error to zero, introduce PI control by making $G_c = K(1 + 5/s)$, with K as in part (a). Modify the Bode plot to account for this, and verify that the phase margin at ω_c of part (a) is reduced by no more than 6°, which will be taken to be acceptable in this case.

   (c) How would the phase margin be affected if the parameter 5 in G_c were increased?

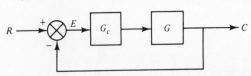

Figure P8.4

8.5. If Fig. P8.4 models an antenna position control with $G = 1/[s(0.1s + 1)]$, $G_c = K$:

   (a) Draw the asymptotic Bode magnitude plot without establishing a 0-dB axis level, and plot the phase-angle curve from points calculated at $\omega = 4, 5, 6, 8, 10, 15, 20,$ and 30.

   (b) Determine the crossover frequencies ω_c required for phase margins ϕ_m of 65°, 55°, 45°, and 35°.

   (c) At the frequencies ω_c of part (b), indicate the approximate deviations of the true curve to the asymptotic magnitude plot.

   (d) Draw the 0-dB axis levels associated with the values ϕ_m specified in part (b) and find the values of K needed to realize these phase margins.

   (e) Predict the closed-loop resonance peaking M_p and bandwidth ω_b for each of the designs from the open-loop response data.

8.6. It is desired to compare the predictions of ω_b and M_p in Problem 8.5 with closed-loop results obtained using the Nichols chart.

   (a) Transfer the Bode plot data for the lowest value of K in Problem 8.5 to the chart.

   (b) How does a curve on the Nichols chart change if K is changed from K_1 to K_2?

   (c) Plot the Nichols chart curves for all values of K in Problem 8.5.

   (d) From the plots, find ω_b and M_p for each case and compare them with the predictions.

8.7. If Fig. P8.4 models a position servo with $G_c = K$ and

$$G = \frac{1}{s(0.1s + 1)(0.01s + 1)}$$

   (a) Draw the asymptotic Bode magnitude plot with undetermined 0-dB axis level.

   (b) Calculate the phase angle at a number of frequencies, plot the phase-angle curve, and find the crossover frequency ω_c if the desired phase margin is 45°.

   (c) Use ω_c to set the 0-dB axis level on the Bode magnitude plot, corrected from the asymptotic plot at several frequencies.

   (d) Replot the results on the Nichols chart and compare bandwidth ω_b and resonance peaking M_p with those predicted from the Bode plot data.

8.8. Figure P8.8 shows the Bode plots for the loop gain function of a unity feedback system.

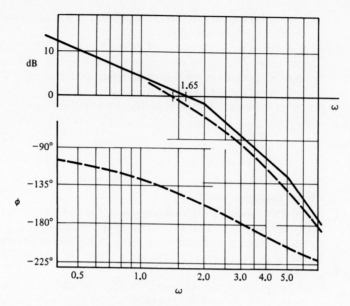

Figure P8.8

(a) What is the loop gain function?

(b) Predict the closed-loop system resonance peaking and bandwidth.

8.9. In Problem 8.8, replot the Bode plot data on the Nichols chart and:

    (a) Compare ω_b and M_p obtained from the Nichols chart with the predictions of Problem 8.8(b).

    (b) Plot the closed-loop frequency response function magnitude versus frequency, with magnitude in decibels, on semilog graph paper.

8.10. Determine a desired Bode plot and corresponding loop gain function of a system with unity feedback that satisfies the following specifications:

    1. The steady-state error for step inputs must be zero.

    2. The steady-state error following unit ramps may not exceed 2%.

    3. The error in response to sine inputs up to 10 rad/sec may not exceed 10%.

    4. To limit the effect of high-frequency noise, the output in response to sine inputs above 250 rad/sec may not exceed 10% of the input.

    5. To ensure adequate bandwidth, the crossover frequency must be about 50 rad/sec.

    6. The phase margin must be at least about 55°.

8.11. If Fig. P8.4 models an electro-hydraulic position control for a robot with

$$G(s) = \frac{1}{s[(s^2/50^2) + 2 \times 0.5 \times (s/50) + 1]}$$

    (a) For $G_c = K$, find K to achieve a phase margin of 50°, and plot the Bode magnitude plot and the phase-angle curve.

    (b) Design phase-lag compensation to reduce steady-state errors for ramp inputs by a factor of 10 while reducing the phase margin by no more than 6°.

8.12. In Problem 8.11:

    (a) If G_c is a phase-lead compensator with a ratio 10 of the break frequencies, find the maximum crossover frequency that can be achieved for a 45° phase margin.

(b) Determine the gain and break frequencies of this compensator and construct the Bode magnitude plot of $G_c G$.

(c) Plot the phase-angle curves of G and $G_c G$ to verify that this compensation is not very effective in raising crossover frequency and bandwidth for under-damped quadratics, because of the relatively fast change of phase angle near the undamped natural frequency.

(d) What is the effect on steady-state errors relative to $G_c = K$?

8.13. If Fig. P8.4 models a position servo with

$$G = \frac{1}{s(0.1s + 1)(0.01s + 1)}$$

design a series compensator G_c to achieve a crossover frequency of 30 rad/sec and a phase margin of about 45°.

8.14. If Fig. P8.4 models a radar tracking system with $G = 1/[(0.1s + 1)s]$, design se-ries compensation to meet the following specifications:

1. The steady-state error following ramp inputs may not exceed 2%.
2. The error in response to sinusoidal inputs up to 5 rad/sec should not exceed about 5%.
3. The crossover frequency should be about 50 rad/sec to meet bandwidth require-ments while limiting the response to high-frequency noise.
4. The ratio of the break frequencies of G_c should not exceed 5 to limit noise effects.
5. The phase margin should be about 50°.

8.15. If Fig. P8.4 models the pressure control of a plant consisting of three simple lags in series

$$G = \frac{1}{(s + 1)(0.25s + 1)(0.1s + 1)}$$

design a PI controller if a crossover frequency of about 2 rad/sec and a phase margin of about 50° are desired.

8.16. For the plant of Problem 8.15, design a controller G_c to satisfy the following speci-fications:

1. The steady-state error for step inputs may not exceed 10%.
2. The crossover frequency of $G_c G$ should be at least 7 rad/sec.
3. The phase margin should be about 45°.

8.17. In Fig. P8.4 with $G = 1/[s(0.5s + 1)]$, design phase-lag compensation to meet the following specifications:

1. The steady-state unit ramp following error may not exceed 10%.
2. The phase margin should be at least 45°.
3. The error in response to sinusoidal inputs up to 0.1 rad/sec may not exceed about 4%.

8.18. In Fig. P8.4 with $G = 1/[(s + 1)(0.25s + 1)^2]$:

(a) If $G_c = K$, find K for a phase margin of about 48°.

(b) Design phase-lag compensation G_c to reduce the steady-state error of part (a) by a factor of 10 for a phase margin of about 42°.

8.19. In Fig. P8.4 with $G = 1/[s(0.2s + 1)(0.05s + 1)]$:

(a) If $G_c = K$, find K for a 45° phase margin and the corresponding crossover frequency.

(b) Design phase-lead compensation with a ratio 10 of the break frequencies to maximize the system bandwidth, maintaining about 45° phase margin.

8.20. In Fig. P8.4 with $G = 100/[s(s + 10)^2]$:
   (a) If $G_c = K$, find K for a unit ramp steady-state following error of 5% and construct the corresponding Bode plot.
   (b) Design phase-lag compensation G_c to achieve at least 55° phase margin without loss of the steady-state accuracy of part (a).

8.21. In Fig. P8.4 with $G = 1/[s(0.1s + 1)(0.02s + 1)]$, design a phase-lag compensator to satisfy the following specifications:
   1. The steady-state unit ramp following error may not exceed 2%.
   2. The phase margin should be at least 50°.

8.22. In Fig. P8.4 with $G = 1/[s(0.05s + 1)(0.01s + 1)]$, design phase-lag compensation to meet the following specifications:
   1. The steady-state ramp input following error may not exceed 5%.
   2. The phase margin must be at least 50°.

8.23. In Fig. P8.4 with $G = 1/[s(0.1s + 1)(0.001s + 1)]$:
   (a) For $G_c = K$ find K and plot KG for 0.1% steady-state error following ramp inputs. Determine the stability.
   (b) With K as above, determine the effect of a lead network

   $$G_c = K\left[\frac{(s/50) + 1}{(s/400) + 1}\right]$$

   on steady-state accuracy and stability and on the bandwidth.
   (c) Could phase-lag compensation have been used if no loss of bandwidth relative to part (a) is permitted?

8.24. In Fig. P8.4 with $G = 1/[(2s + 1)(0.2s + 1)]$:
   (a) If $G_c = K$, find K for 50° phase margin and construct the Bode plot for this gain.
   (b) Design PI control to improve steady-state errors without loss of bandwidth for a 45° phase margin.

8.25. In Fig. P8.4 with $G = 1/[(s + 1)(0.1s + 1)]$, design a controller G_c to meet the following specifications:
   1. Zero steady-state error for step inputs.
   2. The steady-state unit ramp following error may not exceed 2%.
   3. The error in response to sinusoidal inputs up to 5 rad/sec may not exceed 10%.
   4. The phase margin should not be below about 63°.
   Note that a combination of two types of controllers, in series, can be necessary to meet a set of specifications.

8.26. In Fig. P8.4 with $G = 1/[s(0.1s + 1)(0.2s + 1)]$:.
   (a) If $G_c = K$, find K and construct the Bode plot for a steady-state unit ramp following error of 3.3% and determine system stability.
   (b) Design phase-lag compensation to stabilize the system, with about 50° phase margin, without loss of steady-state accuracy.

8.27. In Fig. P8.27, choose K and b to meet the following specifications:
   1. The unit ramp steady-state following error should not exceed 10%.
   2. The phase margin should be about 65°.

8.28. In Fig. P8.4 with

$$G(s) = \frac{1}{s(s + 1)}$$

design phase-lead compensation to meet the following specifications:

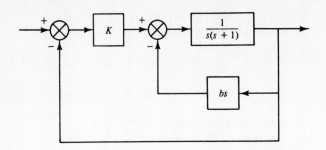

1. The steady-state unit ramp following error may not exceed 10%.
2. The phase margin should be approximately 53°.
3. To limit noise effects, the ratio of the break frequencies of the compensator should not be larger than necessary.

8.29. In Fig. P8.4 with $G = 1/[s(0.1s + 1)]$, design phase-lead compensation to meet the following specifications:
1. The steady-state unit ramp following error may not exceed 2%.
2. The phase margin should be approximately 45°.
3. To limit noise effects, the ratio of the break frequencies of the compensator should not be larger than necessary.

8.30. In Fig. P8.4 with $G = 1/[s(0.2s + 1)(0.05s + 1)]$, design phase-lead compensation to meet the following specifications:
1. The steady-state ramp input following error may not exceed 15%.
2. The phase margin must be approximately 63°.
3. To limit noise effects, the ratio of the break frequencies of the compensator should not be larger than necessary.

8.31. In Fig. P8.4 with

$$G = \frac{1}{s(0.1s + 1)(0.01s + 1)}$$

design phase-lead compensation to meet the following specifications:
1. The steady-state ramp input following error may not exceed 2%.
2. The phase margin must be of the order of 45°.
3. To limit noise effects, the ratio of the break frequencies of the compensator should not be larger than necessary.

8.32. If Fig. P8.4 with $G(s) = 1/(s + 1)$ models a temperature control system, design PI control for a phase margin of 52°, a crossover frequency of 10 rad/sec, and minimum steady-state ramp following error. How large is this error?

8.33. If Fig. P8.4 with $G(s) = 1/[(s + 1)(s + 4)]$ models a motor speed control system, design PI control for a phase margin of 52°, maximum bandwidth, and minimum steady-state ramp following error. Find this error and the crossover frequency, and estimate the bandwidth.

8.34. If Fig. P8.4 with $G(s) = 1/[s(s + 1)(s + 4)]$ models a motor position servo, design PD controllers to obtain a phase margin of 52° at crossover frequencies of about 3 and 4 rad/sec, and find the corresponding steady-state unit ramp following errors.

8.35. In Fig. P8.4 with $G = (1 - 0.1s)/[s(1 + s)]$:
(a) If $G_c = K$, find K for a phase margin of about 55°.

(b) Design a phase-lead network G_c with a ratio 10 of the break frequencies to achieve maximum improvement of crossover frequency of part (a), without loss of phase margin.

(c) Compare the steady-state errors of parts (a) and (b).

8.36. In Fig. P8.4 with $G = (1 - s)/[s(1 + s)]$:

(a) Construct the Bode magnitude plot for $G_c = K$, with K chosen for a phase margin of about 55°.

(b) What is the maximum crossover frequency achievable with phase-lead compensation with ratio 10 of the break frequencies if the phase margin is to equal that of part (a)?

8.37. In Fig. P8.4 with $G_c = K$ and $G = e^{-T_d s}/(s + 1)$, where G is a transfer function frequently used to approximate the dynamic behavior of processes:

(a) Construct Bode magnitude and phase-angle plots for $T_d = 0$, 0.1, and 0.5 sec for $K = 1$.

(b) For $T_d = 0.1$ and 0.5, find the values of K to achieve 50° phase margin and the corresponding crossover frequencies.

8.38. If the combination of a transport lag and a simple lag as in Problem 8.37 cannot model a process adequately, an additional simple lag may often do so. Let

$$G = \frac{e^{-T_d s}}{(s + 1)(0.2s + 1)} \qquad T_d = 0.2 \qquad G_c = K$$

Construct Bode magnitude and phase-angle curves both with and without the transport lag, and find and compare the values of K and the corresponding crossover frequencies for a phase margin of about 55°.

8.39. In Fig. P8.39:

(a) Find K for a phase margin of about 55° and the corresponding steady-state error following a unit step change of disturbance input D in the absence of feedforward control G_f.

(b) Construct the asymptotic Bode magnitude plot of C/D for part (a).

(c) Use Bode plots to choose a constant-gain feedforward control G_f that counterbalances the direct effect of the disturbance over a wide frequency range. Construct the resulting C/D plot and compare with that of part (b) to judge the effect of feedforward on steady-state and dynamic disturbance response.

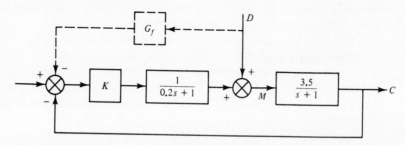

Figure P8.39

8.40. Repeat Problem 8.39 if the time constants, but not the gains, of the two simple lag blocks are interchanged. Compare the effectiveness of feedforward control, chosen for best results at low frequencies, with that in Problem 8.39.

9

Digital Control Systems

9.1 INTRODUCTION

A very strong trend to digital computer control is evident in all areas of application. The introduction of minicomputers after about 1965 and particularly of microcomputers since about 1975, and the greatly increased power and reduced cost of computer hardware have been mainly responsible for this. Applications are found in large processes such as power plants and steel mills, in aircraft control and transportation generally, in manufacturing and robotics, and so on. Microprocessor-based digital single-loop controllers are available for single-loop systems, that is, the equivalent of pneumatic or electronic process controllers.

Digital control offers important advantages in flexibility of modifying controller characteristics or of adapting the controller if plant dynamics change with operating conditions. In multivariable systems, with more than one input and output, modern techniques for optimizing system performance or reducing interactions between feedback loops can be implemented.

It should be emphasized that feedback control in the sense of earlier chapters is only one of the functions of the computer. In fact, most of the information transfer between process and computer is of an on–off nature and exploits the logical decision-making capacity of the computer. A controls practice example will be discussed to illustrate this and to broaden the focus of attention beyond strict feedback control.

To enable the computer to meet the variety of demands imposed on it, it is *time-shared* among its tasks. This inherently requires *sampling* of the variables of interest. The sampling of an output with a sampling interval of T seconds may be visualized by assuming the presence of a relay in the feedback path that closes momentarily every T seconds. Thus feedback is not continuous but only intermittent, at

264

the sampling instants. In the course of the sampling interval, all outputs are sampled, and the computer uses the sample values to calculate actuating signals to the system actuators according to control algorithms. As will be discussed, frequently these are digital implementations of PID control, separate for the individual feedback loops in the process.

Sampling is a fundamental departure from the continuous systems considered thus far, and its effect on system dynamics will require careful attention. A large sampling interval reduces computing and sampling requirements, but too large an interval will not adequately represent actual signal variations and could cause instability.

9.2 COMPONENTS IN A PROCESS CONTROL CONFIGURATION

Large processes with many feedback loops have conventionally frequently been controlled by separate PID controllers for each loop, which are also expected to handle moderate interactions between the loops and under static conditions make each output equal to the corresponding input. To introduce computer control, Fig. 9.1 shows the schematic diagram for a system of which a number of analog outputs must be controlled.

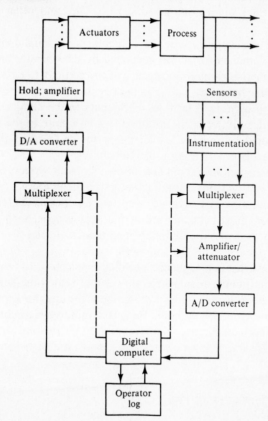

Figure 9.1   Computer control systems.

Sensors measure the system outputs. Those that provide a digital signal are preferable, such as shaft position encoders or turbine flow meters, but usually the sensor outputs are continuous.

Instrumentation can refer to several possibilities. If the sensor output is not electrical, a transducer (TDR) is used to change it to a proportional electrical signal. The sensor or TDR may be followed by a transmitter (TMR) if needed because of the distance to the computer or the quality and strength of the signal. Signal-conditioning filters may be present to improve the signals. Simple lag filters to remove high-frequency noise from the sensor outputs are very common and will be found to have special significance in connection with the effects of sampling.

The *multiplexer* connects each signal in turn, as selected by the computer, to a single *analog-to-digital (A/D) converter.* The *amplifier–attenuator,* its scaling selected by the computer for each signal, may be present to obtain voltage ranges suitable for conversion.

The *computer* applies a scaling inverse to that above and converts to engineering units from calibration data of the sensors and other instrumentation. The system inputs and stored values of present and past output samples are used to calculate signals to the actuators during each sampling interval, on the basis of the computer *control algorithm* selected. Such algorithms are discussed later and often implement PID control for each loop.

In addition to these *cyclic programs,* the computer must handle emergencies and respond to irregular demands for action generated by the operator and the program that use the logical decision-making capacity of the computer. This activity is discussed later. The *priority interrupt* feature is the key in allowing the computer to meet these requirements in organized fashion. The inputs to the computer have assigned priority levels. Unless interrupt is inhibited, which could be the case while cyclic programs are executed, an input will interrupt action on an input of lower priority. The latter is resumed later at the point of interruption.

The *output multiplexer,* on the output side of the computer, directs each digital output to the proper *actuator.* Again, digital actuators are desirable. A very important one is the stepping motor, which advances a certain number of degrees for each pulse it receives. However, most actuators are of the analog type, such as motors, pneumatic valve actuators, or hydraulic cylinders, and require *digital-to-analog conversion (D/A).*

The output of a converter is a sequence of voltage pulses, while the signal to the actuator must be a continuous signal, preferably that of which the pulses are the samples. The *zero-order hold* (ZOH) is a very common technique for approximating this signal reconstruction. This simply holds the last sample voltage constant until the next sample and produces a staircase approximation to the desired signal. Power *amplification* is also necessary, as is an electrical-to-pneumatic transducer for pneumatic actuators.

9.3 FEATURES AND CONFIGURATONS OF COMPUTER CONTROL

Section 9.2 focused on the control of a number of analog outputs. Some of the many other activities of the computer are process monitoring and data logging; alarming and taking appropriate actions when variables exceed permissible limits; sequencing of multiple parallel actuators; process startup and shutdown; sensing the status of contacts, indicating whether on–off valves are open or closed; switching on a motor; or opening a valve. Others, relating more closely to the feedback loops of Section 9.2, are:

1. Provisions for "bumpless transfer" between manual and automatic control to avoid potentially severe switch-over transients
2. Provisions to limit, for reasons of safety, the commanded change of an actuator position in one sampling interval to a given percentage of its value

Before initiating any action, the computer must make sure that it is safe to do so, in that variables are inside safe limits, preliminary commands have been obeyed, and the computer is not in a manual operating mode.

Pending a somewhat detailed example in the next section, a power plant turbine start-up procedure is outlined briefly. When not running, the rotor is rotated slowly by a turning gear driven by an electric motor. This is to prevent thermal sag of the rotor, which would cause destructive unbalance forces. Before rolling the rotor off this gear, the computer checks steam and bearing lubricant pressures. If a specified minimum speed is not reached in a specified time, this phase is repeated after a specified delay. Otherwise, the rotor is accelerated according to a recommended curve that avoids excessive thermal stress. The bearing vibrations are interrogated at given intervals, and if they exceed safe limits, the rotor is held at safe speeds, away from critical values, for a specified delay.

The system configurations of process control have undergone important developments. In the original *centralized configuration* a control computer is located in a central control room, with long individual lines to and from actuators and sensors. In the *star configuration,* indicated in Fig. 9.2(a), a central *supervisory computer* is connected to several unit control rooms, which are much closer to the actuators and sensors. This configuration greatly reduces wiring and installation costs for large installations. Noise pickup and susceptibility to damage are reduced correspondingly. As a logical next step, there is a strong trend toward the true *distributed configuration* illustrated in Fig. 9.2(b). A "data highway," in which multiplexing techniques are used to time-share a pair of wires between many digital signals, in a loop configuration permits bidirectional communication of controllers connected to the data highway with a supervisory computer in the unit control room. The controller for each loop may be mounted adjacent to the corresponding actuator.

This distributed control has been made possible by the availability of general-purpose digital single-loop controllers, microprocessor based, since about 1979. These permit adjustment of controller set points and of the parameters of the control algorithms by a remote supervisory computer. Controllers are available that also allow a choice among a number of different control algorithms, such as:

1. Several forms of PID algorithms.
2. Ratio control, to control, say, the ratio of pulverized coal and airflow for a power plant.
3. Feedforward control, discussed in Section 8.11.
4. Cascade control: Say that to increase a temperature the controller increases a fuel valve opening. The expected larger fuel flow may not result if the fuel

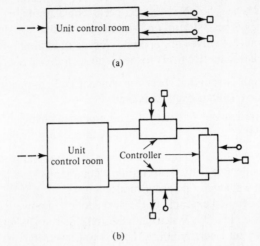

(a)

(b)

Figure 9.2   (a) Star and (b) distributed configurations.

source pressure has dropped due to disturbances. Therefore, the controller output is used instead to provide the set point of a flow controller, which acts quickly to counteract the disturbances and supply the desired flow.

Microprocessor-based *Programmable Controllers* should also be mentioned. These are very widely applied and are available at many levels of sophistication, ranging from control of a simple sequence of operations, and including sequences of which the steps may require closed-loop control of several variables to track variable inputs.

9.4 CONTROL PRACTICE: A LEVEL CONTROL EXAMPLE

This section has been included to illustrate practical issues and to show the extent to which straight engineering considerations of controls practice can complicate a level control. It is not needed for the study of subsequent material. Figure 9.3(a) shows a typical schematic diagram of a large feedwater-level control system for power plants and Fig. 9.3(b) the logic diagram for control valve operation. Although somewhat simplified, the system provides a good illustration of the many considerations that may affect the implementation of a control loop.

A level sensor and transmitter (TMR) provide a feedback signal in the range of 4 to 20 mA. The controller may be analog or digital, or both may be present with the former acting as a backup for the latter. A malfunction of the digital system would then cause automatic switchover to the backup, but manual selection would also be possible. Most controllers have provisions for "bumpless transfer," to avoid severe switch-over transients.

The E/P converter changes the electrical controller output into a proportional 3- to 15-psi pneumatic signal needed for operation of the control valves. The valve system is rather more complicated than the pneumatically actuated valve of smaller systems and is discussed below. A recirculation loop is provided to protect the pumps that supply the water by ensuring that the flow does not fall below a minimum value. LS represents one or more level switches, which give warning signals if the tank level passes set values.

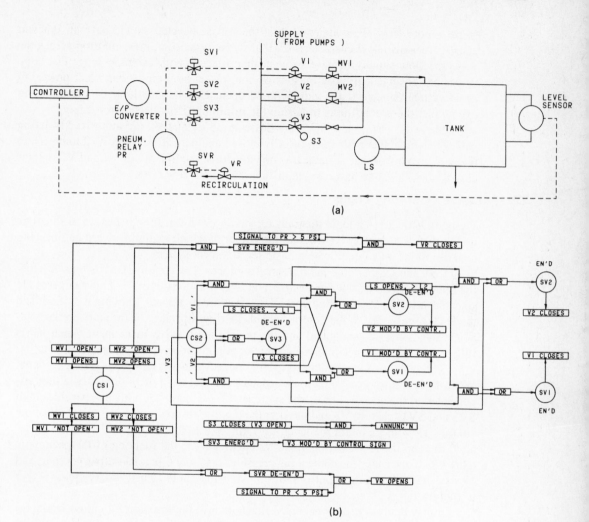

Figure 9.3   Level control example: controls practice.

Control Valve Operation

Valves V1 and V2 are pneumatically actuated control valves of full (100%) capacity. That is, each can supply the maximum tank flow. A small, 5% capacity valve V3 is needed to control the small flows during system start-up and shutdown. The large valves cannot do this adequately. For control-loop linearity, it is desirable to select valves with a linear installed flow characteristic (that is, they provide a linear relation between valve opening and valve flow during operation). To approximate this, the drop in supply pump discharge pressure as pump flow increases must be taken into account.

When only the small valve V3 is in operation, the recirculation valve VR is open to protect the pumps and the discharge pressure is almost constant. Hence a valve with a linear "inherent" flow characteristic will also operate linearly after installation. But with the large valves V1 and V2, pump pressure drops as a valve opens to pass more flow. To counteract this, the valves selected should have an inherent equal-percentage flow characteristic; that is, for a constant pressure drop across the valve the flow should increase stronger than linearly with valve opening. The large valves are arranged to close on loss of control air pressure or electrical power, to avoid system damage if the valves failed open. But the small valve is arranged to open under these conditions to ensure the minimum flow needed for safety.

The valves MV1 and MV2 downstream of V1 and V2 are motorized isolating valves that are either fully open or fully closed and permit V1 and/or V2 to be effectively removed from the system. It is noted that the control valves V1 and V2 in their closed positions may have very significant leakage flow rates.

Valve Operating Logic

Figure 9.3(b) shows a logic diagram for valve operation. It is discussed to illustrate this aspect of computer control and the extent to which straightforward engineering considerations may complicate a basic level control system.

CS1 and CS2 are the main control switches. For start-up, CS1 is closed, closing motorized valves MV1 and MV2, and CS2 is switched to V3. This energizes the solenoid valve SV3, admitting control air pressure to V3 and allowing it to modulate the flow in response to the control signal. During this stage of the start-up the solenoid valve SVr is de-energized and the recirculation valve Vr is open. When V3 is fully open, the switch S3 closes, giving an annunciation to the operator.

The operator then switches CS1 to "open," causing MV1 and MV2 to open. When their "fully open" positions, sensed by the closure of limit switches, are annunciated to the operator, he or she switches CS2 to, say, V1. This de-energizes both SV3 and SV1, closing V3 and passing the control air pressure signal to V1. V1 now controls flow to maintain the level set point. While the flow is still small, it is desired to keep the recirculation valve Vr open. This is the function of the pneumatic relay PR. If its input, the control air pressure, is below 5 psi, its output is 0 psi, and Vr stays open. But if the input is above 5 psi the output is 15 psi, and Vr closes if SVr is energized.

It may be that during system overload or valve malfunction V1 cannot maintain the set-point level. If the level should drop below a certain value l_1, a current relay LS is closed by a level sensor/transmitter. As indicated in Fig. 9.3(b), provided that MV1 is fully open, this de-energizes SV2 and allows V2 to operate in parallel with V1. When the level recovers to l_2, SV2 is again energized and V2 closes.

Figure 9.3(b) is apparently based on straightforward engineering considerations. However, to combine these into a minimal logic diagram such as that shown may require experience and an adequate knowledge of logic circuits.

9.5 SAMPLING CHARACTERISTICS AND SIGNAL RECONSTRUCTION

As a first step toward the analysis and design of digital control systems, it is necessary to consider the effects of sampling. Sampling at intervals of T seconds is indicated in Fig. 9.4(a). The sampler output $f^*(t)$ equals $f(t)$ over the short periods $\tau \ll T$ during each interval when the relay is closed, and $f^*(t)$ is zero between samples. This

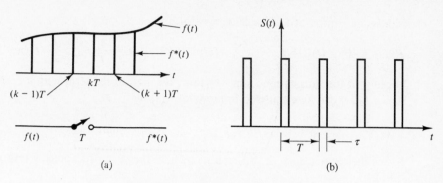

Figure 9.4 (a) Sampling; (b) pulse train.

may be represented mathematically by multiplying $f(t)$ by a train $S(t)$ of pulses, shown in Fig. 9.4(b): $f^*(t) = f(t)S(t)$. To simplify the model for the sampler, this train will be approximated by a train of unit impulses $\delta(t - nT)$ at $t = nT$:

$$S(t) = \sum_{n=-\infty}^{\infty} \delta(t - nT) \tag{9.1}$$

Then, assuming that $f(t) = 0$ for $t < 0$, the mathematical model used to represent the sampled signal is given by

$$f^*(t) = f(t)S(t) = \sum_{n=0}^{\infty} f(nT)\,\delta(t - nT) \tag{9.2}$$

It is clear that this cannot be a correct model of the physical sampling process, because the unit impulse $\delta(t - nT)$ is infinite at $t = nT$. Yet this model can be used for system analysis and design, because the errors are in effect corrected when the impulse series is used to model the input to the system component that follows the sampler.

The Laplace transform of a unit impulse $\delta(t)$ at $t = 0$ is $L[\delta(t)] = 1$, and according to the delay theorem the Laplace transform of a unit impulse, $\delta(t - nT)$ at $t = nT$ is $L[\delta(t - nT)] = 1 \cdot e^{-nTs} = e^{-nTs}$. Therefore, the Laplace transform of the sampled signal $f^*(t)$ in (9.2) is

$$F^*(s) = L[f^*(t)] = \sum_{n=0}^{\infty} f(nT)e^{-nTs} \tag{9.3}$$

An alternative expression for $F^*(s)$ is obtained by noting that, as may be verified from the theory of Fourier series, the impulse train $S(t)$ of (9.1) can be written as a Fourier series

$$S(t) = \frac{1}{T} \sum_{n=-\infty}^{\infty} e^{jn\omega_s t}$$

where

$$\omega_s = \frac{2\pi}{T} = \text{radian sampling frequency} \tag{9.4}$$

It follows that

$$f^*(t) = f(t)S(t) = \frac{1}{T} \sum_{n=-\infty}^{\infty} f(t)e^{jn\omega_s t}$$

and, by definition of the Laplace transform,

$$F^*(s) = L[f^*(t)] = \int_0^\infty \frac{1}{T} \sum_{n=-\infty}^\infty f(t)e^{-(s-jn\omega_s)t}\,dt = \frac{1}{T} \sum_{n=-\infty}^\infty \int_0^\infty f(t)e^{-(s-jn\omega_s)t}\,dt$$

Because of the summation from $-\infty$ to ∞, the sign of the term with n is immaterial. Hence $F^*(s)$ can be expressed in terms of $F(s)$ by

$$F^*(s) = \frac{1}{T} \sum_{n=-\infty}^\infty F(s + jn\omega_s) \qquad (9.5)$$

To see the significance of this result, consider the case of frequency response, when $s = j\omega$:

$$F^*(j\omega) = \frac{1}{T} \sum_{n=-\infty}^\infty F\{j(\omega + n\omega_s)\} \qquad (9.6)$$

This relation is illustrated in Fig. 9.5. If $|F(j\omega)|$ is the frequency spectrum of the input to the sampler, the frequency spectrum $|F^*(j\omega)|$ of the sampler output is periodic. The sampling process introduces unwanted *sidebands*, in addition to the *central band* corresponding to the term for $n = 0$ in (9.6). All bands have the same shape as the input spectrum, except for the factor $1/T$, since the value of each term in (9.6) depends only on the distance of ω to the center of the band. For example, the band for $n = -2$ contributes $(1/T)F(j0)$ at $\omega = 2\omega_s$.

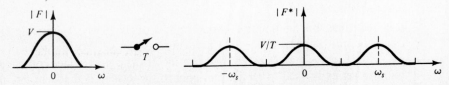

Figure 9.5   Frequency spectrum due to sampling.

Similarly, $F^*(s)$ of (9.5) consists of an infinite sum of transfer functions. The s-plane may be divided as shown in Fig. 9.6 into a *primary strip*, which corresponds to the central band, between $-0.5j\omega_s$ and $+0.5j\omega_s$, and *complementary strips*, which correspond to the sidebands. Two significant properties of $F^*(s)$ are the following.

1. If $F(s)$ has a pole at s_0, then $F^*(s)$ will have poles at $(s_0 + jm\omega_s)$ for integer $m = 0, \pm1, \pm2, \ldots$, so at intervals $j\omega_s$ on a vertical through s_0.

If $F(s)$ has a pole at s_0, whether inside or outside the primary strip, then this will also be a pole of $F^*(s)$, because it is a pole of the term for $n = 0$ in (9.5). But, from the other terms in this series, $F^*(s)$ will also have poles at $s_0 + jn\omega_s$. For example, in the term for $n = -1$, the term s of the factor $(s - s_0)$ in the denominator of $F(s)$ must be replaced by $s - j\omega_s$ so that the root s_0 becomes $s_0 + j\omega_s$. It must be concluded also that the Laplace transform of a sampled signal has an infinite number of poles.

2. $F^*(s)$ is periodic with radian frequency ω_s along any line parallel to the $j\omega$-axis in the s-plane:

$$F^*(s + jm\omega_s) = F^*(s) \qquad (9.7)$$

This may be proved by replacing s in (9.3) by $(s + jm\omega_s)$. This yields, for integer m,

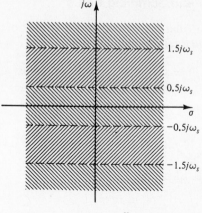

Figure 9.6 Primary and complementary strips.

$$F^*(s + jm\omega_s) = \sum_{n=0}^{\infty} f(nT)e^{-nT(s+jm\omega_s)}$$

$$= \sum_{n=0}^{\infty} f(nT)e^{-nTs}e^{-jnm2\pi} = \sum_{n=0}^{\infty} f(nT)e^{-nTs}$$

By (9.3), this is equal to $F^*(s)$. Note that this property applies to the zeros as well as the poles of $F^*(s)$, but that property 1 applies to poles only.

Zero-Order Hold (ZOH)

The ZOH, mentioned in Section 9.2, reconstructs a staircase approximation to a continuous signal from the sequence of voltage pulses obtained following D/A conversion of the computer output. It does so by "holding" its output constant between samples, equal to the last sample value. There is also an implied ZOH on the input side of the computer, to "hold" a sample long enough for A/D conversion.

The ZOH also acts as a filter that removes most of the sidebands. Filtering out of the sidebands is necessary, for example, when digital data acquisition systems are used to collect experimental data and the frequency spectrum of the original signal is desired. This filtering out of the sidebands is implied physically when the ZOH reconstructs a stepwise continuous signal from a sequence of pulses.

The transfer function, needed later, is derived as follows. As illustrated in Fig. 9.7, for a pulse input $r(t)$ of unit height the output $c(t)$ of the ZOH is a box of unit height between $t = 0$ and $t = T$. This pulse input will be modeled as a unit impulse input $r(t) = \delta(t)[R(s) = 1]$, just as in (9.2) a sequence of pulses was modeled as a train of impulses. In fact, while the model is incorrect in each instance, by using it in both the errors correct each other. The output $c(t)$ of the ZOH is the sum of a unit step at $t = 0$ ($1/s$) and a negative unit step at $t = T(-e^{-Ts}/s)$. So $C(s) = (1 - e^{-Ts})/s$, and the transfer function $G_h(s) = C(s)/R(s)$ is

$$G_h(s) = \frac{1 - e^{-Ts}}{s} \tag{9.8}$$

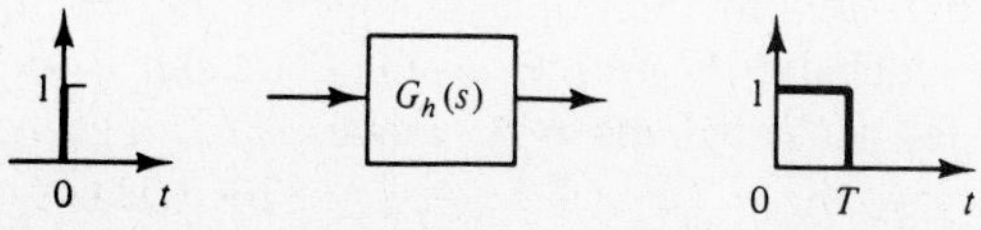

Figure 9.7 Zero-order hold (ZOH).

The Sampling Theorem and Sampling Rates

It is clearly not possible to filter out the sidebands if they overlap with the central band, as in Fig. 9.8(a). Hence follows the

> **Sampling Theorem.** The sampling frequency ω_s must at least equal twice the value of the highest significant frequency in the signal.

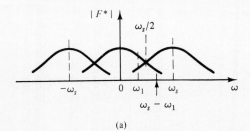

(a)

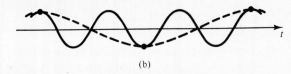

(b)   **Figure 9.8**   Aliasing or folding.

Note that this implies that the sampling theorem is satisfied if all poles of the signal transform $F(s)$ lie inside the primary strip in Fig. 9.6. It is recalled that the total transient is a superposition of components due to all poles of $F(s)$. For complex poles with imaginary part ω_d, the frequency of transient oscillations corresponding to the poles is ω_d. So the highest significant frequency corresponds to the pair with largest ω_d, and if it is inside the primary strip, then $\omega_s > 2\omega_d$, as required. Sampling rates used in practice are generally much higher and may be between 4 and 20 times the system bandwidth, depending on the desired accuracy and the load on and capacity of the computer. One rule of thumb is to choose T as one-tenth of the smallest plant time constant or the desired closed-loop time constant. Another convenient rule suggests sampling at the rate of 6 to 10 times per cycle. Thus, if the largest imaginary part of the significant system poles is 1 rad/sec, which corresponds to transient oscillations with a frequency of $1/6.28$ cycle per second, or a period of 6.28 sec, $T = 1$ sec may be satisfactory.

In Chapter 10 the effect of sampling rate on performance will be considered. It will show, for example, that for exponential transients corresponding to real poles, which largely decay in three time constants, a sampling interval somewhat below the smallest time constant may be adequate.

Aliasing or Folding

These terms are used to describe the consequences if the sampling theorem is not satisfied. In that case the sidebands reach into the primary strip, where their contributions cannot be filtered out. In Fig. 9.8(a), at frequency ω_1 the central band contributes $|F(j\omega_1)|$ and the sideband $|F(j(\omega_s - \omega_1))|$. The latter is an *alias* of the component of the central band an equal distance above $\omega_s/2$. The name *folding* arises because it can also be visualized as that component being folded back into the primary strip about a fold at $\omega_s/2$.

For example, in terms of pole locations, suppose that one pair is located at $-a \pm 0.6j\omega_s$, outside the primary strip. Then sampling will "fold" this pair back into the primary strip, to the locations $-a \pm 0.4j\omega_s$. This also follows from the earlier property 1 that poles repeat at intervals $\pm j\omega_s$, so poles at $-a \pm 0.6j\omega_s$ imply poles at $-a \pm 0.6j\omega_s \mp j\omega_s$.

So $f^*(t)$ contains components inside the primary strip that do not occur in $f(t)$. Note that, as may also be seen in Fig. 9.8(a), the alias occurs at the difference frequency $0.4\omega_s$ between the signal frequency $0.6\omega_s$ and the sampling frequency. Similarly, if a 70 cycles per second (hertz) oscillatory signal is sampled at the rate of 50 Hz, the sidebands reach back into the primary strip and cause an alias at the difference frequency of 20 Hz. Figure 9.8(b) illustrates how such a low-frequency alias of a higher-frequency signal can arise. The sampling frequency is too low to permit the two signals to be distinguished from each other.

Analog Prefilters

These are filters on the sensor outputs before sampling and are very common in digital control systems. They were mentioned earlier and are a direct consequence of the foregoing considerations. Analog prefilters are usually simple lag filters with the time constant chosen to remove signal noise above the frequency range of interest. In continuous systems such noise is usually filtered out by low-pass characteristics of the process or the actuators. However, with sampling this noise would be folded back into the primary strip; that is, it would show up at low frequencies and would not be filtered out. The prefilters avoid this by limiting the bandwidth of the sampler input.

9.6 CONTROL ALGORITHMS AND FINITE DIFFERENCES

In preceding sections a description was given of computer control systems and their components. One function of the computer was found to be the implementation of controller transfer functions by digital *control algorithms*. As will be seen in Chapter 10, there are several techniques for translating controller transfer functions to such algorithms. In this section they will be derived directly by the use of finite-difference approximations of derivatives and integrals. Algorithms for PID control and for phase-lead and phase-lag compensation will be considered as examples. The transfer function $D(s)$ for PID control and the corresponding time domain relation between input $e(t)$ and output $u(t)$ are

$$
\begin{cases}
D(s) = \dfrac{U(s)}{E(s)} = K_p + \dfrac{K_i}{s} + K_d s \\[2mm]
u(t) = K_p e(t) + K_i \displaystyle\int_0^t e\, dt + K_d \dot{e}
\end{cases}
\tag{9.9a}
$$

The finite-difference approximations may also be based on the differential form of (9.9a):

$$
\dot{u} = K_i e + K_p \dot{e} + K_d \ddot{e}
\tag{9.9b}
$$

For phase-lag and phase-lead compensation, the transfer function and the corresponding differential equation, obtained by cross-multiplying and inverting, are

$$\left\{ \begin{array}{l} G_c(s) = \dfrac{U(s)}{E(s)} = \dfrac{K(s + b)}{s + a} \end{array} \right. \qquad (9.10a)$$

$$\dot{u} + au = K\dot{e} + Kbe \qquad (9.10b)$$

The notations $e_k = e(kT)$ and $u_k = u(kT)$ will be used to denote the variables at time $t = kT$, as indicated in Fig. 9.9. In the present context, the sequence e_k, $k = 0, 1, 2, \ldots$, represents the sample values of the error, and u_k represents the signal to the actuator, say a valve.

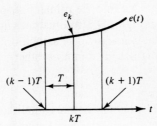

Figure 9.9 Finite differences.

Control algorithms are in general of two kinds:

1. *Position algorithm:* Here the output of the algorithm is u_k, the desired valve position.
2. *Velocity algorithm:* The algorithm supplies $\Delta u_k = u_k - u_{k-1}$, the desired change of valve position between $(k - 1)T$ and kT.

Velocity algorithms are often preferred because they have the fail-safe feature of leaving the valve where it is if computer malfunction causes a zero output. They also facilitate bumpless transfer between manual and automatic control and adapt naturally to actuators such as stepping motors that require such incremental input signals.

Two types of finite-difference approximations are used:

(a) Finite-Difference Approximations of Derivatives

In Fig. 9.9, the following approximations, called *first differences,* can be used for $\dot{e}(t)$, the slope of the $e(t)$ curve at $t = kT$:

$$\frac{e_{k+1} - e_k}{T} \qquad \text{forward difference}$$

$$\frac{e_k - e_{k-1}}{T} \qquad \text{backward difference} \qquad (9.11)$$

$$\frac{e_{k+1} - e_{k-1}}{2T} \qquad \text{central difference}$$

If e_{k+1} is available, the last tends to give the best approximation. *Second differences* to approximate $\ddot{e}$ may be derived as the first difference of first differences:

$$\frac{e_{k+2} - 2e_{k+1} + e_k}{T^2} \qquad \text{forward difference}$$

$$\frac{e_k - 2e_{k-1} + e_{k-2}}{T^2} \qquad \text{backward difference} \qquad (9.12)$$

$$\frac{e_{k+1} - 2e_k + e_{k-1}}{T^2} \qquad \text{central difference}$$

The last, for example, may be derived as

$$\frac{[(e_{k+1} - e_k)/T] - [(e_k - e_{k-1})/T]}{T}$$

(b) Finite-Difference Approximations of Integrals

For the integral term in (9.9a), if in Fig. 9.9 v_{k-1} approximates $K_i \int e\, dt$, (K_i times the area under the curve, up to $(k - 1)T$) then the approximation v_k up to kT may be written alternatively as

$$v_k = v_{k-1} + K_i T e_{k-1} \qquad \text{forward rectangular rule}$$
$$v_k = v_{k-1} + K_i T e_k \qquad \text{backward rectangular rule}$$
$$v_k = v_{k-1} + K_i \frac{T}{2}(e_{k-1} + e_k) \qquad \text{trapezoidal rule} \qquad (9.13)$$

Figure 9.9 shows that the second terms on the right are different approximations to the area between $(k - 1)T$ and kT, the last being the most accurate.

Example 9.6.1   PI Control Algorithms

(a) Using *backward differences:* From (9.9b) with $K_d = 0$,

$$\frac{(u_k - u_{k-1})}{T} = K_i e_k + K_p \frac{e_k - e_{k-1}}{T}$$

or

$$u_k = u_{k-1} + (K_p + K_i T)e_k - K_p e_{k-1} \qquad (9.14a)$$

This is the algorithm for calculating the new signal u_k based on past values of u and present and past values of e and is easily programmed on a computer. The position algorithm (9.14a) can also be written as a velocity algorithm $\Delta u_k = u_k - u_{k-1}$.

(b) Using the *backward rectangular rule:*

$$u_k = K_p e_k + v_k \qquad v_k = v_{k-1} + K_i T e_k \qquad (9.14b)$$

Programming based on these two equations is quite common. In the form of a velocity algorithm, substituting the second equation into the first yields

$$\Delta u_k = u_k - u_{k-1} = K_p(e_k - e_{k-1}) + v_k - v_{k-1}$$
$$= K_p(e_k - e_{k-1}) + K_i T e_k$$
$$= (K_p + K_i T)e_k - K_p e_{k-1} \qquad (9.14c)$$

Note that this is the same as the backward difference algorithm (9.14a).

(c) Using the *trapezoidal rule:*

$$u_k = K_p e_k + v_k \qquad v_k = v_{k-1} + 0.5 K_i T(e_{k-1} + e_k) \qquad (9.14d)$$

This alternative to (9.14b) provides greater accuracy and yields the velocity algorithm

$$u_k - u_{k-1} = K_p(e_k - e_{k-1}) + v_k - v_{k-1}$$
$$= K_p(e_k - e_{k-1}) + 0.5 K_i T(e_{k-1} + e_k)$$
$$= (K_p + 0.5 K_i T)e_k + (0.5 K_i T - K_p)e_{k-1} \qquad (9.14e)$$

Example 9.6.2   PID Control Algorithms

(a) Using *backward differences:* From (9.9b), with (9.11) and (9.12),

$$\frac{u_k - u_{k-1}}{T} = K_p \frac{e_k - e_{k-1}}{T} + K_i e_k + K_d \frac{e_k - 2e_{k-1} + e_{k-2}}{T^2}$$

Rearranging, this becomes

$$u_k = u_{k-1} + \left(K_p + K_i T + \frac{K_d}{T}\right)e_k - \left(K_p + \frac{2K_d}{T}\right)e_{k-1} + \frac{K_d}{T}e_{k-2} \quad (9.15a)$$

(b) Using *trapezoidal integration* and *backward differences* for the derivative: Equation (9.9a), with (9.11) and (9.13), yields

$$u_k = u_{k-1} + K_p(e_k - e_{k-1}) + v_k - v_{k-1} + K_d\left(\frac{e_k - e_{k-1}}{T} - \frac{e_{k-1} - e_{k-2}}{T}\right)$$

$$= u_{k-1} + K_p(e_k - e_{k-1}) + 0.5K_i T(e_{k-1} + e_k) + \frac{K_d}{T}(e_k - 2e_{k-1} + e_{k-2})$$

$$= u_{k-1} + \left(K_p + \frac{K_i T}{2} + \frac{K_d}{T}\right)e_k - \left(K_p + \frac{2K_d}{T} - \frac{K_i T}{2}\right)e_{k-1} + \frac{K_d}{T}e_{k-2} \quad (9.15b)$$

The better algorithm (9.15b) results if in the equation from which (9.15a) was derived the term $K_i e_k$ is replaced by $0.5K_i(e_{k-1} + e_k)$.

Two problems that can arise in both continuous and discrete systems when PID control is present, and were also discussed earlier, are the following:

1. Reset windup or integral windup can occur when integral control is present; this was discussed in Section 5.6. For large changes of input or large disturbances it can lead to severe transient oscillations. To avoid the growth of integrator output that causes this, one solution in control algorithms is to clamp the integrator output to prevent values outside selected HI and LO limits:

$$\text{IF I} > \text{HI THEN I} = \text{HI} \qquad \text{IF I} < \text{LO THEN I} = \text{LO} \qquad (9.16)$$

2. Derivative control can cause impulses in the control signal u, because a step change of the system input implies a step change of the error e, and so an impulse in $\dot{e}$. This is why a derivative of the output is often used instead of a derivative of the error to realize this form of control, as discussed also in the subsection on I-PD control in Section 5.6.

Example 9.6.3   Algorithms for Phase-Lead of Phase-Lag Compensation

(a) Using *backward differences:* From (9.10b), with (9.11),

$$\frac{1}{T}(u_k - u_{k-1}) + au_k = \frac{K}{T}(e_k - e_{k-1}) + Kbe_k$$

or

$$u_k = \frac{u_{k-1} + K(1 + bT)e_k - Ke_{k-1}}{1 + aT} \qquad (9.17a)$$

(b) Using *trapezoidal integration:* Rearrange (9.10b) to $\dot{u} - K\dot{e} = -au + Kbe$ and write it in the integral form:

$$u - Ke = \int_0^t (-au + Kbe)\,d\tau$$

With the last of (9.13), this gives

$$u_k - Ke_k = u_{k-1} - Ke_{k-1} + 0.5T(-au_{k-1} + Kbe_{k-1} - au_k + Kbe_k)$$

which can be arranged to the algorithm

$$u_k = \frac{(1 - 0.5aT)u_{k-1} + K(1 + 0.5bT)e_k + K(0.5bT - 1)e_{k-1}}{1 + 0.5aT} \qquad (9.17b)$$

These examples illustrate the variations possible in algorithms that approximate the same continuous systems.

9.7 *Z* TRANSFORMS

It was found in Section 9.5 that the Laplace transform of a sampled signal in general has an infinite number of poles and zeros. Also, the Laplace transform is not suitable for representing the action of digital control algorithms. As this suggests, this transform is not attractive for analysis and design. The *Z* transform takes its place and is well adapted to represent both computer control algorithms and sampled signals. The Laplace transform of a sampled signal was given in (9.3):

$$F^*(s) = L[f^*(t)] = \sum_{n=0}^{\infty} f(nT)e^{-nTs} \qquad (9.18)$$

The *Z* transform of a sampled signal can be defined as a further transformation:

$$z = e^{Ts} \qquad F(z) = F^*(s)\Big|_{z=e^{Ts}} \qquad (9.19)$$

Substitution into (9.18) then yields

$$F(z) = Z[f(t)] = \sum_{n=0}^{\infty} f(nT)z^{-n} \qquad (9.20a)$$

Only the sequence of samples $f(nT)$, $n = 0, 1, 2, \dots$, is considered in this transform, not the response between samples. It is recalled that e^{-Ts}, and therefore z^{-1}, means a delay of one sampling interval. $F(z)$ in (9.20a) represents a sequence of sample values, with $f(nT)$ occurring after n sampling intervals.

The number sequences f_k, $k = 0, 1, 2, \dots$, in control algorithms are also sequences of values and can be represented by the same transform:

$$F(z) = Z[f_k] = \sum_{k=0}^{\infty} f_k z^{-k} \qquad (9.20b)$$

Table 9.7.1 shows examples derived from the definitions (9.20), by substituting the values of $f(nT)$ or f_k and using known results to write the series in closed form.

  1. *Unit impulse* $\delta(t)$ *or discrete pulse* δ_k:

$$(\delta_k = 1 \qquad \text{for } k = 0, \qquad \delta_k = 0 \quad \text{for } k \neq 0)$$
$$Z[\delta] = 1 + 0 \cdot z^{-1} + \cdots = 1$$

Note that the unit impulse function is to be interpreted as an approximate mathematical model only, in the sense of its use in the sampler model (9.2) and the ZOH model (9.8).

TABLE 9.7.1   *Z* TRANSFORM PAIRS AND THEOREMS

$f(t), f_k$	$L[f(t)]$	$Z[\cdot]$
1. $\delta(t), \delta_k$	1	1
2. $\delta(t - nT), \delta_{k-n}$	e^{-nTs}	z^{-n}
3. $u(t), f_k = 1$	$\dfrac{1}{s}$	$\dfrac{1}{1 - z^{-1}} = \dfrac{z}{z - 1}$
4. $t, f_k = kT$	$\dfrac{1}{s^2}$	$\dfrac{Tz^{-1}}{(1 - z^{-1})^2} = \dfrac{Tz}{(z - 1)^2}$
5. e^{-at}	$\dfrac{1}{s + a}$	$\dfrac{1}{1 - e^{-aT}z^{-1}} = \dfrac{z}{z - e^{-aT}}$
b^k		$\dfrac{z}{z - b}$
$e^{-j\omega_0 t}$	$\dfrac{1}{s + j\omega_0}$	$\dfrac{1}{1 - e^{-j\omega_0 T}z^{-1}}$ $(a = j\omega_0)$

6. Linearity theorem: $Z[af(t) + bg(t)] = aF(z) + bG(z)$

7. Delay theorem: $Z[f(t - nT)]$ or $Z[f_{k-n}] = z^{-n}F(z)$

8. Final value theorem: $\lim\limits_{n \to \infty} f(nT)$ or $f_n = \lim\limits_{z \to 1} (1 - z^{-1})F(z)$

---

2. *Delayed unit pulses* $\delta(t - nT)$ *or* δ_{k-n}:
$$Z[\cdot] = 0 + 0 \cdot z^{-1} + \cdots + 1 \cdot z^{-n} + 0 \cdot z^{-n-1} + \cdots = z^{-n}$$

3. *Unit step* $u(t)$ *or number sequence* $f_k = 1$:
$$Z[\cdot] = 1 + 1 \cdot z^{-1} + 1 \cdot z^{-2} + \cdots = \frac{1}{1 - z^{-1}} = \frac{z}{z - 1}$$

4. *Unit ramp* t *or sequence* $f_k = kT$:
$$Z[\cdot] = 0 + Tz^{-1} + 2Tz^{-2} + \cdots = Tz^{-1}(1 + 2z^{-1} + 3z^{-2} + \cdots)$$
$$= \frac{Tz^{-1}}{(1 - z^{-1})^2} = \frac{Tz}{(z - 1)^2}$$

5. *Exponential decay* e^{-at} *and geometric sequences:*
$$Z[e^{-at}] = 1 + e^{-aT}z^{-1} + e^{-2aT}z^{-2} + \cdots$$
$$= 1 + (e^{aT}z)^{-1} + (e^{aT}z)^{-2} + \cdots = \frac{1}{1 - e^{-aT}z^{-1}} = \frac{z}{z - e^{-aT}}$$

Because e^{-at} is represented by the sequence $(e^{-aT})^k$, its transform implies those of other geometric sequences. For b^{-k}, replace e^{-aT} by b^{-1}, and for b^k replace e^{-aT} by b:
$$Z[b^k] = \frac{z}{z - b} \qquad Z[b^{-k}] = \frac{z}{z - b^{-1}}$$

6. *Linearity:* Proved from definition (9.20).

7. *Delay theorem:*

$$Z[f_{k-n}] = \sum_{k=0}^{\infty} f_{k-n}z^{-k}$$

$$= 0 + 0 \cdot z^{-1} + \cdots + 0 \cdot z^{-n+1} + f_0 z^{-n} + f_1 z^{-n-1} + \cdots$$

$$= z^{-n}[f_0 + f_1 z^{-1} + \cdots] = z^{-n}F(z)$$

since f_{k-n} is assumed to be zero for $k < n$. So delaying a sequence by n intervals is equivalent to multiplying its transform by z^{-n}.

The partial fraction expansion of $F(s)$ is used to find $F(z)$ for transforms not available in the table.

Example 9.7.1

Find the *Z* transform corresponding to

$$G(s) = \frac{a}{s(s+a)} = \frac{1}{s} - \frac{1}{s+a}$$

Using entries 3 and 5 in Table 9.7.1 gives

$$G(z) = \frac{z}{z-1} - \frac{z}{z-e^{-aT}} = \frac{z(1-e^{-aT})}{(z-1)(z-e^{-aT})} \tag{9.21}$$

Example 9.7.2

$$L[\sin \omega t] = \frac{\omega}{s^2 + \omega^2} = \frac{-1/(2j)}{s+j\omega} + \frac{1/(2j)}{s-j\omega}$$

Using entries 5 and 6 in Table 9.7.1,

$$Z[\sin \omega t] = \frac{-1/(2j)}{1-e^{-j\omega T}z^{-1}} + \frac{1/(2j)}{1-e^{j\omega T}z^{-1}} = \frac{z^{-1}\sin \omega T}{1-2z^{-1}\cos \omega T + z^{-2}} \tag{9.22}$$

More examples are given below.

9.8 *Z* TRANSFER FUNCTIONS

Definition. The *Z transfer function* of a system is the ratio of the *Z* transforms of its output and input sequences.

These can be used to describe the action of a control algorithm as well as the relation between sampled inputs and outputs of continuous systems. When used to describe an algorithm, the *Z* transfer function is often called a *discrete transfer function,* and the name *pulse transfer function* is often used for the *Z* transfer function relating input and output samples of continuous systems.

Discrete Transfer Functions and Digital Filters

A computer control algorithm or digital filter can be represented by one or more difference equations of the form

$$y_n = \sum_{i=0}^{p} a_i x_{n-i} - \sum_{j=1}^{q} b_j y_{n-j} \tag{9.23}$$

where y_k and x_k are output and input number sequences. The filter is *nonrecursive* or *transversal* if all b_j are zero. The present output y_n then depends only on present and past inputs. Otherwise, the filter is *recursive,* and the output depends also on past outputs. Using the linearity and delay theorems, the transform of (9.23) is

$$Y(z) = \sum_{i=0}^{p} a_i z^{-i} X(z) - \sum_{j=1}^{q} b_j z^{-j} Y(z) \qquad (9.24)$$

Hence, as illustrated in Fig. 9.10(a), the discrete transfer function $D(z)$ describing the filter is

$$D(z) = \frac{Y(z)}{X(z)} = \frac{\displaystyle\sum_{i=0}^{p} a_i z^{-i}}{1 + \displaystyle\sum_{j=1}^{q} b_j z^{-j}} \qquad (9.25)$$

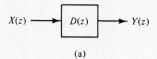

(a)

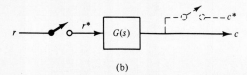

(b)

Figure 9.10   *Z* transfer function.

It is important to note that the algorithm corresponding to a given transfer function $D(z)$ is readily found by using these equations in the reverse order. Cross-multiplication in (9.25) gives (9.24), which, by the linearity and delay theorems, corresponds to the algorithm (9.23).

Example 9.8.1   Delay of One Sampling Interval

$$D(z) = z^{-1}$$
$$Y(z) = z^{-1} X(z)$$

so $y_k = x_{k-1}$, by the delay theorem. The input number sequence is delayed by one interval.

Example 9.8.2   Control Algorithm

$$y_k = a_1 y_{k-1} - a_2 y_{k-2} + b_1 x_k + b_2 x_{k-1}$$
$$Y(z) = a_1 z^{-1} Y(z) - a_2 z^{-2} Y(z) + b_1 X(z) + b_2 z^{-1} X(z)$$
$$D(z) = \frac{Y(z)}{X(z)} = \frac{b_1 + b_2 z^{-1}}{1 - a_1 z^{-1} + a_2 z^{-2}}$$

Example 9.8.3   Integration Algorithms (9.13)

(a) $u_k = u_{k-1} + T e_{k-1}; \; U(z) = z^{-1} U(z) + T z^{-1} E(z);$

$$D(z) = \frac{U(z)}{E(z)} = \frac{T z^{-1}}{1 - z^{-1}} \qquad \text{forward rectangular rule} \qquad (9.26a)$$

(b) $u_k = u_{k-1} + T e_k; \; D(z) = \dfrac{T}{1 - z^{-1}} \quad$ backward rectangular rule $\qquad (9.26b)$

(c) $u_k = u_{k-1} + (T/2)(e_{k-1} + e_k);$

$$D(z) = \frac{T}{2}\frac{1 + z^{-1}}{1 - z^{-1}} = \frac{T}{2}\frac{z + 1}{z - 1} \qquad \text{trapezoidal rule} \qquad (9.26c)$$

Example 9.8.4   PID-Control Algorithms

(a) *PI algorithm* (9.14c)

$$u_k = u_{k-1} + K_p(e_k - e_{k-1}) + K_i Te_k = u_{k-1} + (K_p + K_i T)e_k - K_p e_{k-1}$$

$$D(z) = K_p + \frac{K_i T}{1 - z^{-1}} \tag{9.27a}$$

(b) *PID algorithm* (9.15b)

$$u_k = u_{k-1} + ae_k + be_{k-1} + ce_{k-2}$$

where $c \equiv K_d/T$, $a \equiv K_p + 0.5K_i T + K_d/T$, and $b \equiv 0.5K_i T - K_p - 2K_d/T$.

$$U(z) = z^{-1}U(z) + (a + bz^{-1} + cz^{-2})E(z)$$

$$D(z) = \frac{a + bz^{-1} + cz^{-2}}{1 - z^{-1}} \tag{9.27b}$$

$D(z)$ relates number sequences. But by considering the A/D and D/A conversions on the input and output side of the computer to be included, it is also a pulse transfer function relating sample sequences, and $D(z) = D^*(s)|_{z=e^{Ts}}$ as in (9.19).

Pulse Transfer Functions

In Fig. 9.10(b), a sample sequence r^* is the input to a system with transfer function $G(s)$. Z transforms do not consider the response between samples, and therefore a synchronous fictitious sampler is introduced to produce c^*.

$$C(s) = G(s)R^*(s) \tag{9.28}$$

So $C(s + jn\omega_s) = G(s + jn\omega_s)R^*(s + jn\omega_s) = G(s + jn\omega_s)R^*(s)$, since $R^*(s + jn\omega_s) = R^*(s)$ according to (9.7). Equation (9.5) now yields

$$C^*(s) = \frac{1}{T} \sum_{n=-\infty}^{\infty} C(s + jn\omega_s) = R^*(s)\left[\frac{1}{T} \sum G(s + jn\omega_s)\right]$$

The form in brackets is a pulse transfer function because it is the ratio of the transforms of the sampled output and input signals. It is also equal to $G^*(s)$, by (9.5). With Z transformation, this yields the key relations

$$C^*(s) = G^*(s)R^*(s) \qquad C(z) = G(z)R(z) \tag{9.29}$$

and the definition

The *pulse transfer function $G(z)$* is the Z transform of $G(s)$:

$$G(z) = Z[G(s)] \tag{9.30}$$

The partial fraction expansion of $G(s)$ is used to determine the transform.

Example 9.8.5

(a) $G(s) = \dfrac{a}{s + a}$; $G(z) = \dfrac{az}{z - e^{-aT}}$ $\tag{9.31a}$

(b) $G(s) = \dfrac{a}{s(s + a)}$; $G(z) = Z\left[\dfrac{1}{s} - \dfrac{1}{s + a}\right] = \dfrac{z}{z - 1} - \dfrac{z}{z - e^{-aT}}$;

$$G(z) = \frac{z(1 - e^{-aT})}{(z - 1)(z - e^{-aT})} \tag{9.31b}$$

(c) $G(s) = \dfrac{a}{s^2(s + a)}$; $G(z) = Z\left[\dfrac{1}{s^2} - \dfrac{1/a}{s} + \dfrac{1/a}{s + a}\right]$;

$$G(z) = \frac{Tz^{-1}}{(1 - z^{-1})^2} - \frac{1/a}{1 - z^{-1}} + \frac{1/a}{1 - e^{-aT}z^{-1}} \tag{9.31c}$$

where the partial fraction expansions are found in the usual way.

9.9 SYSTEM CONFIGURATION FOR ANALYSIS AND DESIGN

Figure 9.11(a) shows the most common system configuration, which will also be the focus of the discussion of analysis and design techniques in Chapter 10. The output sampler is modeled in the error path for convenience; in reality it would normally be in the feedback path, and r would be a direct digital input to the controller. Figure 9.11(b) shows the translation of the diagram (a) to Z transforms and Z transfer functions. The controller D is assumed to incorporate A/D and D/A conversion as well as the control algorithm, and so relates input and output pulse sequences. $D(z)$ represents the control algorithm, for example as in (9.25).

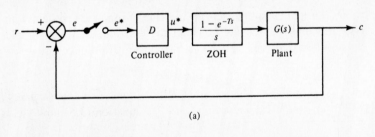

(a)

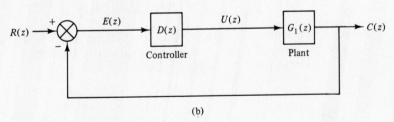

(b)

Figure 9.11 Digital control system.

The transfer function $G_1(z)$ represents the zero-order hold (ZOH) of (9.8) and the plant $G(s)$, which is assumed to include input power amplification and actuator dynamics. $G_1(z)$ relates pulse sequences, because its input is the pulse sequence u^* to the ZOH and its output is provided by the fictitious sampler in the plant output. To derive $G_1(z)$, the general relations (9.28) to (9.30) are used. From Fig. 9.11(a),

$$C(s) = \frac{1}{s}(1 - e^{-Ts})G(s)U^*(s)$$

This is of the form of (9.28). Therefore, since $e^{-Ts} = z^{-1}$, (9.29) and (9.30) yield

$$C(z) = G_1(z)U(z)$$

where

$$G_1(z) = Z\left[\frac{1}{s}(1 - e^{-Ts})G(s)\right] = (1 - z^{-1})Z\left[\frac{G(s)}{s}\right] \tag{9.32}$$

Table 9.9.1 shows the Z transfer functions $G_1(z)$ for some common transfer functions $G(s)$. For the simple lag $G(s) = a/(s + a)$, $G_1(z)$ follows immediately by multiplying $G(z)$ of (9.31b) by $(1 - z^{-1})$. For $G(s) = K/[s(s + a)]$, similarly, $G(z)$ of (9.31c) must be multiplied by $(K/a)(1 - z^{-1})$. This yields

$$\frac{K}{a}\left[\frac{Tz^{-1}}{1 - z^{-1}} - \frac{1}{a} + \frac{(1 - z^{-1})/a}{1 - e^{-aT}z^{-1}}\right]$$

$$= \frac{K}{a}\frac{Tz^{-1}(1 - e^{-aT}z^{-1}) - (1/a)(1 - z^{-1})(1 - e^{-aT}z^{-1}) + (1/a)(1 - z^{-1})^2}{(1 - z^{-1})(1 - e^{-aT}z^{-1})}$$

TABLE 9.9.1 $G_1(z) = (1 - z^{-1})Z\left[\dfrac{G(s)}{s}\right]$

$$G(s) = \frac{a}{s + a}: \quad G_1(z) = \frac{1 - e^{-aT}}{z - e^{-aT}}$$

$$G(s) = \frac{K}{s(s + a)}: \quad G_1(z) = \frac{K}{a}\frac{[T + (1/a)e^{-aT} - 1/a]z + [(1/a) - (1/a)e^{-aT} - Te^{-aT}]}{(z - 1)(z - e^{-aT})}$$

$$G(s) = \frac{K}{(s + a)(s + b)}:$$

$$G_1(z) = \frac{K}{ab(a - b)}\frac{[a - b + be^{-aT} - ae^{-bT}]z + be^{-bT} - ae^{-aT} + (a - b)e^{-(a+b)T}}{(z - e^{-aT})(z - e^{-bT})}$$

Noting that the constant terms in the numerator cancel each other, the result in Table 9.9.1 follows by rearranging this. For $G(s) = K/[(s + a)(s + b)]$,

$$G_1(z) = (1 - z^{-1})Z\left[\frac{K}{s(s + a)(s + b)}\right]$$

$$= \frac{K(1 - z^{-1})}{ab(a - b)}Z\left[\frac{a - b}{s} + \frac{b}{s + a} - \frac{a}{s + b}\right]$$

$$= \frac{K}{ab(a - b)}\left[a - b + \frac{b(1 - z^{-1})}{1 - e^{-aT}z^{-1}} - \frac{a(1 - z^{-1})}{1 - e^{-bT}z^{-1}}\right]$$

The result in Table 9.9.1 is obtained by bringing this on a common denominator and noting that the constant terms in the numerator cancel. The table will prove very useful for the analysis and design problems in Chapter 10.

9.10 CONCLUSION

In this chapter a general introduction has been given to digital computer control, and the foundations were laid for the analysis and design of single-loop digital control systems in Chapter 10. Sampling characteristics were discussed, and digital control algorithms were derived by the use of finite differences. In Chapter 10 this technique will be seen in its proper context, as one of several ways in which a continuous controller can be approximated by a digital algorithm. Z transforms and Z transfer functions were defined, to take the place of the Laplace transform in the study of sampled systems. The basic digital control loop was introduced and will be the focus for the analysis and design techniques of Chapter 10.

PROBLEMS

9.1. The response of a system is dominated by the transient corresponding to a pole at $s = -a$.
  (a) What requirements, if any, does the sampling theorem impose on the size of the sampling interval?
  (b) Give the largest sampling interval that might be adequate to represent the transient.

9.2. The response of a system is dominated by the transient corresponding to a pole pair at $s = -a \pm j$.
  (a) What minimum sampling frequency is required by the sampling theorem?
  (b) Determine the sampling interval that will provide sampling at the rate of six times per cycle of the transient.
  (c) Determine the range of values of a for which the sampling interval found in (b) is unlikely to provide an adequate representation of the transient.

9.3. A unity feedback system has been designed for a crossover frequency of 1 rad/sec on the Bode plot of its loop gain function. The control is to be implemented digitally. Does $T = 1$ appear to be a reasonable choice for the sampling interval?

9.4. Obtain digital filter algorithms to represent the transfer function

$$\frac{U(s)}{E(s)} = \frac{a}{s + a}$$

based on numerical integration by:
  (a) The backward rectangular rule.
  (b) The forward rectangular rule.
  (c) The trapezoidal rule.

9.5. Obtain digital filter algorithms to represent

$$\frac{U(s)}{E(s)} = \frac{a}{s + a}$$

based on finite difference approximations using:
  (a) Backward differences.
  (b) Forward differences.
  (c) Central differences.
  Compare the results with those of Problem 9.4.

9.6. For $\dfrac{U(s)}{E(s)} = \dfrac{s + 1}{5s + 1}$:
  (a) Show that backward differences and the backward rectangular rule lead to the same computer algorithm.

(b) Obtain a control algorithm by the use of central difference approximations.

(c) Find the algorithm based on trapezoidal integration.

9.7. Obtain the discrete transfer functions corresponding to the algorithms in Problem 9.4.

9.8. Obtain the discrete transfer functions corresponding to the algorithms in Problem 9.5.

9.9. Find the discrete transfer functions corresponding to the algorithms in Problem 9.6.

9.10. For $D(s) = \dfrac{U(s)}{E(s)} = \dfrac{s+1}{5s+1}$ as in Problems 9.6 and 9.9:

(a) Obtain a discrete transfer function via partial fraction expansion and Z transformation of $D(s)$ and write the corresponding algorithm.

(b) Show that the same result as in Problem 9.9 corresponding to 9.6(c) is obtained by substituting for s in $D(s)$ the Tustin transformation or bilinear transformation

$$s = \frac{(2/T)(1 - z^{-1})}{1 + z^{-1}} = \frac{(2/T)(z - 1)}{z + 1}$$

9.11. Use the Tustin method, based on trapezoidal integration and consisting of replacing s in $G(s)$ by $(2/T)(1 - z^{-1})/(1 + z^{-1})$, to obtain the Z transfer function corresponding to

$$G(s) = \frac{C(s)}{U(s)} = \frac{K}{s(s + a)}$$

and from it write an algorithm for calculation of the output sequence for a given input sequence.

9.12. Obtain the pulse transfer function corresponding to

$$G(s) = \frac{K}{(s + a)(s + b)} = \frac{C(s)}{U(s)}$$

and from it a recursive algorithm for c_k.

9.13. Obtain the pulse transfer function corresponding to

$$G(s) = \frac{4}{(s + 1)(s + 4)} = \frac{C(s)}{U(s)}$$

and from it a recursive algorithm for c_k.

9.14. Use the Tustin transformation $s = (2/T)(1 - z^{-1})/(1 + z^{-1})$, based on trapezoidal integration, to find the transform $G(z)$ corresponding to $G(s)$ of Problem 9.13, and write the corresponding algorithm.

9.15. Find the Z transfer functions $G(z)$ corresponding to the following transfer functions $G(s)$.

$$\textbf{(a)}\ \frac{K}{s + 5} \qquad \textbf{(b)}\ \frac{K}{s(s + 5)}$$

9.16. Repeat Problem 9.15 for $\dfrac{K}{[s^2(s + 5)]}$.

9.17. Find the Z transfer functions $G(z)$ corresponding to the following transfer functions $G(s)$.

$$\textbf{(a)}\ \frac{K}{(s + 1)(s + 5)} \qquad \textbf{(b)}\ \frac{K(s + 3)}{(s + 1)(s + 5)}$$

9.18. Repeat Problem 9.17 for

$$\frac{K}{s(s + 1)(s + 5)}$$

9.19. Show that the Z transform of

$$\frac{s + a}{(s + a + jb)(s + a - jb)} = \frac{1}{2}\left(\frac{1}{s + a - jb} + \frac{1}{s + a + jb}\right)$$

equals

$$\frac{z^2 - ze^{-aT}\cos bT}{z^2 - 2ze^{-aT}\cos bT + e^{-2aT}}$$

9.20. Show that the Z transform of

$$\frac{Ke^{j\phi}}{s + a - jb} + \frac{Ke^{-j\phi}}{s + a + jb} \quad \text{is} \quad 2K\frac{z^2\cos\phi - ze^{-aT}\cos(\phi - bT)}{z^2 - 2ze^{-aT}\cos bT + e^{-2aT}}$$

9.21. Use the results of Problem 9.20 to find the Z transform corresponding to

$$G(s) = \frac{2}{s^2 + 2s + 2}$$

preceded by a zero-order hold.

9.22. Represent each of the following computer control algorithms by a discrete transfer function.
   (a) $u_n = e_n - 5u_{n-1} - 3u_{n-2}$
   (b) $u_n = 3u_{n-1} - 2u_{n-2} + e_{n-1} + e_{n-2}$
   (c) $u_n = u_{n-1} - 0.5u_{n-2} + e_n$

9.23. Use the series definition of the Z transform to show that the Z transforms of the geometric sequences a^n and $(-a)^n$, where a is positive and real, are as follows:

$$a^n: \quad \frac{z}{z - a} = \frac{1}{1 - az^{-1}}$$

$$(-a)^n: \quad \frac{z}{z + a} = \frac{1}{1 + az^{-1}}$$

9.24. Determine the Z transfer functions $G_1(z)$ that result if the following transfer functions $G(s)$ are preceded by a zero-order hold.

$$\text{(a)} \ \frac{K}{s + 5} \qquad \text{(b)} \ \frac{K}{s(s + 5)} \qquad \text{(c)} \ \frac{K}{(s + 1)(s + 5)}$$

9.25. For satisfactory dynamic modeling of a process, it is often adequate to use a transport lag in combination with one or two simple lags, as given by
   1. $G(s) = Ke^{-T_d s}/(s + 5)$
   2. $G(s) = Ke^{-T_d s}/[(s + 1)(s + 5)]$
   If T_d can be approximated by an integer number of sample intervals, $T_d = mT$, m integer, find the corresponding Z transfer functions $G_1(z)$ if $G(s)$ is preceded by a zero-order hold.

10

Digital Control System Analysis and Design

10.1 INTRODUCTION

This chapter deals with the analysis and design of single-loop digital control systems. Z transforms and Z transfer functions are used, and the system configuration of principal interest will be that in Fig. 9.11(b), repeated in Fig. 10.1. The transfer function $G_1(z)$ in this block diagram is defined by (9.32) and is given in Table 9.9.1 for some common plant transfer functions $G(s)$.

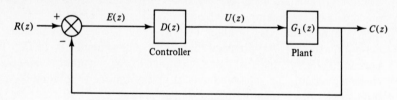

Figure 10.1 System configuration.

10.2 CLOSED-LOOP TRANSFER FUNCTIONS AND BLOCK DIAGRAM REDUCTION

The closed-loop transfer function for the system in Fig. 10.1 is found in the usual way.

$$U(z) = D(z)E(z) \qquad E(z) = R(z) - C(z) \qquad C(z) = G_1(z)U(z)$$

Combining these equations yields

$$C(z) = G_1(z)D(z)[R(z) - C(z)]$$

or

$$\frac{C(z)}{R(z)} = \frac{G_1(z)D(z)}{1 + G_1(z)D(z)} \qquad (10.1)$$

Note that the equation for this configuration is of the same form as for continuous systems, with a loop gain function $G_1(z)D(z)$.

Example 10.2.1

$$G(s) = \frac{a}{s + a} \qquad D(z) = K$$

With this pure gain control, using Table 9.9.1, the loop gain function becomes

$$G_1(z)D(z) = \frac{K(1 - e^{-aT})}{z - e^{-aT}} \qquad (10.2)$$

and the closed-loop transfer function is

$$\frac{C(z)}{R(z)} = \frac{K(1 - e^{-aT})}{z - e^{-aT} + K(1 - e^{-aT})} \qquad (10.3)$$

For $aT = 1$, these equations become $(e^{-1} = 0.3678)$

$$G_1(z)D(z) = \frac{0.6322K}{z - 0.3678}$$

$$\frac{C(z)}{R(z)} = \frac{0.6322K}{z - 0.3678 + 0.6322K} \qquad (10.4)$$

While the familiar technique was used in this example, care is necessary in block diagram reduction, because the usual rules do not always apply. In fact, a transfer function may not exist. Equations (9.28) to (9.30) are the basis of block diagram reduction:

$$C(s) = G(s)R^*(s) \qquad C^*(s) = G^*(s)R^*(s) \qquad C(z) = G(z)R(z) \qquad (10.5)$$

These equations are used for the examples in Fig. 10.2. In part (a), $G_1(s)$ and $G_2(s)$ are separated by a sampler and each block is represented by (10.5), but in part (b) they are not, and the product $G_1(s)G_2(s)$ must be transformed. The notations $G_1G_2^*(s)$ and $G_1G_2(z)$ are used to identify such transforms. In Fig. 10.2(c) it is important to observe that a transfer function $C(z)/R(z)$ cannot be identified. This is because the input is not separated by a sampler from the first block G_1 and therefore the product $G_1(s)R(s)$ must be transformed.

Example 10.2.2

In Fig. 10.2(a) and (b), let

$$G_1(s) = \frac{1}{s} \qquad G_2(s) = \frac{a}{s + a}$$

For Fig. 10.2(a),

$$G_1(z) = Z\left[\frac{1}{s}\right] = \frac{z}{z - 1}$$

$$G_2(z) = Z\left[\frac{a}{s + a}\right] = \frac{az}{z - e^{-aT}}, \quad \text{from (9.31a)}$$

$$G_1(z)G_2(z) = \frac{az^2}{(z - 1)(z - e^{-aT})}$$

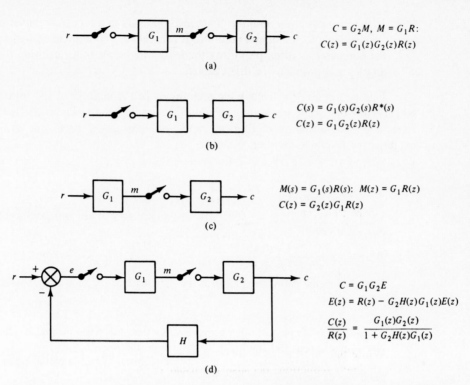

$$C = G_2M, \quad M = G_1R:$$
$$C(z) = G_1(z)G_2(z)R(z)$$

(a)

$$C(s) = G_1(s)G_2(s)R^*(s)$$
$$C(z) = G_1G_2(z)R(z)$$

(b)

$$M(s) = G_1(s)R(s): \quad M(z) = G_1R(z)$$
$$C(z) = G_2(z)G_1R(z)$$

(c)

$$C = G_1G_2E$$
$$E(z) = R(z) - G_2H(z)G_1(z)E(z)$$
$$\frac{C(z)}{R(z)} = \frac{G_1(z)G_2(z)}{1 + G_2H(z)G_1(z)}$$

(d)

Figure 10.2   Examples of block diagram reduction.

For Fig. 10.2(b),

$$G_1G_2(z) = Z\left[\frac{a}{s(s+a)}\right] = \frac{z(1 - e^{-aT})}{(z-1)(z - e^{-aT})}, \quad \text{from (9.31b)}$$

Note that

$$G_1(z)G_2(z) \neq G_1G_2(z)$$

10.3 TRANSIENT RESPONSE OF DIGITAL CONTROL SYSTEMS

Analogous to continuous systems, the determination of the transient response of digital control systems, whether open or closed loop, requires the *inverse Z transformation* of the transform $C(z)$ of the output. For the transform of the output of Fig. 10.1, Eq. (10.1) gives

$$C(z) = \frac{G_1(z)D(z)R(z)}{1 + G_1(z)D(z)} \tag{10.6}$$

Several techniques are available to determine the response $c(nT)$ at the sampling instants:

1. *Long division:* Divide the numerator of $C(z)$ by its denominator to obtain a series expansion in terms of z^{-1}.

2. *Partial fraction expansion:* Probably the most commonly used method.

3. *Difference equation method:* This method leads to a recurrence relation and is very suitable for computer solution. It can also be recommended without use of the computer as often providing the least time-consuming solution for the examples and problems in this chapter.

It is convenient to discuss the techniques in the context of the following example. Note that these methods give only the sample values. The modified Z transform and submultiple sampling methods, not discussed here, are available to examine the response between samples.

Example 10.3.1

$$C(z) = \frac{z(z + 0.4)}{(z - 1)(z - 0.3)(z - 0.8)}$$

(a) *Long-division method:* Write the numerator and denominator as polynomials in z and divide:

$$C(z) = \frac{z^2 + 0.4z}{z^3 - 2.1z^2 + 1.34z - 0.24} = z^{-1} + 2.5z^{-2} + 3.91z^{-3} + 5.101z^{-4} + \cdots$$

Hence, from the definition (9.20) of the Z transform, the sequence of sample values $c_n \equiv c(nT)$ is

$$c_0 = 0, \quad c_1 = 1, \quad c_2 = 2.5, \quad c_3 = 3.91, \quad c_4 = 5.101, \quad \ldots$$

The method is simple and often convenient, for example, when method 2 includes complex pairs of poles. But finding c_n for large n is laborious.

(b) *Partial fraction expansion method:* Expand $C(z)/z$ in partial fractions in the same way as for Laplace transforms:

$$\frac{C(z)}{z} = \frac{z + 0.4}{(z - 1)(z - 0.3)(z - 0.8)} = \frac{A}{z - 1} + \frac{B}{z - 0.3} + \frac{C}{z - 0.8}$$

The residues are

$$A = \frac{z + 0.4}{(z - 0.3)(z - 0.8)}\bigg|_{z=1} = 10 \quad B = 2 \quad C = -12$$

so

$$C(z) = \frac{10z}{z - 1} + \frac{2z}{z - 0.3} + \frac{-12z}{z - 0.8}$$

Hence, from the geometric sequence in item 5 of Table 9.7.1 and from item 3,

$$c_n = 10 + 2(0.3)^n - 12(0.8)^n$$

Note that the expansion is made for $C(z)/z$ instead of $C(z)$. Then the terms of $C(z)$ have the forms available in Table 9.7.1. Otherwise, the technique is analogous to that for Laplace transforms. The advantage of the technique is that it gives a closed-form solution from which c_n can be found directly for any value of n.

(c) *Difference equation method:* Two formulations of this method are considered:

1. Write $C(z)$ in terms of powers of z^{-1} as follows by dividing the numerator and denominator of the expression in (a) by z^3:

$$\frac{C(z)}{1} = \frac{z^{-1} + 0.4z^{-2}}{1 - 2.1z^{-1} + 1.34z^{-2} - 0.24z^{-3}}$$

Cross-multiplication yields

$$(1 - 2.1z^{-1} + 1.34z^{-2} - 0.24z^{-3})C(z) = z^{-1} + 0.4z^{-2}$$

Using Table 9.7.1, the inverse can be written as

$$c_n = 2.1c_{n-1} - 1.34c_{n-2} + 0.24c_{n-3} + \delta_{n-1} + 0.4\delta_{n-2}$$

Solutions of this recurrence relation, assuming $c_i = 0$ for $i < 0$, are

$$c_0 = 0 \qquad c_1 = 0 - 0 + 0 + 1 + 0 = 1$$
$$c_2 = 2.1 - 0 + 0 + 0 + 0.4 = 2.5$$
$$c_3 = 2.1 \times 2.5 - 1.34 \times 1 + 0 + 0 + 0 = 3.91, \ldots$$

2. Write $C(z)/R(z)$, instead of $C(z)$, in terms of powers of z^{-1}. This is the preferred alternative if the transfer function is known. The transform $C(z)$ is in fact the response to a unit step input $R(z) = z/(z - 1)$ of the system

$$\frac{C(z)}{R(z)} = \frac{z + 0.4}{(z - 0.3)(z - 0.8)} = \frac{z + 0.4}{z^2 - 1.1z + 0.24}$$
$$= \frac{z^{-1} + 0.4z^{-2}}{1 - 1.1z^{-1} + 0.24z^{-2}}$$

Cross-multiplication yields

$$(1 - 1.1z^{-1} + 0.24z^{-2})C(z) = (z^{-1} + 0.4z^{-2})R(z)$$

The inverse gives, after rearranging,

$$c_n = 1.1c_{n-1} - 0.24c_{n-2} + r_{n-1} + 0.4r_{n-2}$$

For a unit step input sequence, substitution of $r_i = 1$, $i = 0, 1, \ldots$, leads to the same sequence for c_n as above.

In the next example these methods will be used to examine the proportional control of a simple lag plant. This will illustrate a fundamental difference with continuous control. In contrast to the continuous system, of which the response in this case is always nonoscillatory, the response with digital control becomes severely oscillatory as the gain is increased for constant sampling interval or if the sampling interval is increased for constant gain.

Example 10.3.2   Figure 10.1 with

$$G(s) = \frac{a}{s + a} \qquad D(z) = K$$

From Example 10.2.1, the loop gain function and the closed-loop transfer function are

$$G_1(z)D(z) = \frac{K(1 - e^{-aT})}{z - e^{-aT}} \qquad \frac{C(z)}{R(z)} = \frac{K(1 - e^{-aT})}{z - e^{-aT} + K(1 - e^{-aT})} \qquad (10.7)$$

a) If the difference equation method is to be used, the numerator and denominator are multiplied by z^{-1}, and cross-multiplication and inversion yield

$$c_n = [e^{-aT} - K(1 - e^{-aT})]c_{n-1} + K(1 - e^{-aT})r_{n-1} \qquad (10.8)$$

For a unit step input sequence, $r_i = 1$, $i = 0, 1, \ldots$, is substituted to calculate successive values of c_n.

b) Alternatively, to calculate this response to a unit step by partial fraction expansion, substitute $R(z) = z/(z - 1)$ into (10.7) and determine the residues, to find that

$$C(z) = z \frac{K(1 - e^{-aT})}{(z - 1)[z - e^{-aT} + K(1 - e^{-aT})]}$$
$$= \frac{K}{K + 1} \left[\frac{z}{z - 1} - \frac{z}{z - e^{-aT} + K(1 - e^{-aT})} \right]$$

Hence

$$c_n = \frac{K}{K+1}(1 - [e^{-aT} - K(1 - e^{-aT})]^n) \qquad (10.9)$$

The unit step response of the corresponding continuous system is

$$c(t) = \frac{K}{K+1}(1 - e^{-(K+1)at})$$

The steady-state value, which is the same for both, could have been found directly from the final value theorem in Table 9.7.1:

$$\lim_{n\to\infty} c_n = \lim_{z\to1} (1 - z^{-1})C(z) = \frac{K}{K+1} \qquad (10.10)$$

Figure 10.3 shows the responses as calculated from (10.8) or (10.9) for three values of aT and for the values of gain K indicated. (Ignore for now the other plots in Fig. 10.3.) The time scales of the three response graphs are the same, to facilitate comparison of response speeds. It is evident that, in contrast to the continuous system, the response becomes severely oscillatory as K is increased for constant aT or if T is increased for constant K. For $K = 2$, only the response for $aT = 0.25$ approximates that of the continuous system. The latter has a closed-loop time constant of

$$\frac{1}{(K+1)a} = \frac{1}{3a}$$

so that at time $1/a$ its transient will have largely decayed. For $aT = 0.25$, the sampling interval $T = 1/(4a)$ is somewhat smaller than this time constant and by Section 9.5 can be satisfactory.

The next example presents a proportional control design for a plant that will serve also to illustrate alternative design techniques in subsequent sections.

Example 10.3.3   Figure 10.1 with

$$G(s) = \frac{0.5}{s(s+0.5)} \qquad D(z) = K \qquad T = 1$$

Substituting $K = a = 0.5$, $T = 1$ into Table 9.9.1 gives

$$G_1(z) = 0.213\frac{z + 0.8474}{(z - 1)(z - 0.6065)} \qquad (10.11)$$

So the loop gain function is

$$G_1(z)D(z) = \frac{0.213K(z + 0.8474)}{(z - 1)(z - 0.6065)} \qquad (10.12)$$

For $D(z) = K = 1$, this can be written as

$$G_1(z)D(z) = \frac{0.213z + 0.1805}{z^2 - 1.6065z + 0.6065}$$

and for a unit step $R(z) = z/(z - 1)$ the closed-loop transfer function (10.1) yields

$$C(z) = \frac{0.213z^2 + 0.1805z}{z^3 - 2.3935z^2 + 2.1805z - 0.787}$$
$$= 0.213z^{-1} + 0.690z^{-2} + 1.188z^{-3} + 1.505z^{-4} + 1.557z^{-5} + 1.378z^{-6}$$
$$+ \cdots$$

by long division. So the output sample values are

$$c_0 = 0, \quad c_1 = 0.213, \quad c_2 = 0.690, \quad c_3 = 1.188,$$
$$c_4 = 1.505, \quad c_5 = 1.557, \quad c_6 = 1.378, \quad \cdots$$

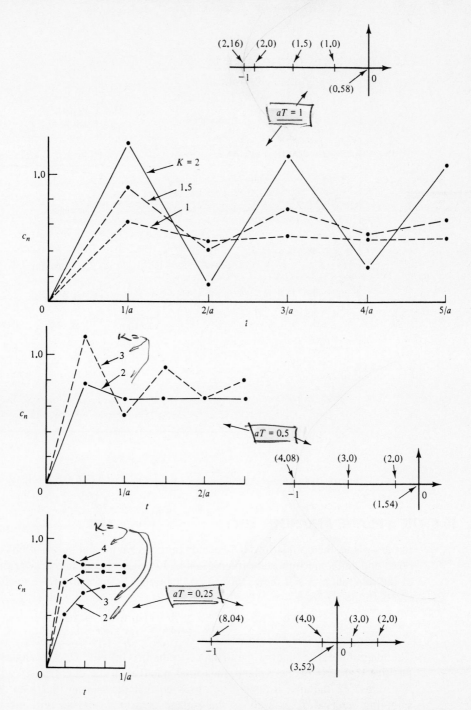

Figure 10.3   Examples 10.3.2 and 10.4.2.

This sample response is plotted on Fig. 10.4, together with those for a number of alternative designs for the same plant, all for sampling interval $T = 1$ sec, discussed in subsequent examples.

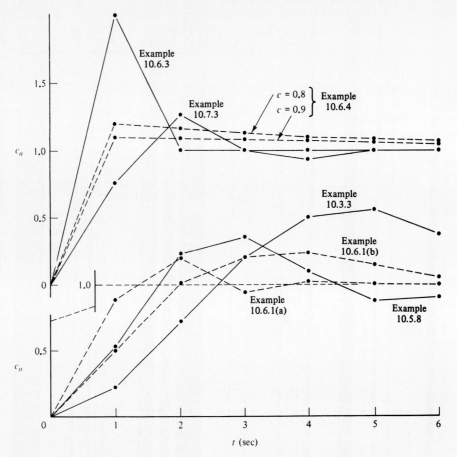

Figure 10.4   Digital control examples for the plant $G(s) = 0.5/[s(s + 0.5)]$.

10.4 THE z-PLANE AND ROOT LOCI

In the s-plane, the correlations between dynamic behavior and the positions of system poles and zeros are important in analysis and design of continuous systems. By use of the relation $z = e^{Ts}$, these correlations can also be used to derive analogous insights in the z-plane.

$$\text{poles or zeros } s = -a \pm jb \quad \text{map to the positions}$$
$$z = e^{(-a \pm jb)T} = e^{-aT}e^{\pm jbT} \tag{10.13}$$

The magnitude of z (that is, the distance to the origin) is e^{-aT}. The angles with the positive real axis of the z-plane, measured positive in the counterclockwise direction, are $\pm bT$ radians. It should be observed that the z-plane locations depend on sampling interval T, as well as the s-plane positions, and so will move if T is changed. Two important special cases are the following:

1. Constant a: A vertical line in the s-plane maps into a circle of radius e^{-aT} in the z-plane as b is varied.

2. *Constant b:* A horizontal line in the s-plane, which means a constant frequency, maps into a radial line at angle bT radians in the z-plane as a is varied.

Figure 10.5 shows the loci of corresponding locations in the upper-half planes, identified by identical number sequences. All loci are symmetrical about the real axes.

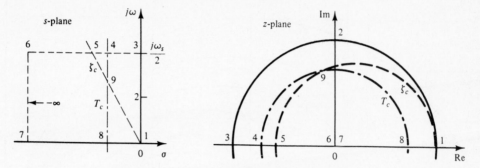

Figure 10.5   Correlations between the s- and z-planes.

1. *Stability:* In the s-plane, the imaginary axis $(a = 0)$ is the boundary of the stable pole region. This axis maps into $z = e^{\pm jbT}$, a circle of unit radius about the origin, called the *unit circle*, as b moves along the imaginary axis. Stability requires that $a > 0$, so $e^{-aT} < 1$. Hence follows the basic theorem for stability of digital control systems.

Stability Theorem.   For stability, all poles of the system Z transfer function must lie inside the unit circle in the z-plane.

2. *Time constant, T_c:* In the s-plane, the poles must lie to the left of a vertical at $-a = -1/T_c$ to achieve a time constant $< T_c$. In the z-plane such a vertical at $-a$ maps into a circle of radius e^{-aT} as b moves along the vertical. So in the z-plane the poles must lie inside a circle of radius e^{-aT}. For constant T, moving the poles closer to the origin of the z-plane increases the speed of response. Note that a constraint on T_c also implies a constraint on the *settling time*, which is given by $4T_c$.

3. *Damping ratio, ζ_c:* Poles $s_{1,2} = -\zeta\omega_n \pm j\omega_n\sqrt{1 - \zeta^2}$ for constant ζ lie along radial lines. In the z-plane

$$z_{1,2} = e^{-\zeta\omega_n T}e^{\pm j\omega_n\sqrt{1-\zeta^2}T}$$

As ω_n increases for constant ζ, so as s moves out along a radial line, the magnitude of z decreases and the phase angle increases, and z describes a logarithmic spiral. Figure 10.6 shows the loci in the upper-half z-plane. A specified minimum damping requires the poles to lie inside the corresponding spiral. As an aid in analysis and design, Table 10.4.1 shows real and imaginary values for points on the curves in Fig. 10.6 where these intersect radial lines at intervals of 10° in the z-plane.

4. *Sampling frequency:* As the imaginary part b of the poles, and thus the frequency of the oscillations, moves closer to the limit $\omega_s/2$ of the primary strip, the number of samples per cycle reduces to the minimum of two permitted by

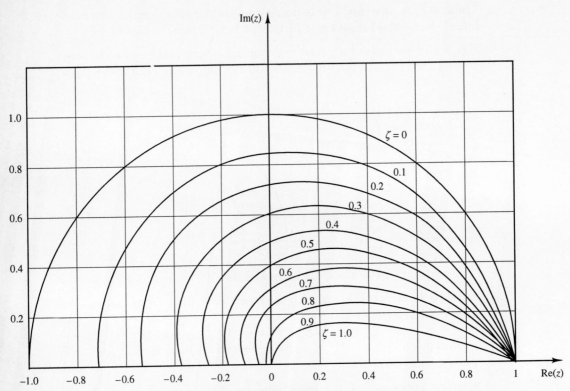

Figure 10.6   Constant ζ loci.

the sampling theorem. In the z-plane, this reduction occurs as the angle $e^{\pm jbT}$ of the poles moves closer to the direction of the negative real axis. Indeed, for $b = \omega_s/2 = \pi/T$ the angle (bT radians) of z is 180°. In design, to ensure an adequate number of samples per cycle, the angle of the closed-loop poles should be well away from 180°. Along the imaginary axis of the z-plane $bT = \pi/2$, so $b = \pi/(2T) = \omega_s/4$. Thus the transients corresponding to poles along the imaginary axis are sampled at the rate of four samples per cycle. To obtain at least eight samples per cycle, the poles should be within 45° of the positive real axis.

5. Horizontal lines in the s-plane, that is constant-frequency loci, map into radial lines in the z-plane. This is implied by point 4 because constant b means a constant angle in the z-plane. There are two special cases:
    a. The positive real axis in the z-plane corresponds to the entire real axis in the s-plane.
    b. The negative real axis in the z-plane corresponds to the boundaries of the primary strip in the s-plane.

Root Loci in the z-Plane

The system characteristic equation for Fig. 10.1 is, from (10.1),

$$1 + G_1(z)D(z) = 0 \qquad (10.14)$$

TABLE 10.4.1   REAL AND IMAGINARY COORDINATES OF POINTS ON CONSTANT-ζ CURVES

		ζ: 0.2	0.3	0.4	0.5	0.6	0.7	0.8
10	Re	0.950	0.932	0.913	0.891	0.864	0.830	0.780
	Im	0.168	0.164	0.161	0.157	0.152	0.146	0.138
20	Re	0.875	0.842	0.807	0.768	0.723	0.668	0.590
	Im	0.319	0.307	0.294	0.280	0.263	0.243	0.215
30	Re	0.778	0.735	0.689	0.640	0.585	0.519	0.431
	Im	0.449	0.424	0.398	0.370	0.338	0.299	0.249
40	Re	0.664	0.615	0.565	0.512	0.454	0.387	0.302
	Im	0.558	0.516	0.474	0.430	0.381	0.324	0.254
50	Re	0.538	0.489	0.439	0.389	0.334	0.273	0.201
	Im	0.641	0.582	0.524	0.463	0.398	0.326	0.239
60	Re	0.404	0.360	0.317	0.273	0.228	0.179	0.124
	Im	0.699	0.623	0.549	0.473	0.395	0.310	0.215
70	Re	0.267	0.233	0.201	0.169	0.137	0.103	0.067
	Im	0.732	0.640	0.552	0.464	0.376	0.284	0.185
80	Re	0.131	0.112	0.094	0.078	0.061	0.044	0.027
	Im	0.741	0.635	0.536	0.440	0.346	0.251	0.153
90	Re	0	0	0	0	0	0	0
	Im	0.726	0.610	0.504	0.404	0.308	0.215	0.123
100	Re	−0.122	−0.100	−0.081	−0.063	−0.047	−0.031	−0.017
	Im	0.690	0.569	0.460	0.360	0.266	0.178	0.096
110	Re	−0.231	−0.187	−0.148	−0.113	−0.081	−0.052	−0.027
	Im	0.635	0.514	0.407	0.310	0.223	0.143	0.073
120	Re	−0.326	−0.259	−0.201	−0.149	−0.104	−0.064	−0.031
	Im	0.565	0.448	0.347	0.259	0.180	0.111	0.053
130	Re	−0.405	−0.315	−0.239	−0.174	−0.117	−0.070	−0.031
	Im	0.482	0.375	0.285	0.207	0.140	0.083	0.037
140	Re	−0.465	−0.355	−0.264	−0.187	−0.123	−0.070	−0.030
	Im	0.391	0.298	0.221	0.157	0.103	0.059	0.025
150	Re	−0.508	−0.380	−0.276	−0.191	−0.122	−0.067	−0.026
	Im	0.293	0.220	0.160	0.110	0.070	0.039	0.015
160	Re	−0.532	−0.391	−0.278	−0.188	−0.116	−0.061	−0.023
	Im	0.194	0.142	0.101	0.068	0.042	0.022	0.008
170	Re	−0.538	−0.388	−0.270	−0.178	−0.107	−0.054	−0.019
	Im	0.095	0.068	0.048	0.031	0.019	0.009	0.003
180	Re	−0.527	−0.373	−0.254	−0.163	−0.095	−0.046	−0.015
	Im	0	0	0	0	0	0	0

This equation is of the same form as that for continuous systems in the s-plane. Therefore, root loci can be constructed to determine closed-loop pole positions in the z-plane by exactly the same rules as were used in the s-plane. The following examples will also illustrate the features of digital control systems outlined earlier. Here, as well as for root locus design in the z-plane discussed in Section 10.6, Fig. 10.6 is useful as a standard graph paper.

Example 10.4.1   Example 10.3.2 for $aT = 1$

Figure 10.7(a) shows the loci constructed from the loop gain function in (10.7) for $aT = 1$,

$$G_1(z)D(z) = \frac{0.6322K}{z - 0.3678}$$

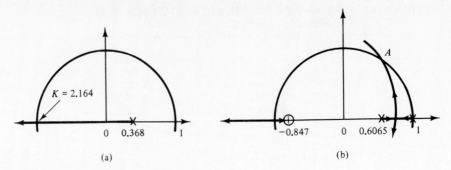

Figure 10.7   Root locus Examples 10.4.1 and 10.4.3.

The root locus is the real axis to the left of the open-loop pole at 0.3678. The graphical application of the magnitude condition yields the root locus gain $0.6322K = 1.3678$, or $K = 2.164$, where the locus crosses the unit circle. This is the stability limit. In contrast to the continuous system, which is always stable, sampling effects cause instability for $K > 2.164$. Of course, root loci are not needed for this example. For example, from (10.14), the system pole is -1 for $K = 1.3678/0.6322 = 2.164$.

Example 10.4.2

Figure 10.3 shows step response sequences for the same loop gain function (10.7) as in the preceding example, but including $aT = 0.25$ and $aT = 0.5$, as well as $aT = 1$.

Figure 10.3 shows also the positions of the closed-loop system pole $z = e^{-aT} - K(1 - e^{-aT})$ for all cases, as well as the values of K for which the pole crosses the unit circle, the stability limit. Note that the limiting value of K for stability decreases as the sampling interval T increases. Furthermore, the values of K are given for which the pole lies at the origin. Then (10.9) reduces to $c_n = K/(K + 1)$, since the second term is zero in this case. This shows and interpolation in Fig. 10.3 confirms that the response then reaches steady state, at the sampling instants, in one sampling interval. This is called *deadbeat response,* which will be discussed later.

It is also important to observe that the oscillatory responses in Fig. 10.3 show only two samples per cycle. This could have been predicted, because for these cases the poles are on the negative real axis of the z-plane and, according to item 4 of the discussion related to Fig. 10.5, these correspond to an oscillatory response with only two samples per cycle.

Example 10.4.3   **Loci for Example 10.3.3**

The loop gain function (10.12) is

$$G_1(z)D(z) = \frac{0.213K(z + 0.8474)}{(z - 1)(z - 0.6065)} \qquad (10.15)$$

Figure 10.7(b) shows a partial root locus sketch, which can be constructed in the usual way. For example, some trial points along the curve in Fig. 10.6 for a damping ratio of interest can be used to determine where the root locus intersects this curve. Similarly, some trial points along the unit circle will serve to determine where the locus intersects it. Application of the magnitude condition to the unit circle crossing A yields the stability limit $K = 2.18$. In the case of this example, these results can also be calculated analytically, because the characteristic equation $1 + G_1(z)D(z) = 0$ is a quadratic equation:

$$z^2 - (1.6065 - 0.213K)z + 0.6065 + 0.1805K = 0$$

On the unit circle the magnitude of each root is 1, so their product, given by the constant term in the characteristic equation, must be 1:

$$0.6065 + 0.1805K = 1.0$$

This yields $K = 2.18$, as before. It is again important to observe that the corresponding continuous design is stable for any K. As illustrated also in the preceding example, a system with sampling generally becomes unstable as K and/or T is increased.

The actual value of gain here and in Example 10.3.3 is $K = 1$. The characteristic equation $1 + G_1(z)D(z) = 0$ is then $z^2 - 1.394z + 0.787 = 0$. Its roots are $z_{1,2} = 0.697 \pm 0.549j$ and correspond to a damping ratio of about 0.2, according to the loci for constant ζ in Fig. 10.6 or the corresponding Table 10.4.1. The step response for Example 10.3.3 in Fig. 10.4 shows a maximum overshoot of about 56%. The characteristic equation of the corresponding continuous system is $s^2 + 0.5s + 0.5 = 0$, which implies a damping ratio $\zeta = 0.35$ and an overshoot of about 32%. For a smaller sampling interval the agreement between the sampled and continuous systems would be better.

10.5 DIGITAL CONTROLLERS BY CONTINUOUS SYSTEM DESIGN

One common approach to the design of digital control systems is to select a compensator $D(s)$ for the equivalent continuous system and approximate it by a digital filter $D(z)$. Implementation as a control algorithm then follows immediately. For example, if

$$D(z) = \frac{U(z)}{E(z)} = \frac{K(1 + az^{-1})}{1 + bz^{-1}}$$

cross-multiplication gives

$$U(z) + bz^{-1}U(z) = KE(z) + Kaz^{-1}E(z)$$

By inversion, this corresponds to the algorithm

$$u_k = -bu_{k-1} + Ke_k + Kae_{k-1}$$

But design verification of both stability and response (for example, by root loci) is always necessary for the chosen sampling interval.

A characteristic that the preceding algorithm has in common with many others is that u_k is implied to be available at the same instant as e_k from which it is calculated. However, a small delay here is usually not objectionable.

Several techniques for translating $D(s)$ to $D(z)$ are available. These will now be discussed and illustrated by application to common forms of compensators, assumed to have been designed previously.

$$\text{PID control:} \quad D(s) = \frac{U(s)}{E(s)} = K_p + \frac{K_i}{s} + K_d s$$

$$\text{phase lead/lag:} \quad D(s) = \frac{U(s)}{E(s)} = \frac{K(s + b)}{s + a} \tag{10.16}$$

1 **Finite-difference approximation; Tustin's method.** This approach was introduced in Chapter 9. Equations (9.14) and (9.15) give PID control algorithms derived using backward differences and the trapezoidal rule. In (9.27) some of these are represented by discrete transfer functions $D(z)$. In (9.17), phase lead/lag is represented in finite-difference forms based on backward differences and the trapezoidal rule. These algorithms can also immediately be written as discrete transfer functions. For example, the algorithm (9.17a), rearranged to the form

$$(1 + aT)u_k - u_{k-1} = K(1 + bT)e_k - Ke_{k-1}$$

yields on transformation

$$D(z) = \frac{U(z)}{E(z)} = K\frac{1 + bT - z^{-1}}{1 + aT - z^{-1}} \tag{10.17}$$

From a more fundamental point of view, (9.26) give $D(z)$ as obtained by different algorithms for the integrator

$$D(s) = \frac{1}{s} \tag{10.18a}$$

$$D(z) = \frac{T}{z - 1} = \frac{Tz^{-1}}{1 - z^{-1}} \qquad \text{forward rectangular rule} \tag{10.18b}$$

$$D(z) = \frac{Tz}{z - 1} = \frac{T}{1 - z^{-1}} \qquad \text{backward rectangular rule} \tag{10.18c}$$

$$D(z) = \frac{T}{2}\frac{z + 1}{z - 1} = \frac{T}{2}\frac{1 + z^{-1}}{1 - z^{-1}} \qquad \text{trapezoidal rule} \tag{10.18d}$$

The inverses of these give the corresponding approximations to a derivative, $D(s) = s$.
One approach to approximating $D(s)$ by $D(z)$ is to replace s by one of these forms. The most common of these is that based on the trapezoidal rule:

$$s = \frac{2}{T}\frac{1 - z^{-1}}{1 + z^{-1}} = \frac{2}{T}\frac{z - 1}{z + 1} \tag{10.19}$$

This is known as the *bilinear transformation* and is also often referred to as *Tustin's method:*

$$D(z) = D(s)\Big|_{s=(2/T)(z-1)/(z+1)} \tag{10.20}$$

A correction of this result may be necessary, because (10.19) is only an approximation to the true relation $z = e^{Ts}$. This will be discussed in Section 10.7, in the context of frequency response considerations.

Example 10.5.1 PID Algorithms

(a) *PI control $D(s) = K_p + K_i/s$:* With the backward rectangular rule,

$$D(z) = K_p + \frac{K_i Tz}{z - 1} \tag{10.21a}$$

This is the same result as derived in (9.27a) by first formulating the algorithm.

(b) *PID control $D(s) = K_p + (K_i/s) + K_d s$:* Using the Tustin approximation for the integral and the backward rectangular approximation for the derivative, this gives

$$D(z) = K_p + \frac{K_i T}{2}\frac{z + 1}{z - 1} + K_d\frac{z - 1}{Tz} \tag{10.21b}$$

which may be verified to be identical to (9.27b).

Example 10.5.2 Phase Lead/Lag (10.16)

Substituting (10.20) gives the following approximation by Tustin's method:

$$D(z) = K\frac{(bT + 2)z + (bT - 2)}{(aT + 2)z + (aT - 2)} \tag{10.22}$$

Example 10.5.3

$$\frac{Y(s)}{X(s)} = \frac{1}{s^2 + as + b}$$

Cross-multiplying gives

$$(s^2 + as + b)Y = X$$

of which the Tustin approximation is

$$\left(\frac{4}{T^2}\frac{1 - 2z^{-1} + z^{-2}}{1 + 2z^{-1} + z^{-2}} + \frac{2a}{T}\frac{1 - z^{-1}}{1 + z^{-1}} + b\right)Y(z) = X(z)$$

This can be arranged as a discrete transfer function $D(z) = Y(z)/X(z)$ or written as a recursive algorithm for y_k.

Impulse-invariant method, or Z-transform method. In this method, $D(z)$ is chosen to be the Z transform of $D(s)$. By definition, this means that the unit pulse response sequence will equal the sampled unit pulse response of $D(s)$.

The actual form used is often

$$D(z) = cZ[D(s)] \tag{10.23}$$

Here c is a constant chosen so that the gain of $D(z)$ will match that of $D(s)$ at a selected frequency, usually $s = 0$. By Fig. 10.5, this corresponds to $z = 1$, so c is chosen to make

$$D(z)\Big|_{z=1} = D(s)\Big|_{s=0} \tag{10.24}$$

As in Example 9.8.5, partial fraction expansion of $D(s)$ is used if necessary to derive $D(z)$. If, as in the following example, the numerator and denominator of $D(s)$ have the same maximum power, a division must be performed first.

Example 10.5.4   Phase Lead/Lag (10.16)

$$Z\left[K\frac{s + b}{s + a}\right] = KZ\left[1 + \frac{b - a}{s + a}\right] = K\left[1 + \frac{b - a}{1 - e^{-aT}z^{-1}}\right]$$

$$= K\frac{1 + b - a - e^{-aT}z^{-1}}{1 - e^{-aT}z^{-1}}$$

To satisfy (10.24) requires $D(z)|_{z=1} = Kb/a$, so

$$D(z) = K\frac{b}{a}\frac{1 - e^{-aT}}{1 + b - a - e^{-aT}}\frac{1 + b - a - e^{-aT}z^{-1}}{1 - e^{-aT}z^{-1}} \tag{10.25}$$

Zero-order-hold equivalence. In this method, $D(z)$ is considered to be generated by preceding $D(s)$ by a ZOH and following it by a sampler, as indicated in Fig. 10.8. From this, as in (9.32),

$$D(z) = (1 - z^{-1})Z\left[\frac{D(s)}{s}\right] \tag{10.26}$$

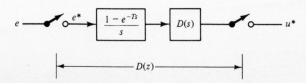

Figure 10.8 Hold equivalence.

Example 10.5.5 Phase Lead/Lag (10.16)

$$D(z) = (1 - z^{-1})Z\left(\frac{K}{s}\frac{s+b}{s+a}\right) = K(1 - z^{-1})Z\left(\frac{b/a}{s} + \frac{1 - b/a}{s+a}\right)$$

$$= K(1 - z^{-1})\left(\frac{b/a}{1 - z^{-1}} + \frac{1 - b/a}{1 - e^{-aT}z^{-1}}\right)$$

$$= K\frac{1 - [1 + (b/a)(e^{-aT} - 1)]z^{-1}}{1 - e^{-aT}z^{-1}} \qquad (10.27)$$

Pole–zero matching. As the name implies, poles and zeros s_i of $D(s)$ are mapped as poles and zeros of $D(z)$ according to $z_i = e^{s_i T}$, and the gain of $D(z)$ is set to satisfy, say, (10.24).

If $D(s)$ has more poles than zeros, it tends to zero as frequency $\omega \to \infty$. This is simulated by making $D^*(s)$ zero at the limits of the primary strip, which by Fig. 10.5 corresponds to making $D(z)$ zero for $z = -1$. In $D(z)$, therefore, the powers of numerator and denominator are made the same by adding factors $(z^{-1} + 1)$ in the numerator. The gains can again be chosen to satisfy (10.24).

Example 10.5.6 $D(s) = a/(s + a)$

$$D(z) = \frac{0.5(1 - e^{-aT})(z + 1)}{z - e^{-aT}} \qquad (10.28)$$

Example 10.5.7 Phase Lead/Lag (10.16)

$$D(z) = K\frac{b}{a}\frac{1 - e^{-aT}}{1 - e^{-bT}}\frac{z - e^{-bT}}{z - e^{-aT}} \qquad (10.29)$$

Tustin's method and pole–zero matching in particular are used exte sively. A design example follows, for a plant considered on previous occasions.

Example 10.5.8 Examples 10.3.3 and 10.4.3 Continued

$$G(s) = \frac{0.5}{s(s + 0.5)}$$

As in (10.11), with $T = 1$,

$$G_1(z) = \frac{0.213(z + 0.847)}{(z - 1)(z - 0.6065)} \qquad (10.30)$$

The step response for Example 3.3 in Fig. 10.4 is for $D(s) = D(z) = K = 1$. This corresponds to a damping ratio $\zeta = 0.2$ for the discrete design and 0.35 for the continuous design. Thus, assuming that accuracy requirements do not permit gain reduction, even the continuous system needs dynamic compensation. Bode plot design leads to the choice

$$D(s) = \frac{6(s + 0.5)}{s + 3} \qquad (10.31)$$

Note that the zero of $D(s)$ has been chosen to cancel the pole of $G(s)$. The closed-loop damping ratio for the continuous design is then found to be $\zeta = 0.866$. Using pole–zero matching yields

$$D(z) = \frac{1 - e^{-3}}{1 - e^{-0.5}}\frac{z - e^{-0.5}}{z - e^{-3}} = 2.415\frac{z - 0.6065}{z - 0.0498} \qquad (10.32)$$

Hence

$$G_1(z)D(z) = \frac{0.514(z + 0.847)}{(z - 1)(z - 0.0498)}$$

Note that if $D(s)$ cancels a pole of $G(s)$ then $D(z)$ obtained by pole–zero matching does so in $G_1(z)$.

Figure 10.9 shows partial loci for K times $D(z)$ of (10.32), and those in Fig. 10.7(b) for $D(z) = K$ for comparison. The root locations indicated for $K = 1$ are in this case also readily found from the system characteristic equation

$$1 + G_1(z)D(z) = z^2 - 0.536z + 0.486 = 0$$

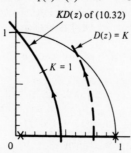

Figure 10.9   Example 10.5.8.

Using the constant ζ loci in Fig. 10.6 or Table 10.4.1, the location corresponds to a damping ratio of only about 0.3, much below the 0.866 of the continuous design.

The sampling interval would have to be reduced for better agreement. The characteristic equation of the continuous design is $1 + G(s)D(s) = s^2 + 3s + 3 = 0$, so $\omega_n = \sqrt{3} = 1.732$ and $\zeta = 0.866$. Therefore, the system time constant is $T_c = 1/(\zeta\omega_n) = 0.67$, and the frequency $(\omega_n\sqrt{1 - \zeta^2})$ of transient oscillations is 0.866 rad/sec, and their period is $2\pi/0.866 = 7.25$ sec. So $T = 1$ gives over seven samples per cycle. But $T > T_c$, so T does not meet the sampling rate recommendations of Section 9.5 in terms of the time constant of the transient. The closed-loop transfer function is

$$\frac{C(z)}{R(z)} = \frac{0.514z + 0.4354}{z^2 - 0.536z + 0.486}$$

and calculating the unit step sample response by the difference equation method, which is faster and easier than long division without a computer, yields

$$c_0 = 0, \quad c_1 = 0.514, \quad c_2 = 1.225, \quad c_3 = 1.356, \quad c_4 = 1.081,$$
$$c_5 = 0.870, \quad c_6 = 0.890, \quad \ldots$$

This is plotted in Fig. 10.4. The overshoot is about 40%, compared to about 0% for the continuous design, as the difference in values of ζ suggests.

10.6 DIRECT DESIGN OF DIGITAL CONTROLLERS

One disadvantage of approximating a continuous design is that this does not fully exploit the flexibility of digital computer control. Direct design can make use of the fact that digital control implementation is subject to relatively few constraints. For example, poles and zeros of phase-lead/lag compensators are on the negative real axis of the s-plane. From Fig. 10.5, this corresponds to the positive real axis of the z-plane. But digital controllers allow simple poles and zeros to be placed on the negative as well as the positive real axis in the z-plane. Furthermore, digital control allows the implementation of a controller that, subject to certain constraints, makes the closed-loop transfer function equal to a desired function $T(z)$.

Direct Root Locus Design in the z-Plane

As was noted earlier, it is convenient to use Fig. 10.6 as a standard graph paper on which the plots are made.

Example 10.6.1   Example 10.5.8 Continued

In Example 10.5.8, the zero of $D(z)$ was chosen at $z = 0.6065$, to cancel a pole of $G_1(z)$, and the pole of $D(z)$ was at $z = 0.0498$. Both are on the positive real axis, a direct consequence of design by "translation" of $D(s)$, which usually has its poles and zeros confined to the negative real axis of the s-plane. To exploit the greater flexibility of digital control, suppose that a pole of $D(z)$ is chosen at -0.6 on the negative real axis of the z-plane.

(a) If the zero of $D(z)$ is still chosen to cancel the pole of $G_1(z)$, then, for $D(z)|_{z=1} = 1$,

$$D(z) = \frac{1 + 0.6}{1 - 0.6065} \frac{z - 0.6065}{z + 0.6} = 4.066 \frac{z - 0.6065}{z + 0.6}$$

$$G_1(z)D(z) = \frac{0.866(z + 0.847)}{(z - 1)(z + 0.6)}$$

Figure 10.10 shows partial loci [curve (a)] for K times this loop gain function with the root locations for $K = 1$ identified. The loci of Fig. 10.9 are shown for comparison. The constant ζ loci in Fig. 10.6 indicate an improvement of ζ from 0.3 to almost 0.5. The poles are also closer to the origin, so the time constant is smaller. But the angle of the poles is undesirably close to the direction of the negative real axis, suggesting a low sampling rate and a poor representation of the continuous output by its samples.

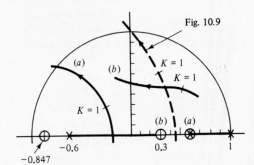

Figure 10.10   Example 10.6.1.

The closed-loop transfer function is

$$\frac{C(z)}{R(z)} = \frac{0.866z + 0.7335}{z^2 + 0.466z + 0.1335}$$

and the difference equation method yields the unit step response sample sequence

$$c_0 = 0, \quad c_1 = 0.866, \quad c_2 = 1.196, \quad c_3 = 0.927, \quad c_4 = 1.008,$$
$$c_5 = 1.006, \quad c_6 = 0.996, \quad \ldots$$

The plot in Fig. 10.4 confirms the smaller overshoot, about 20%, and smaller settling time, and also the small number of samples per cycle of the transient, suggesting that the samples may be a poor representation of the response.

(b) To force the breakaway point of the complex branches to occur between 0.6065 and 1, the zero of $D(z)$ is moved from 0.6065 to 0.3:

$$D(z) = \frac{1 + 0.6}{1 - 0.3} \frac{z - 0.3}{z + 0.6} = 2.286 \frac{z - 0.3}{z + 0.6}$$

Partial loci [curve (b)] are shown in Fig. 10.10. The zero at 0.3 pulls the branches to the left. The roots for $K = 1$ correspond to a damping ratio of almost 0.5, but the settling time for the samples will be larger than for curve (a).

The loop gain function is

$$G_1(z)D(z) = \frac{0.4869(z - 0.3)(z + 0.847)}{(z - 1)(z + 0.6)(z - 0.6065)}$$

and the closed-loop transfer function

$$\frac{C(z)}{R(z)} = \frac{0.4869(z^2 + 0.547z - 0.2541)}{z^3 - 0.5196z^2 - 0.09107z + 0.2402}$$

The difference equation method yields the unit step response sample sequence

$$c_0 = 0, \quad c_1 = 0.487, \quad c_2 = 1.006, \quad c_3 = 1.197, \quad c_4 = 1.226,$$
$$c_5 = 1.134, \quad c_6 = 1.043, \quad \cdots$$

which is plotted as curve (b) in Fig. 10.4. The overshoot is about 25% and the settling time similar to Example 10.5.8, as Fig. 10.10 suggests. The smaller angle of the poles with the positive real axis is reflected by a greater number of samples per cycle than for curve (a).

Deadbeat System Design

An alternative direct design approach is based on designing a controller $D(z)$ such that the closed-loop transfer function $C(z)/R(z)$ becomes equal to a desired function $T(z)$:

$$T(z) = \frac{C(z)}{R(z)} = \frac{G_1(z)D(z)}{1 + G_1(z)D(z)} \tag{10.33a}$$

$$E(z) = R(z) - C(z) = R(z)[1 - T(z)] \tag{10.33b}$$

The controller required to achieve this is found by solving (10.33a) for $D(z)$:

$$D(z) = \frac{1}{G_1(z)} \frac{T(z)}{1 - T(z)} \tag{10.34}$$

But, as one would expect, the choice of $T(z)$ is constrained by characteristics of $G_1(z)$, as well as by desired features of the performance:

1. The method depends on cancellations of zeros and poles of $G_1(z)$ from the loop gain function $G_1(z)D(z)$ by the choice of $D(z)$. This is necessary in order to replace the plant dynamics by the dynamics needed to produce $T(z)$. But for poles and zeros of $G_1(z)$ on or outside the unit circle, this cancellation would, as in the s-plane, result in unstable closed-loop poles. From (10.34), therefore, so that they do not appear in $D(z)$:

   a. Such zeros of $G_1(z)$ must also be chosen to be zeros of $T(z)$.

   b. Such poles of $G_1(z)$ must be included as zeros of $1 - T(z)$.

2. If $T(z) = t_k z^{-k} + t_{k+1} z^{-k-1} + \cdots$, and if dividing numerator and denominator of $G_1(z)$ gives $G_1(z) = g_n z^{-n} + g_{n+1} z^{-n-1} + \cdots$, then, from (10.34),

$$D(z) = \frac{t_k z^{-k} + \cdots}{(g_n z^{-n} + \cdots)(1 - t_k z^{-k} - \cdots)}$$
$$= d_{k-n} z^{-(k-n)} + d_{k-n+1} z^{-(k-n+1)} + \cdots$$

A term in this series with a positive power of z would imply a system that supplies an output before the input is applied. So $k \geq n$ is required for $D(z)$ to be realizable. Therefore, $T(z)$ must be chosen such that the lowest power of z^{-1} is at least as high as in $G_1(z)$.

3. The steady-state error is, from (10.33b) and the final value theorem,

$$e_{ss} = \lim_{z \to 1} (1 - z^{-1})E(z) = \lim_{z \to 1} (1 - z^{-1})R(z)[1 - T(z)]$$

To make this steady-state error zero, the expression must have a factor $(1 - z^{-1})$. For a unit step input $R(z) = 1/(1 - z^{-1})$, this requires that $1 - T(z)$ be of the form, for $m = 1$,

$$1 - T(z) = (1 - z^{-1})^m F(z) \qquad T(z) = 1 - (1 - z^{-1})^m F(z) \qquad (10.35)$$

for which the choice of $F(z)$ is discussed later. For a unit ramp $R(z) = Tz^{-1}/(1 - z^{-1})^2$, this condition applies for $m = 2$.

4. Zero steady-state error at the sampling instants can be achieved in finite time. This is in contrast to a continuous system, where the decay of transients theoretically takes infinite time. The *finite settling time* property applies only at the sampling instants and requires that $T(z)$ in (10.35) have a finite number of terms. For example, if $T(z) = z^{-1}$, the output sequence equals the input sequence with delay T. If the highest power of z^{-1} in $T(z)$ is p, settling requires p sampling intervals.

5. A *deadbeat* design is one that settles to zero error at the sampling instants in a minimum number of sampling intervals. In (10.35), if $F(z) = 1$, $T(z) = z^{-1}$ for step inputs ($m = 1$) and $T(z) = 2z^{-1} - z^{-2}$ for ramp inputs ($m = 2$), so the deadbeat response requires one and two sampling intervals, respectively. The corresponding compensators are, from (10.34),

$$D(z) = \frac{1}{G_1(z)} \frac{z^{-1}}{1 - z^{-1}} \quad (m = 1) \qquad D(z) = \frac{1}{G_1(z)} \frac{2z^{-1} - z^{-2}}{(1 - z^{-1})^2} \qquad (10.36)$$

But, as illustrated by the next example, conditions 1 and 2 may impose a different choice of $F(z)$ than $F(z) = 1$.

Example 10.6.2

$G_1(z)$ has two more poles than zeros, all inside the unit circle. By condition 2, $T(z)$ can at best be z^{-2}. Let

$$T(z) = \alpha_2 z^{-2} + \alpha_3 z^{-3} + \cdots \qquad F(z) = \beta_0 + \beta_1 z^{-1} + \beta_2 z^{-2} + \cdots$$

For zero steady-state error after a step, (10.35) then gives the condition ($m = 1$)

$$1 - (1 - z^{-1})(\beta_0 + \beta_1 z^{-1} + \beta_2 z^{-2} + \cdots) = \alpha_2 z^{-2} + \alpha_3 z^{-3} + \cdots$$

Equating coefficients shows that this equation can be satisfied for $\beta_0 = \beta_1 = \alpha_2 = 1$, $\alpha_3 = \cdots = \beta_2 = \cdots = 0$. Thus $T(z) = z^{-2}$ and $F(z) = 1 + z^{-1}$. Substitution of $T(z)$ into (10.34) yields the controller $D(z) = [1/G_1(z)]z^{-2}/(1 - z^{-2})$.

Example 10.6.3   Examples 10.3.3, 10.5.8, and 10.6.1

$$G_1(z) = \frac{0.213(z + 0.847)}{(z - 1)(z - 0.6065)} \qquad (10.37)$$

Equation (10.35) gives for $F(z) = 1$ and $m = 2$ the following condition for zero steady-state error after a ramp input:

$$1 - T(z) = (1 - z^{-1})^2 \qquad T(z) = 2z^{-1} - z^{-2}$$

This satisfies condition 1, which requires that the pole of $G_1(z)$ on the unit circle be included as a zero of $1 - T(z)$. It also satisfies condition 2, because $T(z)$ has the same lowest power of z^{-1} as $G_1(z)$. The choice $F(z) = 1$ apparently satisfies the conditions. Moreover, from points 4 and 5, this is a deadbeat design and settles, at the sampling instants, in two sampling intervals. Since $T(z) = C(z)/R(z)$,

$$C(z) = 2z^{-1}R(z) - z^{-2}R(z) \qquad c_k = 2r_{k-1} - r_{k-2}$$

For a ramp input, $r_k = kT$, $c_0 = 0$, $c_1 = 0$, $c_2 = 2T$, $c_3 = 3T, \ldots$. For a step input, $r_k = 1$, $c_0 = 0$, $c_1 = 2$, $c_2 = c_3 = \cdots = 1$. So both have zero error after $2T$, but a severe overshoot exists for a step input, for which the sample sequence is plotted in Fig. 10.4.

Example 10.6.3 demonstrates one of several serious disadvantages of deadbeat design in that the response to the step input shows a very large overshoot. The response for inputs other than that for which the design is made often tends to be unsatisfactory. At least as serious is that the output generally oscillates between sampling instants. This occurs because control activity has to be strong to achieve zero error in minimum time. Furthermore, the exact cancellation of plant poles and zeros by $D(z)$ is not possible in practice, and the response is very sensitive to such errors.

A *staleness weighting factor* is commonly introduced to obtain more satisfactory results. It accepts a lower speed of response in return for more damping and better behavior for inputs other than the design input.

Considering the ramp input case, (10.35) is replaced by

$$1 - T_s(z) = \frac{1 - T(z)}{1 - cz^{-1}} = \frac{(1 - z^{-1})^2 F(z)}{1 - cz^{-1}} \tag{10.38}$$

Letting

$$F(z) = \beta_0 + \beta_1 z^{-1} \qquad T_s(z) = \frac{\alpha_1 z^{-1} + \alpha_2 z^{-2}}{1 - cz^{-1}} \tag{10.39}$$

and equating coefficients, it is found that $\beta_0 = 1$ and that the equation can be satisfied with $\beta_1 = 0$ by choosing $\alpha_1 = 2 - c$, $\alpha_2 = -1$:

$$T_s(z) = \frac{C(z)}{R(z)} = \frac{(2 - c)z^{-1} - z^{-2}}{1 - cz^{-1}} \tag{10.40}$$

Cross-multiplication and inversion yield

$$c_k = cc_{k-1} + (2 - c)r_{k-1} - r_{k-2} \tag{10.41}$$

The response sequences for ramp inputs and step inputs can be calculated from this equation. Ramp input $r_k = kT$: $c_0 = 0$, $c_1 = 0$, $c_2 = (2 - c)T$, $c_3 = [c(2 - c) + (2 - c)2 - 1]T = (3 - c^2)T$, $c_4 = (4 - c^3)T, \ldots$. Step input $r_k = 1$: $c_0 = 0$, $c_1 = 2 - c$, $c_2 = c(2 - c) + (2 - c)1 - 1 = 1 + c - c^2$, $c_3 = 1 + c^2 - c^3$, $c_4 = 1 + c^3 - c^4, \ldots$. Both approach zero error, faster for smaller c ($0 < c < 1$). But to limit the step response overshoot in $c_1 = 2 - c$, c must be sufficiently large.

Example 10.6.4   Example 10.6.3 with a Staleness Weighting Factor Included

From the preceding response equations for a step input, Fig. 10.4 shows the response sequences for values $c = 0.9$ and $c = 0.8$ of the staleness weighting factor. These sequences are

$$c = 0.9: \quad c_0 = 0, \quad c_1 = 1.1, \quad c_2 = 1.09, \quad c_3 = 1.081,$$
$$c_4 = 1.073, \quad c_5 = 1.066, \quad c_6 = 1.059, \quad \ldots$$
$$c = 0.8: \quad c_0 = 0, \quad c_1 = 1.2, \quad c_2 = 1.16, \quad c_3 = 1.128,$$
$$c_4 = 1.102, \quad c_5 = 1.082, \quad c_6 = 1.066, \quad \ldots$$

Even allowing for considerable oscillations between samples, the overshoot is strongly reduced, but the approach to equilibrium is slow (that is, settling times are large). Using (10.39), the compensator is

$$D(z) = \frac{1}{G_1(z)} \frac{T_s(z)}{1 - T_s(z)} = \frac{(z - 0.6065)[(2 - c)z - 1]}{0.213(z + 0.847)(z - 1)} \tag{10.42}$$

Overall, it must be concluded that in practice deadbeat response does not live up to its theoretical appeal.

10.7 FREQUENCY RESPONSE AND THE *w* AND BILINEAR FORMS

As has been shown, root locus techniques apply directly to digital control systems. Polar plots and Nyquist diagrams can also be used. It may be noted, incidentally, that plant transport lag factors $e^{-T_d s}$ cause no difficulty if the lag is an integer multiple of the sampling interval, since the transform $z^{-T_d/T}$ represents this delay. For analysis and design by the use of Bode plots, however, a further transformation is necessary. This will serve also to apply the correction to Tustin's method referred to in Section 10.5.

The *bilinear transformation* or Tustin transformation $s = (2/T)(z - 1)/(z + 1)$ was introduced in (10.19) as an approximation to the true relation $z = e^{Ts}$, to translate a continuous system design $D(s)$ to an approximately equivalent rational digital compensator $D(z)$. For $s = j\omega$, this gives, on solving for z,

$$z = \frac{1 + j\omega T/2}{1 - j\omega T/2}$$

The magnitude of z is $|z| = 1$ at all frequencies and agrees with the true relation $z = e^{j\omega T}$. But the phase angle of z increases from $0°$ to $+180°$ as ω increases from 0 to $+\infty$, while for $z = e^{j\omega T}$ this change of phase occurs as ω increases only from 0 to the limit $\omega_s/2 = \pi/T$ of the primary strip. This compression of the entire $j\omega$ axis to the circumference of the unit circle is responsible for a warping problem, which causes deviations at higher frequencies between the responses of $D(s)$ and $D(z)$. For this reason, if $D(z)$ is obtained from $D(s)$ by Tustin's method, it is often desirable to *prewarp* the critical frequencies of $D(s)$ to counteract the warping at higher frequencies.

For this prewarping as well as to provide a basis for the use of Bode plot techniques, consider the *w* transform, defined as

$$w = \frac{2}{T} \frac{z - 1}{z + 1} \qquad z = \frac{1 + wT/2}{1 - wT/2} \tag{10.43}$$

Note that this is identical to the preceding bilinear transformation. But the latter is an approximate relation between s and z to approximate $D(s)$ by $D(z)$, while the w transform is a formal transformation from a transfer function $G(z)$ to a transfer function $G(w)$ in the new variable w.

For frequency response, $s = j\omega$ and $z = e^{j\omega T}$, so

$$w = \frac{2}{T} \frac{e^{j\omega T} - 1}{e^{j\omega T} + 1} = j\frac{2}{T} \frac{(e^{j\omega T/2} - e^{-j\omega T/2})/(2j)}{(e^{j\omega T/2} + e^{-j\omega T/2})/2} = j\frac{2}{T} \frac{\sin(\omega T/2)}{\cos(\omega T/2)}$$

So for frequency response w is imaginary:

$$w = jv \qquad v = \frac{2}{T} \tan \frac{\omega T}{2} \tag{10.44}$$

As ω increases along the imaginary axis of the s-plane from 0 to the limit $\omega_s/2 = \pi/T$ of the primary strip, so as $z = e^{j\omega T}$ moves along the top half of the unit circle from $+1$ to -1, the new frequency variable v increases from 0 to $+\infty$.

Prewarping makes use of (10.44). A typical factor $X(j\omega) = j(\omega/\omega_0) + 1$ of $D(s)$ will equal $X(j\omega_0) = j + 1$ at its critical frequency ω_0. It is desired to make $X(z)$ as obtained by Tustin's method equal to this for $z = e^{j\omega_0 T}$. Because the w transform has the same form as the bilinear transformation, this is achieved by prewarping $X(s)$ to $X(s) = (s/v_0) + 1$, where v_0 and ω_0 are related by (10.44). For example, for the phase lead/lag $D(s) = K(s + b)/(s + a)$ in (10.16),

$$D(s) = \frac{K(b/a)(1 + s/v_b)}{1 + s/v_a} \tag{10.45}$$

The bilinear transformation $s = (2/T)(z - 1)/(z + 1)$ is now applied to this prewarped compensator $D(s)$ and gives

$$D(z) = K \frac{b}{a} \frac{v_a}{v_b} \frac{(v_b T - 2) + (v_b T + 2)z}{(v_a T - 2) + (v_a T + 2)z} \tag{10.46}$$

The w transform is the same as the bilinear transformation. Therefore, although there is normally no reason for such a sequence, the w transform $G(w)$ obtained from $G(z)$ will be the same as $G(s)$ if $G(z)$ has been obtained from $G(s)$ by the bilinear transformation.

Example 10.7.1

$$D(s) = \frac{1}{s^2 + 0.5s + 1} \qquad T = 1$$

The critical frequency $\omega_n = 1$, so from (10.44), $v_n = (2/1) \tan(1/2) = 1.092$ and the prewarped $D(s)$, with the same damping ratio, is

$$D(s) = \frac{v_n^2}{s^2 + 2\zeta v_n s + v_n^2} = \frac{1.1925}{s^2 + 0.546s + 1.1925}$$

and the bilinear transformation (10.19) yields

$$D(z) = 0.190 \frac{(z + 1)^2}{z^2 - 0.893z + 0.652}$$

It may be verified that substituting the w transform (10.43) for z indeed yields a transform $D(w)$ that is the same as that obtained by replacing s in $D(s)$ by w. This is because $D(z)$ was derived by Tustin's method.

Bode Plots

The w transform makes it possible to use Bode plots in the usual way. First, $G_1(z)$ is transformed to $G_1(w)$ by the w transform (10.43). Then, for frequency response, $w = jv$ is substituted, just as $s = j\omega$ in the Laplace domain. This frequency response function $G_1(jv)$ can be plotted on Bode diagrams with the frequency variable v and used for analysis and design in the normal manner. The plots can also be transferred to the Nichols chart. A compensator $D(w)$ can be designed by the usual techniques and transformed to $D(z)$ by substituting (10.43) for w. $D(z)$ is then implemented as an algorithm.

Example 10.7.2   Digital Control of an Integrator

In the system of Fig. 10.1, let

$$G(s) = \frac{1}{s} \qquad D(z) = K$$

Then

$$G_1(z) = (1 - z^{-1})Z\left[\frac{1}{s^2}\right] = \frac{Tz^{-1}}{1 - z^{-1}} = \frac{T}{z - 1}$$

Substituting (10.43) yields

$$G_1(w) = \frac{1 - wT/2}{w} \qquad D(w) = K$$

Figure 10.11 shows the Bode plot of the loop gain function

$$G_1(jv)D(jv) = \frac{K(1 - jvT/2)}{jv}$$

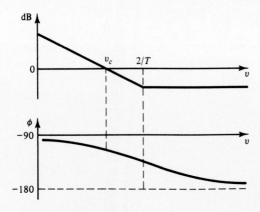

Figure 10.11   Example 10.7.2.

The negative sign reflects a nonminimum-phase zero, and the phase angle of this factor decreases from $0°$ to $-90°$ as v increases from 0 to ∞. As in (7.23), the low-frequency asymptote intersects the 0-dB axis at frequency K. If the crossover frequency $v_c < 2/T$, it must equal K. IF $K > 2/T$, the entire plot is above the 0-dB axis; in effect, $v_c \to \infty$. By (10.44) this means that $\omega_c T/2 = \pi/2$, so $\omega_c = \pi/T = \omega_s/2$, the limit of the primary strip, and hence is too high. If a phase margin of $55°$ is desired, the numerator may contribute $-35°$ at v_c, so

$$\arctan \frac{v_c}{2/T} = 35° \quad \text{or} \quad v_c = \frac{1.4}{T}$$

Substituting this into (10.44) gives the actual crossover frequency $\omega_c = 1.22/T$, and $K = v_c$.

Example 10.7.3   Examples 10.3.3, 10.5.8, 10.6.1, and 10.6.3

$$G_1(z) = \frac{0.213(z + 0.847)}{(z - 1)(z - 0.6065)} \qquad T = 1 \tag{10.47}$$

Substituting (10.43) for z gives the w transform of $G_1(z)$:

$$G_1(w) = \frac{(1 + w/24.144)(1 - w/2)}{w(1 + w/0.49)} \qquad (T = 1) \qquad (10.48)$$

The solid lines in Fig. 10.12 show the Bode plot of $G_1(jv)$. The nonminimum-phase zero is encircled as a reminder that the phase angle contributed by this factor is negative. By inspection, the phase margin is inadequate. It is assumed that the low-frequency asymptote may not be lowered, to avoid loss of accuracy. Phase-lag compensation can be used if the associated reduction of the crossover frequency and speed of response are acceptable. $T = 1$ then also gives more samples per cycle. Here, phase-lead compensation is used, and Fig. 10.12 shows by dashed lines the loop gain function for

$$D(w) = \frac{1 + w/0.5}{1 + w/5} \qquad (10.49)$$

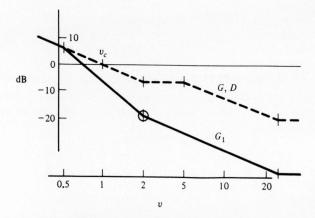

Figure 10.12 Example 10.7.3: Bode plot design.

By (10.44), the crossover frequency $v_c = 1$ corresponds to $\omega_c = 0.927$. The phase angle at $v_c = 1$ is $-125.5°$, so the phase margin $54.5°$. Using (10.43), with $T = 1$, $D(w)$ corresponds to

$$D(z) = \frac{5z - 3}{1.4z + 0.6} = \frac{U(z)}{E(z)}$$

which is equivalent to $1.4U(z) + 0.6z^{-1}U(z) = 5E(z) - 3z^{-1}E(z)$ and is implementable as the algorithm

$$u_k = 0.429u_{k-1} + 3.571e_k - 2.143e_{k-1} \qquad (10.50)$$

The loop gain function $G_1(z)D(z)$ is

$$G_1(z)D(z) = \frac{1.065z^2 + 0.263z - 0.5412}{1.4z^3 - 1.6491z^2 - 0.1148z + 0.3639}$$

and, using (10.33a), $C(z)$ for a unit step $R(z)$ is found to be

$$C(z) = \frac{1.065z^3 + 0.263z^2 - 0.5412z}{1.4z^4 - 1.9841z^3 + 0.7323z^2 - 0.3255z + 0.1773}$$

Long division yields the sample response in Fig. 10.4:

$$c_0 = 0, \quad c_1 = 0.761, \quad c_2 = 1.266, \quad c_3 = 1.002,$$
$$c_4 = 0.935, \quad c_5 = 0.999, \quad c_6 = 1.0$$

It is noted that the overshoot in digital control systems is in general larger than that of continuous systems for the same phase margin. From Table 8.4.1, it would be about 10% for 55° phase margin.

10.8 CONCLUSION

The preceding two chapters were devoted to digital control systems. In Chapter 9 digital computer control was introduced in general terms, and the discussion of sampling characteristics, control algorithms, and Z transforms laid the basis for the present chapter. In this chapter the analysis and design of single-loop digital control systems have been considered in a manner analogous to the earlier work on continuous systems and with attention to the effects of sampling on the dynamic behavior.

PROBLEMS

10.1. In Fig. P10.1, if $G_1(z) = (1 - z^{-1})Z[G(s)/s]$ represents the plant and zero-order hold, prove the following results on the steady-state error sequence $e_{ss} = \lim\limits_{k \to \infty} e_k$ for the inputs specified:

(a) Unit step: $e_{ss} = 1/(1 + K_p)$; $K_p = \lim\limits_{z \to 1} G_1(z)D(z)$

(b) Unit ramp: $e_{ss} = 1/K_v$; $K_v = (1/T)\lim\limits_{z \to 1} (z - 1)G_1(z)D(z)$

These results are analogous to those for continuous systems. K_p and K_v are often called the position and velocity error constants. What are the conditions for zero steady-state errors in (a) and (b)?

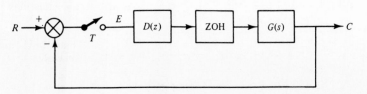

Figure P10.1

10.2. For the position servo model in Fig. P10.1 with $D(z) = K$, $G(s) = 1/[s(s + 1)]$, and $T = 1$:

(a) Determine the loop gain function $G_1(z)D(z)$.

(b) Express the closed-loop transfer function $C(z)/R(z)$.

(c) Find the steady-state value of the output sequence, and hence the steady-state error, for unit step inputs.

10.3. Find the unit step response sequences for the following Z transfer functions by partial fraction expansion.

(a) $\dfrac{0.32}{(z - 0.2)(z - 0.6)}$ (b) $\dfrac{0.32z}{(z - 0.2)(z - 0.6)}$

(c) $\dfrac{0.32}{z(z - 0.2)(z - 0.6)}$

Note that (b) is (a) multiplied by z, and (c) equals (a) multiplied by z^{-1}. Verify the relations between the response sequences that are implied by this.

10.4. Find the unit step response sequence of

$$\frac{C(z)}{R(z)} = \frac{z - 0.5}{z^2 - z + 0.5}$$

Solve

by partial fraction expansion. Use graphical calculation of residues as in Section 1.8, together with the results of Problem 9.23 and the relations $a + jb = Re^{j\phi}$; $R = \sqrt{(a^2 + b^2)}$, $\phi = \tan^{-1}(b/a)$.

10.5. Verify the result of Problem 10.4 by:
   (a) The long-division method.
   (b) The difference equation method.

   Calculate c_k for $k = 0, 1, 2, 3, 4, 5, 6$ by all three methods. What is the advantage of the partial fraction expansion method if solutions at higher sampling instants are of primary interest?

10.6. Determine the response of the system in Problem 10.4 to the input sequence shown in Fig. P10.6. Which technique appears most convenient for this type of problem? What is the Z transform of the sequence?

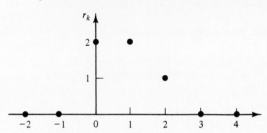

Figure P10.6

10.7. Express the solution of the following difference equations in closed form for $x_k = 1$, $k \geqslant 0$:
   (a) $y_k - 0.7y_{k-1} = 0.3x_k$ **(b)** $y_k - 0.7y_{k-1} + 0.1y_{k-2} = 0.8x_k - 0.4x_{k-1}$
   A closed-form solution allows y_k for a selected value of k to be calculated directly, without first calculating it for all lower k values.

10.8. Determine the inverse transform f_n of

$$F(z) = \frac{z}{z^2 - 2.5z + 1}$$

   (a) By long division.
   (b) By partial fraction expansion.
   (c) By the difference equation method.

10.9. Calculate the unit step response for the system in Problem 10.2 for $K = 1$. Plot the response c_k, $k = 0 - 7$.

10.10. Determine the stability of the algorithms in Problem 9.22.

10.11. Find the conditions, if any, for which the algorithms in Problems 9.4 and 9.5 may be unstable.

10.12. Repeat Problem 9.6 using forward differences and the forward rectangular rule. Is the algorithm stable?

10.13. Obtain a recursive algorithm and pulse transfer function for $G(s)$ of Problem 9.13 based on backward difference approximation of derivatives. Find the conditions, if any, for which the algorithm may be unstable.

10.14. Determine the stability of systems described by the following Z transfer functions by plotting the pole positions on the z-plane:
   (a) $(2z + 6)/(4z^3 - z)$ **(b)** $(0.2z - 0.14)/(z^2 - 0.4z + 0.4)$

10.15. Use the results of Problem 9.23 to interpret the nature of transients corresponding to poles on the positive and negative real axis and at smaller and greater distance to the origin.

10.16. (a) Determine the locus in the s-plane that corresponds to a radial line at angle ϕ radians in the z-plane relative to the positive real axis.

(b) Find the loci in the z-plane for which transient oscillations will be sampled at the rate of, respectively, two, four, six, and eight samples per cycle.

10.17. In Fig. P10.1 with $D(z) = 1$ and $G(s) = 1/(s + 1)$:

(a) Find the loop gain function and the closed-loop Z transfer function.

(b) Determine the Z transform of the output sample sequence for a unit step input and use the final value theorem to find the steady-state response.

(c) Plot the locations of the system poles for $T = 0.1$, $T = 0.693$, $T = 1$, and $T = 2$ sec.

10.18. For the temperature control system model in Fig. P10.1, with $D(z) = K$, $G(s) = 1/(s + 5)$, and $T = 0.2$:

(a) Plot the root loci and use them to find the range of K for which the system is stable.

(b) Calculate and plot the system pole locations for $K = 1$, $K = 2.911$, and $K = 5$.

(c) For these values of K, find the steady-state values of the output sequence for a unit step input, and hence the steady-state errors.

(d) Compare with the steady-state error for the corresponding continuous system and its stability limit on K.

10.19. In Fig. P10.1 with $D(z) = K$, $G(s) = 1/(s + 2)$, and $T = 0.5$:

(a) Express the loop gain function.

(b) Plot the root loci.

(c) Find the limiting value of K for stability.

(d) Express the steady-state sample error for unit step inputs as a function of K.

10.20. In Problem 10.19, calculate and compare the values of K to achieve a system time constant $\tau = 0.25$:

(a) By using the value of K that achieves this for the continuous system. Find the z-plane pole location corresponding to this gain and hence indicate the nature of the transient.

(b) By finding the z-plane pole that corresponds to the s-plane pole for $\tau = 0.25$, and calculating K required to realize this z-plane pole.

(c) Which of these alternatives is closest to the continuous system behavior?

10.21. For the system of Problem 10.2:

(a) Sketch the loci of the z-plane poles for varying K. It can be shown that for this type of open-loop pole–zero configuration the loci include a circle centered on the zero.

(b) Find the limiting value of K for stability.

(c) Find K corresponding to a damping ratio $\zeta = 0.5$. (*Note:* Figure 10.6, with the loci for constant ζ, can be used as a standard graph paper, or a section of the constant-ζ locus where it intersects the root locus can be plotted by using Table 10.4.1.)

10.22. For the positioning servo in Fig. P10.1, with $D(z) = K$, the motor transfer function $G(s) = 1/[s(s + 5)]$ was also considered in Problem 9.24.

(a) Find the loop gain functions for $T = 0.1$ and $T = 1.0$.

(b) Sketch the loci for both cases.

(c) Find for both the limiting value of K for stability.

10.23. In the level control system model in Fig. P10.1, where $D(z) = K$ and $T = 0.2$, the plant $G(s) = 1/[(s + 1)(s + 5)]$ was also considered in Problem 9.24.

(a) Determine the loop gain function.
(b) Sketch the root loci.
(c) Find the limiting value of K for stability.

10.24. In Problem 10.23:
(a) Find K corresponding to a damping ratio $\zeta = 0.5$ (see the note for Problem 10.21).
(b) Find the closed-loop transfer function for K found in part (a) and the steady-state error sequence.

10.25. In Problem 10.17:
(a) From the output transform $C(z)$ in Problem 10.17(b) obtain an algorithm for the output sequence c_k.
(b) Calculate the sample sequence c_k for the four values of T (for $T = 0.1$ up to 1 sec and for the other values up to 5 sec) and plot these sequences.

10.26. In Problem 10.18:
(a) Calculate and plot, by partial fraction expansion, the unit step response sequence c_k, $k = 0, 1, 2, 3$, for the three values of K.
(b) Correlate the nature of the transients with the pole positions, and calculate the closed-loop time constants of the corresponding continuous designs, to compare with continuous system response.

10.27. For Problems 10.17 and 10.25:
(a) What is the closed-loop system time constant without sampling, and hence what is the guide for the maximum sampling interval that might be satisfactory?
(b) Compare the responses with the pole positions in Problem 10.17(c). The response for $T = 0.693$ is called deadbeat response.
(c) Examine the responses in terms of the s-plane to z-plane correlations of Fig. 10.5.

10.28. (a) Repeat Problem 10.20 for a sampling interval $T = 0.1$ instead of $T = 0.5$.
(b) Why do the two approaches agree so much better than in Problem 10.20? With $\tau = 0.25$ desired, what value should serve as an upper limit on the sampling interval selected?
(c) Calculate the unit step response sequence for K as chosen by the equivalent poles approach. Plot c_k for $k = 0, 1, 2, 3, 4, 5$ and show c_∞.

10.29. In Problem 10.24:
(a) Calculate and plot the unit step response c_k, $k \leqslant 8$.
(b) Compare the approximate percentage overshoot with that predicted by Fig. 5.2 for a continuous design with $\zeta = 0.5$.
(c) Use a root locus sketch for a continuous design with $\zeta = 0.5$ to determine its resonant frequency, and hence find the approximate number of samples per cycle to suggest the reason for the result of part (b).

10.30. Determine a digital filter transfer function $D(z)$ to approximate the lead network compensator

$$D(s) = \frac{s + 1}{0.1s + 1}$$

which was designed by continuous system methods, using:
(a) Tustin's method (bilinear transformation).
(b) Impulse-invariant, or Z transform method.
(c) Zero-order-hold equivalence method.
(d) Pole–zero matching method.

10.31. Determine a digital filter transfer function $D(z)$ to approximate the phase-lag compensator

$$D(s) = \frac{0.25s + 1}{s + 1}$$

designed by continuous system methods, using:
(a) Tustin's method (bilinear transformation).
(b) Impulse-invariant or Z transform method.
(c) Zero-order-hold equivalence method.
(d) Pole–zero matching method.

10.32. Use bilinear transformation with prewarping to find a digital approximation $D(z)$ to $D(s)$ of Problem 10.31 for (a) $T = 0.05$ and (b) $T = 0.5$.

10.33. In Fig. P10.1 with $G(s) = 1/(s + 1)$, PI control $D(s) = 3 + 8/s$ has been chosen by continuous system design, to remove steady-state error and obtain a closed-loop time constant 0.5 sec and closed-loop damping ratio 0.707.
(a) Translate $D(s)$ to $D(z)$ by the Tustin method for $T = 0.1$.
(b) Compare the resulting closed-loop poles with those obtained by transforming the s-plane poles of the continuous design (that is, by translating the specifications).

10.34. Repeat Problem 10.33 for $T = 0.5$. Compare T with the desired system time constant.

10.35. In Problem 10.34, with $T = 0.5$, determine $D(z)$ so that the closed-loop poles will correspond to those in the s-plane for a system time constant 0.5 sec and damping ratio 0.707. Compare K_p and K_i with those for the s-plane design ($K_p = 3$, $K_i = 8$).

10.36. In the system of Problem 10.23, PI control is to be used to eliminate the steady-state error. Assuming that $T = 0.2$ and using the Tustin model for integration, design $D(z)$ to cancel one of the plant poles and obtain a system damping ratio of about 0.5. Results and methods in Problems 10.23 and 10.24 will be useful.

10.37. In Fig. P10.1 with $G(s) = 1/[s(s + 1)]$, $D(z) = K(z - z_1)/(z - p_1)$, and $T = 0.5$:
(a) Express the loop gain function.
(b) On a relatively large scale plot, for use in several design problems, construct the upper-half-plane constant ζ locus for $\zeta = 0.5$. (See the note in Problem 10.21.)
(c) Use the result of Problem 10.1 to express the steady-state error for a unit ramp input.

10.38. In Problem 10.37, find the limiting values of K for stability, plot the root loci partially, and find K for a damping ratio of 0.5, for the following three cases:
(a) No compensation ($z_1 = p_1$).
(b) $p_1 = 0.1$ and z_1 cancels a plant pole.
(c) $p_1 = -0.6065$ and z_1 cancels a plant pole.

Also find the steady-state errors following unit ramp inputs for each case when $\zeta = 0.5$.

10.39. Calculate and plot the unit step responses for the three designs with $\zeta = 0.5$ in Problem 10.38 using the difference equation method.

10.40. **(a)** Use s-plane/z-plane correlations to determine the undamped natural frequencies of continuous systems that would correspond to the angular positions of the z-plane system poles for $\zeta = 0.5$ for the three designs in Problem 10.38.

(b) From these, find the numbers of samples per cycle for $T = 0.5$, and correlate these values with the step responses found in Problem 10.39 to determine for which design $T = 0.5$ may be inadequate.

(c) In a general way, correlate the order of magnitude of the settling times found in Problem 10.39 to the system pole positions.

10.41. In Problem 10.38, to obtain an intersection with the $\zeta = 0.5$ locus at a point corresponding to a higher sampling rate, modify the compensator (c) to

$$D(z) = \frac{K(z - 0.2)}{z + 0.6065}$$

(that is, the zero no longer cancels the plant pole).
(a) Express the loop gain function.
(b) Plot the loci in the general range where the intersection with the $\zeta = 0.5$ locus occurs.
(c) Find K at the intersection with this locus and express the steady-state error following unit ramp inputs.
(d) Plot the unit step response on the graph for Problem 10.39.

10.42. (a) Design a deadbeat compensator for step inputs for the system of Problem 10.39 ($T = 0.5$).
(b) Find the characteristic equation and the system pole(s) and plot the step response.

10.43. (a) Repeat Problem 10.42 for a ramp input.
(b) Calculate and plot the response sequences of the resulting system for unit step and unit ramp inputs. What is the overshoot of the step response sequence?

10.44. Introduce a staleness weighting factor into the ramp input design of Problem 10.43 to limit the overshoot of the step response sequence to 20%. Find the compensator $D(z)$ and plot the unit step response sequence.

10.45. In Fig. P10.1 with $G(s) = 1/(s + 1)$, $D(z) = K$, and $T = 0.1$:
(a) Find K for a phase margin of about 67°.
(b) Recommended sampling frequencies may be 4 to 20 times closed-loop system bandwidth. Without calculating the latter, estimate from the crossover frequency corresponding to part (a) whether $T = 0.1$ is small enough.
(c) Calculate the unit step response sequence, and find the overshoot over steady state. Compare it with the 4.5% overshoot for second-order systems when $\zeta = 0.7$ (Fig. 5.2) to comment further on part (b).

10.46. In Fig. P10.1 with $G(s) = 1/[(s + 1)(s + 5)]$, $D(z) = K$, and $T = 0.2$, which was also considered in Problem 10.23:

(a) Determine the w transform of the loop gain function.
(b) Construct the Bode plot and find K for a phase margin of about 67°.
(c) Use the correlations in Table 8.4.1 to find an estimate of the ratio ω_s/ω_b of sampling frequency to closed-loop bandwidth, and hence an indication of whether the sampling rate is adequate.
(d) On the basis of the correlations in Fig. 8.5 and Table 8.4.1, estimate the expected closed-loop resonance peaking M_p.

10.47. Transfer the Bode plot data in Problem 10.46 to the Nichols chart to check the predictions of closed-loop resonance peaking M_p and bandwidth ω_b.

10.48. In Fig. P10.1, with $G(s) = 1/[s(s + 1)]$ and $T = 0.5$ as in Problem 10.37, and with $D(z) = K$:

(a) Find the w transform of the loop gain function.

(b) Using Bode plots, find K for a phase margin of about 67°.

(c) Use Table 8.4.1 to find an estimate of ratio ω_s/ω_b of sampling frequency to closed-loop bandwidth, and hence an indication of whether the sampling rate is satisfactory.

10.49. (a) Repeat Problem 10.48 for 53° phase margin.

(b) Estimate the resonance peaking M_p using Table 8.4.1 and Fig. 8.5, and check peaking and bandwidth estimates on the Nichols chart.

10.50. Prove the following w transforms of common compensators if the bilinear transformation is used to derive the Z transforms:

(a) PI control: $D(s) = K_p + K_i/s$; $D(w) = K_p + K_i/w$.

(b) Phase lag or phase lead:

$$D(s) = \frac{K(s + a)}{s + b} \qquad D(w) = \frac{K(w + a)}{w + b}$$

10.51. PI control must be designed for the system of Problem 10.45 to eliminate the steady-state error ($T = 0.1$).

(a) Express the loop gain function $G_1(w)D(w)$ using results in Problem 10.50.

(b) Using Table 8.4.1, estimate the desired crossover frequency v_c in the w domain if the phase margin is to be about 53° and the ratio of sampling frequency ω_s to closed-loop bandwidth ω_b about 10.

(c) Find the desired ratio K_i/K_p of the controller if the phase margin for v_c in part (b) is to be 53°.

(d) Construct the Bode plot, find the desired gain, and determine K_p and K_i.

(e) Find $D(z)$ from $D(w)$ and hence the control algorithm.

10.52. For the system of Problem 10.46 ($T = 0.2$), and using Problem 10.50, design PI control; that is, find the values of K_p and K_i to achieve a phase margin of about 53° at a crossover frequency for which an estimate of the ratio of sampling frequency to closed-loop bandwidth is about 10. (See Problem 10.51 for a possible sequence of solution.)

10.53. For the system of Problem 10.48 ($T = 0.5$) and using Problem 10.50, design a series compensator $D(z)$ to meet the following specifications:

1. The steady-state ramp input error sequence may not exceed 20%. (Remember that at low frequencies $v \approx \omega$.)

2. The phase margin should not be less than about 53°.

3. The ratio of sampling frequency to estimated closed-loop bandwidth should not be below about 12.

11

State-Space Analysis

11.1 INTRODUCTION

The many books largely devoted to the subject and the continuing emphasis that it receives in the technical literature are evidence of the importance of the state-space approach. The key factor that accounts for this is that the system dynamics are described by a state-space model instead of by transfer functions. A state-space model is a description in terms of a set of first-order differential equations that are written compactly in a standard matrix form. This standard form has permitted the development of general computer programs, which can be used for the analysis and design of even very large systems.

Techniques for modeling and analysis are discussed in this chapter, including a brief section on digital control systems, and design is introduced in Chapter 12. Appendix A reviews the needed topics in matrix analysis.

11.2 STATE-SPACE MODELS

The derivation of state-space models is no different from that of transfer functions in that the differential equations describing the system dynamics are written first. In transfer function models these equations are transformed and variables are eliminated between them to find the relation between selected input and output variables. For state models, instead, the equations are arranged into a set of first-order differential equations in terms of selected state variables, and the outputs are expressed in these same state variables. Because the elimination of variables between equations is not an inherent part of this process, state models can be easier to obtain. State models should normally not be derived from transfer functions, but directly from the original system equations. But in this section examples will be given to relate state models to the transfer functions used thus far.

Consider a system described by the nth-order differential equation

$$\frac{d^n w}{dt^n} + a_n \frac{d^{n-1}w}{dt^{n-1}} + \cdots + a_2 \frac{dw}{dt} + a_1 w = r \tag{11.1}$$

or the equivalent transfer function. A state model for this system is not unique but depends on the choice of a set of *state variables* $x_1(t), x_2(t), \ldots, x_n(t)$. One possible choice is the following:

$$x_1 = w \qquad x_2 = \dot{w} \quad \cdots \quad x_n = \overset{(n-1)}{w} \tag{11.2}$$

Directly from these definitions, the following equations are obtained:

$$\dot{x}_1 = x_2 \qquad \dot{x}_2 = x_3 \quad \cdots \quad \dot{x}_{n-1} = x_n$$

And substitution of the definitions into (11.1) yields

$$\dot{x}_n = -a_1 x_1 - a_2 x_2 - \cdots - a_n x_n + r$$

Together these form a set of n first-order differential equations. The output w can also be expressed in terms of the state variables:

$$w = x_1$$

In vector-matrix form these equations can be arranged as follows:

$$\begin{bmatrix} \dot{x}_1 \\ \dot{x}_2 \\ \vdots \\ \\ \dot{x}_n \end{bmatrix} = \begin{bmatrix} 0 & 1 & 0 & \cdots & 0 \\ 0 & 0 & 1 & & 0 \\ \vdots & & & & \vdots \\ 0 & & & & 1 \\ -a_1 & -a_2 & \cdots\cdots & & -a_n \end{bmatrix} \begin{bmatrix} x_1 \\ x_2 \\ \vdots \\ \\ x_n \end{bmatrix} + \begin{bmatrix} 0 \\ \vdots \\ \\ 0 \\ 1 \end{bmatrix} r$$

$$w = \begin{bmatrix} 1 & 0 & \cdots & 0 \end{bmatrix} \begin{bmatrix} x_1 \\ x_2 \\ \vdots \\ \\ x_n \end{bmatrix}$$

The standard form of a state-space model is as follows:

$$\begin{aligned} \dot{\mathbf{x}} &= \mathbf{Ax} + \mathbf{Bu} \quad \textit{(state equation)} \\ \mathbf{y} &= \mathbf{Cx} + \mathbf{Du} \quad \textit{(output equation)} \end{aligned} \tag{11.3}$$

Here $\mathbf{x}$ is the *state vector,* the vector of the state variables,
   $\mathbf{y}$, the *output vector,*

u, the *control vector*,

A, the *system matrix*.

In the preceding example the control vector is the scalar function r and the output vector the scalar function w. It may be seen that

$$
\mathbf{x} = \begin{bmatrix} x_1 \\ x_2 \\ \vdots \\ \\ x_n \end{bmatrix} = \begin{bmatrix} w \\ \dot{w} \\ \vdots \\ (n-1) \\ w \end{bmatrix} \quad \mathbf{B} = \begin{bmatrix} 0 \\ \vdots \\ \\ 0 \\ 1 \end{bmatrix} \quad \mathbf{A} = \begin{bmatrix} 0 & 1 & 0 & \cdots & 0 \\ 0 & 0 & 1 & & 0 \\ \vdots & & & & \vdots \\ 0 & & & & 1 \\ -a_1 & -a_2 & \cdots & & -a_n \end{bmatrix} \tag{11.4}
$$

$$ \mathbf{C} = [1 \quad 0 \quad \cdots \quad 0] \qquad \mathbf{D} = \mathbf{0} $$

This form of **A** is a *companion matrix*.

Example 11.2.1   A Transfer Function without Zeros

$$ \frac{W}{R} = \frac{5}{s^3 + 6s^2 + 9s + 3} \qquad \dddot{w} + 6\ddot{w} + 9\dot{w} + 3w = 5r $$

Choose state variables $x_1(t)$, $x_2(t)$, and $x_3(t)$ as in (11.2):

$$ x_1 = w \qquad x_2 = \dot{w} \qquad x_3 = \ddot{w} $$

Then a state model representing this transfer function or the corresponding differential equation is obtained as in the general case. The definitions and the differential equation yield

$$ \dot{x}_1 = x_2 \qquad \dot{x}_2 = x_3 \qquad \dot{x}_3 = -3x_1 - 9x_2 - 6x_3 + 5r $$

In matrix form and with the output w expressed also in terms of all state variables,

$$ \mathbf{x} = \begin{bmatrix} x_1 \\ x_2 \\ x_3 \end{bmatrix} \qquad \dot{\mathbf{x}} = \begin{bmatrix} 0 & 1 & 0 \\ 0 & 0 & 1 \\ -3 & -9 & -6 \end{bmatrix} \mathbf{x} + \begin{bmatrix} 0 \\ 0 \\ 5 \end{bmatrix} r \qquad w = [1 \quad 0 \quad 0]\mathbf{x} $$

When the transfer function has zeros or the differential equation contains derivatives of the input, this approach must be amended to avoid the presence of derivatives in the control vector. One way is to derive first a state equation representing only the denominator. Then, as the following example illustrates, the numerator will be reflected only in the output equation; that is, the output equation represents the effect of system zeros, or derivatives of the input.

Example 11.2.2   Transfer Function with Zeros

In Example 11.2.1, introduce a polynomial in the numerator:

$$ \frac{W}{R} = \frac{5s^2 + 2s + 2}{s^3 + 6s^2 + 9s + 3} $$

or

$$ \dddot{w} + 6\ddot{w} + 9\dot{w} + 3w = 5\ddot{r} + 2\dot{r} + 2r $$

First, consider only the denominator:

$$ \frac{V}{R} = \frac{1}{s^3 + 6s^2 + 9s + 3} \qquad \dddot{v} + 6\ddot{v} + 9\dot{v} + 3v = r $$

As in Example 11.2.1,

$$\mathbf{x} = \begin{bmatrix} v \\ \dot v \\ \ddot v \end{bmatrix} \qquad \dot{\mathbf{x}} = \begin{bmatrix} 0 & 1 & 0 \\ 0 & 0 & 1 \\ -3 & -9 & -6 \end{bmatrix}\mathbf{x} + \begin{bmatrix} 0 \\ 0 \\ 1 \end{bmatrix} r$$

But
$$W = (5s^2 + 2s + 2)V \quad \text{or} \quad w = 5\ddot v + 2\dot v + 2v = [2 \; 2 \; 5]\mathbf{x}$$
Hence the output equation, with $y = w$, is
$$y = \mathbf{Cx} \qquad \mathbf{C} = [2 \; 2 \; 5]$$
Observe that the zeros of the transfer function, or the derivatives of the input on the right side of the differential equation, are represented only by $\mathbf{C}$.

If the numerator is of the same order as the denominator, a division must be carried out first for this technique to apply.

Example 11.2.3
$$\frac{Y}{U} = \frac{K(s + a)}{s + b} \quad \text{or} \quad \dot y + by = Kau + K\dot u$$

By the approach of the last example, $V/U = 1/(s + b)$ yields a state equation $\dot v = -bv + u$, with state variable v. But then $Y = K(s + a)V$, so $y = K\dot v + Kav$, and $\dot v$ is now not a state variable. With equal powers in the numerator and denominator, the method of Example 11.2.2 can be used if a division is performed first:
$$Y = K\left(1 + \frac{a - b}{s + b}\right)U = KU + X \qquad \text{where} \quad \frac{X}{U} = \frac{K(a - b)}{s + b}$$
This yields $\dot x = -bx + K(a - b)u, y = Ku + x$.

Note that this case of equal powers in the numerator and denominator is also one where the matrix $\mathbf{D}$ in the output equation (11.3) is not zero. That is, there is a direct link from input to output.

11.3 STATE-SPACE MODELS FOR PHYSICAL SYSTEMS

The choices of state variables $x_1 = w$, $x_2 = \dot w$, $x_3 = \ddot w, \ldots$, made so far have the disadvantage that the variables beyond the second or third have little or no physical meaning and are difficult or impossible to measure. These features are important to identify the behavior of variables of interest and to implement a control that requires feedback from the state variables. Where possible, state variables should be chosen that are measurable and physically meaningful. This implies that state models should not be derived from closed-loop transfer functions. They should be derived directly from the original system equations. This, in fact, tends to simplify the modeling, because these are frequently already first-order equations, and eliminating variables between them is not necessary. Derivation from a block diagram is frequently appropriate, however, since the output signals of the blocks are often measurable physical variables.

If a system block diagram can be separated into simple lag blocks and integrator blocks, a state model can be derived by identifying the output of each as a state variable. For example, a block $X/U = K/(s + b)$ is equivalent to the first-order differential equation $\dot x = -bx + Ku$, so x is a suitable state variable. If a block is described by a quadratic transfer function $K/(s^2 + as + b)$, the output of the block and its derivative are frequently suitable state variables.

The following examples will illustrate this as well as the normal approach for the derivation of state models, that is, directly from the system differential equations, with state variables chosen to be measurable and meaningful. It is not always obvious how many state variables are needed to describe a system. On occasion, the fact that the number of state variables for a system equals the number of independent energy storage elements can be used to advantage.

Example 11.3.1   *RC* Simple Lag Circuit [Example 2.3.1, Fig. 2.8(a)]

Input voltage $e_i(t)$ and output voltage $e_o(t)$ are related by

$$RC\dot{e}_o + e_o = e_i \qquad \frac{E_o}{E_i} = \frac{1}{RCs + 1}$$

The system is represented by a simple lag block, of which the output is measurable and meaningful and is a suitable state variable, identified as x. This is also the output y. Alternatively, the system differential equation is already of first order. Thus, with input $e_i = u$, both the differential equation and the transfer function lead to the state model

$$\dot{x} = \frac{-x}{RC} + \frac{u}{RC} \qquad y = x \qquad (u = e_i)$$

Example 11.3.2   Spring–Mass–Damper (Example 2.2.1, Fig. 2.2)

Mass position x and applied force f are related by

$$m\ddot{x} + c\dot{x} + kx = f \qquad \frac{X(s)}{F(s)} = \frac{1}{ms^2 + cs + k}$$

The output of the quadratic lag block and its derivative are mass position and velocity and so are suitable state variables, also suggested by the differential equation:

$$x_1 = x \qquad x_2 = \dot{x} \qquad y = x_1 \qquad u = f$$

Therefore, here also the transfer function and differential equation descriptions will yield the same state model

$$\dot{\mathbf{x}} = \begin{bmatrix} 0 & 1 \\ -\dfrac{k}{m} & -\dfrac{c}{m} \end{bmatrix} \mathbf{x} + \begin{bmatrix} 0 \\ \dfrac{1}{m} \end{bmatrix} u \qquad y = \begin{bmatrix} 1 & 0 \end{bmatrix} \mathbf{x}$$

Example 11.3.3   Field-Controlled DC Motor (Example 2.4.1, Fig. 2.9)

The original equations for field voltage e_f, field current i_f, and shaft position θ_o are

$$e_f = R_f i_f + L_f \dot{i}_f \qquad K_T i_f = J\ddot{\theta}_o + B\dot{\theta}_o$$

Shaft position, shaft velocity, and field current are suitable state variables, and with input and output

$$u = e_f \qquad y = \theta_o$$

the equations yield the state model

$$\mathbf{x} = \begin{bmatrix} \theta_o \\ \dot{\theta}_o \\ i_f \end{bmatrix} \qquad \dot{\mathbf{x}} = \begin{bmatrix} 0 & 1 & 0 \\ 0 & -\dfrac{B}{J} & \dfrac{K_T}{J} \\ 0 & 0 & -\dfrac{R_f}{L_f} \end{bmatrix} \mathbf{x} + \begin{bmatrix} 0 \\ 0 \\ \dfrac{1}{L_f} \end{bmatrix} u \qquad (11.5a)$$

$$y = \begin{bmatrix} 1 & 0 & 0 \end{bmatrix} \mathbf{x}$$

This direct model can be compared with other types of models that can be derived from transfer function descriptions. Figure 11.1 shows the original equations in block diagram form. Two of the blocks are simple lags and one is an integrator. Identifying their outputs as state variables yields the same state model as above.

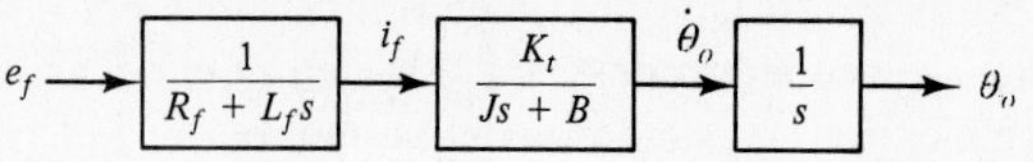

$$e_f \longrightarrow \boxed{\dfrac{1}{R_f + L_f s}} \xrightarrow{i_f} \boxed{\dfrac{K_t}{Js + B}} \xrightarrow{\dot{\theta}_o} \boxed{\dfrac{1}{s}} \longrightarrow \theta_o$$

Figure 11.1   Field-controlled dc motor.

A different state model results by first deriving the overall transfer function. For $B = J = L_f = 1$, $R_f = 6$, and $K_t = 30$, the overall transfer function is

$$\frac{\theta_o}{E_f} = \frac{30}{s(s+1)(s+6)} = \frac{30}{s^3 + 7s^2 + 6s}$$

Analogous to Example 11.2.1, cross-multiplication and inversion using the second form yields

$$\dddot{\theta}_o + 7\ddot{\theta}_o + 6\dot{\theta}_o = 30 e_f$$

and leads to the state model

$$\mathbf{x} = \begin{bmatrix} \theta_o \\ \dot{\theta}_o \\ \ddot{\theta}_o \end{bmatrix} \qquad \dot{\mathbf{x}} = \begin{bmatrix} 0 & 1 & 0 \\ 0 & 0 & 1 \\ 0 & -6 & -7 \end{bmatrix}\mathbf{x} + \begin{bmatrix} 0 \\ 0 \\ 30 \end{bmatrix}u \qquad (11.5b)$$

$$y = \theta_o = [1 \quad 0 \quad 0]\mathbf{x}$$

Another special state model is obtained from the partial fraction expansion of θ_o/E_f:

$$\frac{\theta_o}{E_f} = \frac{5}{s} - \frac{6}{s+1} + \frac{1}{s+6}$$

This relation is illustrated in Fig. 11.2, which shows a parallel connection of simple lag and integrator blocks, instead of the series connection in Fig. 11.1. Identifying the outputs of the blocks as state variables yields the model

$$\mathbf{x} = \begin{bmatrix} x_1 \\ x_2 \\ x_3 \end{bmatrix} \qquad \dot{\mathbf{x}} = \begin{bmatrix} 0 & 0 & 0 \\ 0 & -1 & 0 \\ 0 & 0 & -6 \end{bmatrix}\mathbf{x} + \begin{bmatrix} 5 \\ 6 \\ 1 \end{bmatrix}u \qquad (11.5c)$$

$$y = [1 \quad -1 \quad 1]\mathbf{x}$$

While the significance of the state variables is not clear, it is interesting to note that the system matrix is diagonal. This form will receive much attention later.

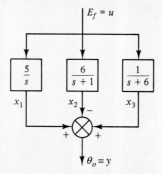

Figure 11.2   Diagonal state model.

In the examples so far, single-input, single-output systems were considered; that is, the input u and output y were scalars. However, state models are well adapted to the representation of multivariable control systems, where in general both u and y are vectors.

Example 11.3.4   Multivariable Control System (Fig. 11.3)

A control system with two inputs and two outputs is shown in Fig. 11.3. It could be a system in which both the flow rate and temperature of a liquid are controlled by the adjustment of valves in hot and cold supply lines. The outputs of the simple lag blocks are again defined as state variables. The output equation is seen to be

$$\mathbf{y} = \begin{bmatrix} y_1 \\ y_2 \end{bmatrix} = \begin{bmatrix} 1 & 1 & 0 & 0 \\ 0 & 0 & 1 & 1 \end{bmatrix} \mathbf{x}$$

and from equations such as $\dot{x}_1 = -x_1 + u_1$ and $u_1 = K_1(r_1 - y_1) = K_1(r_1 - x_1 - x_2)$, which can be seen by inspection of Fig. 11.3, the state equation is found to be

$$\dot{\mathbf{x}} = \begin{bmatrix} -1 - K_1 & -K_1 & 0 & 0 \\ 0 & -5 & -5K_2 & -5K_2 \\ -0.4K_1 & -0.4K_1 & -0.5 & 0 \\ 0 & 0 & -4K_2 & -2 - 4K_2 \end{bmatrix} \mathbf{x} + \begin{bmatrix} K_1 & 0 \\ 0 & 5K_2 \\ 0.4K_1 & 0 \\ 0 & 4K_2 \end{bmatrix} \mathbf{u}$$

where $\mathbf{u} = \begin{bmatrix} r_1 \\ r_2 \end{bmatrix}$.

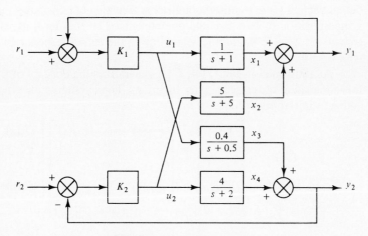

Figure 11.3   A 2 × 2 system.

11.4 TRANSFER FUNCTION MATRICES AND STABILITY

With a state model description of the system dynamics, the first question to be answered is how stability may be determined. To derive the stability criterion, the generalization of the concept of a transfer function is considered. A matrix of transfer functions or transfer function matrix is found that corresponds to the state model. This requires Laplace transformation of the state model equations. The Laplace transform of a vector is the vector of the Laplace transforms of its elements, so the transforms of $\mathbf{x}$ and $\dot{\mathbf{x}}$ are as follows:

$$\mathbf{X}(s) = L[\mathbf{x}(t)] = \begin{bmatrix} L[x_1] \\ \vdots \\ L[x_n] \end{bmatrix} = \begin{bmatrix} X_1(s) \\ \vdots \\ X_n(s) \end{bmatrix} \qquad L[\dot{\mathbf{x}}] = \begin{bmatrix} L[\dot{x}_1] \\ \vdots \\ L[\dot{x}_n] \end{bmatrix}$$

where $L[\dot{x}_i] = sX_i(s) - x_i(0)$. Hence the transform of $\dot{\mathbf{x}} = \mathbf{A}\mathbf{x} + \mathbf{B}\mathbf{u}$ is

$$s\mathbf{X}(s) - \mathbf{x}_o = \mathbf{A}\mathbf{X}(s) + \mathbf{B}\mathbf{U}(s)$$

or

$$(s\mathbf{I} - \mathbf{A})\mathbf{X}(s) = \mathbf{B}\mathbf{U}(s) + \mathbf{x}_o$$

where $\mathbf{I}$ is the unit matrix. Note that $s\mathbf{X}(s) = s\mathbf{I}\mathbf{X}(s)$ and that $(s - \mathbf{A})$ is incorrect since s is a scalar and $\mathbf{A}$ is not. Then, if the system output is $\mathbf{y} = \mathbf{C}\mathbf{x}$, so $\mathbf{Y}(s) = \mathbf{C}\mathbf{X}(s)$, the state response $\mathbf{X}(s)$ to initial conditions alone, and the output response $\mathbf{Y}(s)$ to both initial conditions and input are

$$\begin{aligned} \mathbf{X}(s) &= (s\mathbf{I} - \mathbf{A})^{-1}\mathbf{x}_o \qquad \text{\textit{state response to IC alone}} \\ \mathbf{Y}(s) &= \mathbf{G}(s)\mathbf{U}(s) + \mathbf{C}(s\mathbf{I} - \mathbf{A})^{-1}\mathbf{x}_o \end{aligned} \tag{11.6}$$

where

$$\mathbf{G}(s) = \mathbf{C}(s\mathbf{I} - \mathbf{A})^{-1}\mathbf{B} \tag{11.7}$$

is called the *transfer function matrix* because it relates the transforms of the input and output vectors for zero initial conditions. Equation (11.6) shows the total response as a superposition of two separate components. The first term gives the input–output response, for $\mathbf{x}_o = \mathbf{0}$, and the second gives the output response to initial conditions, with $\mathbf{u} = \mathbf{0}$.

As illustrated in Fig. 11.4, $\mathbf{G}(s)$ generalizes the concept of a transfer function. It is a matrix of ordinary transfer functions. For a system with r inputs and m outputs and zero initial conditions,

$$\mathbf{Y} = \mathbf{G}\mathbf{U}: \qquad \begin{bmatrix} Y_1(s) \\ \vdots \\ Y_m(s) \end{bmatrix} = \begin{bmatrix} g_{11}(s) & \cdots & g_{1r}(s) \\ \vdots & & \vdots \\ g_{m1}(s) & \cdots & g_{mr}(s) \end{bmatrix} \begin{bmatrix} U_1(s) \\ \vdots \\ U_r(s) \end{bmatrix} \tag{11.8}$$

$$Y_i(s) = Y_{i1}(s) + \cdots + Y_{ir}(s) \qquad Y_{ij}(s) = g_{ij}(s)U_j(s)$$

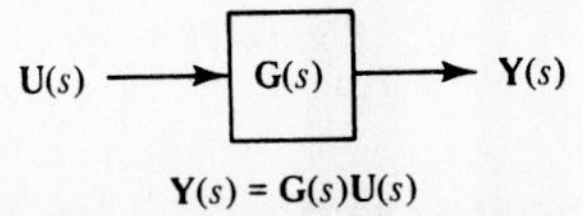

$$U(s) \longrightarrow \boxed{G(s)} \longrightarrow Y(s)$$

$$Y(s) = G(s)U(s)$$

Figure 11.4　Transfer function matrix.

The element $g_{ij}(s)$ is an ordinary transfer function that gives the part Y_{ij} of Y_i that is due to input U_j. It should be recognized that this transfer function matrix may also arise directly; that is, it is not necessarily linked to a state model. For example, Fig. 11.5 shows the two-input two-output feedback control system of Fig. 11.3 expressed in terms of vectors and transfer matrices $\mathbf{G}_o$ and $\mathbf{K}$ given by

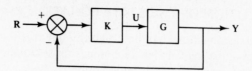

Figure 11.5   Multivariable control system.

$$\mathbf{r} = \begin{bmatrix} r_1 \\ r_2 \end{bmatrix} \qquad \mathbf{u} = \begin{bmatrix} u_1 \\ u_2 \end{bmatrix} \qquad \mathbf{y} = \begin{bmatrix} y_1 \\ y_2 \end{bmatrix} \qquad \mathbf{K} = \begin{bmatrix} K_1 & 0 \\ 0 & K_2 \end{bmatrix}$$

$$\mathbf{G}_o(s) = \begin{bmatrix} 1/(s+1) & 5/(s+5) \\ 0.4/(s+0.5) & 4/(s+2) \end{bmatrix} \tag{11.9}$$

For stability, all poles of all the transfer functions $g_{ij}(s)$ in (11.8) must lie in the left-half s-plane. To determine this, the transfer function matrix in (11.7) is written as follows, using the fact that the inverse equals the adjoint divided by the determinant:

$$\mathbf{G}(s) = \frac{\mathbf{C} \text{ adj } (s\mathbf{I} - \mathbf{A})\mathbf{B}}{|s\mathbf{I} - \mathbf{A}|} \tag{11.10}$$

The numerator is a matrix of polynomials, and all elements of $\mathbf{G}$ have the same denominator. This denominator is the polynomial $|s\mathbf{I} - \mathbf{A}|$. Hence

Stability Theorem 11.4.1   The system described by the state model (11.3) is stable if and only if the *eigenvalues* of the system matrix $\mathbf{A}$, that is, the roots of the system

$$\textit{characteristic equation} \quad |s\mathbf{I} - \mathbf{A}| = 0 \tag{11.11}$$

all lie in the left-half s-plane.

The ability to exploit standard computer routines available to determine the eigenvalues of even very large matrices $\mathbf{A}$ is a major advantage of the state-space formulation.

Example 11.4.1

$$\mathbf{C} = \begin{bmatrix} 1 & 0 \end{bmatrix} \qquad \mathbf{A} = \begin{bmatrix} 0 & 1 \\ -2 & -3 \end{bmatrix} \qquad s\mathbf{I} - \mathbf{A} = \begin{bmatrix} s & -1 \\ 2 & s+3 \end{bmatrix} \qquad \mathbf{B} = \begin{bmatrix} 1 \\ 0 \end{bmatrix}$$

$$(s\mathbf{I} - \mathbf{A})^{-1} = \frac{1}{s^2 + 3s + 2} \begin{bmatrix} s+3 & 1 \\ -2 & s \end{bmatrix}$$

$$G = \mathbf{C}(s\mathbf{I} - \mathbf{A})^{-1}\mathbf{B} = \frac{\begin{bmatrix} 1 & 0 \end{bmatrix}}{(s+1)(s+2)} \begin{bmatrix} s+3 & 1 \\ -2 & s \end{bmatrix} \begin{bmatrix} 1 \\ 0 \end{bmatrix} = \frac{s+3}{(s+1)(s+2)}$$

$|s\mathbf{I} - \mathbf{A}| = s^2 + 3s + 2$. The eigenvalues are -1 and -2, so the system is stable.

Example 11.4.2

$$\mathbf{A} = \begin{bmatrix} -1 & 0 \\ 0 & -2 \end{bmatrix} \qquad s\mathbf{I} - \mathbf{A} = \begin{bmatrix} s+1 & 0 \\ 0 & s+2 \end{bmatrix}$$

$|s\mathbf{I} - \mathbf{A}| = s^2 + 3s + 2 = (s+1)(s+2)$

The eigenvalues are -1 and -2, so the system is stable. Note for future reference that $\mathbf{A}$ is a diagonal matrix and that the eigenvalues are equal to the diagonal elements of $\mathbf{A}$.

11.5 SOLUTION OF THE STATE EQUATION $\dot{x}$ = Ax + Bu

As for transfer functions, not only stability but also transient responses must be determined from the state model. Techniques for solution of the state-space equation toward this end are discussed in this section.

Solution by Laplace Transforms

Based on (11.6) and (11.7), this method is often useful for smaller systems, as well as for larger ones if only certain input–output relations need to be calculated.

Example 11.5.1

$$A = \begin{bmatrix} 0 & 1 \\ -2 & -3 \end{bmatrix} \qquad B = \begin{bmatrix} 0 \\ 2 \end{bmatrix} \qquad C = [1 \ \ 0] \qquad x_0 = \begin{bmatrix} -1 \\ 0 \end{bmatrix}$$

$$sI - A = \begin{bmatrix} s & -1 \\ 2 & s+3 \end{bmatrix} \qquad (sI - A)^{-1} = \frac{1}{s^2 + 3s + 2} \begin{bmatrix} s+3 & 1 \\ -2 & s \end{bmatrix}$$

since $|sI - A| = s^2 + 3s + 2 = (s+1)(s+2)$. Equation (11.7) gives

$$G = C(sI - A)^{-1}B = \frac{[1 \ \ 0]}{(s+1)(s+2)} \begin{bmatrix} s+3 & 1 \\ -2 & s \end{bmatrix} \begin{bmatrix} 0 \\ 2 \end{bmatrix} = \frac{2}{(s+1)(s+2)}$$

This is an ordinary transfer function, since the system is single input/single output, and responses to inputs can be calculated in the usual way. From (11.6), the state response to x_0 is

$$X(s) = (sI - A)^{-1} \begin{bmatrix} -1 \\ 0 \end{bmatrix} = \begin{bmatrix} \dfrac{-2}{s+1} + \dfrac{1}{s+2} \\ \dfrac{2}{s+1} - \dfrac{2}{s+2} \end{bmatrix} \qquad x(t) = \begin{bmatrix} -2e^{-t} + e^{-2t} \\ 2e^{-t} - 2e^{-2t} \end{bmatrix}$$

where partial fraction expansion was used for $X(s)$.

Example 11.5.2

$$A = \begin{bmatrix} -1 & 0 \\ 0 & -2 \end{bmatrix} \qquad B = \begin{bmatrix} 0 \\ 1 \end{bmatrix} \qquad C = \begin{bmatrix} 1 & 0 \\ 0 & 1 \end{bmatrix} \qquad x_0 = \begin{bmatrix} 2 \\ -3 \end{bmatrix}$$

Since

$$sI - A = \begin{bmatrix} s+1 & 0 \\ 0 & s+2 \end{bmatrix}$$

and C is a unit matrix, the output response to x_0 and a step input $U = 1/s$ is

$$Y = CX = C(sI - A)^{-1}(BU + x_0) = \begin{bmatrix} \dfrac{1}{s+1} & 0 \\ 0 & \dfrac{1}{s+2} \end{bmatrix} \left(\begin{bmatrix} 0 \\ 1 \end{bmatrix} \dfrac{1}{s} + \begin{bmatrix} 2 \\ -3 \end{bmatrix} \right)$$

Then, using partial fractions,

$$Y = \begin{bmatrix} \dfrac{1}{s+1} & 0 \\ 0 & \dfrac{1}{s+2} \end{bmatrix} \begin{bmatrix} 2 \\ -3 + \dfrac{1}{s} \end{bmatrix} = \begin{bmatrix} \dfrac{2}{s+1} \\ \dfrac{1-3s}{s(s+2)} \end{bmatrix}$$

$$y(t) = \begin{bmatrix} 2e^{-t} \\ 0.5(1 - 7e^{-2t}) \end{bmatrix}$$

It is useful to observe that for problems where the size of $\mathbf{A}$ is greater than 2×2, direct solution from the simultaneous equations is often more convenient than formal matrix inversion if only few input–output relations are of interest.

Formal Solution: The Transition Matrix

The solution of the scalar equation $\dot{x} = ax + bu$, $x(0) = x_0$ is known to be $x(t) = e^{at}x_0 + \int_0^t e^{a(t-\tau)}bu(\tau)\,d\tau$, where $e^{at} = 1 + at + (1/2!)a^2t^2 + (1/3!)a^3t^3 + \cdots$. By analogy, for the matrix case

$$\begin{cases} \dot{\mathbf{x}} = \mathbf{Ax} \quad \text{(homogeneous equation):} \quad \mathbf{x}(t) = e^{\mathbf{A}t}\mathbf{x}_0 & (11.12a) \\[2mm] \dot{\mathbf{x}} = \mathbf{Ax} + \mathbf{Bu}: \quad \mathbf{x}(t) = e^{\mathbf{A}t}\mathbf{x}_0 + \int_0^t e^{\mathbf{A}(t-\tau)}\mathbf{Bu}(\tau)\,d\tau & (11.12b) \end{cases}$$

where the matrix exponential is defined by

$$e^{\mathbf{A}t} = \mathbf{I} + \mathbf{A}t + \left(\frac{1}{2!}\right)\mathbf{A}^2t^2 + \left(\frac{1}{3!}\right)\mathbf{A}^3t^3 + \cdots = \sum_{k=0}^{\infty} \frac{\mathbf{A}^k t^k}{k!} \qquad (11.13)$$

11.2.a   To prove (11.12a), the derivative of this matrix is required.

$$\begin{aligned} \frac{de^{\mathbf{A}t}}{dt} &= \left(\frac{d}{dt}\right)\left(\mathbf{I} + \mathbf{A}t + \left(\frac{1}{2!}\right)\mathbf{A}^2t^2 + \cdots\right) \\[2mm] &= \mathbf{A} + \left(\frac{2}{2!}\right)\mathbf{A}^2t + \left(\frac{3}{3!}\right)\mathbf{A}^3t^2 + \cdots \\[2mm] &= \mathbf{A}\left(\mathbf{I} + \mathbf{A}t + \left(\frac{1}{2!}\right)\mathbf{A}^2t^2 + \cdots\right) \\[2mm] &= \mathbf{A}e^{\mathbf{A}t} \quad \text{or} \quad e^{\mathbf{A}t}\mathbf{A} \end{aligned} \qquad (11.14)$$

Note that this is analogous to the derivative of a scalar exponential. Substituting the solution (11.12a) into the differential equation now yields

$$\dot{\mathbf{x}} = \frac{d}{dt}e^{\mathbf{A}t}\mathbf{x}_0 = \mathbf{A}e^{\mathbf{A}t}\mathbf{x}_0 = \mathbf{Ax}$$

This satisfies the differential equation. The initial condition is also satisfied:

$$\mathbf{x}(0) = e^{\mathbf{A}.0}\mathbf{x}_0 = (\mathbf{I} + \mathbf{A}.0 + \cdots)\mathbf{x}_0 = \mathbf{x}_0$$

The matrix $e^{\mathbf{A}t}$ relates the state at t to that at time zero and is called the

$$\text{\textit{transition matrix}} \quad \boldsymbol{\phi}(t) = e^{\mathbf{A}t} \qquad (11.15)$$

11.2.b   For the proof of (11.12b), some properties analogous to those for scalar exponentials are needed. Because $\mathbf{x}(t_2 + t_1) = e^{\mathbf{A}(t_2+t_1)}\mathbf{x}_0$, and also $\mathbf{x}(t_2 + t_1) = e^{\mathbf{A}t_2}\mathbf{x}(t_1) = e^{\mathbf{A}t_2}e^{\mathbf{A}t_1}\mathbf{x}_0$, it follows that

$$e^{\mathbf{A}(t_1+t_2)} = e^{\mathbf{A}t_1}e^{\mathbf{A}t_2} \qquad \boldsymbol{\phi}(t_1 + t_2) = \boldsymbol{\phi}(t_1)\boldsymbol{\phi}(t_2) \qquad (11.16)$$

With $t_2 = -t_1 = t$, this yields the additional relation

$$\mathbf{I} = e^{-\mathbf{A}t}e^{\mathbf{A}t} \qquad (e^{\mathbf{A}t})^{-1} = e^{-\mathbf{A}t} \qquad (11.17)$$

To prove (11.12b), assume a solution

$$\mathbf{x}(t) = e^{\mathbf{A}t}\mathbf{f}(t) \qquad \dot{\mathbf{x}} = \mathbf{A}e^{\mathbf{A}t}\mathbf{f} + e^{\mathbf{A}t}\dot{\mathbf{f}} = \mathbf{A}\mathbf{x} + e^{\mathbf{A}t}\dot{\mathbf{f}}$$

For this to satisfy $\dot{\mathbf{x}} = \mathbf{A}\mathbf{x} + \mathbf{B}\mathbf{u}$ requires that $e^{\mathbf{A}t}\dot{\mathbf{f}} = \mathbf{B}\mathbf{u}$ or, using (11.17), $\dot{\mathbf{f}} = e^{-\mathbf{A}t}\mathbf{B}\mathbf{u}$. Integration yields $\mathbf{f}(t) = \mathbf{f}(0) + \int_0^t e^{-\mathbf{A}\tau}\mathbf{B}\mathbf{u}(\tau)\,d\tau$, and substitution then gives the solution (11.12b), since $\mathbf{x}(0) = e^{\mathbf{A}\cdot 0}\mathbf{f}(0) = \mathbf{f}(0)$.

An alternative expression for the transition matrix follows from the equation $\mathbf{X}(s) = (s\mathbf{I} - \mathbf{A})^{-1}\mathbf{x}_0$ in (11.6) for the state response to initial conditions. From comparison with (11.12a), this yields a useful equation for determining the transition matrix for smaller systems:

$$e^{\mathbf{A}t} = L^{-1}[(s\mathbf{I} - \mathbf{A})^{-1}] \tag{11.18}$$

Discrete-Time Solution

As illustrated in Fig. 11.6, the time axis is discretized into intervals of width T, and $u_i(t)$ is approximated by a staircase function, constant over the intervals. Using (11.12b), for $t = T$,

$$\mathbf{x}(T) = e^{\mathbf{A}T}\mathbf{x}(0) + e^{\mathbf{A}T}\int_0^T e^{-\mathbf{A}\tau}\,d\tau\,\mathbf{B}\mathbf{u}(0)$$

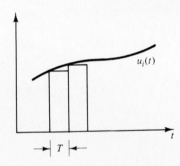

Figure 11.6   Time discretization.

or

$$\mathbf{x}(T) = \boldsymbol{\phi}\mathbf{x}(0) + \boldsymbol{\Delta}\mathbf{u}(0)$$

where

$$\boldsymbol{\phi} = e^{\mathbf{A}T} \qquad \boldsymbol{\Delta} = e^{\mathbf{A}T}\int_0^T e^{-\mathbf{A}\tau}\,d\tau\,\mathbf{B} \tag{11.19}$$

This can be repeated for the second interval, with $\mathbf{x}(T)$ as initial condition, to yield $\mathbf{x}(2T) = \boldsymbol{\phi}\mathbf{x}(T) + \boldsymbol{\Delta}\mathbf{u}(T)$, and then for the succeeding intervals. For the kth interval, with $\mathbf{x}_k = \mathbf{x}(kT)$,

$$\mathbf{x}_{k+1} = \boldsymbol{\phi}\mathbf{x}_k + \boldsymbol{\Delta}\mathbf{u}_k \tag{11.20}$$

This permits the solution to be calculated forward in time.

The series (11.13) is used to compute $\boldsymbol{\phi}$, adding terms until the next one does not change the result by more than a specified tolerance. To maintain accuracy, it turns out to be desirable to calculate $\boldsymbol{\phi}(T)$ as $[\boldsymbol{\phi}(T/n)]^n$, where n is a positive even integer chosen so that the elements of $\mathbf{A}T/n$ will be "sufficiently small."

To find $\boldsymbol{\Delta}$, (11.19) is integrated:

$$\boldsymbol{\Delta} = e^{\mathbf{A}T}(-e^{-\mathbf{A}T}\mathbf{A}^{-1} + \mathbf{A}^{-1})\mathbf{B} = (\boldsymbol{\phi} - \mathbf{I})\mathbf{A}^{-1}\mathbf{B} \tag{11.21}$$

If **A** is singular, this solution is invalid. An alternative solution is obtained if, analogous to (11.13), $e^{A(T-\tau)}$ in (11.19) is written as a series:

$$\Delta = \int_0^T \sum_{k=0}^{\infty} \frac{A^k(T-\tau)^k}{k!} \, d\tau \, B = \sum_{k=0}^{\infty} \frac{A^k T^{k+1}}{(k+1)!} B \qquad (11.22)$$

Approximate Discrete-Time Solution

The vector finite-difference approximation to $\dot{x}$ is of the same form as a scalar finite difference:

$$\dot{x} = \frac{1}{T}[x(t+T) - x(t)]$$

The equation $\dot{x} = Ax + Bu$ is then approximated by

$$x(t+T) = (I + AT)x(t) + TBu(t) \qquad (11.23a)$$

This is of the form of (11.20) with

$$\phi = I + AT \qquad \Delta = TB \qquad (11.23b)$$

Only the first terms of the previous series for ϕ and Δ are included, so T must be chosen sufficiently small, often as one-tenth of the smallest system time constant. But this approach is convenient for nonlinear and time-varying parameter systems, where (11.19) does not apply.

Example 11.5.3

$$A = \begin{bmatrix} -3/4 & -1/4 \\ -1/4 & -3/4 \end{bmatrix} \qquad (sI - A) = \begin{bmatrix} s+3/4 & 1/4 \\ 1/4 & s+3/4 \end{bmatrix}$$

$$|sI - A| = (s+3/4)^2 - (1/4)^2 = (s+1)(s+0.5)$$

$$(sI - A)^{-1} = \frac{1}{(s+1)(s+0.5)} \begin{bmatrix} s+3/4 & -1/4 \\ -1/4 & s+3/4 \end{bmatrix}$$

Using (11.18) and partial fraction expansion yields the transition matrix

$$\phi(t) = L^{-1}[(sI - A)^{-1}] = \frac{1}{2} \begin{bmatrix} e^{-t} + e^{-0.5t} & e^{-t} - e^{-0.5t} \\ e^{-t} - e^{-0.5t} & e^{-t} + e^{-0.5t} \end{bmatrix}$$

Discrete-time solutions ϕ as calculated from (11.19) and its approximation (11.23) for intervals $T = 1$ and $T = 0.1$ are as follows:

	$T = 1$	$T = 0.1$
(11.19):	$\begin{bmatrix} 0.4872 & -0.1193 \\ -0.1193 & 0.4872 \end{bmatrix}$	$\begin{bmatrix} 0.9280 & -0.0232 \\ -0.0232 & 0.9280 \end{bmatrix}$
(11.23):	$\begin{bmatrix} 0.25 & -0.25 \\ -0.25 & 0.25 \end{bmatrix}$	$\begin{bmatrix} 0.9250 & -0.0250 \\ -0.0250 & 0.9250 \end{bmatrix}$

It is seen that the approximation (11.23) is very poor for $T = 1$. This is not surprising, because the plant time constants are 1 and 2 seconds. For $T = 0.1$ the interval is one-tenth of the smallest time constant, and reasonable accuracy may be expected.

Other Solutions

Standard routines such as the Runge–Kutta or predictor–corrector methods are quite satisfactory for smaller systems, but become very time consuming for, say, the large matrices in power system dynamics. The discrete-time solution or other available routines are needed for such cases.

$2^{\circledast}$ A further method of solution is based on the <u>use of coordinate transformations,</u> a very important subject in its own right, and considered next.

11.6 EIGENVALUES, EIGENVECTORS, AND MODES

It is recalled that the state model for a system is not unique, but depends on the choice of a set of state variables. To simplify analysis and design for a system

$$\dot{\mathbf{x}} = \mathbf{Ax} + \mathbf{Bu} \qquad \mathbf{y} = \mathbf{Cx} \qquad \mathbf{x}(0) = \mathbf{x}_0 \qquad (11.24)$$

it is often beneficial to define a new state vector $\mathbf{z}$ by a *coordinate transformation*

$$\mathbf{x} = \mathbf{Tz} \qquad (11.25)$$

The state model corresponding to these new state variables is found by substituting this into (11.24):

$$\mathbf{T\dot{z}} = \mathbf{ATz} + \mathbf{Bu} \qquad \mathbf{y} = \mathbf{CTz} \qquad \mathbf{x}_0 = \mathbf{Tz}_0$$

If $\mathbf{T}^{-1}$ exists, the transformed system is

$$\dot{\mathbf{z}} = \hat{\mathbf{A}}\mathbf{z} + \hat{\mathbf{B}}\mathbf{u} \qquad \mathbf{y} = \hat{\mathbf{C}}\mathbf{z} \qquad \mathbf{z}(0) = \mathbf{z}_0$$

$$\hat{\mathbf{A}} = \mathbf{T}^{-1}\mathbf{AT} \qquad \hat{\mathbf{B}} = \mathbf{T}^{-1}\mathbf{B} \qquad \hat{\mathbf{C}} = \mathbf{CT} \qquad \mathbf{z}_0 = \mathbf{T}^{-1}\mathbf{x}_0 \qquad (11.26)$$

The definition of a new set of internal state variables should evidently not affect the eigenvalues or input–output behavior. This may be verified by evaluating the characteristic equation and the transfer function matrix of the transformed system:

1.
$$|s\mathbf{I} - \hat{\mathbf{A}}| = |s\mathbf{I} - \mathbf{T}^{-1}\mathbf{AT}| = |\mathbf{T}^{-1}(s\mathbf{I} - \mathbf{A})\mathbf{T}|$$
$$= |\mathbf{T}^{-1}||s\mathbf{I} - \mathbf{A}||\mathbf{T}| = |s\mathbf{I} - \mathbf{A}| \qquad (11.27)$$

So the characteristic equation does not change.

2. The transfer function matrix of (11.7) becomes

$$\hat{\mathbf{G}} = \hat{\mathbf{C}}(s\mathbf{I} - \hat{\mathbf{A}})^{-1}\hat{\mathbf{B}} = \mathbf{CT}(s\mathbf{I} - \mathbf{T}^{-1}\mathbf{AT})^{-1}\mathbf{T}^{-1}\mathbf{B}$$

$$= \mathbf{CT}[\mathbf{T}^{-1}(s\mathbf{I} - \mathbf{A})\mathbf{T}]^{-1}\mathbf{T}^{-1}\mathbf{B} = \mathbf{CTT}^{-1}(s\mathbf{I} - \mathbf{A})^{-1}\mathbf{TT}^{-1}\mathbf{B} \qquad (11.28)$$

$$= \mathbf{C}(s\mathbf{I} - \mathbf{A})^{-1}\mathbf{B} = \mathbf{G}$$

So $\mathbf{G}$ is also unaffected by the transformation.

The particular transformation of interest in this section is that for which $\hat{\mathbf{A}}$ is a diagonal matrix $\mathbf{\Lambda}$:

$$\mathbf{T} = \mathbf{U} \qquad \hat{\mathbf{A}} = \mathbf{U}^{-1}\mathbf{AU} = \mathbf{\Lambda} = \mathrm{diag}(\lambda_1, \lambda_2, \ldots, \lambda_n) \qquad (11.29)$$

Since, from (11.27),

$$|s\mathbf{I} - \mathbf{A}| = |s\mathbf{I} - \mathbf{\Lambda}| = \det\{\mathrm{diag}(s - \lambda_1, \ldots, s - \lambda_n)\}$$
$$= (s - \lambda_1)(s - \lambda_2)\ldots(s - \lambda_n) \qquad (11.30)$$

it follows that:

The elements λ_i of the diagonal matrix $\mathbf{\Lambda}$ are the roots of the characteristic equation $|s\mathbf{I} - \mathbf{\Lambda}| = |s\mathbf{I} - \mathbf{A}| = 0$ and so are the eigenvalues.

This was illustrated by Example 11.4.2.

Let the transformation matrix $\mathbf{U}$ required to transform $\mathbf{A}$ to $\mathbf{\Lambda}$ be of the form

$$\mathbf{U} = [\mathbf{u}_1, \mathbf{u}_2, \ldots, \mathbf{u}_n]; \qquad \mathbf{u}_i = \begin{bmatrix} u_{1i} \\ \vdots \\ u_{ni} \end{bmatrix} = i\text{th column of } \mathbf{U} \qquad (11.31)$$

Equation (11.29) shows that $\mathbf{AU} = \mathbf{U\Lambda}$:

$$\mathbf{A}[\mathbf{u}_1, \mathbf{u}_2, \ldots, \mathbf{u}_n] = [\mathbf{u}_1, \mathbf{u}_2, \ldots, \mathbf{u}_n] \, \mathrm{diag}(\lambda_1, \ldots, \lambda_n)$$

By equating the ith columns,

$$\mathbf{Au}_i = \lambda_i \mathbf{u}_i \qquad (\lambda_i \mathbf{I} - \mathbf{A})\mathbf{u}_i = \mathbf{0} \qquad (11.32)$$

This is a set of homogeneous equations for $u_{1i}, \ldots, u_{ni}$. For this set to have a non-trivial solution, the determinant $|\lambda_i \mathbf{I} - \mathbf{A}|$ of the coefficients must be zero. This condition is satisfied by virtue of the fact that λ_i is an eigenvalue.

The solution $\mathbf{u}_i$ of (11.32) is the ith (right) *eigenvector* of $\mathbf{A}$. Use of the *modal matrix* $\mathbf{U} = [\mathbf{u}_1, \ldots, \mathbf{u}_n]$ as a transformation matrix diagonalizes $\mathbf{A}$.

Because $|\lambda_i \mathbf{I} - \mathbf{A}| = 0$, only the ratios of the elements of eigenvectors are fixed, and one element can always be chosen. Thus any multiples of $\mathbf{u}_1, \ldots, \mathbf{u}_n$ are also eigenvectors.

Example 11.6.1

$$\mathbf{A} = \begin{bmatrix} -3 & 2 \\ -1 & 0 \end{bmatrix} \qquad |\lambda \mathbf{I} - \mathbf{A}| = \lambda^2 + 3\lambda + 2$$

The eigenvalues are $\lambda_1 = -1$, $\lambda_2 = -2$. Eigenvector $\mathbf{u}_i$ is the solution of $\mathbf{Au}_i = \lambda_i \mathbf{u}_i$:

$$\begin{bmatrix} -3 & 2 \\ -1 & 0 \end{bmatrix} \begin{bmatrix} u_{11} \\ u_{21} \end{bmatrix} = -1 \begin{bmatrix} u_{11} \\ u_{21} \end{bmatrix}: \quad u_{11} = u_{21} \qquad \mathbf{u}_1 = \begin{bmatrix} 1 \\ 1 \end{bmatrix}$$

$$\begin{bmatrix} -3 & 2 \\ -1 & 0 \end{bmatrix} \begin{bmatrix} u_{12} \\ u_{22} \end{bmatrix} = -2 \begin{bmatrix} u_{12} \\ u_{22} \end{bmatrix}: \quad u_{12} = 2u_{22} \qquad \mathbf{u}_2 = \begin{bmatrix} 1 \\ \frac{1}{2} \end{bmatrix}$$

Hence

$$\mathbf{U} = \begin{bmatrix} 1 & 1 \\ 1 & \frac{1}{2} \end{bmatrix}$$

The first and second rows are dependent in each case and give the same relation between u_{1i} and u_{2i}. Thus any multiples of $\mathbf{u}_1$ and $\mathbf{u}_2$ are also eigenvectors.

Numerous computer routines are available to compute both the eigenvalues and eigenvectors of large matrices.

Complex Pairs of Eigenvalues

The case when $\mathbf{A}$ has repeated eigenvalues will not be considered. Transformation to diagonal form may then not be possible. The case of distinct complex pairs of eigenvalues is a direct generalization of that for a single pair discussed next.

If $\mathbf{A}$ has a pair of complex eigenvalues, then

$$\mathbf{\Lambda} = \mathrm{diag}\,(\lambda_1, \lambda_2, \ldots, \sigma + j\omega, \sigma - j\omega, \ldots, \lambda_n) \qquad (11.33)$$

and the corresponding adjacent columns of $\mathbf{U}$ are also complex conjugates. Real arithmetic is often preferable and can be achieved by a further transformation:

$$\dot{\mathbf{x}} = \mathbf{A}\mathbf{x} \qquad \mathbf{x} = \mathbf{U}\mathbf{z} \qquad \dot{\mathbf{z}} = \mathbf{\Lambda}\mathbf{z} \qquad \mathbf{\Lambda} = \mathbf{U}^{-1}\mathbf{A}\mathbf{U}$$

$$\dot{\mathbf{z}} = \mathbf{\Lambda}\mathbf{z} \qquad \mathbf{z} = \mathbf{K}\mathbf{v} \qquad \dot{\mathbf{v}} = \hat{\mathbf{\Lambda}}\mathbf{v} \qquad \hat{\mathbf{\Lambda}} = \mathbf{K}^{-1}\mathbf{\Lambda}\mathbf{K} = \mathbf{K}^{-1}\mathbf{U}^{-1}\mathbf{A}\mathbf{U}\mathbf{K}$$
$$= \mathbf{T}^{-1}\mathbf{A}\mathbf{T}$$

where

$$\mathbf{T} = \mathbf{U}\mathbf{K} \tag{11.34}$$

and where $\mathbf{K}$ and the resulting matrix $\hat{\mathbf{\Lambda}}$ are

$$\mathbf{K} = \begin{bmatrix} 1 & & & & & & 0 \\ & \ddots & & & & & \\ & & 0.5 & -0.5_j & & & \\ & & 0.5 & 0.5_j & & & \\ & & & & \ddots & & \\ 0 & & & & & & 1 \end{bmatrix} \qquad \hat{\mathbf{\Lambda}} = \begin{bmatrix} \lambda_1 & & & & & 0 \\ & \ddots & & & & \\ & & \sigma & \omega & & \\ & & -\omega & \sigma & & \\ & & & & \ddots & \\ 0 & & & & & \lambda_n \end{bmatrix} \tag{11.35}$$

$\mathbf{T}$ turns out to be real, with one column equal to the real part of the corresponding eigenvector and the next one to the imaginary part.

Left Eigenvectors and Eigenvector Normalization

In addition to the (right) eigenvectors above, the left eigenvectors are also used for analysis and design. The left eigenvectors $\mathbf{v}_i$ of $\mathbf{A}$ satisfy the equation and its transpose below, where the prime identifies a transpose:

$$\mathbf{v}_i' \mathbf{A} = \lambda_i \mathbf{v}_i' \qquad \mathbf{A}'\mathbf{v}_i = \lambda_i \mathbf{v}_i \tag{11.36}$$

The second form implies that $\mathbf{v}_i$ is also the right eigenvector of $\mathbf{A}'$. Note that, since $|\lambda\mathbf{I} - \mathbf{A}| = |\lambda\mathbf{I} - \mathbf{A}'|$, λ_i is also an eigenvalue of $\mathbf{A}'$.

Theorem 11.6.1

$$\mathbf{U} = [\mathbf{u}_1, \ldots, \mathbf{u}_n] \qquad \mathbf{V} = [\mathbf{v}_1, \ldots, \mathbf{v}_n] \tag{11.37}$$

(i) $\mathbf{u}_j$ and $\mathbf{v}_i, j \neq i$, are orthogonal, that is, their inner product is zero: $\mathbf{v}_i' \mathbf{u}_j = 0$.
(ii) $\mathbf{u}_i$ and $\mathbf{v}_i$ can be normalized so that $\mathbf{v}_i' \mathbf{u}_i = 1$. Then

$$\mathbf{V}'\mathbf{U} = \mathbf{I} \qquad \mathbf{V}' = \mathbf{U}^{-1} \tag{11.38}$$

Proof. The scalar or inner product of $\mathbf{v}_i$ and $\mathbf{u}_j$ is defined by

$$\mathbf{v}_i'\mathbf{u}_j(= \mathbf{u}_j'\mathbf{v}_i) = [v_{1i} \quad v_{2i} \quad \cdots \quad v_{ni}] \begin{bmatrix} u_{1j} \\ \cdot \\ \cdot \\ \cdot \\ u_{nj} \end{bmatrix} = \sum_{k=1}^{n} v_{ki}u_{kj} \tag{11.39}$$

The vectors are orthogonal if this product is zero.

Premultiplying $\mathbf{A}\mathbf{u}_j = \lambda_j\mathbf{u}_j$ by $\mathbf{v}_i'$ yields $\mathbf{v}_i'\mathbf{A}\mathbf{u}_j = \lambda_j\mathbf{v}_i'\mathbf{u}_j$.
Postmultiplying $\mathbf{v}_i'\mathbf{A} = \lambda_i\mathbf{v}_i'$ by $\mathbf{u}_j$ yields $\mathbf{v}_i'\mathbf{A}\mathbf{u}_j = \lambda_i\mathbf{v}_i'\mathbf{u}_j$.

The left sides of the two forms are the same, so subtracting both yields $(\lambda_j - \lambda_i)\mathbf{v}_i'\mathbf{u}_j = 0$. If, as will be assumed, the eigenvalues are all distinct, this proves part (i).

One element of each of $\mathbf{v}_i$ and $\mathbf{u}_i$ can be chosen, so the condition $\mathbf{v}_i'\mathbf{u}_i = 1$ of part (ii) can always be satisfied. Then, since $\mathbf{v}_i'\mathbf{u}_j = 0$,

$$\mathbf{V}'\mathbf{U} = [\mathbf{v}_1 \cdots \mathbf{v}_n]'[\mathbf{u}_1 \cdots \mathbf{u}_n] = \begin{bmatrix} \mathbf{v}_1' \\ \vdots \\ \mathbf{v}_n' \end{bmatrix} [\mathbf{u}_1 \cdots \mathbf{u}_n] \tag{11.40}$$

$$= \begin{bmatrix} \mathbf{v}_1'\mathbf{u}_1 & \cdots & \mathbf{v}_1'\mathbf{u}_n \\ & & \\ & & \\ \mathbf{v}_n'\mathbf{u}_1 & \cdots & \mathbf{v}_n'\mathbf{u}_n \end{bmatrix} = \mathbf{I}$$

Example 11.6.2   Example 11.6.1 Continued

$$\lambda_1 = -1 \qquad \lambda_2 = -2$$

$\mathbf{A}'\mathbf{v}_i = \lambda_i \mathbf{v}_i$, with $\mathbf{A}' = \begin{bmatrix} -3 & -1 \\ 2 & 0 \end{bmatrix}$ yields $\mathbf{v}_1 = \begin{bmatrix} -1 \\ 2 \end{bmatrix}$, $\mathbf{v}_2 = \begin{bmatrix} -1 \\ 1 \end{bmatrix}$, or any multiples. Here $\mathbf{v}_1'\mathbf{u}_1 = -1$, $\mathbf{v}_2'\mathbf{u}_2 = \frac{1}{2}$, and to make $\mathbf{v}_i'\mathbf{u}_i = 1$, $\mathbf{v}_1$ and $\mathbf{v}_2$ are rescaled to

$$\mathbf{v}_1 = \begin{bmatrix} -1 \\ 2 \end{bmatrix} \qquad \mathbf{v}_2 = \begin{bmatrix} 2 \\ -2 \end{bmatrix} \qquad (\mathbf{U}^{-1}\mathbf{A}\mathbf{U} = \mathbf{V}'\mathbf{A}\mathbf{U} = \mathbf{\Lambda})$$

Modal Decomposition and the Transition Matrix e^{At}

With the modal transformation $\mathbf{x} = \mathbf{U}\mathbf{z}$, the state equation $\dot{\mathbf{x}} = \mathbf{A}\mathbf{x} + \mathbf{B}\mathbf{u}$ becomes

$$\dot{\mathbf{z}} = \mathbf{\Lambda}\mathbf{z} + (\mathbf{U}^{-1}\mathbf{B})\mathbf{u} = \mathbf{\Lambda}\mathbf{z} + (\mathbf{V}'\mathbf{B})\mathbf{u} \tag{11.41}$$

The homogeneous equation $\dot{\mathbf{x}} = \mathbf{A}\mathbf{x}$, $\mathbf{x}(0) = \mathbf{x}_0$, becomes

$$\dot{\mathbf{z}} = \mathbf{\Lambda}\mathbf{z} \qquad \dot{z}_i = \lambda_i z_i \qquad i = 1, \ldots, n \tag{11.42}$$

Since $\mathbf{\Lambda}$ is diagonal, the system is represented by n independent differential equations. The initial condition vector is

$$\mathbf{z}(0) = \mathbf{U}^{-1}\mathbf{x}(0) = \mathbf{V}'\mathbf{x}(0) = \begin{bmatrix} \mathbf{v}_1' \\ \vdots \\ \mathbf{v}_n' \end{bmatrix} \mathbf{x}(0) \qquad z_i(0) = \mathbf{v}_i'\mathbf{x}_0 \tag{11.43}$$

The modal response to this initial condition is

$$\begin{cases} \mathbf{z}(t) = e^{\mathbf{\Lambda}t}\mathbf{z}_0 = e^{\mathbf{\Lambda}t}\mathbf{U}^{-1}\mathbf{x}_0 = e^{\mathbf{\Lambda}t}\mathbf{V}'\mathbf{x}_0 \\ z_i(t) = z_i(0)e^{\lambda_i t} = (\mathbf{v}_i'\mathbf{x}_0)e^{\lambda_i t} \end{cases} \tag{11.44}$$

The response $\mathbf{x}$ is the sum of modal components and can be expressed in many different forms.

$$\mathbf{x}(t) = \mathbf{U}\mathbf{z}(t) = \mathbf{U}e^{\mathbf{\Lambda}t}\mathbf{U}^{-1}\mathbf{x}_0 = \mathbf{U}e^{\mathbf{\Lambda}t}\mathbf{V}'\mathbf{x}_0 \tag{11.45}$$

$$\mathbf{x}(t) = [\mathbf{u}_1 \quad \cdots \quad \mathbf{u}_n]\mathbf{z} = \mathbf{u}_1 z_1 + \mathbf{u}_2 z_2 + \cdots + \mathbf{u}_n z_n$$

$$\mathbf{x}(t) = (\mathbf{v}_1'\mathbf{x}_0)e^{\lambda_1 t}\mathbf{u}_1 + \cdots + (\mathbf{v}_n'\mathbf{x}_0)e^{\lambda_n t}\mathbf{u}_n \tag{11.46}$$

Equation (11.46) shows the *modal decomposition* of $\mathbf{x}$. The ith row of $\mathbf{V}'$, or ith column of $\mathbf{V}$, is used to find the ith initial (scalar) mode size $\mathbf{v}_i'\mathbf{x}_0$ from $\mathbf{x}_0$. The (scalar) factor $e^{\lambda_i t}$ then gives the response in the mode, and $\mathbf{u}_i$ is the ith mode shape, which shows how the modal response is distributed over the elements of $\mathbf{x}$. The modal decomposition shows how the total response consists of a sum of responses in the individual modes. It also shows that if, for example, the initial condition is confined to the ith mode, that is, only $\mathbf{v}_i' \mathbf{x}_0$ is nonzero, the response will be only in the ith mode. This means that the ratios of the elements of $\mathbf{x}$ will equal those in $\mathbf{u}_i$.

Example 11.6.3   Examples 11.6.1 and 11.6.2 Continued

The modal decomposition is

$$\mathbf{x}(t) = [-1 \quad 2]\mathbf{x}_0 e^{-t}\begin{bmatrix} 1 \\ 1 \end{bmatrix} + [2 \quad -2]\mathbf{x}_0 e^{-2t}\begin{bmatrix} 1 \\ \frac{1}{2} \end{bmatrix}$$

The second mode decays faster than the first, so as t increases, the ratio of the elements of $\mathbf{x}(t)$ will increasingly approximate the first mode shape.

Since $\mathbf{x}(t)$ also equals $e^{\mathbf{A}t}\mathbf{x}_0$, it follows from (11.45) that

$$e^{\mathbf{A}t} = \mathbf{U}e^{\mathbf{\Lambda}t}\mathbf{U}^{-1} = \mathbf{U}e^{\mathbf{\Lambda}t}\mathbf{V}' \tag{11.47}$$

provides an additional way for calculation of the transition matrix. Here $e^{\mathbf{\Lambda}t}$ is a diagonal matrix. Using (11.18) gives

$$e^{\mathbf{\Lambda}t} = L^{-1}[(s\mathbf{I} - \mathbf{\Lambda})^{-1}] = L^{-1}[(\text{diag}\{s - \lambda_1, \ldots, s - \lambda_n\})^{-1}]$$

$$= L^{-1}\left[\text{diag}\left\{ \frac{1}{s - \lambda_1}, \ldots, \frac{1}{s - \lambda_n} \right\} \right] \tag{11.48}$$

$$= \text{diag}(e^{\lambda_1 t}, e^{\lambda_2 t}, \ldots, e^{\lambda_n t})$$

11.7 CONTROLLABILITY, OBSERVABILITY, AND STABILIZABILITY

The preceding sections have been concerned with methods to determine the stability and transient response of systems described by given state models. Before considering design, that is, how this behavior may be changed, some important new concepts will be introduced. Generally, there are fewer control variables than state variables. Controllability is concerned with the question of whether it is at all possible to control all states, disregarding how this might be done. Similarly, the number of output variables measured is usually smaller than the number of state variables. In some very important design techniques it is desirable to use feedback from all the state variables. Observability is concerned with the question of whether it is at all possible to find all states from the measured outputs, regardless of the method used. Assuming distinct eigenvalues, let the system be given by

$$\dot{\mathbf{x}} = \mathbf{A}\mathbf{x} + \mathbf{B}\mathbf{u} \qquad \mathbf{y} = \mathbf{C}\mathbf{x} + \mathbf{D}\mathbf{u} \tag{11.49}$$

Definition.   The system (11.49), or the pair $(\mathbf{A}, \mathbf{B})$, is state *controllable* if and only if there exists a control $\mathbf{u}$ that will transfer any initial state $\mathbf{x}(0)$ to any state $\mathbf{x}(T)$ in finite time T. The system (11.49), or the pair $(\mathbf{C}, \mathbf{A})$, is *observable* if the state $\mathbf{x}(0)$ can be determined from knowledge of $\mathbf{u}$ and $\mathbf{y}$ over a finite time interval $0 < t < T$.

Controllability and observability are important concepts in the theory and design of multivariable systems. One method for determining whether a system is controllable or observable is based on the modal approach. With $\mathbf{x} = \mathbf{U}\mathbf{z}$, (11.49) becomes, as in (11.41),

$$\dot{\mathbf{z}} = \mathbf{\Lambda}\mathbf{z} + (\mathbf{V}'\mathbf{B})\mathbf{u} \qquad \mathbf{y} = (\mathbf{C}\mathbf{U})\mathbf{z} + \mathbf{D}\mathbf{u} \qquad (11.50)$$

Suppose that the ith row of $\mathbf{V}'\mathbf{B}$ has only zero elements. Then, from (11.50), $\dot{z}_i = \lambda_i z_i$. Therefore, the ith mode is uncontrollable, since it is not affected by the control. Similarly, suppose that the jth column of $\mathbf{C}\mathbf{U}$ has only zero elements. Then the jth mode is unobservable, because it does not show up in the output. This proves the "only if" parts of the following theorem.

> **Theorem 11.7.1.** The system (11.49) is controllable if and only if $\mathbf{V}'\mathbf{B}$ has no rows consisting entirely of zero elements. The system is observable if and only if $\mathbf{C}\mathbf{U}$ has no columns consisting entirely of zero elements.

The proof of this and the following alternative theorem will be omitted. This alternative does not require the modal transformation, but also does not identify the uncontrollable or unobservable modes. The theorem expresses controllability in terms of the matrices $\mathbf{A}$ and $\mathbf{B}$ and observability in terms of $\mathbf{A}$ and $\mathbf{C}$.

> **Theorem 11.7.2.** The pair $(\mathbf{A}, \mathbf{B})$ is controllable if and only if the rank of the *controllability matrix*
>
> $$\mathbf{\Gamma}_c = [\mathbf{B} \quad \mathbf{A}\mathbf{B} \quad \mathbf{A}^2\mathbf{B} \quad \cdots \quad \mathbf{A}^{n-1}\mathbf{B}] \qquad (11.51)$$
>
> is n; that is, it contains an $n \times n$ nonsingular matrix or has n linearly independent columns. The pair $(\mathbf{C}, \mathbf{A})$ is observable if and only if the rank of the *observability matrix*
>
> $$\mathbf{\Gamma}_o = [\mathbf{C}' \quad A'\mathbf{C}' \quad A'^2\mathbf{C}' \quad \cdots \quad A'^{n-1}\mathbf{C}'] \qquad (11.52)$$
>
> is n, where $\mathbf{A}'$ and $\mathbf{C}'$ are the transposes of $\mathbf{A}$ and $\mathbf{C}$.

Stabilizability

If an uncontrollable mode is stable, it will decay in any case. This is clearly much less serious than an uncontrollable unstable mode, because in this case the system cannot be stabilized.

> **Theorem 11.7.3** The pair $(\mathbf{A}, \mathbf{B})$ is *stabilizable* if it is controllable, or if it is uncontrollable but the uncontrollable modes are stable.

Thus, if a system has unstable eigenvalues and Theorem 11.7.2 shows it to be uncontrollable, the unstable modes can be checked one at a time to determine whether the system is stabilizable. In effect this means that Theorem 11.7.1 is used under certain conditions.

Example 11.7.1

$$\mathbf{A} = \begin{bmatrix} 0 & -1 \\ -3 & 2 \end{bmatrix} \qquad \mathbf{B} = \begin{bmatrix} 1 \\ -3 \end{bmatrix} \qquad \mathbf{C} = [1 \quad 0]$$

The controllability matrix and the observability matrix are

$$[\mathbf{B} \quad \mathbf{A}\mathbf{B}] = \begin{bmatrix} 1 & 3 \\ -3 & -9 \end{bmatrix} \qquad [\mathbf{C}' \quad A'\mathbf{C}'] = \begin{bmatrix} 1 & 0 \\ 0 & -1 \end{bmatrix}$$

The system is observable, but it is not controllable, because the controllability matrix has a rank of 1 since the determinant is zero (that is, the matrix is singular). $|s\mathbf{I} - \mathbf{A}| = s^2 - 2s - 3$, and the eigenvalues are $\lambda_1 = -1$, $\lambda_2 = +3$. Because the second mode is unstable, to determine whether this uncontrollable system is stabilizable, the left eigenvector $\mathbf{v}_2$ is found from $\mathbf{A}'\mathbf{v}_2 = +3\mathbf{v}_2$. It is found that $\mathbf{v}_2' = [1 \quad -1]$, so $\mathbf{v}_2'\mathbf{B} = 4 \neq 0$. Hence the unstable mode is controllable, so the system is stabilizable.

Controllability, Observability, and Transfer Functions

In the state model (11.49) with $\mathbf{D} = 0$, if $\mathbf{c}$ is the ith row of $\mathbf{C}$ and $\mathbf{b}$ the jth column of $\mathbf{B}$, then the ith output y and jth input u are related by

$$\dot{\mathbf{x}} = \mathbf{A}\mathbf{x} + \mathbf{b}u \qquad y = \mathbf{c}\mathbf{x} \tag{11.53}$$

and, by (11.10), the corresponding transfer function is

$$G(s) = \frac{\mathbf{c} \ \text{adj}(s\mathbf{I} - \mathbf{A})\mathbf{b}}{|s\mathbf{I} - \mathbf{A}|} = \frac{p(s)}{q(s)} \tag{11.54}$$

where $q(s) = |s\mathbf{I} - \mathbf{A}|$ is of order n.

> **Theorem 11.7.4.** The system (11.49) is controllable and observable if and only if no cancellations occur between numerators and denominators of the transfer function or transfer function matrix.

Thus, if a state model is made of a transfer function $G(s)$ that has common factors in the numerator and denominator, this state model will be found to be uncontrollable or unobservable or both. And if the common factors in $p(s)$ and $q(s)$ are canceled and a state model made of the resulting transfer function, this model will hide the modes corresponding to the canceled factors. This demonstrates that transfer functions may not represent all system dynamics, since uncontrollable and unobservable parts of the system do not show up in this model.

The state model (11.53), with n state variables, is called a *realization* of the transfer function $G(s)$ of (11.54). If $p(s)$ and $q(s)$ have common factors (that is, if they are not relatively prime or coprime), then (11.53) is not a *minimal realization*. A realization $\{\mathbf{A}, \mathbf{b}, \mathbf{c}\}$ of $G(s)$ is minimal if it has the smallest possible number of state variables.

Earlier results can now be stated as follows.

1. A realization $\{\mathbf{A}, \mathbf{b}, \mathbf{c}\}$ of $G(s)$ is minimal if and only if $q(s) = |s\mathbf{I} - \mathbf{A}|$ and $p(s) = \mathbf{c} \ \text{adj}(s\mathbf{I} - \mathbf{A})\mathbf{b}$ are relatively prime (that is, have no common factors).
2. If $q(s)$ is of order n, then an nth-order realization $\{\mathbf{A}, \mathbf{b}, \mathbf{c}\}$ of $G(s)$ will be controllable and observable if and only if $G(s) = p(s)/q(s)$ is irreducible, that is, if $p(s)$ and $q(s)$ have no common factors.
3. A realization $\{\mathbf{A}, \mathbf{b}, \mathbf{c}\}$ of $G(s)$ is minimal if and only if $\{\mathbf{A}, \mathbf{b}\}$ is controllable and $\{\mathbf{c}, \mathbf{A}\}$ is observable.

11.8 STATE-SPACE METHODS FOR DIGITAL SIMULATION AND CONTROL

This chapter is concluded with an introduction into the use of state-space methods in digital simulation and control. The brief discussion of this important area will demonstrate a strong analogy with the methods for continuous systems.

The discrete-time and approximate discrete-time solutions of $\dot{x} = Ax + Bu$, $y = Cx$ in Section 11.5 have in effect also provided digital simulations of this state model:

$$x_{k+1} = Px_k + Qu_k \qquad y_k = Cx_k \qquad (11.55)$$

Here $P = \phi$ and $Q = \Delta$, where ϕ and Δ are given in (11.19) in terms of the time interval T and the matrices A and B. Such discrete state models may also describe single-variable or multivariable control algorithms. For example, for a higher-order difference equation

$$x_{k+n} + a_1 x_{k+n-1} + \cdots + a_n x_k = bu_k$$

a possible state vector and associated discrete state model are as follows:

$$\mathbf{x}_k = \begin{bmatrix} x_{1,k} \\ x_{2,k} \\ \vdots \\ x_{n,k} \end{bmatrix} = \begin{bmatrix} x_k \\ x_{k+1} \\ \vdots \\ x_{k+n-1} \end{bmatrix} \qquad \mathbf{x}_{k+1} = \begin{bmatrix} 0 & 1 & 0 & \cdots & & 0 \\ 0 & & 0 & 1 & & \\ \vdots & & & & \ddots & \\ & & & & & 1 \\ -a_n & \cdot & \cdot & \cdots & & -a_1 \end{bmatrix} \mathbf{x}_k + \begin{bmatrix} 0 \\ \cdot \\ \cdot \\ 0 \\ b \end{bmatrix} u_k \qquad (11.56)$$

Note that P is a companion matrix in this case. Discrete state models may also be derived from Z transfer functions as for the continuous case.

Example 11.8.1   Discrete state model of a Z transfer function

$$\frac{C(z)}{R(z)} = \frac{0.514z^{-1} + 0.4354z^{-2}}{1 - 0.536z^{-1} + 0.486z^{-2}} = \frac{0.514z + 0.4354}{z^2 - 0.536z + 0.486}$$

First, consider only the denominator.

$$\frac{W(z)}{R(z)} = \frac{1}{z^2 - 0.536z + 0.486} \qquad w_{k+2} - 0.536w_{k+1} + 0.486w_k = r_k$$

Define the state as in (11.56)

$$x_k = \begin{bmatrix} x_{1,k} \\ x_{2,k} \end{bmatrix} = \begin{bmatrix} w_k \\ w_{k+1} \end{bmatrix}$$

so that $x_{1,k+1} = x_{2,k}$. Then from the difference equation

$$x_{2,k+1} = -0.486x_{1,k} + 0.536x_{2,k} + r_k$$

and the state equation becomes

$$\begin{bmatrix} x_{1,k+1} \\ x_{2,k+1} \end{bmatrix} = \begin{bmatrix} 0 & 1 \\ -0.486 & 0.536 \end{bmatrix} \begin{bmatrix} x_{1,k} \\ x_{2,k} \end{bmatrix} + \begin{bmatrix} 0 \\ 1 \end{bmatrix} r_k$$

To obtain the output equation, it is seen that

$$C(z) = (0.514z + 0.4354)W(z) \qquad c_k = 0.514w_{k+1} + 0.4354w_k$$

In terms of the state vector, this yields

$$c_k = \begin{bmatrix} 0.4354 & 0.514 \end{bmatrix} \begin{bmatrix} x_{1,k} \\ x_{2,k} \end{bmatrix}$$

A Z transfer function matrix $G(z)$ can be derived from (11.55) as for the continuous case. From the definition of the Z transform, $Z[\mathbf{x}_{k+1}] = \sum_{k=0}^{\infty} \mathbf{x}_{k+1}z^{-k}$. Let $n = k + 1$ to obtain

$$Z[\mathbf{x}_{k+1}] = \sum_{n=1}^{\infty} \mathbf{x}_n z^{-n+1} = z \sum_{n=1}^{\infty} \mathbf{x}_n z^{-n} = z \left(\sum_{n=0}^{\infty} \mathbf{x}_n z^{-n} - \mathbf{x}_0 \right)$$

Using this, the transforms of (11.55) are

$$z\mathbf{X}(z) - z\mathbf{x}_0 = \mathbf{P}\mathbf{X}(z) + \mathbf{Q}\mathbf{U}(z) \qquad \mathbf{Y}(z) = \mathbf{C}\mathbf{X}(z)$$

Thus, analogous to (11.6) and (11.7),

$$\mathbf{Y}(z) = \mathbf{G}(z)\mathbf{U}(z) + \mathbf{C}(z\mathbf{I} - \mathbf{P})^{-1}z\mathbf{x}_0 \qquad (11.57)$$

where

$$\mathbf{G}(z) = \mathbf{C}(z\mathbf{I} - \mathbf{P})^{-1}\mathbf{Q} \qquad (11.58)$$

is the Z transfer function matrix and gives the input–output response for zero initial conditions, while the second term in (11.57) gives the response to initial conditions. For $\mathbf{C} = \mathbf{I}$, the state response is obtained.

$\mathbf{G}(z)$ consists of ordinary Z transfer functions and represents a stable system if all poles of all these transfer functions lie inside the unit circle of the z-plane. But all have the same poles, which are the roots of the characteristic equation

$$|z\mathbf{I} - \mathbf{P}| = 0 \qquad (11.59)$$

Hence follows the

Stability Theorem 11.8.1.   The system (11.55) is stable if and only if all eigenvalues [that is, all roots of (11.59)] lie inside the unit circle.

It is noted that, analogous to the Routh–Hurwitz criterion, the Jury theorem is available to determine whether any eigenvalues lie outside the unit circle.

Transient responses are also calculated by techniques equivalent to those for continuous systems. Solution of the discrete state model equation (11.55) gives

$$\mathbf{x}_1 = \mathbf{P}\mathbf{x}_0 + \mathbf{Q}\mathbf{u}_0, \qquad \mathbf{x}_2 = \mathbf{P}\mathbf{x}_1 + \mathbf{Q}\mathbf{u}_1 = \mathbf{P}^2\mathbf{x}_0 + \mathbf{P}\mathbf{Q}\mathbf{u}_0 + \mathbf{Q}\mathbf{u}_1, \dots$$

$$\mathbf{x}_k = \mathbf{P}^k\mathbf{x}_0 + \sum_{i=0}^{k-1} \mathbf{P}^{k-1-i}\mathbf{Q}\mathbf{u}_i \qquad (11.60)$$

The first term represents the response to initial conditions, and comparison with (11.57) yields the following analog of (11.18):

$$\mathbf{P}^k = Z^{-1}[(z\mathbf{I} - \mathbf{P})^{-1}z] \qquad (11.61)$$

The techniques for Z transform inversion can be used as well, in the manner of the use of Laplace transforms in Section 11.5. Solutions based on transformation of $\mathbf{P}$ to diagonal form are also applicable. Design techniques for discrete-time systems are in essence quite analogous to those for continuous systems, discussed in Chapter 12.

11.9 CONCLUSION

In this chapter the modeling and analysis of systems in state space has been introduced, and several new concepts have been defined that are important for both analysis and design. This material is the subject of a very extensive literature, ranging from the practical to the highly theoretical. The great advantage of the state-space approach is that the standard matrix formulation has allowed the development of standard computer programs suitable for the analysis and design of large systems.

PROBLEMS

11.1. Derive state-space models for systems described by the following transfer functions.

(a) $\dfrac{C(s)}{R(s)} = \dfrac{6}{s^3 + 3s^2 + 5s + 1}$ (b) $\dfrac{C(s)}{R(s)} = \dfrac{6(s + 1)}{s^3 + 3s^2 + 5s + 1}$

11.2. Derive the state-space models for systems described by the following transfer functions.

(a) $\dfrac{C(s)}{R(s)} = \dfrac{5}{(s + 1)(s^2 + 3s + 5)}$ (b) $\dfrac{C(s)}{R(s)} = \dfrac{s + 2}{s(s^2 + 6s + 1)}$

11.3. Derive state models for the following systems.

(a) $\dfrac{C(s)}{R(s)} = \dfrac{4s + 3}{3s^2 + 5s + 4}$ (b) $\dfrac{C(s)}{R(s)} = 2\dfrac{s + 3}{s + 10}$ (c) $\dfrac{C(s)}{R(s)} = \dfrac{s^2 + 3s + 4}{s^2 + 7s + 9}$

11.4. Obtain three alternative state models for the system in Fig. P11.4 with $G_1 = 5/(s + 1)$, $G_2 = 1/(s + 5)$, and $G_3 = 1/(s + 10)$:

(a) One where the system matrix **A** is a companion matrix.

(b) One where **A** is a diagonal matrix. (Use partial fraction expansion of $C(s)/R(s)\,[=(X_1/R) + (X_2/R) + (X_3/R)]$.)

(c) One where the physically meaningful and measurable outputs of the simple lag blocks are chosen to be the state variables.

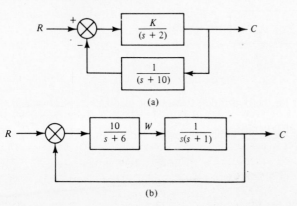

R → G_1 → G_2 → G_3 → C

Figure P11.4

11.5. Repeat Problem 11.4 for $G_1 = 1/(s + 3)$, $G_2 = 1/(s + 1)$, and $G_3 = 1/(s + 2)$.

11.6. Obtain state models for the system in Fig. P11.4, using a meaningful selection of state variables, if $G_3 = 1/[(s + 1)(s + 2)]$ is the plant and $G_1 G_2$ is a controller D given by:

(a) $D(s) = 1 + \dfrac{1.2}{s}$ (b) $D(s) = \dfrac{s + 3}{s + 10}$

11.7. Selecting meaningful state variables, derive state models for the systems in Fig. P11.7.

R → ⊕ (+, −) → $\dfrac{K}{(s + 2)}$ → C

$\dfrac{1}{(s + 10)}$

(a)

R → ⊗ → $\dfrac{10}{s + 6}$ → W → $\dfrac{1}{s(s + 1)}$ → C

(b)   **Figure P11.7**

11.8. Figure P11.8 shows a motor position servo with velocity feedback and a disturbance input D in addition to the reference input R. Obtain a state model.

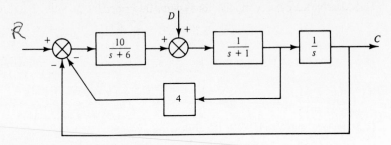

Figure P11.8

11.9. Determine state models for the systems shown in Fig. P11.9.

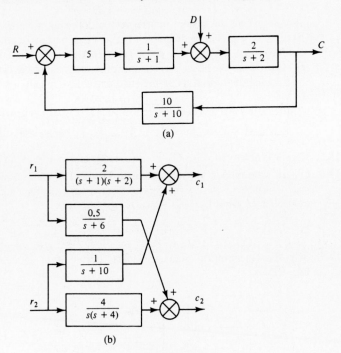

(a)

(b)

Figure P11.9

11.10. Derive a state model for the two-input, two-output feedback control system shown in Fig. P11.10.

11.11. The point model equations for a nuclear reactor with allowance for six groups of delayed neutrons are

$$\dot{n} = \frac{\delta k - \beta}{\Lambda} n + \sum_{i=1}^{6} \lambda_i C_i \qquad \dot{C}_i = \frac{\beta_i}{\Lambda} n - \lambda_i C_i \qquad i = 1, \ldots, 6$$

where n is the neutron density, C_i the concentration of the ith group precursor, and $\lambda_i C_i$ the growth rate of delayed neutrons due to decay of the ith group precursor. Put the equations in the form of a state model.

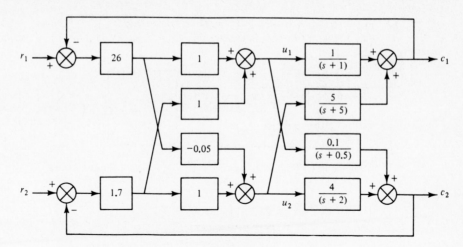

Figure P11.10

11.12. For the two-tank system in Fig. P11.12, q_1 and q_2 are liquid volume flow rates, h_1 and h_2 are liquid levels, where h_2 is the system output, and A_1 and A_2 are tank areas equal to $A_1 = 8, A_2 = 4$. The linearized models for flow through the resistors R_1 and R_2 can be taken to be $w_1 = (h_1 - h_2)/3$, $w_2 = h_2/2$. Derive a state model, selecting meaningful and measurable state variables.

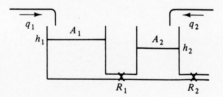

Figure P11.12

11.13. Derive a state model, with $y(t)$ as output, for the representation of a train shown in Fig. P11.13.

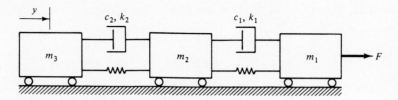

Figure P11.13

11.14. Derive a linearized state model for the system in Fig. P11.14, where h_1 and h_2 are levels, with h_2 the output, q_1 and q_2 are volume flow rates, A_1 and A_2 are areas, and R_1 and R_2 are linearized valve resistances.

11.15. Derive a state model to represent the field-controlled dc servomotor shown in Fig. P11.15. Choose θ, $\dot{\theta}$, and i_f as state variables. θ is the output.

11.16. Derive a state model representation, with displacement y of mass m as output, for the mechanical system in Fig. P11.16.

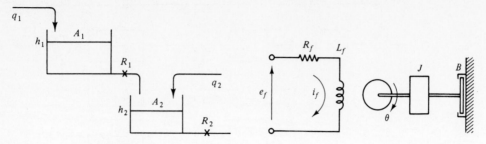

Figure P11.14   **Figure P11.15**

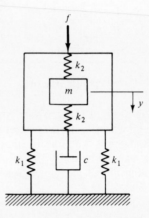

Figure P11.16

11.17. Derive two state models for the system with transfer function

$$\frac{C(s)}{R(s)} = \frac{s + 3}{s(s + 1)(s + 2)}$$

(a) One for which the system matrix is a companion matrix.

(b) One for which the system matrix is diagonal.
[Use partial fraction expansion of $C(s)/R(s)$ and define $C/R = X_1/R + X_2/R + X_3/R$.]

11.18. Determine the transfer function matrices $\mathbf{G}(s)$ corresponding to the following state models:

(a) $\dot{\mathbf{x}} = \begin{bmatrix} 0 & -1 \\ 0 & -2 \end{bmatrix} \mathbf{x} + \begin{bmatrix} 0 \\ 1 \end{bmatrix} u;$ $\mathbf{y} = \begin{bmatrix} 1 & 0 \\ 0 & 1 \end{bmatrix} \mathbf{x}$

(b) $\dot{\mathbf{x}} = \begin{bmatrix} 0 & 1 \\ -2 & -3 \end{bmatrix} \mathbf{x} + \begin{bmatrix} 0 & 1 \\ 1 & 0 \end{bmatrix} \mathbf{u};$ $\mathbf{y} = \begin{bmatrix} 1 & 0 \\ 1 & -1 \end{bmatrix} \mathbf{x}$

11.19. For

$$\dot{\mathbf{x}} = \begin{bmatrix} -3/4 & -1/4 \\ -1/4 & -3/4 \end{bmatrix} \mathbf{x} + \begin{bmatrix} 0 \\ 1 \end{bmatrix} u$$

(a) Determine stability.

(b) Find the transfer function relating x_1 and u.

11.20. A single-input, single-output system is described by the state model

$$\dot{\mathbf{x}} = \begin{bmatrix} -2 & 0 & 0 \\ -1 & -4 & 0 \\ -1 & -1 & -5 \end{bmatrix} \mathbf{x} + \begin{bmatrix} 1 \\ 1 \\ 1 \end{bmatrix} u \qquad y = [0 \quad 0 \quad 1] \mathbf{x}$$

Find the corresponding transfer function $Y(s)/U(s)$. Note that, instead of by inverting $s\mathbf{I} - \mathbf{A}$, this transfer function can be found also by transforming the individual state equations and eliminating variables between them.

11.21. In Problem 11.19:

(a) Evaluate the matrices $\boldsymbol{\phi}$ and $\boldsymbol{\Delta}$ in the discrete-time solution (11.19) for $T = 1$.

(b) Use them to find $\mathbf{x}(1)$, $\mathbf{x}(2)$, and $\mathbf{x}(3)$ for a unit step input with zero initial state.

11.22. For $\dot{\mathbf{x}} = \mathbf{Ax} + \mathbf{b}u$, $\mathbf{A} = \begin{bmatrix} 0 & -1 \\ 0 & -2 \end{bmatrix}$, $\mathbf{b} = \begin{bmatrix} 0 \\ 1 \end{bmatrix}$:

(a) Determine the transition matrix $\boldsymbol{\phi}(t) = e^{\mathbf{A}t} = L^{-1}(s\mathbf{I} - \mathbf{A})^{-1}$ by Laplace transforms.

(b) Verify part (a) using the series $\Sigma \mathbf{A}^k t^k / k!$.

(c) Find the state response to a unit step input.

11.23. For the system

$$\dot{\mathbf{x}} = \begin{bmatrix} 1 & -2 \\ 2 & -3 \end{bmatrix} \mathbf{x}$$

(a) Determine stability.

(b) Find the transition matrix $\boldsymbol{\phi} = e^{\mathbf{A}t} = L^{-1}(s\mathbf{I} - \mathbf{A})^{-1}$.

(c) Find the state response to initial conditions $x_1(0) = 2$, $x_2(0) = 1$.

11.24. For the system

$$\dot{\mathbf{x}} = \begin{bmatrix} -1 & 0 \\ 1 & -2 \end{bmatrix} \mathbf{x}$$

(a) Determine the transition matrix.

(b) Use it to evaluate matrix $\boldsymbol{\phi}$ in the discrete-time solution (11.19) for $T = 1$ and express the state response $\mathbf{x}(1)$, $\mathbf{x}(2)$, $\mathbf{x}(3)$ for an arbitrary initial condition $\mathbf{x}_0$.

11.25. Determine the output response of the system

$$\dot{\mathbf{x}} = \begin{bmatrix} -1 & 1 \\ 0 & -2 \end{bmatrix} \mathbf{x} + \begin{bmatrix} 0 \\ 1 \end{bmatrix} u \qquad y = \begin{bmatrix} 1 & 0 \\ 0 & 1 \end{bmatrix} \mathbf{x} \qquad \mathbf{x}_0 = \begin{bmatrix} +1 \\ +1 \end{bmatrix}$$

to the initial conditions and a unit step input.

11.26. Determine the response of the system

$$\dot{\mathbf{x}} = \begin{bmatrix} 0 & 1 \\ -6 & -5 \end{bmatrix} \mathbf{x} + \begin{bmatrix} 0 \\ 1 \end{bmatrix} u \qquad y = [1 \quad 0] \mathbf{x}$$

to a unit step input.

11.27. For the system

$$\dot{\mathbf{x}} = \begin{bmatrix} 0 & 1 \\ -6 & -5 \end{bmatrix} \mathbf{x}$$

determine the eigenvalues and the eigenvectors, and use these results to find the transition matrix.

11.28. For the system

$$\dot{\mathbf{x}} = \begin{bmatrix} 0 & 1 \\ -2 & -3 \end{bmatrix} \mathbf{x} + \begin{bmatrix} 0 \\ 2 \end{bmatrix} u \qquad y = \begin{bmatrix} 3 & 1 \end{bmatrix} \mathbf{x}$$

(a) Determine the eigenvalues and right and left (normalized) eigenvectors.
(b) Transform the state model to diagonal form.
(c) Express the modal decomposition of the state response to initial condition

$$\mathbf{x}_0 = \begin{bmatrix} 2 \\ 3 \end{bmatrix}.$$

(d) Verify part (c) by calculation of the transition matrix.

11.29. For the system

$$\dot{\mathbf{x}} = \begin{bmatrix} 0 & -1 \\ 1 & 0 \end{bmatrix} \mathbf{x}$$

determine the transition matrix by the use of eigenvalues and eigenvectors:

(a) By the standard method, using complex arithmetic.
(b) Using the approach of (11.35), with real arithmetic.

11.30. For the system of Problem 11.24:

(a) Find the eigenvalues and normalized right and left eigenvectors.
(b) Use them to find the transition matrix.
(c) Find the modal form of the system model.
(d) Find the modal decomposition of the response to initial conditions $\mathbf{x}_0$ and use it to verify part (b).

11.31. For the system

$$\dot{\mathbf{x}} = \begin{bmatrix} -3 & 2 \\ 4 & -5 \end{bmatrix} \mathbf{x} \qquad \mathbf{x}_0 = \begin{bmatrix} 5 \\ 3 \end{bmatrix}$$

(a) Find the eigenvalues and normalized right and left eigenvectors.
(b) Express the response to $\mathbf{x}_0$ by modal decomposition.

11.32. Verify the solution in Problem 11.31(b) by calculation of the transition matrix by two different techniques.

11.33. Determine the transition matrix of each of the following systems by the use of eigenvalues and eigenvectors.

(a) $\dot{\mathbf{x}} = \begin{bmatrix} 0 & 1 \\ 0 & -2 \end{bmatrix} \mathbf{x}$ (b) $\dot{\mathbf{x}} = \begin{bmatrix} -1 & 1 \\ 0 & -2 \end{bmatrix} \mathbf{x}$

11.34. For a system with transfer function

$$\frac{Y(s)}{W(s)} = \frac{1}{s(s+1)(s+2)}$$

(a) Obtain a state model.
(b) Find normalized right and left eigenvectors.
(c) Determine the transition matrix.

11.35. Check the controllability of the following pairs $(\mathbf{A}, \mathbf{B})$.

(a) $\begin{bmatrix} 0 & 1 \\ -2 & -3 \end{bmatrix}, \begin{bmatrix} 0 \\ 2 \end{bmatrix}$ (b) $\begin{bmatrix} 1 & 1 \\ 1 & 0 \end{bmatrix}, \begin{bmatrix} 1 \\ 0 \end{bmatrix}$

(c) $\begin{bmatrix} 0 & 1 & 0 \\ 0 & 0 & 1 \\ 0 & -1 & -2 \end{bmatrix}, \begin{bmatrix} 0 \\ 1 \\ 1 \end{bmatrix}$ **(d)** $\begin{bmatrix} 1 & 0 & 0 \\ 0 & 0 & 1 \\ 1 & -2 & -1 \end{bmatrix}, \begin{bmatrix} 0 \\ 0 \\ 1 \end{bmatrix}$

11.36. Check the controllability and stabilizability of the following pairs $(\mathbf{A}, \mathbf{B})$.

(a) $\begin{bmatrix} -3 & 1 \\ -2 & 1.5 \end{bmatrix}, \begin{bmatrix} 0 \\ 1 \end{bmatrix}$ **(b)** $\begin{bmatrix} 0 & 1 \\ 2 & -1 \end{bmatrix}, \begin{bmatrix} 1 \\ 1 \end{bmatrix}$

11.37. Check the controllability and stabilizability of the following pairs $(\mathbf{A}, \mathbf{B})$.

(a) $\begin{bmatrix} -3 & 1 \\ -2 & 1.5 \end{bmatrix}, \begin{bmatrix} 1 \\ 4 \end{bmatrix}$ **(b)** $\begin{bmatrix} 0 & 1 & 0 \\ 0 & 0 & 1 \\ a & b & c \end{bmatrix}, \begin{bmatrix} 0 \\ 0 \\ 1 \end{bmatrix}$

11.38. Check the observability of the model with state equation

$$\dot{\mathbf{x}} = \begin{bmatrix} 0 & 1 & 0 \\ 0 & 0 & 1 \\ -6 & -11 & -6 \end{bmatrix} \mathbf{x} + \begin{bmatrix} 0 \\ 0 \\ 1 \end{bmatrix} u$$

for the following output equations.
(a) $y = [3 \quad 2 \quad 1]\mathbf{x}$ **(b)** $y = [4 \quad 5 \quad 1]\mathbf{x}$

11.39. Noting that the system matrix in Problem 11.38 is a companion matrix and that $s^3 + 6s^2 + 11s + 6 = (s + 1)(s + 2)(s + 3)$:
(a) Derive the system transfer function for the output equation in Problem 11.38(b). Remember that the output equation shows numerator dynamics.
(b) Is this transfer function reducible?
(c) Does the state model represent a minimal realization of this transfer function?
(d) Relate these results to an earlier statement that pole–zero cancellation can be used to improve output response but does not consider the response to initial conditions.

11.40. A system has the discrete transfer function

$$\frac{C(z)}{R(z)} = \frac{2z + 6}{4z^3 - z}$$

Derive an equivalent difference state model.

11.41. Determine the stability of a system described by the difference state model $\mathbf{x}_{k+1} = \mathbf{P}\mathbf{x}_k + \mathbf{Q}\mathbf{u}_k$ if

$$\mathbf{P} = \begin{bmatrix} 0.4 & 0.2 \\ 0.3 & -0.1 \end{bmatrix}$$

11.42. Derive a difference state model to represent the algorithm $x_{k+2} = 3x_{k+1} - 2x_k + u_k$ and use it to determine the stability of this algorithm.

11.43. Derive a difference equation state model to represent the control algorithm

$$x_k = -4x_{k-1} - 6x_{k-2} + 2u_{k-1} + u_{k-2}$$

and use it to determine its stability.

11.44. Derive a difference state model to represent the algorithm $x_k = -5x_{k-1} - 3x_{k-2} + u_k$ and use it to determine its stability.

11.45. Determine the vector difference state model for digital simulation of the continuous system

$$\dot{\mathbf{x}} = \begin{bmatrix} 0 & 1 \\ -2 & -3 \end{bmatrix} \mathbf{x} + \begin{bmatrix} 0 \\ 1 \end{bmatrix} u$$

if the computation interval $T = 1$. Also determine the stability of the simulation.

11.46. Referring to Problem 11.21, determine a difference state model for digital simulation of the continuous system

$$\dot{\mathbf{x}} = \begin{bmatrix} -3/4 & -1/4 \\ -1/4 & -3/4 \end{bmatrix} \mathbf{x} + \begin{bmatrix} 0 \\ 1 \end{bmatrix} u$$

if the computation interval $T = 1$, and check the stability of the simulation.

11.47. Derive a digital simulation, in the form of a difference state model, for the system $G(s) = K/[(s + a)(s + b)] = C(s)/U(s)$ by finding first an equivalent continuous state model.

11.48. Determine a closed-form solution of the difference equation $x_{n+2} - 3x_{n+1} + 2x_n = u_n$ for $u_n = 0$ and the initial conditions $x_0 = 1$, $x_1 = 0.5$ by the use of a difference state model and Z transforms.

12

Introduction to
State-Space Design

12.1 INTRODUCTION

First, an example will be given to illustrate how state-space design relates to design based on transfer functions. Then the two major approaches to design in state space are discussed:

1. *Pole assignment:* The closed-loop eigenvalues are placed in specified locations.
2. *Optimal control:* A specified mathematical performance criterion is minimized.

Several techniques for pole assignment will be discussed, including *modal control,* based on the modal form in Chapter 11. Optimal control is introduced, including the famous optimal regulator problem.

In the original formulations, only the regulator problem is considered, to bring nonzero states caused by transient disturbances to the origin of the state space. Then, for systems with disturbances and inputs that have constant nonzero steady-state values, multivariable generalizations are given of *integral control* and *feedforward control.* In this introduction to a very large area, one goal is to provide a framework for further study. Such a framework must provide for the possibility that a fundamental assumption of the basic pole assignment and optimal control techniques is not satisfied. This assumption is that all states are available for feedback. If this is not the case, or if it is simply not practical to measure all states, one possibility is to design a *dynamic observer* to obtain an estimate of the state vector from the available output vector. Alternatively, constant or dynamic output feedback can be designed, that is, feedback from the output instead of the state.

12.2 STATE FEEDBACK AND POLE ASSIGNMENT

To illustrate the relation between state-space design and the techniques of earlier chapters, the motor position servo in Figure 12.1 will be considered. It uses the field-controlled dc motor of Example 11.3.3 and Fig. 11.1. The variables θ_o, $\dot{\theta}_o$, i_f, and e_f are shaft position, shaft velocity, field current, and field voltage. The reference input is taken to be zero, and the purpose of control is to return the output to zero after deviations due to unspecified transient disturbances.

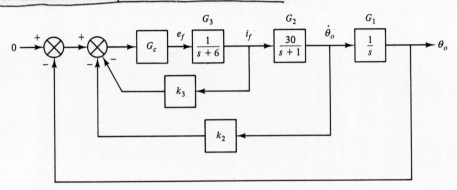

Figure 12.1 Motor position servo with state feedback.

First, let the feedback links k_2 and k_3 be absent. Then this is a classical unity feedback system with compensator G_c and plant transfer function

$$G = \frac{\theta_o}{E_f} = \frac{30}{s(s + 1)(s + 6)} = \frac{30}{s^3 + 7s^2 + 6s} \tag{12.1}$$

Root loci or Bode plots can be used to design a P control $G_c(s) = K_c$ or PD control $G_c(s) = K_c + K_d s$. The system characteristic equation $1 + G_c G = 0$ is then

$$s^3 + 7s^2 + (6 + 30K_d)s + 30K_c = 0$$

K_c and K_d provide a certain freedom in choosing locations for the closed-loop poles, but according to (6.9) the sum of these poles is constrained to be -7.

Now let the feedback loops with k_2 and k_3 be introduced. The system characteristic equation may be found, for example, by noting that the three feedback loops can be replaced by a single feedback from i_f:

$$k_3 + k_2 \frac{30}{s + 1} + \frac{30}{s(s + 1)} = \frac{k_3 s^2 + (k_3 + 30k_2)s + 30}{s(s + 1)}$$

Using this, the system characteristic equation is found to be

$$s^3 + 7s^2 + 6s + G_c[k_3 s^2 + (k_3 + 30k_2)s + 30]$$
$$= s^3 + (7 + G_c k_3)s^2 + (6 + 30G_c k_2 + G_c k_3)s + 30G_c = 0 \tag{12.2}$$

The effect of adding the feedback loops with k_2 and k_3 can now be seen:

1. Constant values of G_c, k_2, and k_3 can always be found to make the coefficients in this equation equal to arbitrary preassigned values.
2. Thus the closed-loop poles can be placed at arbitrary preassigned real and complex conjugate locations by constant values of G_c, k_2, and k_3.

For example, let the desired pole positions be -7 and $-3 \pm 3j$. The desired characteristic equation is then

$$(s + 7)(s + 3 - 3j)(s + 3 + 3j) = s^3 + 13s^2 + 60s + 126 = 0 \qquad (12.3)$$

because the desired poles or eigenvalues are its roots. Equating coefficients in (12.2) and (12.3) yields

$$7 + G_c k_3 = 13 \qquad 6 + 30 G_c k_2 + G_c k_3 = 60 \qquad 30 G_c = 126$$

or

$$G_c = 4.2 \qquad k_3 = 1/0.7 \qquad k_2 = 0.8/2.1 \qquad (12.4)$$

Thus a feedback configuration using only constant gains has been found for which the closed-loop poles have assigned values.

Observe that this was achieved without the use of state-space concepts. However, a state-space formulation and solution help to clarify the nature of this feedback and to generalize the result.

In Fig. 12.1, θ_o, $\dot{\theta}_o$, and i_f are the state variables in the state model (11.5a) for the field-controlled dc motor:

$$\mathbf{x} = \begin{bmatrix} \theta_o \\ \dot{\theta}_o \\ i_f \end{bmatrix} \qquad \dot{\mathbf{x}} = \begin{bmatrix} 0 & 1 & 0 \\ 0 & -1 & 30 \\ 0 & 0 & -6 \end{bmatrix} \mathbf{x} + \begin{bmatrix} 0 \\ 0 \\ 1 \end{bmatrix} e_f \qquad (12.5)$$

Examination of Fig. 12.1 shows that the control (12.4) is in fact a constant gain feedback from all state variables:

$$\begin{aligned} e_f &= -G_c(\theta_o + k_2 \dot{\theta}_o + k_3 i_f) \\ &= -[G_c \quad G_c k_2 \quad G_c k_3]\mathbf{x} = -[4.2 \quad 1.6 \quad 6]\mathbf{x} \end{aligned} \qquad (12.6)$$

In the general case, the state equation (12.5) and the *state feedback* control (12.6) are

$$\dot{\mathbf{x}} = \mathbf{A}\mathbf{x} + \mathbf{B}\mathbf{u} \qquad \mathbf{u} = -\mathbf{K}\mathbf{x} \qquad (12.7)$$

and the system can be represented by the classical feedback control configuration in Fig. 12.2. As in Fig. 12.1, the system inputs are taken to be zero, and the purpose of control is to change the initial state $\mathbf{x}_0$ to $\mathbf{x} = \mathbf{0}$. The plant is described by its state-space model, and it is assumed that all state variables are available for use in feedback.

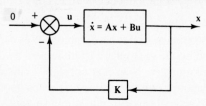

Figure 12.2   State feedback.

The closed-loop system equation corresponding to (12.7) is

$$\dot{\mathbf{x}} = (\mathbf{A} - \mathbf{B}\mathbf{K})\mathbf{x} \qquad (12.8)$$

So the *closed-loop matrix* is $(\mathbf{A} - \mathbf{B}\mathbf{K})$ and the closed-loop characteristic equation is

$$|s\mathbf{I} - (\mathbf{A} - \mathbf{B}\mathbf{K})| = 0 \qquad (12.9)$$

The desired characteristic equation is found, like (12.3), from the desired closed-loop eigenvalues. If these are $\hat{\lambda}_1, \ldots, \hat{\lambda}_n$, then the desired characteristic equation is

$$(s - \hat{\lambda}_1)(s - \hat{\lambda}_2) \cdots (s - \hat{\lambda}_n) = s^n + a_{dn}s^{n-1} + \cdots + a_{d2}s + a_{d1} = 0 \qquad (12.10)$$

The elements of $\mathbf{K}$ are found by equating coefficients in (12.9) and (12.10).

Example 12.2.1   Motor Position Servo

Equation (12.5) gives a state model of the field-controlled motor in terms of measurable and meaningful state variables:

$$\mathbf{x} = \begin{bmatrix} \theta_o \\ \dot\theta_o \\ i_f \end{bmatrix} \qquad \dot{\mathbf{x}} = \begin{bmatrix} 0 & 1 & 0 \\ 0 & -1 & 30 \\ 0 & 0 & -6 \end{bmatrix} \mathbf{x} + \begin{bmatrix} 0 \\ 0 \\ 1 \end{bmatrix} u = \mathbf{A}\mathbf{x} + \mathbf{b}u$$

Now assume a state feedback

$$u = -\mathbf{k}'\mathbf{x} = -[k_1 \quad k_2 \quad k_3]\mathbf{x}$$

Then

$$\mathbf{A} - \mathbf{b}\mathbf{k}' = \begin{bmatrix} 0 & 1 & 0 \\ 0 & -1 & 30 \\ -k_1 & -k_2 & -6 - k_3 \end{bmatrix}$$

and the actual closed-loop characteristic equation is

$$|s\mathbf{I} - (\mathbf{A} - \mathbf{b}\mathbf{k}')| = \begin{vmatrix} s & -1 & 0 \\ 0 & s+1 & 30 \\ k_1 & k_2 & s+6+k_3 \end{vmatrix}$$

$$= s^3 + (7 + k_3)s^2 + (6 + 30k_2 + k_3)s + 30k_1 = 0$$

As in (12.3), the desired characteristic equation is

$$s^3 + 13s^2 + 60s + 126 = 0$$

Equating coefficients in both equations yields

$$k_1 = 4.2 \qquad k_2 = 1.6 \qquad k_3 = 6$$

This agrees with (12.6).

Constant-gain state feedback proved enough to assign the closed-loop system eigenvalues to arbitrary locations. The following theorem shows the generality of this result and is of fundamental importance.

Feedback Control Theorem 12.2.1   Any specified set of closed-loop eigenvalues can be obtained by state feedback with $\mathbf{K}$ consisting of constant gains if and only if the pair $(\mathbf{A}, \mathbf{B})$ is controllable.

The "only if" part is obvious, because uncontrollable modes cannot be changed. The "if" part, already illustrated by the example, will be proved by design of $\mathbf{K}$ using the techniques presented in this chapter. Note that constant gains can achieve complete pole assignment, while dynamic compensation was often found necessary just for adequate performance in classical single-input, single-output design, because there is feedback from all states instead of only from a single output.

The direct technique of equating coefficients in (12.9) and (12.10) becomes laborious beyond four or five state variables. The alternative techniques presented next make use of special *canonical forms* of system matrices, such as the companion and diagonal forms. First, some remarks are in order.

1. It should be recognized that a feedback matrix $\mathbf{K}$ in $\mathbf{u} = -\mathbf{K}\mathbf{x}$ can be implemented in a system with unity feedback from an output. For example, the state feedback (12.6) is realized by a unity feedback system in Fig. 12.1. This is done by incorporating the controller $G_c = k_1$ from the state that represents the output in the forward path and scaling the other gains accordingly.

2. For *multiple-input systems,* pole assignment by modal control, based on transformation to the diagonal canonical form, will provide a method for design. A common alternative here is to reduce them in effect to single-input systems by fixing the ratios of the inputs according to $\mathbf{u} = \boldsymbol{\alpha}w$, where $\boldsymbol{\alpha}$ is a vector of constants and w the new single input. Then

$$\dot{\mathbf{x}} = \mathbf{A}\mathbf{x} + \mathbf{B}\mathbf{u} = \mathbf{A}\mathbf{x} + (\mathbf{B}\boldsymbol{\alpha})w = \mathbf{A}\mathbf{x} + \mathbf{b}w$$

and $\boldsymbol{\alpha}$ must be chosen such that $(\mathbf{A}, \mathbf{b})$ is controllable.

3. The eigenvalues in pole assignment design could be chosen in locations in the left-half plane far from the imaginary axis, making the speed of response arbitrarily fast. But this would require large control inputs and actuator capacities. A tank, say, can be filled arbitrarily fast if the supply flow rate and the rate at which it can be changed are large enough. However, this implies a high cost of control. The choice of desired eigenvalues in pole assignment techniques implies a trade-off between performance and cost of control. One approach is to locate a dominating pair for satisfactory bandwidth and damping and choose the other eigenvalues to the left of this pair. It is necessary to observe that the choice of desired eigenvalues for more complex systems generally involves considerable trial and error until the resulting system is satisfactory from the point of view of the transient responses of states and control inputs to, say, step inputs.

4. For simpler systems, Graham and Lathrop (1953) have proposed coefficients for a closed-loop transfer function

$$\frac{C(s)}{R(s)} = \frac{a_1}{s^n + a_n s^{n-1} + \cdots + a_2 s + a_1}$$

which will result in a well-damped response and a small overshoot for step inputs. Note that the form of the transfer function implies zero steady-state errors after step inputs and so a type 1 system. Table 12.2.1 shows the proposed denominator polynomials and their roots, the desired eigenvalues.

TABLE 12.2.1   DESIRED CLOSED-LOOP CHARACTERISTIC POLYNOMIALS AND THEIR FACTORED FORMS

$s + \omega_0$

$s^2 + 1.4\omega_0 s + \omega_0^2$

$s^3 + 1.75\omega_0 s^2 + 2.15\omega_0^2 s + \omega_0^3$

$s^4 + 2.1\omega_0 s^3 + 3.4\omega_0^2 s^2 + 2.7\omega_0^3 s + \omega_0^4$

$s^5 + 2.8\omega_0 s^4 + 5.0\omega_0^2 s^3 + 5.5\omega_0^3 s^2 + 3.4\omega_0^4 s + \omega_0^5$

$s^6 + 3.25\omega_0 s^5 + 6.6\omega_0^2 s^4 + 8.6\omega_0^3 s^3 + 7.45\omega_0^4 s^2 + 3.95\omega_0^5 s + \omega_0^6$

$d + 1$

$d + 0.707 \pm j0.707$

$(d + 0.7081)(d + 0.521 \pm j1.068)$

$(d + 0.424 \pm j1.263)(d + 0.626 \pm j0.4141)$

$(d + 0.8955)(d + 0.376 \pm j1.292)(d + 0.5758 \pm j0.5339)$

$(d + 0.3099 \pm j1.263)(d + 0.5805 \pm j0.7828)(d + 0.7346 \pm j0.2873)$

ω_0 is a free parameter, chosen to satisfy a specification on bandwidth.

$d = s/\omega_0 \qquad (d + a \pm jb) \equiv (d + a + jb)(d + a - jb)$

12.3 POLE ASSIGNMENT USING COMPANION MATRICES

Consider a single-input plant described by the nth-order differential equation

$$\frac{d^n y}{dt^n} + a_n \frac{d^{n-1} y}{dt^{n-1}} + \cdots + a_2 \frac{dy}{dt} + a_1 y = c_m \frac{d^{m-1} u}{dt^{m-1}} + \cdots + c_2 \frac{du}{dt} + c_1 u$$

$$(12.11)$$

As in Example 11.2.2, this can be represented by the state model

$$\begin{cases} \dot{\mathbf{x}} = \mathbf{A}\mathbf{x} + \mathbf{b}u \\ y = \mathbf{C}\mathbf{x} \end{cases}$$

with :

$$\mathbf{C} = [c_1 \quad c_2 \quad \cdots \quad c_m \quad 0 \quad \cdots \quad 0]$$

$$\mathbf{A} = \begin{bmatrix} 0 & 1 & \cdots & 0 \\ \vdots & & & \vdots \\ 0 & & & 1 \\ -a_1 & -a_2 & \cdots & -a_n \end{bmatrix} \qquad \mathbf{b} = \begin{bmatrix} 0 \\ \vdots \\ 0 \\ 1 \end{bmatrix} \qquad (12.12)$$

The open-loop eigenvalues are the eigenvalues of the companion matrix $\mathbf{A}$. From (12.11), they are also the roots of the plant characteristic equation

$$s^n + a_n s^{n-1} + \cdots + a_2 s + a_1 = 0 \qquad (12.13)$$

This has the same coefficients as the last row of $\mathbf{A}$, with opposite signs and in reverse order. Now assume a state feedback

$$u = -\mathbf{k}'\mathbf{x} = -[k_1 \quad k_2 \quad \cdots \quad k_n]\mathbf{x} \qquad (12.14)$$

where the prime identifies the transpose of a column vector $\mathbf{k}$. The closed-loop system matrix $\mathbf{A} - \mathbf{b}\mathbf{k}'$ is then again a companion matrix:

$$\mathbf{A} - \begin{bmatrix} 0 \\ \vdots \\ 0 \\ 1 \end{bmatrix} [k_1 \quad \cdots \quad k_n] = \begin{bmatrix} 0 & 1 & \cdots & 0 \\ \vdots & & & \vdots \\ & & & 1 \\ -a_1 - k_1 & -a_2 - k_2 & \cdots & -a_n - k_n \end{bmatrix} \qquad (12.15)$$

So, by analogy with (12.12) and (12.13), the closed-loop eigenvalues must be the roots of the closed-loop characteristic equation

$$s^n + (a_n + k_n)s^{n-1} + \cdots + (a_2 + k_2)s + (a_1 + k_1) = 0 \qquad (12.16)$$

Let the desired closed-loop eigenvalues be $\hat{\lambda}_1, \ldots, \hat{\lambda}_n$, so the desired characteristic equation is

$$(s - \hat{\lambda}_1)(s - \hat{\lambda}_2) \cdots (s - \hat{\lambda}_n) = s^n + a_{dn} s^{n-1} + \cdots + a_{d2} s + a_{d1} = 0$$

$$(12.17)$$

Equation (12.16) will be equal to this if

$$\mathbf{k}' = [a_{d1} - a_1 \quad a_{d2} - a_2 \quad \cdots \quad a_{dn} - a_n] \qquad (12.18)$$

If the state model is in companion form, this provides a simple design technique for pole assignment.

Example 12.3.1

$$\dot{\mathbf{x}} = \begin{bmatrix} 0 & 1 & 0 \\ 0 & 0 & 1 \\ -6 & -11 & -6 \end{bmatrix} \mathbf{x} + \begin{bmatrix} 0 \\ 0 \\ 3 \end{bmatrix} u$$

With $u = -\mathbf{k}'\mathbf{x}$, the elements in the last row of $\mathbf{A} - \mathbf{bk}'$ are $-6 - 3k_1$, $-11 - 3k_2$, and $-6 - 3k_3$. If $\hat{\lambda}_1 = -2$, $\hat{\lambda}_2 = -3$, and $\hat{\lambda}_3 = -4$, the desired characteristic equation is $(s + 2)(s + 3)(s + 4) = s^3 + 9s^2 + 26s + 24 = 0$, so the desired elements in the last row of $\mathbf{A} - \mathbf{bk}'$ are -24, -26, and -9. Hence

$$\mathbf{k}' = [6 \quad 5 \quad 1]$$

If, as is usual, the state model is not in companion form, computer algorithms are available for a coordinate transformation to this form and for transforming back after the feedback gains have been found. A method in which such a transformation arises implicitly is to derive a state model in companion form directly, design the state feedback, and then replace the feedback from less desirable state variables by feedback from more attractive ones. This is illustrated next for the motor control example in Section 12.2.

Example 12.3.2   Motor Position Servo

Equation (12.5) gives a state model of the field-controlled motor in terms of measurable and meaningful state variables:

$$\mathbf{x} = \begin{bmatrix} \theta_o \\ \dot{\theta}_o \\ i_f \end{bmatrix} \qquad \dot{\mathbf{x}} = \begin{bmatrix} 0 & 1 & 0 \\ 0 & -1 & 30 \\ 0 & 0 & -6 \end{bmatrix} \mathbf{x} + \begin{bmatrix} 0 \\ 0 \\ 1 \end{bmatrix} e_f$$

A model in <u>companion form</u>, or as it is frequently called the *phase-variable canonical form,* is given in (11.5b) of Example 11.3.3:

$$\hat{\mathbf{x}} = \begin{bmatrix} \theta_o \\ \dot{\theta}_o \\ \ddot{\theta}_o \end{bmatrix} \qquad \dot{\hat{\mathbf{x}}} = \begin{bmatrix} 0 & 1 & 0 \\ 0 & 0 & 1 \\ 0 & -6 & -7 \end{bmatrix} \hat{\mathbf{x}} + \begin{bmatrix} 0 \\ 0 \\ 30 \end{bmatrix} e_f$$

With the desired characteristic equation (12.3), the design equations are

$$e_f = -[\hat{k}_1 \quad \hat{k}_2 \quad \hat{k}_3]\hat{\mathbf{x}}$$
$$30\hat{k}_1 = 126 \qquad 6 + 30\hat{k}_2 = 60 \qquad 7 + 30\hat{k}_3 = 13$$

yielding

$$[\hat{k}_1 \quad \hat{k}_2 \quad \hat{k}_3] = \frac{1}{30}[126 \quad 54 \quad 6]$$

To replace the feedback from $\ddot{\theta}_o$ by an equivalent feedback from i_f, the second equation in the state model for $\dot{\mathbf{x}}$ gives

$$\dot{x}_2 = \ddot{\theta}_o = \hat{x}_3 = -\dot{\theta}_o + 30i_f = -x_2 + 30x_3$$

The state vectors $\hat{\mathbf{x}}$ and $\mathbf{x}$ are therefore related by

$$\hat{\mathbf{x}} = \begin{bmatrix} 1 & 0 & 0 \\ 0 & 1 & 0 \\ 0 & -1 & 30 \end{bmatrix} \mathbf{x}$$

It follows that the feedback from $\mathbf{x}$ is

$$[k_1 \quad k_2 \quad k_3] = \frac{1}{30}[126 \quad 54 \quad 6]\begin{bmatrix} 1 & 0 & 0 \\ 0 & 1 & 0 \\ 0 & -1 & 30 \end{bmatrix}$$
$$= [4.2 \quad 1.6 \quad 6]$$

This agrees with the result in (12.6).

12.4 OPTIMAL CONTROL AND THE OPTIMAL REGULATOR PROBLEM

Optimal control and pole assignment are the principal techniques for design in state space. In optimal control, the control is sought that gives the best trade-off between performance and cost of control. Optimal design can be carried out for both open- and closed-loop systems. A standard form of the closed-loop case is to seek the control that minimizes the value of a *performance index J* of the form

$$J = \frac{1}{2} \int_0^\infty (\mathbf{x}'\mathbf{Q}\mathbf{x} + \mathbf{u}'\mathbf{R}\mathbf{u}) \, dt \tag{12.19}$$

$$\mathbf{Q} = \text{diag}(q_i) \qquad \mathbf{R} = \text{diag}(r_i)$$

Q and **R** are *weighting matrices* and are usually diagonal. The terms $\mathbf{x}'\mathbf{Q}\mathbf{x}$ and $\mathbf{u}'\mathbf{R}\mathbf{u}$ of the integrand are *quadratic forms* that measure, respectively, the performance and the cost of control. These terms have scalar values:

$$\mathbf{x}'\mathbf{Q}\mathbf{x} = [x_1 \quad \cdots \quad x_n] \begin{bmatrix} q_1 & \cdots & 0 \\ \vdots & & \vdots \\ 0 & \cdots & q_n \end{bmatrix} \begin{bmatrix} x_1 \\ \vdots \\ x_n \end{bmatrix} = [x_1 \quad \cdots \quad x_n] \begin{bmatrix} q_1 x_1 \\ \vdots \\ q_n x_n \end{bmatrix}$$

$$= \sum_{i=1}^n q_i x_i^2 \qquad \mathbf{u}'\mathbf{R}\mathbf{u} = \sum_{j=1}^r r_j u_j^2 \tag{12.20}$$

$$J = \frac{1}{2} \int_0^\infty \left(\sum_i q_i x_i^2 + \sum_j r_j u_j^2 \right) dt \tag{12.21}$$

Thus the optimal control minimizes a weighted sum of areas under x_i^2 and u_j^2 curves such as that shown in Fig. 12.3. In the present formulation, $\mathbf{x} = \mathbf{0}$ is the desired state, and the best response is equivalent to the minimum weighted area under the x_i^2 curves. The choice of the elements of **Q** and **R** allows the relative weighting of individual state variables and individual control inputs, as well as the relative weighting of state variables and control inputs. Increasing the r_j relative to the q_i increases the weighting on the control and has the effect of reducing the control inputs at the expense of the response.

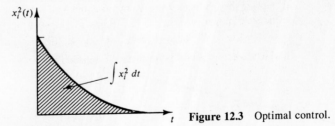

Figure 12.3 Optimal control.

The criterion (12.19) could weight the output response instead of the state response if **Q** is redefined as follows:

$$\mathbf{y} = \mathbf{C}\mathbf{x} \qquad \mathbf{Q} = \mathbf{C}'\mathbf{Q}_o\mathbf{C} \qquad \mathbf{x}'\mathbf{Q}\mathbf{x} = \mathbf{y}'\mathbf{Q}_o\mathbf{y} \tag{12.22}$$

It should be noted that the choice of **Q** and **R** can require considerable trial and error until the transient responses are satisfactory.

Consider now the design of state feedback to minimize J for the system

$$\dot{\mathbf{x}} = \mathbf{A}\mathbf{x} + \mathbf{B}\mathbf{u} \qquad \mathbf{y} = \mathbf{C}\mathbf{x} \tag{12.23}$$

The control equation and the closed-loop system model are

$$\mathbf{u} = -\mathbf{K}\mathbf{x} \qquad \dot{\mathbf{x}} = (\mathbf{A} - \mathbf{B}\mathbf{K})\mathbf{x} \qquad \mathbf{K} = \{k_{ij}\} \tag{12.24}$$

Substituting the control (12.24) into J of (12.19) yields

$$J = \frac{1}{2} \int_0^\infty \mathbf{x}'(\mathbf{Q} + \mathbf{K}'\mathbf{R}\mathbf{K})\mathbf{x} \, dt \tag{12.25}$$

It can be shown that if the closed-loop system is stable [that is, $\mathbf{x}(\infty) = \mathbf{0}$], then there exists a constant, real, symmetric, and positive definite matrix $\mathbf{P}$ such that

$$\frac{d}{dt}(\mathbf{x}'\mathbf{P}\mathbf{x}) = -\mathbf{x}'(\mathbf{Q} + \mathbf{K}'\mathbf{R}\mathbf{K})\mathbf{x} \qquad \mathbf{P} = \{p_{ij}\} \tag{12.26}$$

Positive definite forms are discussed in Chapter 14. $\mathbf{P}$ is positive definite if all its principal minors are greater than zero; that is,

$$p_{11} > 0 \qquad \begin{bmatrix} p_{11} & p_{12} \\ p_{12} & p_{22} \end{bmatrix} > 0 \quad \cdots \quad \begin{bmatrix} p_{11} & p_{12} & \cdots & p_{1n} \\ \vdots & & & \vdots \\ p_{1n} & p_{2n} & & p_{nn} \end{bmatrix} > 0 \tag{12.27}$$

Then, substituting (12.26) into (12.25) gives

$$J = -\frac{1}{2} \int_0^\infty \frac{d}{dt}(\mathbf{x}'\mathbf{P}\mathbf{x}) \, dt = -\frac{1}{2} \mathbf{x}'\mathbf{P}\mathbf{x} \Big|_0^\infty = -\frac{1}{2} \mathbf{x}'(\infty)\mathbf{P}\mathbf{x}(\infty) + \frac{1}{2} \mathbf{x}'(0)\mathbf{P}\mathbf{x}(0)$$

or

$$J = \frac{1}{2} \mathbf{x}'(0)\mathbf{P}\mathbf{x}(0) \tag{12.28}$$

$\mathbf{P}$ may be found from (12.26), (12.24), and the property that the transpose of a product equals the product of the transposes in reverse order:

$$\frac{d}{dt}(\mathbf{x}'\mathbf{P}\mathbf{x}) = \dot{\mathbf{x}}'\mathbf{P}\mathbf{x} + \mathbf{x}'\mathbf{P}\dot{\mathbf{x}} = \mathbf{x}'[(\mathbf{A} - \mathbf{B}\mathbf{K})'\mathbf{P} + \mathbf{P}(\mathbf{A} - \mathbf{B}\mathbf{K})]\mathbf{x}$$

For this to be equal to the right side of (12.26) for all $\mathbf{x}$ requires that

$$(\mathbf{A} - \mathbf{B}\mathbf{K})'\mathbf{P} + \mathbf{P}(\mathbf{A} - \mathbf{B}\mathbf{K}) = -\mathbf{Q} - \mathbf{K}'\mathbf{R}\mathbf{K} \tag{12.29}$$

This is a matrix *Liapunov equation*, of which the symmetric and positive definite solution $\mathbf{P}$ is required. The minimum of J in (12.28) is found by substituting the solution $\mathbf{P}$ and using the conditions

$$\frac{\partial J}{\partial k_{ij}} = 0 \tag{12.30}$$

where the k_{ij} are the elements of $\mathbf{K}$.

A modified criterion, to remove the dependence on initial conditions, is

$$J = \text{tr } \mathbf{P} \tag{12.31}$$

discussed in Section 12.8. This criterion aims to minimize the trace of $\mathbf{P}$, that is, the sum of its diagonal elements (Appendix A).

Analytical solution is practical only for the very simplest problems and when the number of gains k_{ij} to be optimized is small. Usually, numerical parameter optimization routines are required and algorithms to solve the Liapunov equation at each stage of the hillclimbing process. This approach can also be used with output feedback, discussed later, when only certain outputs are available for feedback.

For full state feedback, a preferred alternative to solution via the Liapunov equation is that based on the matrix Riccati equation. There are several ways, omitted here, of proving the following key result from the Liapunov equation (12.29).

> **Optimal Regulator Theorem 12.4.1**   The optimal control is a constant-gain state feedback
>
> $$\mathbf{u}_{\text{opt}} = -\mathbf{Kx} \qquad \mathbf{K} = \mathbf{R}^{-1}\mathbf{B}'\mathbf{P} \tag{12.32}$$
>
> where $\mathbf{P}$ is a symmetric and positive definite matrix obtained by solution of the algebraic matrix *Riccati equation*
>
> $$\mathbf{PA} + \mathbf{A}'\mathbf{P} + \mathbf{Q} - \mathbf{PBR}^{-1}\mathbf{B}'\mathbf{P} = \mathbf{0} \tag{12.33}$$
>
> and, under mild restrictions, gives a stable closed-loop system.

In very simple cases the Riccati equation can be solved directly, but usually computer solution is required. A number of computer routines for this purpose are available. This optimal solution is independent of the initial conditions.

Example Based on the Riccati Equation

For illustration, consider the position control system in Fig. 12.4. The plant transfer function $G(s)$ and a state model with state variables x_1 and x_2 are

$$G(s) = \frac{A}{s(s+a)} \qquad \mathbf{x} = \begin{bmatrix} x_1 \\ x_2 \end{bmatrix} \qquad \dot{\mathbf{x}} = \begin{bmatrix} 0 & 1 \\ 0 & -a \end{bmatrix}\mathbf{x} + \begin{bmatrix} 0 \\ A \end{bmatrix}u \tag{12.34}$$

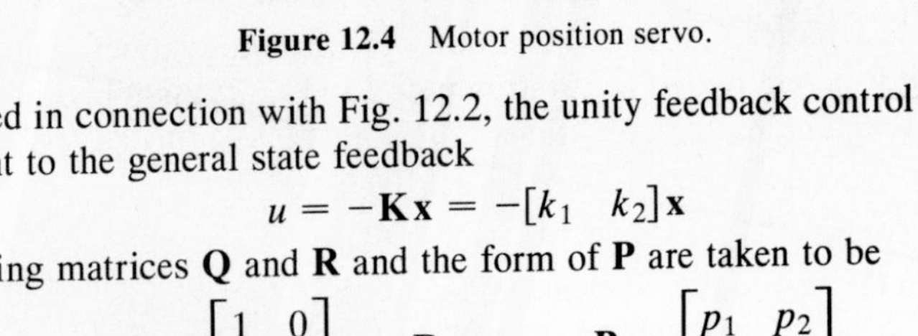

Figure 12.4   Motor position servo.

As discussed in connection with Fig. 12.2, the unity feedback control configuration is equivalent to the general state feedback

$$u = -\mathbf{Kx} = -[k_1 \quad k_2]\mathbf{x} \tag{12.35}$$

The weighting matrices $\mathbf{Q}$ and $\mathbf{R}$ and the form of $\mathbf{P}$ are taken to be

$$\mathbf{Q} = \begin{bmatrix} 1 & 0 \\ 0 & q \end{bmatrix} \qquad \mathbf{R} = r \qquad \mathbf{P} = \begin{bmatrix} p_1 & p_2 \\ p_2 & p_3 \end{bmatrix} \tag{12.36}$$

The Riccati equation becomes

$$\mathbf{P}\begin{bmatrix} 0 & 1 \\ 0 & -a \end{bmatrix} + \begin{bmatrix} 0 & 0 \\ 1 & -a \end{bmatrix}\mathbf{P} + \begin{bmatrix} 1 & 0 \\ 0 & q \end{bmatrix} - \mathbf{P}\begin{bmatrix} 0 \\ A \end{bmatrix}r^{-1}[0 \quad A]\mathbf{P} = 0$$

or

$$\begin{bmatrix} 0 & p_1 - ap_2 \\ p_1 - ap_2 & 2p_2 - 2ap_3 \end{bmatrix} + \begin{bmatrix} 1 & 0 \\ 0 & q \end{bmatrix} - \begin{bmatrix} A^2 p_2^2 & A^2 p_2 p_3 \\ A^2 p_2 p_3 & A^2 p_3^2 \end{bmatrix} = \begin{bmatrix} 0 & 0 \\ 0 & 0 \end{bmatrix}$$

This represents four scalar equations, of which, due to symmetry, only three are different. These are shown in Table 12.4.1, which also gives their solutions p_1, p_2, and p_3 and the optimal control (12.32). Only the positive root $p_2 = \sqrt{r}/A$ of the first equation is admissible, because the negative root leads to a solution for p_1 that does not satisfy (12.27). The last of the three equations then yields a quadratic for p_3:

$$p_3^2 + \frac{2ar}{A^2} p_3 - \frac{2r\sqrt{r}}{A^3} - \frac{qr}{A^2} = 0$$

Only the positive root satisfies (12.27). Substituting p_2 and p_3 into the second Riccati equation gives

$$p_1 = \frac{a}{A}\sqrt{r} + \frac{A}{\sqrt{r}}\left(-\frac{ar}{A^2} + \frac{\sqrt{r}}{A} \sqrt{\frac{a^2 r}{A^2} + \frac{2\sqrt{r}}{A} + q} \right)$$

This simplifies to the expression in Table 12.4.1. The optimal control (12.32) yields

$$\mathbf{K} = r^{-1}\mathbf{B'P} = \frac{A}{r}[\, p_2 \quad p_3 \,]$$

This leads to the state feedback in the table. It is evident that for more complex problems numerical solution becomes necessary.

Example Based on the Liapunov Equation

Again for the system of Fig. 12.4 and equations (12.34) to (12.36), the closed-loop system matrix is

$$(\mathbf{A} - \mathbf{BK}) = \begin{bmatrix} 0 & 1 \\ 0 & -a \end{bmatrix} - \begin{bmatrix} 0 \\ A \end{bmatrix}[k_1 \quad k_2] = \begin{bmatrix} 0 & 1 \\ -Ak_1 & -Ak \end{bmatrix}$$

where

$$Ak = a + Ak_2 \qquad k = k_2 + \frac{a}{A}$$

The Liapunov equation (12.29) becomes

$$\begin{bmatrix} -Ak_1 p_2 & -Ak_1 p_3 \\ p_1 - Akp_2 & p_2 - Akp_3 \end{bmatrix} + \begin{bmatrix} -Ak_1 p_2 & p_1 - Akp_2 \\ -Ak_1 p_3 & p_2 - Akp_3 \end{bmatrix}$$

$$= -\begin{bmatrix} 1 + k_1^2 r & k_1 k_2 r \\ k_1 k_2 r & q + k_2^2 r \end{bmatrix}$$

This represents three different scalar equations, given in Table 12.4.1 with their solutions for p_1, p_2, and p_3. Here p_2 is obtained from the first equation, followed by solution of p_3 from the last equation. Substituting p_2 and p_3 into the second equation gives

$$p_1 = \frac{k}{2k_1} + \frac{1}{2}\left(k_2 + \frac{a}{A} \right)k_1 r + \frac{k_1 q}{2k} + \frac{k_1 k_2^2 r}{2k} + \frac{1}{2Ak} + \frac{k_1^2 r}{2Ak} - k_1 k_2 r$$

TABLE 12.4.1 OPTIMAL CONTROL, FIGURE 12.4

Scalar Riccati equations:

$$1 - A^2 r^{-1} p_2^2 = 0 \qquad p_1 - ap_2 - A^2 r^{-1} p_2 p_3 = 0$$
$$2p_2 - 2ap_3 + q - A^2 r^{-1} p_3^2 = 0$$

Solutions of scalar Riccati equations:

$$p_1 = \sqrt{\frac{a^2 r}{A^2} + \frac{2r^{1/2}}{A} + q} \qquad p_2 = \frac{1}{A} \sqrt{r}$$

$$p_3 = -\frac{ar}{A^2} + \frac{\sqrt{r}}{A} \sqrt{\frac{a^2 r}{A^2} + \frac{2r^{1/2}}{A} + q}$$

Riccati optimal control $u = -\mathbf{K}\mathbf{x}$:

$$u = -\left[\frac{1}{\sqrt{r}} \qquad -\frac{a}{A} + \frac{1}{\sqrt{r}} \sqrt{\frac{a^2 r}{A^2} + \frac{2r^{1/2}}{A} + q} \right] \mathbf{x}$$

Scalar Liapunov equations:

$$2Ak_1 p_2 = 1 + k_1^2 r \qquad p_1 - Akp_2 - Ak_1 p_3 = -k_1 k_2 r$$
$$2p_2 - 2Akp_3 = -q - k_2^2 r \qquad (Ak \equiv a + Ak_2)$$

Solutions of scalar Liapunov equations:

$$p_1 = \frac{k_1 q}{2k} + \frac{k_1 k_2^2 r}{2k} + \frac{1}{2Ak} + \frac{k_1^2 r}{2Ak} + \frac{k}{2k_1} + \frac{ak_1 r}{2A} - \frac{k_1 k_2 r}{2}$$
$$p_2 = \frac{1}{2Ak_1} + \frac{k_1 r}{2A} \qquad p_3 = \frac{q}{2Ak} + \frac{k_2^2 r}{2Ak} + \frac{1}{2A^2 kk_1} + \frac{k_1 r}{2A^2 k}$$

Partial derivatives for $a = 0$ $(k = k_2)$:

$$\frac{dp_2}{dk_2} = 0 \qquad \frac{dp_2}{dk_1} = \frac{1}{2A}\left(r - \frac{1}{k_1^2} \right) \qquad \frac{dp_3}{dk_1} = \frac{1}{2A^2 k_2}\left(r - \frac{1}{k_1^2} \right)$$

$$\frac{dp_3}{dk_2} = \frac{1}{2A}\left(r - \frac{q}{k_2^2} \right) - \frac{1}{2A^2 k_2^2}\left(k_1 r + \frac{1}{k_1} \right)$$

$$\frac{dp_1}{dk_1} = \frac{q}{2k_2} + \frac{k_1 r}{Ak_2} - \frac{k_2}{2k_1^2} \qquad \frac{dp_1}{dk_2} = \frac{1}{2k_1} - \frac{1}{2Ak_2^2}\left(Ak_1 q + 1 + k_1^2 r \right)$$

from which the entry in the table follows. In the Liapunov approach, these solutions must be substituted into J of (12.28) or (12.31) and the optimal gains k_1 and k_2 found from the conditions $\partial J/\partial k_i = 0$. In general, this requires the partial derivatives $\partial p_i/\partial k_j$. Table 12.4.1 shows these for the special case $a = 0$. As these suggest, the expressions for $a \neq 0$ are laborious. The need for numerical methods is indicated even for this relatively simple system, unless the plant or the control is constrained to simplify the scalar Liapunov equations.

Example 12.4.1   Liapunov Approach for $G(s) = 1/s^2$

This is a special case $a = 0$, $A = 1$ in Table 12.4.1. Let the initial conditions be $\mathbf{x}_0' = \begin{bmatrix} 1 & 0 \end{bmatrix}$. Then, ignoring the factor $\frac{1}{2}$ in (12.28), it is found that

$$J = \begin{bmatrix} 1 & 0 \end{bmatrix} \mathbf{P} \begin{bmatrix} 1 \\ 0 \end{bmatrix} = p_1$$

So the conditions $\partial J/\partial k_i = 0$ for the optimum are, using Table 12.4.1,

$$\frac{dp_1}{dk_1} = \frac{q}{2k_2} + \frac{k_1 r}{k_2} - \frac{k_2}{2k_1^2} = 0$$

$$\frac{dp_1}{dk_2} = \frac{1}{2k_1} - \frac{1}{2k_2^2}(1 + qk_1 + rk_1^2) = 0$$

These reduce to

$$k_2^2 = k_1^2(q + 2rk_1) \qquad k_2^2 = k_1(1 + qk_1 + rk_1^2)$$

Eliminating k_2 yields k_1, and substituting k_1 then gives k_2:

$$k_1 = \sqrt{1/r} \qquad k_2 = \sqrt{(q + 2\sqrt{r})/r} \tag{12.37}$$

Example 12.4.2   Constrained Optimum for Example 12.4.1

Suppose that a constraint $k_2 = k_1$ is imposed. From Table 12.4.1 with $a = 0$, $A = 1$, and $k = k_2 = k_1$, J and the condition for the optimum are

$$J = p_1 = \frac{q}{2} + \frac{1}{2} + \frac{1}{2k_1} + \frac{1}{2}k_1 r \qquad \frac{dJ}{dk_1} = \frac{dp_1}{dk_1} = \frac{1}{2}r - \frac{1}{2k_1^2} = 0$$

So the constrained optimum is $k_1 = k_2 = \sqrt{1/r}$.

12.5 MODAL CONTROL FOR POLE ASSIGNMENT USING STATE FEEDBACK

In this alternative to the pole assignment techniques of Sections 12.2 and 12.3, use is made of the relations associated with the modal transformation in Section 11.6:

$$\mathbf{x} = \mathbf{U}\mathbf{z} \qquad \mathbf{A}\mathbf{u}_i = \lambda_i \mathbf{u}_i$$
$$\boxed{\mathbf{v}_i'\mathbf{u}_j = 0 \quad (i \neq j) \qquad \mathbf{v}_i'\mathbf{u}_i = 1 \qquad \mathbf{U}^{-1} = \mathbf{V}'}$$
$$\mathbf{z} = \mathbf{U}^{-1}\mathbf{x} = \mathbf{V}'\mathbf{x} \qquad z_i = \mathbf{v}_i'\mathbf{x} \tag{12.38}$$

Here $\mathbf{u}_i$ is the ith column of $\mathbf{U}$, $\mathbf{v}_i'$ is the ith row of $\mathbf{V}'$, and z_i is the ith modal variable. Now consider a single-input system with state feedback

$$\dot{\mathbf{x}} = \mathbf{A}\mathbf{x} + \mathbf{b}w \qquad w = -\mathbf{k}'\mathbf{x} \tag{12.39}$$

and let the control signal w be generated by feeding back the ith and jth modal variables z_i, z_j, with gains k_i, k_j:

$$w = -k_i z_i - k_j z_j \tag{12.40}$$

$w = -k_i \mathbf{v}_i' \mathbf{x} - k_j \mathbf{v}_j' \mathbf{x}$

Using (12.38), this may be written as

$$w = -\mathbf{k}'\mathbf{x} \qquad \mathbf{k}' = k_i\mathbf{v}_i' + k_j\mathbf{v}_j' \tag{12.41}$$

The closed-loop system equation is then

$$\dot{\mathbf{x}} = (\mathbf{A} - k_i\mathbf{b}\mathbf{v}_i' - k_j\mathbf{b}\mathbf{v}_j')\mathbf{x} \tag{12.42}$$

Postmultiply the closed-loop system matrix in this equation by $\mathbf{u}_k$ with $k \neq i, j$. Using the modal properties in (12.38) then yields

$$(\mathbf{A} - k_i\mathbf{b}\mathbf{v}_i' - k_j\mathbf{b}\mathbf{v}_j')\mathbf{u}_k = \mathbf{A}\mathbf{u}_k = \lambda_k\mathbf{u}_k \qquad k \neq i, j \tag{12.43}$$

By definition, this means that the eigenvalues and eigenvectors λ_k and $\mathbf{u}_k$, $k \neq i, j$, of the open-loop system are also eigenvalues and eigenvectors of the closed loop and have not been affected by the feedback from the ith and jth modal variables. (This property of $\mathbf{u}_k$ is not true for $\mathbf{v}_k$.)

The importance of this result is that it shows that *modal control* is possible; that is, feedback from selected modal variables can be used to change only these eigenvalues to more desirable locations, without affecting the others.

To determine the feedback gains needed, it is noted first that the new eigenvalues and eigenvectors $\hat{\lambda}_i$, $\hat{\lambda}_j$, $\hat{\mathbf{u}}_i$, $\hat{\mathbf{u}}_j$ satisfy, by definition,

$$\begin{aligned}(\mathbf{A} - k_i\mathbf{b}\mathbf{v}_i' - k_j\mathbf{b}\mathbf{v}_j')\hat{\mathbf{u}}_i = \hat{\lambda}_i\hat{\mathbf{u}}_i \\ (\mathbf{A} - k_i\mathbf{b}\mathbf{v}_i' - k_j\mathbf{b}\mathbf{v}_j')\hat{\mathbf{u}}_j = \hat{\lambda}_j\hat{\mathbf{u}}_j\end{aligned} \tag{12.44}$$

Since any n-dimensional vector can be written as a series in the n modal components, let

$$\hat{\mathbf{u}}_i = \sum_{k=1}^{n} q_{ik}\mathbf{u}_k \qquad \hat{\mathbf{u}}_j = \sum_{k=1}^{n} q_{jk}\mathbf{u}_k \qquad \mathbf{b} = \sum_{k=1}^{n} p_k\mathbf{u}_k \tag{12.45}$$

The terms in the first of (12.44) can then be expressed as follows:

$$\mathbf{A}\hat{\mathbf{u}}_i = \mathbf{A}\sum q_{ik}\mathbf{u}_k = \sum q_{ik}\mathbf{A}\mathbf{u}_k = \sum \lambda_k q_{ik}\mathbf{u}_k$$

$$-k_i\mathbf{b}\mathbf{v}_i'\hat{\mathbf{u}}_i = -k_i\mathbf{b}\sum q_{ik}\mathbf{v}_i'\mathbf{u}_k = -k_i\mathbf{b}q_{ii} = -k_i q_{ii}\sum p_k\mathbf{u}_k$$

$$-k_j\mathbf{b}\mathbf{v}_j'\hat{\mathbf{u}}_i = -k_j\mathbf{b}\sum q_{ik}\mathbf{v}_j'\mathbf{u}_k = -k_j\mathbf{b}q_{ij} = -k_j q_{ij}\sum p_k\mathbf{u}_k$$

$$\hat{\lambda}_i\hat{\mathbf{u}}_i = \hat{\lambda}_i\sum q_{ik}\mathbf{u}_k$$

Substituting these into the first of (12.44) and premultiplying by $\mathbf{v}_i'$ yields the first equation below, and premultiplying by $\mathbf{v}_j'$ yields the second if the properties $\mathbf{v}_i'\mathbf{u}_j = 0$ and $\mathbf{v}_i'\mathbf{u}_i = 1$ are used:

$$\begin{aligned}(\lambda_i - \hat{\lambda}_i - k_i p_i)q_{ii} - k_j p_i q_{ij} = 0 \\ -k_i p_j q_{ii} + (\lambda_j - \hat{\lambda}_i - k_j p_j)q_{ij} = 0\end{aligned}$$

Nontrivial solutions for q_{ii} and q_{ij} from this set require that the determinant of the coefficients be zero:

$$(\lambda_i - \hat{\lambda}_i)(\lambda_j - \hat{\lambda}_i) - k_i p_i(\lambda_j - \hat{\lambda}_i) - k_j p_j(\lambda_i - \hat{\lambda}_i) = 0$$

Rearranging this yields the first of (12.46). The second of (12.46) is obtained by an analogous derivation from the second of (12.44).

$$\boxed{\frac{k_i p_i}{\lambda_i - \hat{\lambda}_i} + \frac{k_j p_j}{\lambda_j - \hat{\lambda}_i} = 1} \qquad \boxed{\frac{k_i p_i}{\lambda_i - \hat{\lambda}_j} + \frac{k_j p_j}{\lambda_j - \hat{\lambda}_j} = 1} \tag{12.46}$$

Solving this set for k_i and k_j and using $\mathbf{v}_i'\mathbf{b} = \sum p_k\mathbf{v}_i'\mathbf{u}_k = p_i$, $\mathbf{v}_j'\mathbf{b} = p_j$ yields the following theorem.

Modal Control Theorem 12.5.1

(i) In (12.41), the gains k_i, k_j to move λ_i, λ_j to $\hat{\lambda}_i$, $\hat{\lambda}_j$ are

$$k_i = \frac{(\lambda_i - \hat{\lambda}_i)(\lambda_i - \hat{\lambda}_j)}{(\mathbf{v}_i'\mathbf{b})(\lambda_i - \lambda_j)} \qquad k_j = \frac{(\lambda_j - \hat{\lambda}_j)(\lambda_j - \hat{\lambda}_i)}{(\mathbf{v}_j'\mathbf{b})(\lambda_j - \lambda_i)} \qquad (12.47)$$

(ii) To move only λ_i to $\hat{\lambda}_i$ ($\hat{\lambda}_j = \hat{\lambda}_j$, $k_j = 0$):

$$k_i = \frac{\lambda_i - \hat{\lambda}_i}{\mathbf{v}_i'\mathbf{b}} \qquad (12.48)$$

(iii) It can be shown that the generalized form of (12.47) for moving $\lambda_1, \ldots,$ λ_m to $\hat{\lambda}_1, \ldots, \hat{\lambda}_m$, using feedback from $z_1, \ldots, z_m$, is

$$k_i = \frac{\displaystyle\prod_{j=1}^{m}(\lambda_i - \hat{\lambda}_j)}{(\mathbf{v}_i'\mathbf{b})\displaystyle\prod_{j=1,\neq i}^{m}(\lambda_i - \lambda_j)} \qquad i = 1, \ldots, m \qquad (12.49)$$

Real poles can be moved simultaneously or sequentially. If λ_i has been moved to $\hat{\lambda}_i$ and λ_j is to be moved next, the eigenvector $\hat{\mathbf{v}}_j$ for the system resulting from the move of λ_i must be determined first; only $\mathbf{u}_j$ is unaffected. This will be illustrated in Example 12.5.3. However, as illustrated in Example 12.5.2, eigenvalue moves to or from complex conjugate locations can only occur in pairs.

Example 12.5.1

$$\dot{\mathbf{x}} = \begin{bmatrix} -3/4 & -1/4 \\ -1/4 & -3/4 \end{bmatrix}\mathbf{x} + \begin{bmatrix} 0 \\ 1 \end{bmatrix}u$$

$|\lambda\mathbf{I} - \mathbf{A}| = \lambda^2 + 1.5\lambda + 0.5$ yields the open-loop eigenvalues $\lambda_1 = -1$, $\lambda_2 = -0.5$. Let it be desired to speed up the response by designing state feedback that will make the (closed-loop) system eigenvalues equal to $\hat{\lambda}_1 = -1$, $\hat{\lambda}_2 = -2$. Hence only $z_2 = -\mathbf{v}_2'\mathbf{x}$ is required since only the second eigenvalue needs to be moved.

$$\mathbf{A}'\mathbf{v}_2 = \lambda_2\mathbf{v}_2 \qquad \begin{bmatrix} -3/4 & -1/4 \\ -1/4 & -3/4 \end{bmatrix}\begin{bmatrix} v_{12} \\ v_{22} \end{bmatrix} = -0.5\begin{bmatrix} v_{12} \\ v_{22} \end{bmatrix} \qquad \mathbf{v}_2 = \begin{bmatrix} 1 \\ -1 \end{bmatrix}$$

or any multiple. Then, using (12.48),

$$k_2 = \frac{-0.5 + 2}{[1 \quad -1]\begin{bmatrix} 0 \\ 1 \end{bmatrix}} = -1.5$$

and $\mathbf{k}' = k_2\mathbf{v}_2' = [-1.5 \quad 1.5]$. Verify that -1 and -2 are the eigenvalues of $(\mathbf{A} - \mathbf{b}\mathbf{k}')$. Note that normalization of $\mathbf{v}_2$ is not necessary since it occurs in the numerator and denominator of $\mathbf{k}'$.

Example 12.5.2

$$\dot{\mathbf{x}} = \begin{bmatrix} 0 & -2 \\ 1 & -2 \end{bmatrix}\mathbf{x} + \begin{bmatrix} 0 \\ 1 \end{bmatrix}u$$

$|\lambda\mathbf{I} - \mathbf{A}| = \lambda^2 + 2\lambda + 2$ yields the open-loop eigenvalues $\lambda_1 = -1 + j$, $\lambda_2 = -1 - j$. This is a complex pair, so both must be moved simultaneously. Let the desired locations for the closed-loop system be $\hat{\lambda}_1 = -4 + 4j$, $\hat{\lambda}_2 = -4 - 4j$. The equations

$\mathbf{A}'\mathbf{v}_1 = (-1 + j)\mathbf{v}_1$, $\mathbf{A}'\mathbf{v}_2 = (-1 - j)\mathbf{v}_2$ yield the eigenvectors $\mathbf{v}'_1 = \begin{bmatrix} 1 & -1 + j \end{bmatrix}$, $\mathbf{v}'_2 = \begin{bmatrix} 1 & -1 - j \end{bmatrix}$. Then, from (12.47), the required gains are

$$k_1 = \frac{-1 + j + 4 - 4j}{-1 + j}\frac{-1 + j + 4 + 4j}{-1 + j + 1 + j} = -3\,\frac{3 + 5j}{2j}\frac{j}{j} = -7.5 + 4.5j$$

$$k_2 = -7.5 - 4.5j$$

Equation (12.41) now gives the feedback

$$\begin{aligned}
\mathbf{k}' &= k_1\mathbf{v}'_1 + k_2\mathbf{v}'_2 \\
&= (-7.5 + 4.5j)\begin{bmatrix} 1 & -1 + j \end{bmatrix} + (-7.5 - 4.5j)\begin{bmatrix} 1 & -1 - j \end{bmatrix} = \begin{bmatrix} -15 & 6 \end{bmatrix}
\end{aligned}$$

The closed-loop system matrix

$$\mathbf{A} - \mathbf{b}\mathbf{k}' = \begin{bmatrix} 0 & -2 \\ 1 & -2 \end{bmatrix} - \begin{bmatrix} 0 \\ 1 \end{bmatrix}\begin{bmatrix} -15 & 6 \end{bmatrix} = \begin{bmatrix} 0 & -2 \\ 16 & -8 \end{bmatrix}$$

may be verified to have the desired eigenvalues.

Multiple Input Systems: $\dot{\mathbf{x}} = \mathbf{A}\mathbf{x} + \mathbf{B}\mathbf{u}$

One approach for systems with more than one input is to reduce them in effect to single-input systems, as discussed in the last part of Section 12.2.

A feature of systems with multiple inputs is that the design that places all eigenvalues in specified locations is not unique. For a plant with r inputs, $\mathbf{B}$ consists of r columns and may be written as $\mathbf{B} = \begin{bmatrix} \mathbf{b}_1 & \mathbf{b}_2 & \cdots & \mathbf{b}_r \end{bmatrix}$. The state equation and the state feedback control can then be expressed as

$$\begin{aligned}
\dot{\mathbf{x}} = \mathbf{A}\mathbf{x} + \mathbf{B}\mathbf{u} &= \mathbf{A}\mathbf{x} + \mathbf{b}_1 u_1 + \cdots + \mathbf{b}_r u_r \\
\mathbf{u} = -\mathbf{K}\mathbf{x} \qquad u_i &= -\mathbf{k}'_i\mathbf{x} \qquad i = 1, \ldots, r
\end{aligned} \tag{12.50}$$

where u_i is the ith element of $\mathbf{u}$ and $\mathbf{k}'_i$ the ith row of $\mathbf{K}$. From Theorems 12.2.1 and 11.7.1, input u_i can be used to place the eigenvalues for which $\mathbf{v}'_j\mathbf{b}_i \neq 0$. A mode is usually controllable from more than one input, so the design is not unique. The alternatives have the same closed-loop poles, but the transient responses may differ widely. This can be interpreted in terms of different zero locations, and more advanced techniques seek to optimize these locations.

An approach to modal control that is sometimes convenient and can be used also for single-input systems is to move (real) eigenvalues one at a time, as illustrated in the following example of a system with two inputs.

Example 12.5.3 Two-Tank Level Control System

Figure 12.5 shows a process consisting of two interconnected tanks. R_0, R_1, and R_2 are linearized valve resistance values and h_1, h_2 and A_1, A_2 represent tank levels and surface

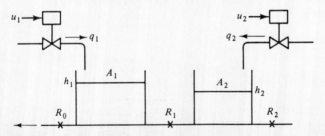

Figure 12.5 Two-tank system.

areas, respectively. The flow rates q_1 and q_2 into the tanks are controlled by the signals u_1 and u_2 via valves and actuators and are modeled by the linearized relations

$$q_1 = K_1 u_1 \qquad q_2 = K_2 u_2$$

From the results in Chapter 2, the differential equations for the levels in the tanks are:

$$\dot{h}_1 = -\frac{1}{A_1}\left(\frac{1}{R_1} + \frac{1}{R_0}\right)h_1 + \frac{1}{A_1 R_1}h_2 + \frac{1}{A_1}q_1$$

$$\dot{h}_2 = \frac{1}{A_2 R_1}h_1 - \frac{1}{A_2}\left(\frac{1}{R_1} + \frac{1}{R_2}\right)h_2 + \frac{1}{A_2}q_2$$

Let $A_1 = 1, A_2 = 0.5, R_0 = 1, R_1 = 0.5, R_2 = 2, K_1 = 1$, and $K_2 = 0.5$. Then, eliminating q_1 and q_2, the equations become

$$\dot{h}_1 = -3h_1 + 2h_2 + u_1 \qquad \dot{h}_2 = 4h_1 - 5h_2 + u_2$$

These correspond to the following state model:

$$\mathbf{x} = \begin{bmatrix} x_1 \\ x_2 \end{bmatrix} = \begin{bmatrix} h_1 \\ h_2 \end{bmatrix} \qquad \mathbf{u} = \begin{bmatrix} u_1 \\ u_2 \end{bmatrix} \qquad \dot{\mathbf{x}} = \begin{bmatrix} -3 & 2 \\ 4 & -5 \end{bmatrix}\mathbf{x} + \begin{bmatrix} 1 & 0 \\ 0 & 1 \end{bmatrix}\mathbf{u}$$

$|\lambda\mathbf{I} - \mathbf{A}| = \lambda^2 + 8\lambda + 7$ yields the open-loop eigenvalues $\lambda_1 = -1, \lambda_2 = -7$. The eigenvectors $\mathbf{v}_i$ are found from $\mathbf{A}'\mathbf{v}_1 = -1\mathbf{v}_1, \mathbf{A}'\mathbf{v}_2 = -7\mathbf{v}_2$ to be $\mathbf{v}_1' = [2 \quad 1]$, $\mathbf{v}_2' = [1 \quad -1]$. Since $\mathbf{v}_i'\mathbf{b}_j \neq 0$ for all i, j, either input can be used to place either or both eigenvalues. Let u_1 be used to move only λ_1 to a desired position $\hat{\lambda}_1 = -4$ for the closed-loop system. Using (12.48), the required control is $u_1 = -\mathbf{k}_1'\mathbf{x}$, with

$$\mathbf{k}_1' = k_1\mathbf{v}_1' = \frac{-1 + 4}{[2 \quad 1]\begin{bmatrix} 1 \\ 0 \end{bmatrix}}[2 \quad 1] = [3 \quad 1.5]$$

The new system is $\dot{\mathbf{x}} = \mathbf{A}\mathbf{x} + \mathbf{b}_1 u_1 + \mathbf{b}_2 u_2 = \hat{\mathbf{A}}\mathbf{x} + \mathbf{b}_2 u_2$, where

$$\hat{\mathbf{A}} = \mathbf{A} - \mathbf{b}_1\mathbf{k}_1' = \begin{bmatrix} -3 & 2 \\ 4 & -5 \end{bmatrix} - \begin{bmatrix} 1 \\ 0 \end{bmatrix}[3 \quad 1.5] = \begin{bmatrix} -6 & 0.5 \\ 4 & -5 \end{bmatrix}$$

This has the eigenvalues $\hat{\lambda}_1 = -4, \lambda_2 = -7$, as expected. Input u_2 is chosen next to move λ_2 to a desired position $\hat{\lambda}_2 = -9$. As was noted earlier, $\mathbf{u}_2$ is not affected but the vector $\mathbf{v}_2$ changes when λ_1 is moved. Therefore, the eigenvector $\hat{\mathbf{v}}_2$ of $\hat{\mathbf{A}}$ is found from $\hat{\mathbf{A}}'\hat{\mathbf{v}}_2 = -7\hat{\mathbf{v}}_2$: $\hat{\mathbf{v}}_2' = [-4 \quad 1]$. Using (12.48), the required control is $u_2 = -\mathbf{k}_2'\mathbf{x}$, with

$$\mathbf{k}_2' = k_2\hat{\mathbf{v}}_2' = \frac{-7 + 9}{[-4 \quad 1]\begin{bmatrix} 0 \\ 1 \end{bmatrix}}[-4 \quad 1] = [-8 \quad 2]$$

The total feedback $\mathbf{u} = -\mathbf{K}\mathbf{x}$ and the closed-loop matrix, which may be verified to have the eigenvalues -4 and -9, are given by

$$\mathbf{K} = \begin{bmatrix} \mathbf{k}_1' \\ \mathbf{k}_2' \end{bmatrix} = \begin{bmatrix} 3 & 1.5 \\ -8 & 2 \end{bmatrix} \qquad \mathbf{A} - \mathbf{B}\mathbf{K} = \begin{bmatrix} -3 & 2 \\ 4 & -5 \end{bmatrix} - \begin{bmatrix} 3 & 1.5 \\ -8 & 2 \end{bmatrix}$$

$$= \begin{bmatrix} -6 & 0.5 \\ 12 & -7 \end{bmatrix}$$

12.6 MULTIVARIABLE INTEGRAL CONTROL

Thus far, only the response to an initial condition $\mathbf{x}(0)$ has been considered, with $\mathbf{x} = \mathbf{0}$ as the desired state. Reference inputs $\mathbf{y}_{ref}$ (that is, the desired output $\mathbf{y}$) or disturbance inputs $\mathbf{w}$ were implied to be transient only, with zero steady-state values. In practice, however, this is generally not the case. Now, in single-variable design the use of integral control proved very effective to obtain zero steady-state error (that is, $y \to y_{ref}$ as $t \to \infty$) if y_{ref} and w have constant steady-state values. The generalization of integral control to multivariable systems is of great practical importance and is considered in this section.

First, it is necessary to augment the state-space model to allow for the presence of disturbance inputs $\mathbf{w}$:

$$\dot{\mathbf{x}} = \mathbf{Ax} + \mathbf{Bu} + \mathbf{Ew} \qquad \mathbf{y} = \mathbf{Cx} + \mathbf{Du} + \mathbf{Fw} \qquad (12.51)$$

Assume that $\mathbf{y}_{ref}$ and $\mathbf{w}$ are vectors of constants, of dimensions m and q. The design problem can now be stated as follows:

Design $\mathbf{u}$ such that

$$\left\{ \begin{array}{ll} \dot{\mathbf{x}} \longrightarrow \mathbf{0} \text{ as } t \longrightarrow \infty & \text{for stability} \\ \mathbf{y} \longrightarrow \mathbf{y}_{ref} \text{ as } t \longrightarrow \infty & \text{for zero steady-state errors} \end{array} \right. \qquad (12.52)$$

The last condition in (12.52) means that the steady-state value of each element of $\mathbf{y}$ must be made equal to the corresponding element of $\mathbf{y}_{ref}$. This implies not only zero steady-state errors but also that each output can be set independently by manipulation of the corresponding reference input. This is often required, even if limited interaction during transients may be quite acceptable. Such a system is said to be *statically* or *asymptotically decoupled* or *noninteracting* since each reference input affects only the corresponding output for $t \to \infty$.

To solve this design problem, it will be reformulated to be in the same form as those of the preceding sections. First define m new state variables:

$$\mathbf{p} = \int_0^t (\mathbf{y} - \mathbf{y}_{ref})\, dt \qquad (12.53)$$

Differentiating this and using (12.51) yields the differential equations

$$\left\{ \begin{array}{l} \dot{\mathbf{x}} = \mathbf{Ax} + \mathbf{Bu} + \mathbf{Ew} \\ \dot{\mathbf{p}} = \mathbf{y} - \mathbf{y}_{ref} = \mathbf{Cx} + \mathbf{Du} + \mathbf{Fw} - \mathbf{y}_{ref} \end{array} \right. \qquad (12.54)$$

Written in matrix form, these equations provide the augmented state model

$$\begin{bmatrix} \dot{\mathbf{x}} \\ \dot{\mathbf{p}} \end{bmatrix} = \begin{bmatrix} \mathbf{A} & \mathbf{0} \\ \mathbf{C} & \mathbf{0} \end{bmatrix} \begin{bmatrix} \mathbf{x} \\ \mathbf{p} \end{bmatrix} + \begin{bmatrix} \mathbf{B} \\ \mathbf{D} \end{bmatrix} \mathbf{u} + \begin{bmatrix} \mathbf{E} & \mathbf{0} \\ \mathbf{F} & -\mathbf{I} \end{bmatrix} \begin{bmatrix} \mathbf{w} \\ \mathbf{y}_{ref} \end{bmatrix} \qquad (12.55)$$

Since $\mathbf{w}$ and $\mathbf{y}_{ref}$ are constant, in the steady state $\dot{\mathbf{x}} = \mathbf{0}$, $\dot{\mathbf{p}} = \mathbf{0}$, provided that the system is stable. This means that the steady-state solutions $\mathbf{x}_s$, $\mathbf{p}_s$, and $\mathbf{u}_s$ must satisfy the equation

$$\begin{bmatrix} \mathbf{E} & \mathbf{0} \\ \mathbf{F} & -\mathbf{I} \end{bmatrix} \begin{bmatrix} \mathbf{w} \\ \mathbf{y}_{ref} \end{bmatrix} = -\begin{bmatrix} \mathbf{A} & \mathbf{0} \\ \mathbf{C} & \mathbf{0} \end{bmatrix} \begin{bmatrix} \mathbf{x}_s \\ \mathbf{p}_s \end{bmatrix} - \begin{bmatrix} \mathbf{B} \\ \mathbf{D} \end{bmatrix} \mathbf{u}_s \qquad (12.56)$$

Substituting this for the last term in (12.55) gives

$$\begin{bmatrix} \dot{\mathbf{x}} \\ \dot{\mathbf{p}} \end{bmatrix} = \begin{bmatrix} \mathbf{A} & \mathbf{0} \\ \mathbf{C} & \mathbf{0} \end{bmatrix} \begin{bmatrix} \mathbf{x} - \mathbf{x}_s \\ \mathbf{p} - \mathbf{p}_s \end{bmatrix} + \begin{bmatrix} \mathbf{B} \\ \mathbf{D} \end{bmatrix} (\mathbf{u} - \mathbf{u}_s) \tag{12.57}$$

Now define new variables as follows, representing the deviations from this steady state:

$$\mathbf{z} = \begin{bmatrix} \mathbf{z}_1 \\ \mathbf{z}_2 \end{bmatrix} = \begin{bmatrix} \mathbf{x} - \mathbf{x}_s \\ \mathbf{p} - \mathbf{p}_s \end{bmatrix} \quad \left(\dot{\mathbf{z}} = \begin{bmatrix} \dot{\mathbf{x}} \\ \dot{\mathbf{p}} \end{bmatrix} \right) \qquad \mathbf{v} = \mathbf{u} - \mathbf{u}_s \tag{12.58}$$

With these, (12.57) becomes

$$\dot{\mathbf{z}} = \hat{\mathbf{A}}\mathbf{z} + \hat{\mathbf{B}}\mathbf{v} \qquad \hat{\mathbf{A}} = \begin{bmatrix} \mathbf{A} & \mathbf{0} \\ \mathbf{C} & \mathbf{0} \end{bmatrix} \qquad \hat{\mathbf{B}} = \begin{bmatrix} \mathbf{B} \\ \mathbf{D} \end{bmatrix} \tag{12.59}$$

The significance of this result is that, by defining the deviations from steady state as state and control variables, the design problem has been reformulated to be in the same form as those of the preceding sections, with $\mathbf{z} = \mathbf{0}$ as the desired state. This means that the pole assignment or optimal control techniques can be used to design a constant-gain state feedback matrix $\mathbf{K}$ from $\mathbf{z}$:

$$\mathbf{v} = -\mathbf{K}\mathbf{z} \tag{12.60}$$

In terms of the original variables, with $\mathbf{K}$ partitioned appropriately,

$$\mathbf{K} = \begin{bmatrix} \mathbf{K}_1 & \mathbf{K}_2 \end{bmatrix} \qquad \mathbf{u} - \mathbf{u}_s = -\begin{bmatrix} \mathbf{K}_1 & \mathbf{K}_2 \end{bmatrix} \begin{bmatrix} \mathbf{x} - \mathbf{x}_s \\ \mathbf{p} - \mathbf{p}_s \end{bmatrix}$$
$$= -\mathbf{K}_1(\mathbf{x} - \mathbf{x}_s) - \mathbf{K}_2(\mathbf{p} - \mathbf{p}_s) \tag{12.61}$$

The steady-state terms must balance each other. Therefore, replacing $\mathbf{p}$ by the integral expression (12.53), the control $\mathbf{u}$ equals

$$\mathbf{u} = -\mathbf{K}_1\mathbf{x} - \mathbf{K}_2\mathbf{p} = -\mathbf{K}_1\mathbf{x} - \mathbf{K}_2 \int_0^t (\mathbf{y} - \mathbf{y}_{\text{ref}}) \, dt \tag{12.62}$$

This important result represents a multivariable generalization of PI control. The control consists of proportional state feedback and integral control of output error.

Figure 12.6 illustrates the feedback configuration. In the steady state, $\dot{\mathbf{p}} = \mathbf{y} - \mathbf{y}_{\text{ref}} = \mathbf{0}$, so the errors are zero. The control is called *robust* because this remains true for any plant parameter variations as long as the system remains stable.

The controllability of the pair $(\mathbf{A}, \mathbf{B})$ is not sufficient to ensure that of the pair $(\hat{\mathbf{A}}, \hat{\mathbf{B}})$ in (12.59). This controllability can be examined using the following theorem. The number of plant inputs is r and the number of outputs is m.

Theorem 12.6.1.   The pair $(\hat{\mathbf{A}}, \hat{\mathbf{B}})$ is controllable if and only if $(\mathbf{A}, \mathbf{B})$ is controllable and

$$\mathbf{G} = \begin{bmatrix} \mathbf{A} & \mathbf{B} \\ \mathbf{C} & \mathbf{D} \end{bmatrix} \tag{12.63}$$

has full rank $(n + m)$, equal to the number of rows.

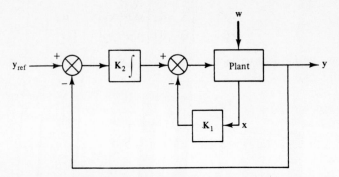

Figure 12.6 Multivariable PI control.

Since the number of columns of $\mathbf{G}$ is $(n + r)$, $r \geq m$ is necessary; that is, there must be at least as many plant inputs as there are outputs to be regulated to zero errors.

Example 12.6.1 Two-Tank Process of Example 12.5.3

In Fig. 12.5, assume that the second input is absent $(u_2 = 0)$, and that a disturbance flow rate w enters the first tank via a return line from elsewhere in the process. This adds a term $w/A_1 = w$ to the right side of the equation for $\dot{h}_1$ in Example 12.5.3, and the equations for the levels h_1 and h_2 become

$$\dot{h}_1 = -3h_1 + 2h_2 + u_1 + w \qquad \dot{h}_2 = 4h_1 - 5h_2$$

Let it be desired to control the level in the second tank with zero steady-state error for reference inputs and disturbances with constant steady-state values. In equations (12.51) then

$$\mathbf{x} = \begin{bmatrix} h_1 \\ h_2 \end{bmatrix} \qquad \mathbf{A} = \begin{bmatrix} -3 & 2 \\ 4 & -5 \end{bmatrix} \qquad \mathbf{B} = \begin{bmatrix} 1 \\ 0 \end{bmatrix} \qquad \mathbf{E} = \begin{bmatrix} 1 \\ 0 \end{bmatrix} \qquad \begin{matrix} \mathbf{C} = [0 \quad 1] \\ \mathbf{D} = \mathbf{F} = \mathbf{0} \end{matrix}$$

Here the output matrix $\mathbf{C}$ reflects that the system output is

$$y = h_2 = x_2 = [0 \quad 1]\mathbf{x}$$

The matrices for integral control design are, from (12.59) and (12.63),

$$\mathbf{G} = \begin{bmatrix} -3 & 2 & 1 \\ 4 & -5 & 0 \\ 0 & 1 & 0 \end{bmatrix} \qquad \hat{\mathbf{A}} = \begin{bmatrix} -3 & 2 & 0 \\ 4 & -5 & 0 \\ 0 & 1 & 0 \end{bmatrix}$$

$$\hat{\mathbf{B}} = \begin{bmatrix} 1 \\ 0 \\ 0 \end{bmatrix} \qquad \lambda\mathbf{I} - \hat{\mathbf{A}} = \begin{bmatrix} \lambda + 3 & -2 & 0 \\ -4 & \lambda + 5 & 0 \\ 0 & -1 & \lambda \end{bmatrix}$$

The pair $(\mathbf{A}, \mathbf{B})$ may be verified to be controllable and $|\mathbf{G}| = 4$, so $\mathbf{G}$ has full rank $n + m = 3$, and the controllability theorem 12.6.1 is satisfied. The eigenvalues of $\hat{\mathbf{A}}$ are those of $\mathbf{A}$ as in Example 12.5.3 and an open-loop eigenvalue at the origin due to the integral control: $\lambda_1 = 0$, $\lambda_2 = -1$, $\lambda_3 = -7$. Let it be sufficient to move the eigenvalue λ_1 to $\hat{\lambda}_1 = -2$, leaving the others unchanged. Hence only the eigenvector $\mathbf{v}_1$ is needed, and is found from $\hat{\mathbf{A}}'\mathbf{v}_1 = 0 \cdot \mathbf{v}_1$ to be

$$\mathbf{v}_1' = [1 \quad 0.75 \quad 1.75]$$

Then, using (12.48), the feedback matrix in (12.60) and (12.61) is found to be

$$\mathbf{K} = \frac{0 + 2}{\mathbf{v}_1'\hat{\mathbf{B}}} \mathbf{v}_1' = [2 \quad 1.5 \quad 3.5] = [\mathbf{K}_1 \quad \mathbf{K}_2]$$

so $\mathbf{K}_1 = [2 \quad 1.5]$, $K_2 = 3.5$, and (12.62) gives

$$u = -2x_1 - 1.5x_2 - 3.5 \int_0^t (y - y_{\text{ref}}) \, dt$$

To evaluate this design, the response of the level h_2 in the second tank to unit step inputs of w and y_{ref} is calculated. This may be done by substituting the control into the original state model equations (12.55). Redefining $\mathbf{z}$ to represent the original state variables yields

$$\mathbf{z} = \begin{bmatrix} \mathbf{x} \\ \mathbf{p} \end{bmatrix} = \begin{bmatrix} x_1 \\ x_2 \\ p \end{bmatrix} = \begin{bmatrix} h_1 \\ h_2 \\ p \end{bmatrix} = \begin{bmatrix} z_1 \\ z_2 \\ z_3 \end{bmatrix} \qquad u = -[2 \quad 1.5 \quad 3.5]\mathbf{z}$$

$$\dot{\mathbf{z}} = \hat{\mathbf{A}}\mathbf{z} + \hat{\mathbf{B}}u + \begin{bmatrix} \mathbf{E} & \mathbf{0} \\ \mathbf{F} & -\mathbf{I} \end{bmatrix} \begin{bmatrix} w \\ y_{\text{ref}} \end{bmatrix}$$

$$= \begin{bmatrix} -3 & 2 & 0 \\ 4 & -5 & 0 \\ 0 & 1 & 0 \end{bmatrix} \mathbf{z} - \begin{bmatrix} 1 \\ 0 \\ 0 \end{bmatrix} [2 \quad 1.5 \quad 3.5]\mathbf{z} + \begin{bmatrix} 1 & 0 \\ 0 & 0 \\ 0 & -1 \end{bmatrix} \begin{bmatrix} w \\ y_{\text{ref}} \end{bmatrix}$$

$$= \begin{bmatrix} -5 & 0.5 & -3.5 \\ 4 & -5 & 0 \\ 0 & 1 & 0 \end{bmatrix} \mathbf{z} + \begin{bmatrix} 1 & 0 \\ 0 & 0 \\ 0 & -1 \end{bmatrix} \begin{bmatrix} w \\ y_{\text{ref}} \end{bmatrix} = \overline{\mathbf{A}}\mathbf{z} + \overline{\mathbf{B}}\overline{\mathbf{r}}$$

Any of the techniques in Section 11.5 can be used to calculate responses. Since only one output is of interest, Laplace transforms will be used.

Only z_2 is needed, so the output matrix can be taken to be $\overline{\mathbf{C}} = [0 \quad 1 \quad 0]$, and the transfer function matrix from $\overline{\mathbf{r}}$ to z_2 is

$$\mathbf{G}(s) = \overline{\mathbf{C}}(s\mathbf{I} - \overline{\mathbf{A}})^{-1}\overline{\mathbf{B}}$$

This involves the inverse of a 3×3 matrix, but because of the form of $\overline{\mathbf{C}}$, only the second row of this inverse

$$(s\mathbf{I} - \overline{\mathbf{A}})^{-1} = \begin{bmatrix} s+5 & -0.5 & 3.5 \\ -4 & s+5 & 0 \\ 0 & -1 & s \end{bmatrix}^{-1}$$

is required. From Appendix A3, this row is $[c_{12} \quad c_{22} \quad c_{32}]/\Delta$, where $\Delta = |s\mathbf{I} - \overline{\mathbf{A}}| = (s+1)(s+2)(s+7)$ is known from the assigned poles, and where c_{i2} are the cofactors of the elements in the second column of $(s\mathbf{I} - \overline{\mathbf{A}})$. These cofactors are $c_{i2} = (-1)^{i+2}m_{i2}$, where the minor m_{i2} is the determinant of the matrix obtained by removing column 2 and row i of the matrix. It may be seen that

$$c_{12} = (-1)^3(-4s) = 4s \qquad c_{22} = (-1)^4 s(s+5) = s(s+5)$$
$$c_{32} = (-1)^5[0 - 3.5(-4)] = -14$$

The transfer function matrix now becomes

$$\mathbf{G}(s) = \frac{1}{\Delta}[4s \quad s(s+5) \quad -14] \begin{bmatrix} 1 & 0 \\ 0 & 0 \\ 0 & -1 \end{bmatrix}$$

and the transform $Z_2(s)$ of $z_2(t)$ is $Z_2(s) = \mathbf{G}(s)\overline{\mathbf{R}}(s)$, or

$$Z_2(s) = \left[\frac{4s}{(s+1)(s+2)(s+7)} \quad \frac{14}{(s+1)(s+2)(s+7)} \right] \begin{bmatrix} W(s) \\ Y_{\text{ref}}(s) \end{bmatrix}$$

Responses to unit steps in w and y_{ref} in turn can now be calculated by partial fraction expansion and are shown in Fig. 12.7, identified by *ic*. As expected, h_2 has zero steady-state error in response to y_{ref} and a transient deviation in response to w.

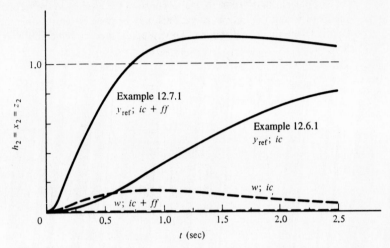

Figure 12.7 Examples 12.6.1 and 12.7.1: integral control and feedforward.

It is useful to observe that the design process itself occupies less than half of the space devoted to this example. Also, instead of by modal control, this design could have been done by equating coefficients in actual and desired characteristic equations based on equations (12.59) and (12.60).

12.7 FEEDFORWARD CONTROL FOR MEASURABLE DISTURBANCES

While with integral control the static errors can be made zero, the errors during the transients may be large. Feedforward control, discussed in Section 8.11, is an important technique in practice to reduce the effect of disturbances if these are measurable, that is, if signals can be obtained to represent them. In this section the concept is formulated in state space, and the control equations are derived for feedforward from both reference inputs and disturbance inputs. The system is again described by the augmented state equations (12.51)

$$\dot{\mathbf{x}} = \mathbf{A}\mathbf{x} + \mathbf{B}\mathbf{u} + \mathbf{E}\mathbf{w} \qquad \mathbf{y} = \mathbf{C}\mathbf{x} + \mathbf{D}\mathbf{u} + \mathbf{F}\mathbf{w} \qquad (12.64)$$

The disturbances $\mathbf{w}$ and reference inputs $\mathbf{y}_{ref}$ are assumed to have constant steady-state values. Subtracting $\mathbf{y}_{ref}$ from both sides of the second equation, they can be written as

$$\begin{bmatrix} \dot{\mathbf{x}} \\ \mathbf{y} - \mathbf{y}_{ref} \end{bmatrix} = \begin{bmatrix} \mathbf{A} & \mathbf{B} \\ \mathbf{C} & \mathbf{D} \end{bmatrix} \begin{bmatrix} \mathbf{x} \\ \mathbf{u} \end{bmatrix} + \begin{bmatrix} \mathbf{E} & \mathbf{0} \\ \mathbf{F} & -\mathbf{I} \end{bmatrix} \begin{bmatrix} \mathbf{w} \\ \mathbf{y}_{ref} \end{bmatrix}$$

or

$$\begin{bmatrix} \dot{\mathbf{x}} \\ \mathbf{y} - \mathbf{y}_{ref} \end{bmatrix} = \mathbf{G} \begin{bmatrix} \mathbf{x} \\ \mathbf{u} \end{bmatrix} + \mathbf{H} \begin{bmatrix} \mathbf{w} \\ \mathbf{y}_{ref} \end{bmatrix} \qquad \mathbf{G} = \begin{bmatrix} \mathbf{A} & \mathbf{B} \\ \mathbf{C} & \mathbf{D} \end{bmatrix} \qquad \mathbf{H} = \begin{bmatrix} \mathbf{E} & \mathbf{0} \\ \mathbf{F} & -\mathbf{I} \end{bmatrix} \qquad (12.65)$$

The purpose of control, as in (12.52), is to ensure that $\dot{\mathbf{x}} \to \mathbf{0}$ and $\mathbf{y} \to \mathbf{y}_{\text{ref}}$. The left side of (12.65) is then zero, so the steady-state solution $\mathbf{x}_s, \mathbf{u}_s$ must satisfy

$$\mathbf{G} \begin{bmatrix} \mathbf{x}_s \\ \mathbf{u}_s \end{bmatrix} = -\mathbf{H} \begin{bmatrix} \mathbf{w} \\ \mathbf{y}_{\text{ref}} \end{bmatrix} \qquad \begin{bmatrix} \mathbf{x}_s \\ \mathbf{u}_s \end{bmatrix} = -\mathbf{G}^{-1} \mathbf{H} \begin{bmatrix} \mathbf{w} \\ \mathbf{y}_{\text{ref}} \end{bmatrix} \tag{12.66}$$

if $\mathbf{G}$ is square and nonsingular. (For rectangular matrices, pseudoinverse matrices can be used.)

When no feedback is used together with the feedforward control, the solution $\mathbf{u}_s$ is the desired feedforward. In matrix form, the solution $\mathbf{u}_s$ in (12.66) can be expressed as

$$\mathbf{u} = -[\mathbf{0} \quad \mathbf{I}] \mathbf{G}^{-1} \mathbf{H} \begin{bmatrix} \mathbf{w} \\ \mathbf{y}_{\text{ref}} \end{bmatrix} \tag{12.67}$$

Here the dimensions and partitioning of the first matrix follow from (12.66), to make $\mathbf{u}$ equal to $\mathbf{u}_s$.

This feedforward control from $\mathbf{y}_{\text{ref}}$ and $\mathbf{w}$ can be implemented if $\mathbf{w}$ is measurable. While feedforward may be used without the use of feedback, it should be emphasized that feedforward control is not robust; that is, it does not provide protection against plant parameter variations. Therefore, feedback is usually also included. To design the combination of feedback and feedforward, substituting the first form in (12.66) for the last term in (12.65) yields

$$\begin{bmatrix} \dot{\hat{\mathbf{x}}} \\ \mathbf{y} - \mathbf{y}_{\text{ref}} \end{bmatrix} = \mathbf{G} \begin{bmatrix} \mathbf{x} - \mathbf{x}_s \\ \mathbf{u} - \mathbf{u}_s \end{bmatrix} = \mathbf{G} \begin{bmatrix} \hat{\mathbf{x}} \\ \hat{\mathbf{u}} \end{bmatrix} \tag{12.68}$$

Here, as for integral control, new variables are defined for the deviations from steady state:

$$\hat{\mathbf{x}} = \mathbf{x} - \mathbf{x}_s \qquad (\dot{\hat{\mathbf{x}}} = \dot{\mathbf{x}}) \qquad \hat{\mathbf{u}} = \mathbf{u} - \mathbf{u}_s \tag{12.69}$$

Substituting for $\mathbf{G}$ from (12.65), (12.68) can be written

$$\dot{\hat{\mathbf{x}}} = \mathbf{A}\hat{\mathbf{x}} + \mathbf{B}\hat{\mathbf{u}} \qquad \mathbf{y} - \mathbf{y}_{\text{ref}} = \mathbf{C}\hat{\mathbf{x}} + \mathbf{D}\hat{\mathbf{u}} \tag{12.70}$$

As in the case of integral control, by defining the deviations from steady state as state and control variables, the problem has been reformulated to be in standard form, with desired state $\hat{\mathbf{x}} = \mathbf{0}$. Hence previous optimal control or pole assignment techniques can be used to design a state feedback:

$$\hat{\mathbf{u}} = -\mathbf{K}\hat{\mathbf{x}} \qquad \mathbf{u} - \mathbf{u}_s = -\mathbf{K}(\mathbf{x} - \mathbf{x}_s) \qquad \mathbf{u} = -\mathbf{K}\mathbf{x} + [\mathbf{K} \quad \mathbf{I}] \begin{bmatrix} \mathbf{x}_s \\ \mathbf{u}_s \end{bmatrix} \tag{12.71}$$

Substituting (12.66) now gives the control:

$$\mathbf{u} = -\mathbf{K}\mathbf{x} - [\mathbf{K} \quad \mathbf{I}] \mathbf{G}^{-1} \mathbf{H} \begin{bmatrix} \mathbf{w} \\ \mathbf{y}_{\text{ref}} \end{bmatrix} \tag{12.72}$$

This is a proportional state feedback plus feedforward control from $\mathbf{y}_{\text{ref}}$ and $\mathbf{w}$.

If zero steady-state errors are specified for constant steady-state values of $\mathbf{y}_{\text{ref}}$ and $\mathbf{w}$, integral control may be incorporated in addition to feedforward and state feedback. Putting $\mathbf{K} = \mathbf{0}$ in (12.72) again yields the control (12.67) with feedforward alone.

The solution (12.66) requires the inversion of **G**. It is often more convenient to solve the set of equations (12.66) directly instead of by the use of matrices. These scalar equations may be written from the original state model (12.64) by putting $\dot{\mathbf{x}} = \mathbf{0}$, $\mathbf{y} - \mathbf{y}_{\text{ref}} = \mathbf{0}$:

$$0 = \mathbf{Ax}_s + \mathbf{Bu}_s + \mathbf{Ew} \qquad 0 = \mathbf{Cx}_s + \mathbf{Du}_s + \mathbf{Fw} - \mathbf{y}_{\text{ref}} \qquad (12.73)$$

The scalar forms of these equations are then solved for the elements of $\mathbf{x}_s$ and $\mathbf{u}_s$ in terms of those of **w** and $\mathbf{y}_{\text{ref}}$, as will be illustrated in the next example.

Example 12.7.1   Example 12.6.1 with Feedforward

From the data given in Example 12.6.1, the matrices **G** and **H** in (12.65) are

$$\mathbf{G} = \begin{bmatrix} -3 & 2 & 1 \\ 4 & -5 & 0 \\ 0 & 1 & 0 \end{bmatrix} \qquad \mathbf{H} = \begin{bmatrix} 1 & 0 \\ 0 & 0 \\ 0 & -1 \end{bmatrix}$$

Formal matrix inversion yields the following result for $\mathbf{G}^{-1}\mathbf{H}$ in (12.66):

$$\mathbf{G}^{-1} = \frac{1}{4} \begin{bmatrix} 0 & 1 & 5 \\ 0 & 0 & 4 \\ 4 & 3 & 7 \end{bmatrix} \qquad \mathbf{G}^{-1}\mathbf{H} = \frac{1}{4} \begin{bmatrix} 0 & -5 \\ 0 & -4 \\ 4 & -7 \end{bmatrix}$$

As was suggested above, direct solution from the scalar form of (12.73) may be preferred. The scalar equations are

$$-3x_{s1} + 2x_{s2} + u_s + w = 0 \qquad 4x_{s1} - 5x_{s2} = 0 \qquad x_{s2} = y_{\text{ref}}$$

These yield

$$x_{s2} = y_{\text{ref}} \qquad x_{s1} = 1.25y_{\text{ref}} \qquad u_s = 1.75y_{\text{ref}} - w$$

So the feedforward control without state feedback or integral control is

$$u_s = -w + 1.75y_{\text{ref}}$$

To show that this gives the same result for $\mathbf{G}^{-1}\mathbf{H}$ as obtained by matrix inversion, the solutions can be arranged in matrix form as

$$\begin{bmatrix} x_{s1} \\ x_{s2} \\ u_s \end{bmatrix} = -\frac{1}{4} \begin{bmatrix} 0 & -5 \\ 0 & -4 \\ 4 & -7 \end{bmatrix} \begin{bmatrix} w \\ y_{\text{ref}} \end{bmatrix}$$

which agrees with the preceding result. Using this form, the formal solution (12.67) in turn yields the same scalar solution for the feedforward control as was obtained previously:

$$u = -\frac{1}{4} [0 \quad 0 \quad 1] \begin{bmatrix} 0 & -5 \\ 0 & -4 \\ 4 & -7 \end{bmatrix} \begin{bmatrix} w \\ y_{\text{ref}} \end{bmatrix} = -w + 1.75y_{\text{ref}}$$

The feedforward can also be superimposed on the earlier design with state feedback and integral control. With the same assigned poles, the feedback control (12.62) is

$$u = -\mathbf{K}_1 \mathbf{x} - \mathbf{K}_2 \int_0^t (\mathbf{y} - \mathbf{y}_{\text{ref}})\, dt = -2x_1 - 1.5x_2 - 3.5 \int_0^t (y - y_{\text{ref}})\, dt$$

The feedback matrix **K** in (12.72) corresponds to the feedback from the original states, so $\mathbf{K} = \mathbf{K}_1 = [2 \quad 1.5]$ and

$$[\mathbf{K} \quad \mathbf{I}]\mathbf{G}^{-1}\mathbf{H} = \frac{1}{4}[2 \quad 1.5 \quad 1] \begin{bmatrix} 0 & -5 \\ 0 & -4 \\ 4 & -7 \end{bmatrix} = \frac{1}{4}[4 \quad (-10 - 6 - 7)] = [1 \quad -5.75]$$

With integral control included, (12.72) becomes

$$u = -2x_1 - 1.5x_2 - 3.5 \int_0^t (y - y_{\text{ref}}) \, dt - w + 5.75 y_{\text{ref}}$$

and the control of Example 12.6.1 is augmented to that illustrated in Fig. 12.8.

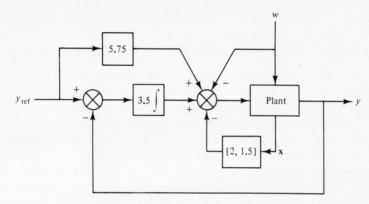

Figure 12.8   Feedforward control plus PI control.

To evaluate the effect of feedforward control, responses to unit step inputs of w and y_{ref} are calculated when this control is added to that in Example 12.6.1. The control u can be expressed in terms of the state vector z of that example as

$$u = -[2 \quad 1.5 \quad 3.5] z + [-1 \quad 5.75] \begin{bmatrix} w \\ y_{\text{ref}} \end{bmatrix}$$

and when this is substituted, the equation for $\dot{z}$ in the example becomes

$$\dot{z} = \begin{bmatrix} -3 & 2 & 0 \\ 4 & -5 & 0 \\ 0 & 1 & 0 \end{bmatrix} z - \begin{bmatrix} 1 \\ 0 \\ 0 \end{bmatrix} \left\{ [2 \quad 1.5 \quad 3.5] z - [-1 \quad 5.75] \begin{bmatrix} w \\ y_{\text{ref}} \end{bmatrix} \right\}$$

$$+ \begin{bmatrix} 1 & 0 \\ 0 & 0 \\ 0 & -1 \end{bmatrix} \begin{bmatrix} w \\ y_{\text{ref}} \end{bmatrix}$$

This reduces to

$$\dot{z} = \begin{bmatrix} -5 & 0.5 & -3.5 \\ 4 & -5 & 0 \\ 0 & 1 & 0 \end{bmatrix} z + \begin{bmatrix} 0 & 5.75 \\ 0 & 0 \\ 0 & -1 \end{bmatrix} \begin{bmatrix} w \\ y_{\text{ref}} \end{bmatrix}$$

The first column contains only zeros, so feedforward control has eliminated the effect of the disturbance input. Analogous to the last example, the transfer function relating y_{ref} and the level z_2 in the second tank is given by

$$\frac{Z_2(s)}{Y_{\text{ref}}(s)} = \frac{1}{\Delta} [4s \quad s(s+5) \quad -14] \begin{bmatrix} 5.75 \\ 0 \\ -1 \end{bmatrix}$$

$$= \frac{14 + 23s}{(s+1)(s+2)(s+7)}$$

From this, partial fraction expansion yields the unit step response identified by $ic + ff$ in Fig. 12.7. Feedforward of the reference input is responsible for an overshoot of about 17.5%, but also for a major improvement of the speed of response. Feedforward has also eliminated the effect of the disturbance input. As may be seen in Fig. 12.8, a second path is constructed from w to y, which counteracts the direct effect of the disturbance on the output.

12.8 DYNAMIC OBSERVERS AND OUTPUT FEEDBACK DESIGN

Constant-gain state feedback $\mathbf{u} = -\mathbf{Kx}$ can provide specified eigenvalues or an optimal and stable design. But it assumes that all states are available for feedback. Even if all states can be measured, the use of many sensors for this purpose is not an acceptable solution. In practice, often only the output $\mathbf{y}$ can be considered to be available for feedback. To enable the use of the design techniques presented earlier in this chapter, which require all states, the solution is then to use a *dynamic observer,* implemented as a computer algorithm. For observable systems, the observer determines an estimate $\hat{\mathbf{x}}$ of the state $\mathbf{x}$ from the measured output $\mathbf{y}$. Constant-gain feedback from $\hat{\mathbf{x}}$ is then again used for optimal control or pole assignment.

Frequently, however, this would be considered too complex, and a simpler design would be sought by the use of *output feedback.* This output feedback may be constrained to be constant or of low dynamic order. It should be recognized that the classical design techniques presented earlier in this book are in fact examples of output feedback.

Like the others in this chapter, the topics of this section are the subjects of an extensive literature.

Dynamic Observers

It would appear that an estimate $\hat{\mathbf{x}}$ of the state $\mathbf{x}$ can be obtained by an analog or digital simulation $\dot{\hat{\mathbf{x}}} = \mathbf{A}\hat{\mathbf{x}} + \mathbf{Bu}$ of the plant dynamics $\dot{\mathbf{x}} = \mathbf{Ax} + \mathbf{Bu}$. But $\hat{\mathbf{x}}$ cannot match $\mathbf{x}$ for all t, because this requires $\hat{\mathbf{x}}(0) = \mathbf{x}(0)$, and $\mathbf{x}(0)$ is not known. However, by subtracting the two equations, the error $\mathbf{x}_e$ of the estimate is seen to satisfy

$$\mathbf{x}_e = \mathbf{x} - \hat{\mathbf{x}} \qquad \dot{\mathbf{x}}_e = \mathbf{Ax}_e \qquad \mathbf{x}_e(0) = \mathbf{x}(0) - \hat{\mathbf{x}}(0)$$

If $\mathbf{A}$ is stable, $\mathbf{x}_e(t)$ will tend to zero, so the estimate $\hat{\mathbf{x}}$ will approach $\mathbf{x}$. This is as desired, except that the dynamics of this estimate are those of $\mathbf{A}$; that is, the estimator has the same speed of response as the plant. Because the estimates are to be used to control the plant, the observer dynamics should really be faster than the controller dynamics. This can be achieved by adding to the model equation a forcing term proportional to the difference between the actual output $\mathbf{y} = \mathbf{Cx}$ and the estimated output $\mathbf{C}\hat{\mathbf{x}}$:

$$\dot{\hat{\mathbf{x}}} = \mathbf{A}\hat{\mathbf{x}} + \mathbf{Bu} + \mathbf{L}(\mathbf{y} - \mathbf{C}\hat{\mathbf{x}}) \qquad (12.74)$$

This is the state equation for the dynamic observer, with $\mathbf{L}$ a constant matrix. An alternative form is

$$\dot{\hat{\mathbf{x}}} = (\mathbf{A} - \mathbf{LC})\hat{\mathbf{x}} + \mathbf{Bu} + \mathbf{Ly} \qquad (12.75)$$

To show that the estimate $\hat{\mathbf{x}}$ approaches $\mathbf{x}$, subtract (12.74) from the state model $\dot{\mathbf{x}} = \mathbf{Ax} + \mathbf{Bu}$, $\mathbf{y} = \mathbf{Cx}$. For the error $\mathbf{x}_e$ of the estimate, this yields

$$\mathbf{x}_e = \mathbf{x} - \hat{\mathbf{x}} \qquad \dot{\mathbf{x}}_e = (\mathbf{A} - \mathbf{LC})\mathbf{x}_e \tag{12.76}$$

If $(\mathbf{A} - \mathbf{LC})$ is stable, $\mathbf{x}_e$ indeed approaches zero, so $\hat{\mathbf{x}}$ approaches $\mathbf{x}$. Moreover, through $\mathbf{L}$ the opportunity exists to speed up the dynamics of the estimator.

In fact, design techniques for $\mathbf{L}$ exist that are very similar to those used for the design of the state feedback matrix $\mathbf{K}$. For example, $\mathbf{L}$ may be chosen to place the ob-server eigenvalues in specified locations. These locations would be chosen to the left of those used for design of $\mathbf{K}$, in order that these controller eigenvalues will domi-nate the dynamics. This is called a *full-order observer,* because it is of the same order as the plant and estimates all states, including those that may already be available by measurement. In *reduced-order observers,* only those states that are not available are estimated. The full-order observer is usually preferred when considerable measure-ment noise is present. The famous Kalman filter, derived by optimal control tech-niques, gives an optimal estimate if $\mathbf{y}$ is contaminated by noise.

Example 12.8.1   Observer for Example 12.5.3

In Example 12.5.3 a state feedback $u = -\mathbf{Kx}$ was designed to place the closed-loop system eigenvalues at -4 and -9. Suppose that the state variable x_1, the level h_1 in the first tank, is not available by measurement and must be estimated. The system matrices are then

$$\mathbf{A} = \begin{bmatrix} -3 & 2 \\ 4 & -5 \end{bmatrix} \qquad \mathbf{B} = \begin{bmatrix} 1 & 0 \\ 0 & 1 \end{bmatrix} \qquad \mathbf{C} = [0 \ \ 1] \qquad \mathbf{L} = \begin{bmatrix} l_1 \\ l_2 \end{bmatrix}$$

and the observer characteristic equation is

$$|s\mathbf{I} - (\mathbf{A} - \mathbf{LC})| = \det \begin{bmatrix} s + 3 & -2 + l_1 \\ -4 & s + 5 + l_2 \end{bmatrix}$$
$$= s^2 + (8 + l_2)s + (7 + 4l_1 + 3l_2)$$

Let the observer eigenvalues be placed at -12 and -15, about a factor of 3 to the left of the controller eigenvalues. Then the desired characteristic equation is

$$(s + 12)(s + 15) = s^2 + 27s + 180$$

and equating coefficients yields

$$l_1 = 29 \qquad l_2 = 19$$

Note that the observer can be designed only if the system is observable, that is, if the rank of the observability matrix (11.52) equals the dimension n of $\mathbf{A}$.

Output Feedback

The dynamic observer may also be viewed as a form of output feedback, although a rather special and complex one. In most applications this complexity is not justified. Classical control is witness that usually quite satisfactory dynamics can be achieved using feedback from the available output. This output feedback may be constant or dynamic. For the system

$$\dot{\mathbf{x}} = \mathbf{Ax} + \mathbf{Bu} \qquad \mathbf{y} = \mathbf{Cx} \tag{12.77}$$

these alternatives lead to the following equations:

1. *Constant-gain output feedback:*
$$\mathbf{u} = -\mathbf{Ky} = -\mathbf{KCx} \qquad \dot{\mathbf{x}} = (\mathbf{A} - \mathbf{BKC})\mathbf{x} \tag{12.78}$$

2. *Dynamic output feedback:* Like the plant, the dynamic compensator can be represented by a state model:

$$\mathbf{u} = \mathbf{Hz} + \mathbf{Ny} \qquad \dot{\mathbf{z}} = \mathbf{Fz} + \mathbf{Gy} \qquad (12.79)$$

The vectors $\mathbf{x}$, $\mathbf{y}$, $\mathbf{u}$, and $\mathbf{z}$ have dimensions n, m, r, and p. These equations can be combined as follows, where the definitions of $\mathbf{v}$, $\hat{\mathbf{A}}$, and so on, are implied by the equations.

$$\begin{bmatrix} \dot{\mathbf{x}} \\ \dot{\mathbf{z}} \end{bmatrix} = \begin{bmatrix} \mathbf{A} & \mathbf{0} \\ \mathbf{0} & \mathbf{0} \end{bmatrix} \begin{bmatrix} \mathbf{x} \\ \mathbf{z} \end{bmatrix} + \begin{bmatrix} \mathbf{B} & \mathbf{0} \\ \mathbf{0} & \mathbf{I} \end{bmatrix} \begin{bmatrix} \mathbf{u} \\ \dot{\mathbf{z}} \end{bmatrix} : \dot{\mathbf{v}} = \hat{\mathbf{A}}\mathbf{v} + \hat{\mathbf{B}}\hat{\mathbf{u}} \qquad (12.80a)$$

$$\begin{bmatrix} \mathbf{u} \\ \dot{\mathbf{z}} \end{bmatrix} = \begin{bmatrix} \mathbf{N} & \mathbf{H} \\ \mathbf{G} & \mathbf{F} \end{bmatrix} \begin{bmatrix} \mathbf{y} \\ \mathbf{z} \end{bmatrix} \qquad \begin{bmatrix} \mathbf{y} \\ \mathbf{z} \end{bmatrix} = \begin{bmatrix} \mathbf{C} & \mathbf{0} \\ \mathbf{0} & \mathbf{I} \end{bmatrix} \begin{bmatrix} \mathbf{x} \\ \mathbf{z} \end{bmatrix} : \hat{\mathbf{y}} = \hat{\mathbf{C}}\mathbf{v} \qquad \hat{\mathbf{u}} = \mathbf{P}\hat{\mathbf{y}}$$

$$(12.80b)$$

Substitution of (12.80b) into (12.80a) yields the closed-loop equation

$$\dot{\mathbf{v}} = (\hat{\mathbf{A}} + \hat{\mathbf{B}}\mathbf{P}\hat{\mathbf{C}})\mathbf{v} \qquad (12.80c)$$

Comparison of (12.80) with (12.77) and (12.78) shows that the problem has been reformulated into one with constant-gain feedback. The unknown matrices $\mathbf{N}$, $\mathbf{H}$, $\mathbf{G}$, and $\mathbf{F}$ are combined in the single matrix $\mathbf{P}$, which is designed for the $(n + p)$th-order system. So techniques for the design of constant-gain output feedback apply also to the design of dynamic output feedback compensation.

Both pole assignment and optimal control design procedures are available, but not all eigenvalues can be placed in specified locations. If a permissible number is placed, the remainder could be anywhere, including in the right-half plane. In fact, a stable design may not be possible below a certain dynamic order of feedback.

Design is often carried out by starting with constant-gain feedback and increasing the assumed order of compensation a step at a time until the results are satisfactory. For illustration, consider the design of an optimal constant-gain output feedback (12.78) which minimizes the standard performance index

$$J = \frac{1}{2} \int_0^\infty (\mathbf{x}'\mathbf{Qx} + \mathbf{u}'\mathbf{Ru}) \, dt \qquad (12.81)$$

From the derivation of the matrix Liapunov equation (12.29), it may be seen that this equation with $\mathbf{K}$ replaced by $\mathbf{KC}$ applies also for output feedback:

$$(\mathbf{A} - \mathbf{BKC})'\mathbf{V} + \mathbf{V}(\mathbf{A} - \mathbf{BKC}) = -\mathbf{Q} - \mathbf{C}'\mathbf{K}'\mathbf{RKC} \qquad (12.82)$$

This is completely equivalent to imposing the constraint implied by $\mathbf{KC}$ on the state feedback matrix $\mathbf{K}$ in (12.29). As in (12.28) to (12.31), and as illustrated by Examples 12.4.1 and 12.4.2, the optimal control minimizes $J = (1/2)\mathbf{x}'(0)\mathbf{Vx}(0)$ and can be found from the conditions $dJ/dk_{ij} = 0$, where $\mathbf{K} = \{k_{ij}\}$. However, it was found in Section 12.4 that this approach soon becomes laborious. The problem of minimizing J subject to (12.82) is a *parameter-optimization* problem. Standard computer routines for numerical parameter optimization, such as the Newton–Raphson method or Rosenbrock's method, are often used to find the best parameter values in $\mathbf{K}$. The Liapunov equation is solved at each stage of the *hillclimbing* procedure.

As noted earlier, the optimal solution depends on the initial condition $\mathbf{x}(0)$. A common change to overcome this disadvantage is to minimize J for a random initial condition. It can be shown that

The average value of J in (12.81) for a random initial condition $\mathbf{x}_0$ is minimized by choosing $\mathbf{K}$ to minimize

$$\hat{J} = \text{trace } (\mathbf{V}) \tag{12.83}$$

where $\mathbf{V}$ is the symmetric and positive definite solution of the Liapunov equation (12.82).

A condition for the procedure to converge is that the closed-loop system matrix $(\mathbf{A} - \mathbf{BKC})$ have stable eigenvalues. To achieve this, a preliminary optimization may be used, which is also useful as an output feedback pole assignment technique. The eigenvalue with largest real part is moved as far to the left as possible by

$$\text{minimizing max } \text{Re}\{\lambda_i(\mathbf{A} - \mathbf{BKC}), i = 1, \ldots, n\} \tag{12.84}$$

Alternatively, a measure of the distances of the eigenvalues to selected desired locations can be minimized. Modal control has also been applied, and techniques are available to control selected dominant modes, using measurement of the outputs to approximate the corresponding modal variables.

12.9 CONCLUSION

Chapters 11 and 12 have provided an introduction to analysis and design in state space, the subject of many books and innumerable technical papers. These techniques provide the only feasible approach for large systems studies and are used extensively for the modeling, analysis, and design of such systems. Examples are electrical power systems, electrical power plants, chemical processes and refineries, aerospace systems, transportation systems, and process control generally, including multivariable control systems.

The great advantage is that the standard matrix formulation has allowed the development of standard computer programs suitable for the analysis and design of large systems. The choice of desired locations in pole assignment and of the weighting matrices in optimal control, however, frequently requires considerable trial-and-error adjustment to obtain a satisfactory transient response. Extensive work has been done to improve this, for example, by considering the system zeros in addition to the poles, and on the problem of output feedback both with and without dynamic compensation.

PROBLEMS

12.1. In Fig. P12.1 with $G_1 = 1/s$ and $G_2 = 4/(s + 4)$, find G_c and k_1 to place the closed-loop poles at $-4 \pm 4j$, without using state-space concepts.

12.2. Verify Problem 12.1 via a state-space model with state feedback, using meaningful and measurable state variables.

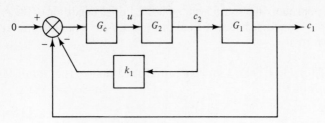

Figure P12.1

12.3. In Fig. P12.1 with $G_1 = 1/(s + 1)$ and $G_2 = 4/(s + 4)$, find G_c and k_1 to place the closed-loop poles at $-4 \pm 4j$, without using state-space concepts.

12.4. Verify Problem 12.3 via a state-space model with state feedback, using meaningful and measurable state variables.

12.5. In Fig. P12.5 with $G_1 = 1/s$, $G_2 = 1/(0.2s + 1)$, and $G_3 = 4/(0.05s + 1)$, find G_c, k_2, and k_1 to place the closed-loop poles at -1 and $-2 \pm 4j$, without using state-space concepts.

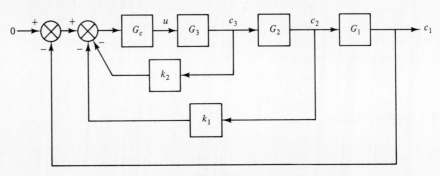

Figure P12.5

12.6. Verify Problem 12.5 via a state-space model with state feedback, using meaningful and measurable state variables.

12.7. In Fig. P12.5 with $G_1 = 1/(s + 1)$, $G_2 = 1/(s + 3)$, and $G_3 = 4/(s + 5)$, find G_c, k_1, and k_2 to place the closed-loop poles at -3, -4, and -5, without using state-space concepts.

12.8. Verify Problem 12.7 via a state-space model with state feedback, using meaningful and measurable state variables.

12.9. For a system with plant transfer function $G(s) = 2/(s^2 + 4s + 5)$, design a state feedback control $u = -\mathbf{k}'\mathbf{x}$ to place the closed-loop eigenvalues at $-3 \pm 2j$.

12.10. For the plant

$$\dot{\mathbf{x}} = \begin{bmatrix} 0 & 1 \\ -2 & -3 \end{bmatrix} \mathbf{x} + \begin{bmatrix} 0 \\ 2 \end{bmatrix} u$$

design a state feedback control to place the closed-loop eigenvalues at $-2 \pm 2j$.

12.11. A single-input, single-output plant has the transfer function

$$G(s) = \frac{4}{s(0.2s + 1)(0.05s + 1)}$$

(a) Obtain a state model in companion form.

(b) Design a state feedback that will place the closed-loop poles at -1 and $-2 \pm 4j$.

12.12. Noting that $s^3 + 9s^2 + 23s + 15 = (s + 1)(s + 3)(s + 5)$, determine a state feedback control for the system

$$\dot{\mathbf{x}} = \begin{bmatrix} 0 & 1 & 0 \\ 0 & 0 & 1 \\ -15 & -23 & -9 \end{bmatrix} \mathbf{x} + \begin{bmatrix} 0 \\ 0 \\ 4 \end{bmatrix} u$$

that will place the closed-loop poles at -3, -4, and -5.

12.13. Verify Problem 12.4 by a state model for which the plant matrix is a companion matrix and by then transforming the state feedback to the state variables in Problem 12.4.

12.14. Noting that Problems 12.6 and 12.11 represent state feedback control of the same plant, verify Problem 12.6 by transforming the control of Problem 12.11.

12.15. Noting that Problems 12.8 and 12.12 represent state feedback control of the same plant, verify Problem 12.8 by transforming the control of Problem 12.12.

12.16. Determine the optimal state feedback for the plant $G(s) = 1/s^2$ and weighting matrices $\mathbf{Q} = \text{diag}\{1, q\}$, $\mathbf{R} = r$. Compare the result with that in (12.37).

12.17. In the Riccati optimal control of Fig. 12.4 and Table 12.4.1:
  (a) Verify the expected effect of an increase of weighting r.
  (b) What should be the effect of increasing the weighting q on k_2 and on system damping?
  (c) Verify (b) by inspection of the characteristic equation of Fig. 12.4.

12.18. Repeat Example 12.4.1 for the initial condition $\mathbf{x}'(0) = \begin{bmatrix} 0 & 1 \end{bmatrix}$ and compare the results.

12.19. Repeat Example 12.4.2, again for the initial condition $\mathbf{x}'(0) = \begin{bmatrix} 1 & 0 \end{bmatrix}$, if a constraint $k_1 = 10$ is imposed.

12.20. For the system in Fig. 12.4 and Table 12.4.1 with $a = A = k_2 = 1$, show that the optimal gain k_1 for the initial condition $\mathbf{x}'(0) = \begin{bmatrix} 1 & 0 \end{bmatrix}$ is the solution of the cubic equation

$$rk_1^3 + 0.5(q + r)k_1^2 - 2 = 0$$

12.21. Verify the solution of Problem 12.10 by modal control design.

12.22. For a system with plant transfer function $G(s) = 2/(s^2 + 3s + 2)$, use modal control to design a state feedback that places the closed-loop eigenvalues at $-2 \pm 2j$.

12.23. Verify the solution of Problem 12.12 by modal control design.

12.24. For Example 12.3.1, perform the design also by modal control, if it is given that the roots of $s^3 + 6s^2 + 11s + 6 = 0$ are -1, -2, and -3. Move only one pole to get the desired result.

12.25. Repeat Problem 12.24 by moving all three poles to get the desired result.

12.26. Figure P12.26 shows a plant model of an aircraft roll control system. Using modal control and state variables as indicated, design state feedback to place the closed-loop eigenvalues at -5, -5, and -50.

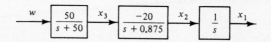

Figure P12.26

12.27. The two-tank hydraulic system in Fig. P11.12 is represented by the state model

$$\dot{\mathbf{x}} = \begin{bmatrix} -1 & 1 \\ 1 & -2 \end{bmatrix} \mathbf{x} + \begin{bmatrix} 0.1 & 0 \\ 0 & 0.1 \end{bmatrix} \mathbf{w}$$

Use modal control to design state feedback that will place the closed-loop eigenvalues at -10. Use w_1 to assign the smallest open-loop eigenvalue and w_2 for the largest.

12.28. For the system with

$$\mathbf{A} = \begin{bmatrix} -2 & 5 \\ -1 & -5 \end{bmatrix} \quad \mathbf{B} = \begin{bmatrix} 0 \\ 5 \end{bmatrix} \quad \mathbf{C} = \begin{bmatrix} 1 & 0 \end{bmatrix} \quad \mathbf{D} = 0$$

design state feedback plus integral control so that the system characteristic equation will be

$$s^3 + 12s^2 + 50s + 100 = 0$$

Partition the feedback appropriately to identify the state feedback and integral control terms.

12.29. In the notations of Eqs. (12.51) let $A = 0$, $B = 1$, $C = 1$, $D = 0$, $E = 1$, and $F = 0$. Determine whether the conditions for integral control hold, and design state feedback plus integral control to place the poles of the closed-loop system at $-p$.

12.30. For the system

$$\dot{\mathbf{x}} = \begin{bmatrix} -1 & 0 \\ 0 & -2 \end{bmatrix} \mathbf{x} + \begin{bmatrix} 1 \\ 1 \end{bmatrix} u + \begin{bmatrix} 1 & 0 \\ 0 & 1 \end{bmatrix} \begin{bmatrix} w_1 \\ w_2 \end{bmatrix} \quad y = \begin{bmatrix} 1 & 0 \end{bmatrix} \mathbf{x}$$

where w_1 and w_2 are constant disturbances, with a constant reference input y_{ref}, check whether the conditions for integral control are satisfied, and design a control system with integral control and state feedback that places the closed-loop eigenvalues at -1, -2, and -3.

12.31. Repeat Problem 12.30 for desired eigenvalues -10, -10, and -10.

12.32. For the system

$$\dot{\mathbf{x}} = \begin{bmatrix} -3 & 1 \\ 3 & -5 \end{bmatrix} \mathbf{x} + \begin{bmatrix} 1 \\ 0 \end{bmatrix} u$$

subject to constant disturbances, design state feedback plus integral control so that no closed-loop eigenvalues will be to the right of -1, and the steady-state error between x_2 and a constant reference input will be zero.

12.33. The plant in a dc motor speed control system is described by the state model

$$\dot{\mathbf{x}} = \begin{bmatrix} 0 & 1 \\ -2 & -3 \end{bmatrix} \mathbf{x} + \begin{bmatrix} 0 \\ 2 \end{bmatrix} u + \begin{bmatrix} -1 \\ 0 \end{bmatrix} w \quad y = \begin{bmatrix} 1 & 0 \end{bmatrix} \mathbf{x}$$

where x_1 is the motor speed, x_2 the armature current, u the armature voltage, and w the load torque. Check whether the conditions for integral control are satisfied, and design a state feedback plus integral control so that the closed-loop poles will be located at -1, -1, and -2.

12.34. For a system

$$\dot{\mathbf{x}} = \begin{bmatrix} 0 & 1 \\ -6 & -5 \end{bmatrix} \mathbf{x} + \begin{bmatrix} 0 \\ 1 \end{bmatrix} u$$

subject to constant disturbances, check whether the conditions for integral control are satisfied, and design a state feedback plus integral control so that the steady-state error between x_1 and a constant reference input will be zero, and the closed-loop eigenvalues will be -3 and $-2 \pm 2j$.

12.35. The system

$$\dot{\mathbf{x}} = \begin{bmatrix} 0 & 1 \\ -12 & -8 \end{bmatrix} \mathbf{x} + \begin{bmatrix} 1 \\ 0 \end{bmatrix} u \qquad y = [1 \quad 0] \mathbf{x}$$

is subject to constant disturbances. Check whether the conditions for integral control are satisfied, and design state feedback plus integral control for constant reference inputs and such that the closed-loop eigenvalues are -6 and $-4 \pm 4j$.

12.36. For the system

$$\dot{\mathbf{x}} = \begin{bmatrix} -3 & 1 \\ 3 & -5 \end{bmatrix} \mathbf{x} + \begin{bmatrix} 1 \\ 0 \end{bmatrix} u + \begin{bmatrix} 1 \\ 2 \end{bmatrix} d$$

where d is a constant measurable disturbance, design feedforward control to minimize the effect of d on the error between x_2 and a constant reference input y_{ref}.

12.37. Assuming that in Problem 12.33 the load torque w is measured,
  (a) Design feedforward control to reduce transient speed errors,
  (b) Express the total control, consisting of this feedforward and the state feedback and integral control of Problem 12.33.

12.38. Add a term

$$\begin{bmatrix} 1 \\ 1 \end{bmatrix} d$$

where d is a constant measurable disturbance, to the right side of the state equation in Problem 12.35, and design feedforward control.

12.39. **(a)** Repeat Problem 12.38 for Problem 12.34.
  (b) Add the above control to the state feedback with integral control of Problem 12.34.

12.40. The linearized model of a distillation column is given by

$$\dot{\mathbf{x}} = \begin{bmatrix} -0.8 & 0.1 \\ -0.1 & -1.2 \end{bmatrix} \mathbf{x} + \begin{bmatrix} -1 & -0.7 \\ 0.2 & 0.3 \end{bmatrix} \mathbf{u} + \begin{bmatrix} 2 \\ 0.5 \end{bmatrix} d \qquad y = \mathbf{I}\mathbf{x}$$

where the time unit is in hours. Design a feedforward plus state feedback controller to minimize the effect of the constant disturbance d on the error between $\mathbf{y}$ and the constant $\mathbf{y}_{\text{ref}}$ and to achieve a dominant time constant not exceeding 2 hours.

12.41. Design the observer matrix $\mathbf{L}$ in (12.74) to estimate the state of the system

$$\dot{\mathbf{x}} = \begin{bmatrix} -1 & 1 \\ 0 & -4 \end{bmatrix} \mathbf{x} + \begin{bmatrix} 0 \\ 4 \end{bmatrix} u \qquad y = [1 \quad 0] \mathbf{x}$$

from the output y. Place the observer eigenvalues at $-10 \pm 10j$.

12.42. For the system

$$\dot{\mathbf{x}} = \begin{bmatrix} 0 & 1 \\ -5 & -4 \end{bmatrix} \mathbf{x} + \begin{bmatrix} 0 \\ 2 \end{bmatrix} u \qquad y = [1 \quad 0] \mathbf{x}$$

a state feedback has been designed to place the system eigenvalues at $-3 \pm 3j$. However, the state x_2 needed to implement this control is not available. Find $\mathbf{L}$ for an observer with eigenvalues three times as far from the origin as those of the controlled system.

12.43. For a system

$$\dot{\mathbf{x}} = \begin{bmatrix} 0 & 1 & 0 \\ 0 & -5 & 5 \\ 0 & 0 & -20 \end{bmatrix} \mathbf{x} + \begin{bmatrix} 0 \\ 0 \\ 80 \end{bmatrix} u \qquad y = \begin{bmatrix} 1 & 0 & 0 \end{bmatrix} \mathbf{x}$$

state feedback has been designed to place the closed-loop eigenvalues at -1 and $-2 \pm 4j$. Since only y is available, find $\mathbf{L}$ for an observer with eigenvalues at -6 and $-6 \pm 6j$.

12.44. For a system

$$\dot{\mathbf{x}} = \begin{bmatrix} -1 & 1 & 0 \\ 0 & -3 & 1 \\ 0 & 0 & -5 \end{bmatrix} \mathbf{x} + \begin{bmatrix} 0 \\ 0 \\ 4 \end{bmatrix} u \qquad y = \begin{bmatrix} 1 & 0 & 0 \end{bmatrix} \mathbf{x}$$

design $\mathbf{L}$ for an observer that will provide an estimate of the state from the available output y. Place the observer eigenvalues at -10.

12.45. Try to design an observer in Problem 12.44 if y is changed to $y = \begin{bmatrix} 0 & 1 & 0 \end{bmatrix} \mathbf{x}$. What is the cause of the problem you will encounter?

12.46. For the system of Fig. 12.4, with scalar Liapunov equations as in Table 12.4.1, find the optimal gain k_1 for the initial condition $\mathbf{x}'(0) = \begin{bmatrix} 0 & 1 \end{bmatrix}$, if $k_2 = 0$, so if the feedback from x_2 is removed (that is, for output feedback $u = k_1 y, y = \begin{bmatrix} 1 & 0 \end{bmatrix} \mathbf{x}$).

12.47. In Problem 12.46, show that for the initial condition $\mathbf{x}'(0) = \begin{bmatrix} 1 & 0 \end{bmatrix}$ the optimal output feedback gain k_1 is the solution of the cubic equation

$$\frac{r}{a} k_1^3 + \left(\frac{Aq}{2a} + \frac{ar}{2A} \right) k_1^2 - \frac{a}{2A} = 0$$

12.48. In Problem 12.46, show that the output feedback gain k_1 to minimize the criterion $J = \mathrm{tr}\, \mathbf{V}$ of (12.31) or (12.83) satisfies the cubic equation

$$\frac{r}{a} k_1^3 + \left(\frac{Aq}{2a} + \frac{ar}{2A} + \frac{r}{2aA} \right) k_1^2 - \frac{a}{2A} - \frac{1}{2aA} = 0$$

12.49. For the difference state model used in Problem 11.48, design state feedback so that the system eigenvalues will be located at the positions $(0.5 \pm 0.2j)$ in the z-plane.

13

Multivariable Systems in the Frequency Domain

13.1 INTRODUCTION

The presence of interaction is the key problem in the design of multivariable control systems. Manipulation of each plant input in general affects all outputs. To allow system outputs to be adjusted independently, it is often desirable that, at least in the steady state, manipulation of a system reference input i affect only the corresponding output i. If, as is frequently the case, the interactions are small, this can be achieved by separate controllers for each variable to be controlled. The interactions are viewed as disturbances on these independent control loops. However, with larger interactions or to improve the performance of systems with smaller interactions, the multivariable nature of the system must be accounted for in design. This is done implicitly in the state-space design techniques of the preceding chapter. But little explicit attention was given to the severity of interaction in the resulting closed-loop system. The exception is the multivariable PI controller in Section 12.6. It provides exact asymptotic or static decoupling; that is, under static conditions each output is equal to the corresponding reference input.

Frequency domain techniques provide an alternative to the state-space approach. The system dynamics are described by transfer function matrices, and interaction considerations figure prominently in design. This approach is well adapted for the design of partially or completely decoupled or noninteracting control systems. Complete decoupling translates to the natural requirement of a diagonal transfer function matrix.

The purposes of this brief introduction are as follows:

- To provide a perspective on the interaction problem
- To present the basic stability theorem for multivariable systems in the frequency domain
- To outline the very successful Nyquist array techniques for design.

13.2 SYSTEM CONFIGURATION AND EQUATIONS

The system configuration of principal interest in this chapter is the unity feedback system shown in Fig. 13.1(a). Figure 13.1(b) shows the explicit form for the case of two inputs and two outputs. $\mathbf{G}$, $\mathbf{K}$, and $\mathbf{K}_d$ are transfer function matrices, as defined for the discussion of stability in state space, and $\mathbf{r}$, $\mathbf{c}$, $\mathbf{e}$, and $\mathbf{u}$ are vectors of transforms of inputs, outputs, errors, and actuating signals. $\mathbf{G}$ is the plant transfer function matrix with elements $g_{ij}(s)$, which are ratios of polynomials $n_{ij}(s)$ and $d_{ij}(s)$, and $\mathbf{K}_d$ is a diagonal matrix of controller transfer functions $k_{di}(s)$. For 2×2 systems,

$$\mathbf{G} = \{g_{ij}(s)\} = \begin{bmatrix} g_{11} & g_{12} \\ g_{21} & g_{22} \end{bmatrix} = \begin{bmatrix} \dfrac{n_{11}}{d_{11}} & \dfrac{n_{12}}{d_{12}} \\ \dfrac{n_{21}}{d_{21}} & \dfrac{n_{22}}{d_{22}} \end{bmatrix} \qquad \mathbf{K}_d = \begin{bmatrix} k_{d1}(s) & 0 \\ 0 & k_{d2}(s) \end{bmatrix} \qquad (13.1)$$

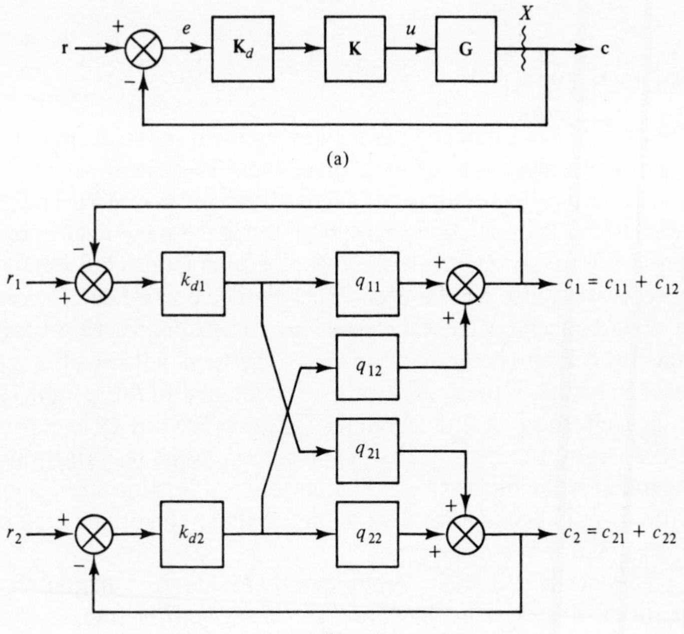

(a)

(b)

Figure 13.1   System configuration.

$\mathbf{K}$ is a decoupling matrix designed to reduce interactions in $\mathbf{Q} = \mathbf{GK}$:

$$\mathbf{K} = \begin{bmatrix} k_{11} & k_{12} \\ k_{12} & k_{22} \end{bmatrix} \qquad \mathbf{Q} = \mathbf{GK} = \begin{bmatrix} q_{11} & q_{12} \\ q_{21} & q_{22} \end{bmatrix} \qquad (13.2)$$

The plant transfer matrix $\mathbf{G}$ may be derived from the system equations or from the transfer functions of subsystems. A common approach is also to obtain $\mathbf{G}$ from a state model of the plant by using (11.7).

Example 13.2.1   Two-Tank Level Control System

Figure 13.2 shows again the two-tank system with two control inputs of Fig. 12.5, for which Example 12.5.3 gives the following state model:

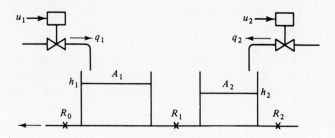

Figure 13.2   Two-tank system.

$$\mathbf{x} = \begin{bmatrix} x_1 \\ x_2 \end{bmatrix} = \begin{bmatrix} h_1 \\ h_2 \end{bmatrix} \qquad \mathbf{u} = \begin{bmatrix} u_1 \\ u_2 \end{bmatrix} \qquad \dot{\mathbf{x}} = \begin{bmatrix} -3 & 2 \\ 4 & -5 \end{bmatrix} \mathbf{x} + \begin{bmatrix} 1 & 0 \\ 0 & 1 \end{bmatrix} \mathbf{u}$$

Let it now be desired to control the levels h_1 and h_2 in both tanks instead of only h_2. Since these are also the state variables, the output equation is

$$\mathbf{y} = \mathbf{Cx} \qquad \mathbf{C} = \mathbf{I}$$

The transfer function matrix **G**, from

$$\mathbf{G}(s) = \mathbf{C}(s\mathbf{I} - \mathbf{A})^{-1}\mathbf{B} \qquad \mathbf{A} = \begin{bmatrix} -3 & 2 \\ 4 & -5 \end{bmatrix} \qquad \mathbf{C} = \mathbf{B} = \mathbf{I}$$

is then found to be

$$\mathbf{G} = \begin{bmatrix} s + 3 & -2 \\ -4 & s + 5 \end{bmatrix}^{-1} = \frac{1}{(s + 1)(s + 7)} \begin{bmatrix} s + 5 & 2 \\ 4 & s + 3 \end{bmatrix} \qquad (13.3)$$

The closed-loop transfer function matrix **R** in Fig. 13.1(a) can be derived exactly as for single-variable systems, except that of course the order of matrix multiplications may not be changed. The system equations

$$\mathbf{c} = \mathbf{Gu} = \mathbf{GKK}_d\mathbf{e} = \mathbf{QK}_d\mathbf{e} \qquad \mathbf{e} = \mathbf{r} - \mathbf{c}$$

yield

$$\mathbf{c} = \mathbf{QK}_d(\mathbf{r} - \mathbf{c}) \qquad \mathbf{c} = (\mathbf{I} + \mathbf{QK}_d)^{-1}\mathbf{QK}_d\mathbf{r}$$

so the closed-loop system is described by

$$\mathbf{c} = \mathbf{Rr} \qquad \mathbf{R} = closed\text{-}loop\ transfer\ matrix \qquad (13.4)$$

R can be expressed in several alternative ways:

$$\mathbf{R} = (\mathbf{I} + \mathbf{QK}_d)^{-1}\mathbf{QK}_d \qquad (13.5)$$
$$\mathbf{R} = (\mathbf{I} + \mathbf{T})^{-1}\mathbf{T} \qquad (13.6)$$
$$\mathbf{R} = \mathbf{F}^{-1}\mathbf{T} \qquad (13.7)$$

where **T** and **F** are defined by

$$\mathbf{T} = \mathbf{GKK}_d = \mathbf{QK}_d \qquad loop\ gain\ matrix \qquad (13.8)$$
$$\mathbf{F} = \mathbf{I} + \mathbf{T}$$
$$= \mathbf{I} + \mathbf{QK}_d$$
$$= (\mathbf{Q} + \mathbf{K}_d^{-1})\mathbf{K}_d \qquad return\ difference\ matrix \qquad (13.9)$$

The reason for the name of the matrix **F** will become clear during the discussion of stability, where these alternative forms are used.

13.3   INTERACTION AND DECOUPLING

Figure 13.1(b) helps to visualize the interaction problem at the heart of multivariable system design. The off-diagonal elements q_{12} and q_{21} of **Q** cause coupling between two separate single-input/single-output (SISO) *diagonal control loops,* shown in Fig. 13.3. If $\mathbf{Q} = \mathbf{GK}$ is diagonal, the system will consist of a number of independent SISO diagonal control loops, each of which can be designed independently by classical techniques. The closed-loop transfer matrix $\mathbf{R}$ would be diagonal and the system noninteracting or decoupled. The system would have the desirable characteristic that each output can be adjusted by manipulating the corresponding reference input without affecting the other outputs. However, aside from conditions on its existence, the decoupling controller $\mathbf{K}$ required to achieve exact decoupling tends to be of the same order of complexity as the plant $\mathbf{G}$ itself. This may be seen from (13.2). $\mathbf{K}$ must satisfy $\mathbf{K} = \mathbf{G}^{-1}\mathbf{Q}$ for a diagonal matrix $\mathbf{Q}$. Since the elements of $\mathbf{G}$ usually have more poles than zeros, $\mathbf{G}^{-1}$ also implies potential problems of physical realizability of $\mathbf{K}$. Moreover, exact decoupling, if indeed it were possible in the presence of the usual approximations and parameter variations in $\mathbf{G}$, means that $\mathbf{K}$ is used to cancel dynamics of $\mathbf{G}$. These canceled modes will still be present in disturbance input responses and could be uncontrollable.

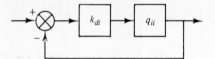

Figure 13.3   Diagonal control loop.

In view of these difficulties, it is fortunate that limited interaction is generally quite acceptable and can usually be achieved by a matrix $\mathbf{K}$ consisting of constant elements. For example, as discussed later, the important Nyquist array (NA) design techniques originated by Rosenbrock are based on the principle of *diagonal dominance.* This represents a condition under which interaction has been reduced sufficiently that the overall control can still be obtained by independent SISO design of diagonal loops.

Numerous techniques are available, both numerical hillclimbing procedures and interactive routines with computer graphics, for the design of the constant-gain elements of $\mathbf{K}$. For example, they may be based on solution of the following problem:

Design $\mathbf{K}$, with constant-gain elements, to "minimize" the

$$\text{column dominance ratios} \quad r_{cj} = \frac{q_{ij}}{q_{jj}} \tag{13.10}$$

in each column j of $\mathbf{Q}$.

These techniques generally allow for a *postcompensator* $\mathbf{L}$, indicated in Fig. 13.4, in addition to the *precompensator* $\mathbf{K}$, to reduce the dominance ratios of $\mathbf{Q} = \mathbf{LGK}$:

$$\mathbf{Q} = \begin{bmatrix} l_{11} & l_{12} \\ l_{21} & l_{22} \end{bmatrix} \begin{bmatrix} g_{11} & g_{12} \\ g_{21} & g_{22} \end{bmatrix} \begin{bmatrix} k_{11} & k_{12} \\ k_{21} & k_{22} \end{bmatrix} \tag{13.11}$$

It should be noted that $\mathbf{L}$ is in the plant output. Certain special forms of $\mathbf{L}$ can be implemented here, as discussed later, but a general form of $\mathbf{L}$ cannot, and must therefore be moved around the loop after design to be combined with $\mathbf{K}$. This does not affect stability, but it does change interactions.

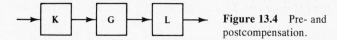

Figure 13.4 Pre- and postcompensation.

For a perspective, it is important to observe that **G** is just a matrix of transfer functions relating input and output variables numbered in specified orders and expressed in given physical units. One is evidently free to change these orders or to use other units. If input 1 affects output 2 most strongly, and input 2 similarly affects output 1, dominance is improved, that is, column dominance ratios are reduced, by interchanging the columns of **G**. This renumbers the inputs to achieve the desired pairing of inputs and outputs. Interchanging rows renumbers the outputs and has the same effect and gives the same column dominance ratios. Similarly, if an input is expressed in units that are half the previous size, the numerical value of this input doubles, so the elements in the corresponding column of **G** must be halved in order for the outputs to remain unchanged. This does not affect the column dominance ratio in **G**. But expressing an output in different units does affect these ratios, because it changes the elements in a row of **G**. However, instead of making such changes directly in **G**, it is preferable to apply them by use of **K** and **L**, because this allows for a much broader range of improvements.

It is noted first that, with or without **L**, only the jth column of **K** appears in the jth column of **Q**, and the use of column dominance, instead of row dominance, permits the columns of **K** to be found independently. **K** performs column operations on **LG** (that is, the columns of **Q** are linear combinations of those of **LG**). Similarly, **L** performs row operations on **GK**. The changes of **G** discussed previously can be realized by what are known as *elementary column operations* and *elementary row operations*.

K is a *permutation matrix* if $k_{11} = k_{22} = 0$ and $k_{12} = k_{21} = 1$. This interchanges the columns of **LG**, an *elementary column operation*. A permutation matrix **L** interchanges the rows of **GK**.

A diagonal matrix **K** multiplies the first column of **G** by k_{11} and the second by k_{22}. This is equivalent to expressing the inputs in different units and does not change the column dominance ratios. A diagonal **L** rescales the outputs and can be very useful. If, say, the elements in the second row of **G** are much smaller than those in the first row, then **K**, which only combines columns of **G**, is not well suited to achieve small dominance ratios in the second column. But the first column is likely to be highly dominant, and **L** permits this to be shared with the second column. If **K** = **I** and $l_{11} = 1$, the dominance ratios are $l_{22}g_{21}/g_{11}$ in column 1 and $g_{12}/(l_{22}g_{22})$ in column 2. Increasing $l_{22} > 1$ implies *dominance sharing* to improve the second column at the expense of the first.

Example 13.3.1   Interactions in G of (13.3).

K required to make **Q** = **GK** diagonal is given by

$$\mathbf{K} = \mathbf{G}^{-1}\mathbf{Q} = \begin{bmatrix} s+3 & -2 \\ -4 & s+5 \end{bmatrix} \begin{bmatrix} q_{11} & 0 \\ 0 & q_{22} \end{bmatrix}$$

$$= \begin{bmatrix} (s+3)q_{11} & -2q_{22} \\ -4q_{11} & (s+5)q_{22} \end{bmatrix}$$

This is realizable, for example, for

$$q_{11} = \frac{k_1}{s} \qquad q_{22} = \frac{k_2}{s}$$

But, from (13.3), the column dominance ratios of **G** itself are

$$r_{c1} = \frac{4}{s+5} \qquad r_{c2} = \frac{2}{s+3} \tag{13.12}$$

Thus, even for **K** = **I** the ratios in **Q** are less than 1 at all frequencies.

13.4 BASIC STABILITY THEOREM

In this section, the general problem of stability in the frequency domain will be considered. From (13.4) to (13.9), the loop gain matrix **T**, the return difference matrix **F**, and the closed-loop transfer matrix **R** are

$$\begin{aligned} \mathbf{T} = \mathbf{GKK}_d = \mathbf{QK}_d \qquad \mathbf{F} = \mathbf{I} + \mathbf{T} \\ \mathbf{R} = (\mathbf{I} + \mathbf{QK}_d)^{-1}\mathbf{QK}_d = (\mathbf{I} + \mathbf{T})^{-1}\mathbf{T} = \mathbf{F}^{-1}\mathbf{T} \end{aligned} \tag{13.13a}$$

Use will be made also of a modified loop gain matrix **T̂**, a diagonal constant matrix **H**, and a corresponding alternative expression for **F**:

$$\begin{aligned} \mathbf{T} = \{t_{ij}\} = \mathbf{\hat{T}H} \qquad \mathbf{\hat{T}} = \{\hat{t}_{ij}\} \\ \mathbf{H} = \mathrm{diag}\{h_i\} \qquad h_i = \text{constant} \\ \mathbf{F} = \mathbf{I} + \mathbf{T} = \mathbf{I} + \mathbf{\hat{T}H} = (\mathbf{\hat{T}} + \mathbf{H}^{-1})\mathbf{H} \end{aligned} \tag{13.13b}$$

The h_i may be considered to represent the gain factors of the elements k_{di} of $\mathbf{K}_d$, and $\hat{t}_{ii} = t_{ii}h_i^{-1}$.

For *single-input/single-output* (SISO) *systems,* $R = T/(1 + T) = T/F$. The numerator of F is the closed-loop characteristic polynomial (clcp) because its roots are roots of the denominator of R and so are the closed-loop system poles. The denominator of F is the open-loop characteristic polynomial (olcp) because the roots equal those of T, the open-loop poles. Hence the relations

$$\frac{\text{clcp}}{\text{olcp}} = F = 1 + T = (\hat{T} + H^{-1})H = \frac{T}{R} \tag{13.14}$$

For *multiple-input/multiple-output* (MIMO) *systems,* as will be shown later, each transfer function in (13.14) must be replaced by the determinant of the corresponding matrix:

$$\frac{\text{clcp}}{\text{olcp}} = \det \mathbf{F} = \det(\mathbf{I} + \mathbf{T}) = \det[(\mathbf{\hat{T}} + \mathbf{H}^{-1})\mathbf{H}] = \frac{\det \mathbf{T}}{\det \mathbf{R}} \tag{13.15}$$

It is only necessary to prove the first of these equalities, because the others then follow from (13.13) and the fact that the determinant of a product of matrices equals the product of their determinants. The denominator of det **F** is indeed the olcp because it is the product of the denominators of the elements of **T**. To show that the numerator of det **F** is the clcp, let the loops in Fig. 13.1(a) be broken at × and a vector signal **a**(s) injected to the right. The "returned" signal at the left is $-\mathbf{Ta}$, and the difference between injected and returned signals is

$$\mathbf{a} - (-\mathbf{Ta}) = (\mathbf{I} + \mathbf{T})\mathbf{a} = \mathbf{Fa}$$

Hence the name of **F** in (13.9). Closing the break forces this difference to be zero:
$$\mathbf{Fa} = (\mathbf{I} + \mathbf{T})\mathbf{a} = 0 \tag{13.16}$$
For SISO systems the roots of $1 + T = 0$ are the closed-loop poles because, for these values of s, (13.16) can have a nontrivial (nonzero) solution for a. In the MIMO case, the condition for a nontrivial solution is known to be that the determinant of the coefficient matrix be zero:
$$\det \mathbf{F} = 0 \tag{13.17}$$
So the numerator of $\det \mathbf{F}$ is indeed the clcp.

If X is any one of the expressions (13.14) for SISO systems and (13.15) for MIMO systems, these equations can be written as follows:
$$\frac{\text{clcp}}{\text{olcp}} = X = K\frac{(s + z_1)(s + z_2)\cdots(s + z_n)}{(s + p_1)(s + p_2)\cdots(s + p_m)} \tag{13.18}$$
X is a ratio of polynomials. Its denominator, the olcp, can usually be made available in a factored form so that $-p_1, \ldots, -p_m$ can be assumed to be known. But the closed-loop system eigenvalues $-z_1, \ldots, -z_n$, the roots of the clcp, are not known. For stability, these must all lie in the left-half s-plane.

To determine this, a plot of X is made on a complex plane, calculated from its numerator and denominator polynomials, in unfactored form, as s travels once clockwise around the Nyquist contour D in Fig. 13.5. D consists of the imaginary axis and a semicircle of radius $R \to \infty$ and, in effect, encloses the entire right-half s-plane. Poles on the imaginary axis are excluded by semicircular indentations of radius $r \to 0$, as indicated for a pole at the origin.

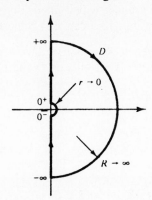

Figure 13.5   Nyquist contour D.

As discussed in Section 7.3, each pole $-p_i$ inside D contributes one counter-clockwise encirclement of the origin in the plot of X as s travels once clockwise around D, and each zero $-z_i$ inside D, a clockwise encirclement. Hence the

Principle of the Argument.   If
$$p_c,\, p_o = \text{number of roots of clcp, olcp inside } D \tag{13.19}$$
then the plot of X will encircle its origin ($p_c - p_o$) times clockwise.

Hence follows, since $p_c = 0$ is required for stability, the basic stability theorem:

Theorem 13.4.1. If a plot of X as s travels once clockwise around D encircles the origin N_X times clockwise, the system is stable if and only if

$$N_X = -p_o \qquad \text{(negative = counterclockwise)} \qquad (13.20a)$$

Instead of from a plot of X, N_X may be found also as the difference of the clockwise encirclements of plots of its numerator and denominator. Thus, depending on which plots are used, if N_F, N_R, and N_T are the clockwise origin encirclements of plots of, respectively, det $\mathbf{F}$, det $\mathbf{R}$, and det $\mathbf{T}$, then

$$N_X = N_F \qquad N_X = N_T - N_R \qquad (13.20b)$$

This is a generalization of the Nyquist stability criterion presented in Chapter 7. Application requires the evaluation of the determinants of matrices. This is acceptable for use in analysis when these matrices are known, but is a serious disadvantage for design when they are not known. Rosenbrock's widely used Nyquist array techniques represent one solution to this problem.

13.5 DESIGN BY NYQUIST ARRAY TECHNIQUES

As indicated previously, the difficulty in applying the Stability Theorem 13.4.1 to the design of multivariable systems is that determination of N_X requires finding the origin encirclements of plots of the determinants of matrices as defined in (13.20) during the design process. Rosenbrock's Nyquist array techniques overcome this problem if the matrices are *diagonally dominant*. As claimed in Section 13.3, this represents a condition under which interactions have been reduced sufficiently that the diagonal loops in Fig. 13.3 can still be designed independently.

Definition. An $m \times m$ matrix $\mathbf{Z}(s) = \{z_{ij}(s)\}$ is diagonally dominant on the Nyquist contour D if for all s on D and for all i either

$$|z_{ii}(s)| > d_{ir}(s) = \sum_{j=1, \neq i}^{m} |z_{ij}(s)| \qquad \text{(row dominance)} \qquad (13.21a)$$

or

$$|z_{ii}(s)| > d_{ic}(s) = \sum_{j=1, \neq i}^{m} |z_{ji}(s)| \qquad \text{(column dominance)} \qquad (13.21b)$$

Thus the magnitude of the diagonal element z_{ii} is larger than the sum d_{ir} or d_{ic} of the magnitudes of the off-diagonal elements in the row or the column. For 2×2 systems this is satisfied if the column dominance ratios r_{cj} of (13.10) are less than 1 in both columns. As seen from (13.12), $\mathbf{G}$ for the two-tank system satisfies this condition.

A graphical interpretation of this condition is based on the construction of *Gershgorin bands,* as follows:

1. Construct Nyquist plots of the diagonal elements $z_{ii}(s)$, as illustrated in Fig. 13.6 for one element.
2. For each, draw circles of radii d_{ir} or d_{ic} calculated from (13.21), with their centers on the plot of z_{ii} at the corresponding frequency. The bands swept out by

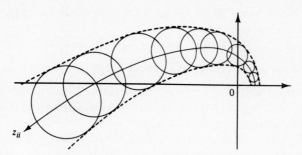

Figure 13.6 Gershgorin band.

these circles, illustrated in Fig. 13.6 for z_{ii}, are the Gershgorin bands. Evidently, if for all i these bands exclude the origin, then (13.21) is satisfied and **Z** is diagonally dominant. It is noted that interactive computer graphics are used for computation and display of such plots.

The exploitation of this concept depends on the following theorem, given here without proof.

Theorem 13.5.1.   For diagonally dominant matrices **Z**, the origin encirclements N_Z of det **Z** as s travels once clockwise around Nyquist contour D equal the sum of the encirclements N_{zi} of the diagonal elements z_{ii}:

$$N_Z = \sum_{i=1}^{m} N_{zi} \qquad (13.22)$$

This is evident for diagonal matrices, for which the determinant equals the product of the diagonal elements.

Rosenbrock's *inverse Nyquist array (INA) technique* will not be discussed here, although it is used extensively for multivariable system design. It employs and would require discussion of Nyquist plots of inverse transfer functions. Also, the direct technique outlined next is quite similar.

The *direct Nyquist array (DNA) technique* is based on the condition that, from (13.13b),

$$\mathbf{F} = \mathbf{I} + \mathbf{T} = \mathbf{I} + \hat{\mathbf{T}}\mathbf{H} = (\hat{\mathbf{T}} + \mathbf{H}^{-1})\mathbf{H} \qquad (13.23)$$

be diagonally dominant. If this is satisfied, then, by (13.22), the condition $N_X = N_F = -p_0$ of the basic stability theorem can be replaced by the condition that the sum of the origin encirclements of the diagonal elements of these matrices be equal to $-p_0$.

One form of the DNA technique is based on the last expression for **F** in (13.23). Since **H** is diagonal and constant, $(\hat{\mathbf{T}} + \mathbf{H}^{-1})$ must be diagonally dominant. This will be so if the Gershgorin bands based on the diagonal elements $(\hat{t}_{ii} + h_i^{-1})$ exclude the origin. Several points should now be noted:

1. The encirclements of $(\hat{t}_{ii} + h_i^{-1})$ about the origin equal those of $\hat{t}_{ii}$ about $-h_i^{-1}$.

2. Since $\mathbf{H}^{-1}$ is diagonal, the Gershgorin circle radii of $(\hat{\mathbf{T}} + \mathbf{H}^{-1})$ are the same as those of $\hat{\mathbf{T}}$.

3. For dominance, the bands based on $(\hat{t}_{ii} + h_i^{-1})$ should exclude the origin, so those based on $\hat{t}_{ii}$ should exclude $-h_i^{-1}$.

4. Using point 1, if point 3 is satisfied the stability condition becomes

$$\Sigma_i \text{ (clockwise encirclements of } \hat{t}_{ii} \text{ about } -h_i^{-1}) = -p_0$$

Hence both dominance and stability can be found from plots of the $\hat{t}_{ii}$ with their Gershgorin bands. It is recalled that $\hat{t}_{ii}$ is the element t_{ii} of loop gain matrix **T** with what could be the gain of $k_{di}(s)$ removed: $\hat{t}_{ii} = t_{ii}h_i^{-1}$.

Figure 13.7 shows $\hat{t}_{11}$ and $\hat{t}_{22}$ with their Gershgorin bands for a 2×2 system. From points 3 and 4, assuming open-loop stability ($p_0 = 0$), the system is stable if $-1/h_1$ and $-1/h_2$ lie to the left of A_1 and A_2, respectively.

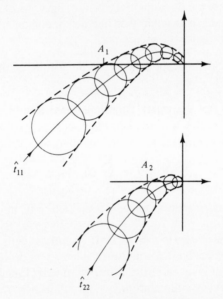

Figure 13.7   DNA example.

An alternative form of the DNA technique is based on the condition that $(\mathbf{I} + \mathbf{T})$ in (13.23) must be diagonally dominant. This form shows more clearly the links with classical SISO design based on the loop gain function. Analogous to the preceding discussion, $(\mathbf{I} + \mathbf{T})$ is dominant if the Gershgorin bands centered on $t_{ii} + 1$ exclude the origin, and so if those centered on t_{ii} exclude -1. Also, as in point 2, the circle radii of $(\mathbf{I} + \mathbf{T})$ and **T** are the same. Then the system is stable if and only if the sum of the clockwise encirclements of t_{ii} about -1 is $-p_0$. In the open-loop stable case, this is equivalent to the -1 point lying to the left of each polar plot t_{ii}, a direct extension of the condition for SISO systems. These results can be summarized as in the following theorem:

DNA Stability Theorem 13.5.2

1. If the Gershgorin bands centered on the $\hat{t}_{ii}$ exclude the points $-h_i^{-1}$, then the system is stable if and only if

$$\sum_i \text{(clockwise encirclements of } \hat{t}_{ii} \text{ about } -h_i^{-1}) = -p_0 \qquad (13.24)$$

2. If the bands centered on the t_{ii} exclude -1, the system is stable if and only if

$$\sum_i (\text{clockwise encirclements of } t_{ii} \text{ about } -1) = -p_0 \qquad (13.25)$$

As was noted, the second form of this theorem closely resembles the classical Nyquist criterion for SISO systems. The difference is that, assuming open-loop stability, the curve that must pass on the right side of the -1 point for stability for SISO systems has broadened for MIMO systems into a band to allow for the effects of interaction.

It is useful to emphasize that diagonal dominance is only a sufficient condition. So the system will not necessarily become unstable as soon as the critical point enters the band. For example, consider the matrices

$$\begin{bmatrix} 1 & 20 \\ 0.1 & 10 \end{bmatrix} \qquad \begin{bmatrix} 1 & 20 \\ 0.5 & 50 \end{bmatrix}$$

The first form is not diagonally dominant, since the ratio in the second column exceeds 1. The second form is diagonally dominant. Yet it is clear from earlier discussions that the second form can be obtained from the first simply by expressing the second output in different units. Clearly, such a decision should not have such momentous effects. Indeed, the diagonal dominance condition has been generalized to include matrices such as the first as being equivalent to diagonally dominant. Only the original diagonal dominance conditions have been presented. These are rather severe in that they should be satisfied over the entire frequency range. There have been extensive and very successful developments to weaken the conditions that must be met to satisfy (13.22).

13.6 CONCLUSION

Frequency domain techniques for multivariable systems have been introduced to indicate an alternative to the state-space approach, which has proved quite useful for smaller systems. Interaction and decoupling were discussed after formulation of the system equations. Then the basic frequency domain stability theorems were developed. Finally, Rosenbrock's well-known Nyquist array techniques were outlined, with the associated stability theorems and diagonal dominance conditions.

PROBLEMS

13.1. In the configuration of Fig. 13.1, let

$$\mathbf{K}_d = \mathbf{I} \qquad \mathbf{K} = \begin{bmatrix} 5 & 0 \\ 0 & 4 \end{bmatrix} \qquad \mathbf{G} = \begin{bmatrix} \dfrac{1}{s+2} & \dfrac{0.25}{s+2} \\[2mm] \dfrac{0.2}{s+1} & \dfrac{1}{s+1} \end{bmatrix}$$

(a) Find the matrices $\mathbf{Q}$, $\mathbf{T}$, and $\mathbf{F}$.
(b) Find the closed-loop transfer function matrix.

13.2. In the configuration of Fig. 13.1, let

$$\mathbf{K}_d = \mathbf{I} \qquad \mathbf{K} = \begin{bmatrix} 1.0 & 1.0 \\ -0.05 & -1.0 \end{bmatrix} \qquad \mathbf{G} = \begin{bmatrix} \dfrac{1}{s+1} & \dfrac{5}{s+5} \\ \dfrac{0.1}{s+0.5} & \dfrac{4}{s+2} \end{bmatrix}$$

Formulate **Q**, **T**, and **F**.

13.3. Use the results in Problem 13.1 to calculate the responses of outputs c_1 and c_2 to a unit step input of r_1.

13.4. Determine the matrices **G**, **K**, and **K**$_d$ in Fig. 13.1 to represent the system in Fig. P11.10.

13.5. (a) Determine the transfer function matrix **G** corresponding to the state model

$$\dot{\mathbf{x}} = \begin{bmatrix} 0 & 1 \\ -2 & -3 \end{bmatrix}\mathbf{x} + \begin{bmatrix} 0 & 1 \\ 1 & 0 \end{bmatrix}\mathbf{u} \qquad \mathbf{y} = \begin{bmatrix} 1 & 0 \\ 1 & -1 \end{bmatrix}\mathbf{x}$$

    (b) Verify the element (1, 1) of **G** directly from the scalar state equations, without using matrix inversion.

13.6. Find the transfer function $Y(s)/U(s)$ corresponding to the state model

$$\dot{\mathbf{x}} = \begin{bmatrix} 0 & 1 & 0 \\ 0 & 0 & 1 \\ 0 & -1 & -2 \end{bmatrix}\mathbf{x} + \begin{bmatrix} 0 \\ 1 \\ 1 \end{bmatrix}u \qquad y = [0 \quad 0 \quad 1]\mathbf{x}$$

Instead of by inverting $(s\mathbf{I} - \mathbf{A})$, find this transfer function by transforming the individual state equations and eliminating variables between them.

13.7. In Fig. 13.1, let

$$\mathbf{G} = \begin{bmatrix} 1 & 6 \\ 5 & 2 \end{bmatrix}$$

Show that for dominance of $\mathbf{Q} = \mathbf{GK}$ the use of column dominance simplifies the design of **K**, by setting up the pertinent ratios for both row and column dominance in terms of the elements of **K**.

13.8. In Problem 13.7:
    (a) Design **K** with only two nonzero elements to achieve dominance ratios of less than 1 in both columns.
    (b) What is the type of matrix **K** designed in (a), and what is its physical significance?

13.9. In Fig. P13.9, mass flow rate and temperature of a flow are controlled by valve openings in "hot" and "cold" supply lines.
    (a) What are the dominance ratios in the columns of **G**?
    (b) Give the forms of **G** and the corresponding column dominance ratios for each of the following:

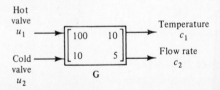

Figure P13.9

        (i)   If the order of the inputs is reversed.

       (ii) If the order of the outputs is reversed.

      (iii) If u_2 is measured in units one-tenth of those in Fig. P13.9.

      (iv) If c_2 is measured in units one-fifth of those in Fig. P13.9.

  (c) Give the pre- and/or postcompensators **K** and **L** that represent the actions of part (b) with the form of **G** as given in Fig. P13.9.

  (d) Choose **K** and **L** for **Q** = **LGK** and note conditions under which small column dominance ratios are difficult to achieve using **K** alone.

13.10. **(a)** Noting that in general the outputs of **G** cannot be manipulated, how is a permutation matrix **L** implemented?

  (b) To implement a diagonal **L**, is it in effect sufficient to multiply the rows of **G** by the diagonal elements of **L**?

13.11. **(a)** Verify that Fig. P13.11 could be a refinement of Fig. P13.9, with allowance for the fact that temperature generally responds more slowly than flow rate, and with c_1 and c_2 expressed in units, respectively, 10 and 5 times those in Fig. P13.9.

  (b) Make Bode plots of the column dominance ratios to show behavior over the frequency range.

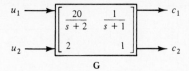

$$\mathbf{G}$$

                                            Figure P13.11

13.12. For the system of Fig. P13.20, make asymptotic Bode plots of the column dominance ratios $r_{cj} = g_{ij}/g_{jj}$ to examine dominance of **G** over the entire frequency range.

13.13. Repeat Problem 13.12 for **G** of Problem 13.1.

13.14. Repeat Problem 13.12 for **G** of Problem 13.2.

13.15. For the plant

$$\mathbf{G} = \begin{bmatrix} \dfrac{5}{s+5} & \dfrac{-1}{s+1} \\[2mm] \dfrac{4}{s+2} & \dfrac{0.1}{s+0.5} \end{bmatrix}$$

determine a very simple compensator **K** and/or **L** that, even if it does not make the dominance ratios less than 1 at all frequencies, represents a major improvement.

13.16. In Fig. 13.4, use elementary column and/or row operations to obtain a **K** and **L** for "minimum" column dominance ratios in **Q** = **LGK**. The numerator and denominator polynomials n_{ij} and d_{ij} of the elements g_{ij} of **G** are

$$n_{11} = 2 \quad d_{11} = s+2 \quad\quad n_{12} = 1 \quad d_{12} = s+1$$
$$n_{21} = 0.4 \quad d_{21} = s+5 \quad\quad n_{22} = 0.01 \quad d_{22} = s+0.5$$

13.17. Determine the open-loop and closed-loop characteristic polynomials and the closed-loop eigenvalues for the system of Problem 13.1.

13.18. For the system of Problem 13.2, find the characteristic polynomial and determine system stability. (Remember that a necessary condition is that all coefficients have the same sign.)

13.19. Determine the combinations of h_1 and h_2 for which the system given in Fig. P13.19 in terms of polar plots of the diagonal elements with their Gershgorin bands will be stable. The system is known to be open-loop stable.

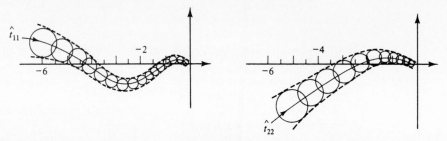

Figure P13.19

13.20. In Fig. P13.20, obtain the polar plot of the diagonal element g_{11} of $\mathbf{G}$ together with its Gershgorin band in terms of the column dominance ratio. (It is sufficient to consider the frequencies $\omega = 4$, 6, and 8.)

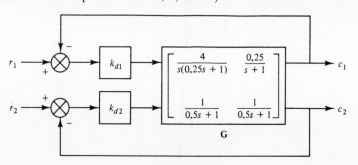

Figure P13.20

13.21. In Problem 13.20:
  (a) Also plot g_{22} with its Gershgorin band.
  (b) What type of matrix $\mathbf{K}$ is represented by the constant gain elements k_{d1} and k_{d2}? Are the column dominance ratios of $\mathbf{Q} = \mathbf{GK}$ different from those of $\mathbf{G}$?
  (c) Is $\mathbf{G}$ or $\mathbf{Q}$ column diagonally dominant, and is this needed for application of an appropriate stability theorem?
  (d) Determine the ranges of values of k_{d1} and k_{d2} for which this theorem guarantees stability.

13.22. In Fig. 13.1, let $\mathbf{K}$ have been designed for a given plant $\mathbf{G}$ such that the numerator and denominator polynomials n_{ij} and d_{ij} of the elements q_{ij} of $\mathbf{Q} = \mathbf{GK}$ are as follows: $n_{11} = 1, d_{11} = s + 1; n_{12} = 3, d_{12} = s + 2; n_{21} = 1.5, d_{21} = s + 1;$ $n_{22} = 1.5, d_{22} = s + 2.$
  (a) Determine whether $\mathbf{Q}$ is column diagonally dominant.
  (b) Determine whether the condition for application of the DNA stability theorem is satisfied for $\mathbf{K}_d = \mathbf{I}$.

14

Nonlinear Control Systems

14.1 INTRODUCTION

As discussed in Section 1.5, most systems are nonlinear for large enough variations about the operating point, and linearization is based on the assumption that these variations are sufficiently small. But this cannot be satisfied, for example, for systems that include relays, which can switch position for very small changes. Start-up and shutdown also frequently require the consideration of nonlinear effects, because of the size of the transients.

A differential equation $A\ddot{x} + B\dot{x} + Cx = f(t)$ is nonlinear if one or more of A, B, or C is a function of the dependent variable x or its derivatives. For example, $a\ddot{x} + b\dot{x}^2 + cx^3 = 0$ could represent a spring–mass–damper system of which the damping coefficient $B = b\dot{x}$ depends on $\dot{x}$ and the spring constant $C = cx^2$ on x. Note that A, B, or C of a linear system may be functions of the independent variable t. Such linear time-varying parameter systems have their own, very considerable problems, which are not discussed in this book.

The principle of superposition does not apply to nonlinear systems. Thus, if input x_1 yields output y_1 and x_2 yields y_2, it is no longer true that for an input $(c_1x_1 + c_2x_2)$ the output will be $(c_1y_1 + c_2y_2)$. This has serious consequences. In fact, the analysis and design techniques discussed so far, including the use of transfer functions and Laplace transforms, are no longer valid. Worse, there is no general equivalent technique to replace them. Instead, a number of techniques exist, each of limited purpose and limited applicability. An extensive literature exists on the subject. In this chapter only the well-known phase plane and describing function methods are discussed and an introduction given to the Liapunov, Popov, and circle criteria for the stability of nonlinear systems.

14.2 NONLINEAR BEHAVIOR AND COMMON NONLINEARITIES

As a minimum, it is important to be aware of the main characteristics of nonlinear behavior, if only to permit recognition if these are encountered experimentally or in system simulations:

1. The nature of the response depends on input and initial conditions. For example, a nonlinear system can change from stable to unstable, or vice versa, if, say, the size of the step input is doubled.
2. Instability shows itself frequently in the form of *limit cycles*. These are oscillations of fixed amplitude and frequency, which can be sustained in the feedback loop even if the system input is zero. In linear systems an unstable transient grows theoretically to infinite amplitude, but nonlinear effects limit this growth.
3. The steady-state response to a sinusoidal input can contain harmonics and subharmonics of the input frequency.
4. The *jump phenomenon* is illustrated by the frequency response plot in Fig. 14.1. If the frequency of the input is reduced from high values, the amplitude of the response drops suddenly at the vertical tangent point C to the value at D. Whether or not such jumps occur depends on the size of the input, the degree of peaking, and the nonlinearity.

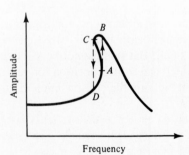

Figure 14.1   Jump phenomenon.

Figure 14.2 shows common types of nonlinearities, with x as input and y as output.

1. *Nonlinear gain:* Very common, for example, valve flow versus pressure drop or valve opening, or force versus deflection for rubber springs.
2. *Saturation:* The output levels off to a constant limit beyond a certain value of the input. Amplifiers saturate, and valve flow cannot rise beyond pump capacity.
3. *Deadband:* An insensitive zone, for example, in instruments or relays, or due to overlap of the lands on a hydraulic control valve spool over the ports to the cylinder.
4. *Backlash:* Due to play in mechanical connections.
5. *Hysteresis:* In electromagnetic circuits; in materials.
6. *Coulomb friction* or *dry friction:* The friction force depends only on the direction of velocity.

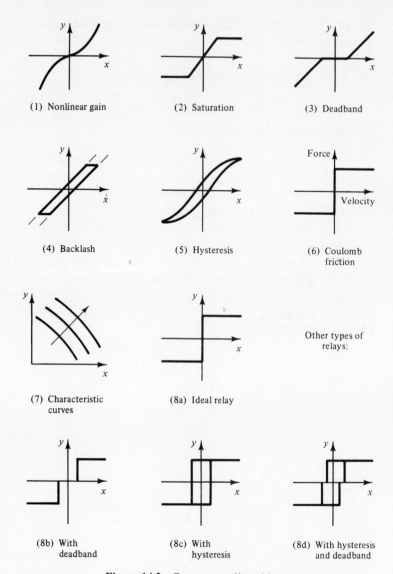

Figure 14.2   Common nonlinearities.

7. *Nonlinear characteristic curves:* The torque–speed curves of motors or the flow–pressure curves of valves.

8. *Relays, with various imperfections:* A very important class of nonlinearities.

14.3 THE PHASE-PLANE METHOD

The phase-plane method is a graphical method for finding the transient response of first- or second-order systems to initial conditions or simple inputs. Despite these restrictions, it is useful because of the insight it provides and because many systems approximate second-order responses. Consider the nonlinear equation

$$\ddot{x} + g(x, \dot{x})\dot{x} + h(x, \dot{x})x = 0 \qquad (14.1)$$

Substitute into this a new function y defined by

$$\dot{x} = y \qquad \ddot{x} = \dot{y} = \frac{dy}{dx}\dot{x} = y\frac{dy}{dx} \qquad (14.2)$$

Then the equation reduces to one of first order:

$$y\left(\frac{dy}{dx}\right) + g(x, y)y + h(x, y)x = 0 \qquad (14.3)$$

Rearranging this yields the *phase-plane equation,*

$$\frac{dy}{dx} = \frac{-g(x, y)y - h(x, y)x}{y} \qquad (14.4)$$

The phase plane is a plot of y versus x as shown in Fig. 14.3. At each point (x, y), dy/dx is the slope of the *phase-plane trajectory* through that point. In Fig. 14.3 these slopes are indicated by the short line segments drawn at points along the *isoclines.* Isoclines are loci along which the trajectory slopes are constant and can be plotted from the *isocline equation.* The equation of the isocline for trajectory slope m is found by substituting $dy/dx = m$ into (14.4) and solving for y:

$$y = \frac{-h(x, y)x}{g(x, y) + m} \qquad (14.5)$$

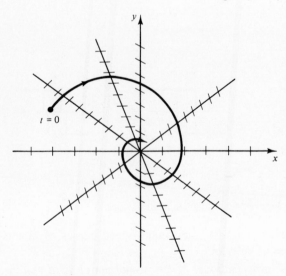

Figure 14.3 Phase plane; focus.

Example 14.3.1   Underdamped System, $\zeta < 1$

$$\ddot{x} + 2\zeta\omega_n\dot{x} + \omega_n^2 x = 0 \qquad (14.6)$$

Phase-plane equation: $\dfrac{dy}{dx} = \dfrac{-(2\zeta\omega_n y + \omega_n^2 x)}{y}$

$$(14.7)$$

Isocline equation: $y = \dfrac{-\omega_n^2 x}{2\zeta\omega_n + m}$

The isoclines are straight lines through the origin, shown in Fig. 14.3 for several values of m. On each isocline, short line segments are drawn at the corresponding slope m. On $y = 0$, dy/dx is infinite, and $dy/dx = 0$ on the isocline $y = -\omega_n x/(2\zeta)$. On the y-axis, $x = 0$ and $dy/dx = -2\zeta\omega_n$. The initial condition $(x, \dot{x})(0)$ is represented by a point on the plane, and the corresponding trajectory is drawn by following the slope segments.

Above the x-axis, $y = \dot{x} > 0$, so x increases. Hence the direction of motion along trajectories must be to the right above the x-axis, and the left below it.

The graphical technique is called the *isocline method* and is useful to sketch the nature of the *phase-plane portrait*, as in the following examples. For numerical work, this and alternative graphical techniques have been largely replaced by computer methods.

Example 14.3.2   Overdamped System, $\zeta > 1$

Figure 14.3 represents an underdamped system, because x oscillates between positive and negative values. For the case $\zeta > 1$, consider when the slope of the isocline equals that of the trajectory (that is, when the isocline is $y = mx$). Substituting this into the second of (14.7) yields a quadratic equation for m of which the roots are

$$m_{1,2} = \omega_n(-\zeta \pm \sqrt{\zeta^2 - 1}) \qquad (14.8)$$

For $\zeta > 1$ this gives two real slopes, both negative, as indicated in Fig. 14.4. These isoclines satisfy the differential equation and so are also possible trajectories. They cannot be crossed, and trajectories that approach them follow these straight lines into the origin.

In Fig. 14.3 the loci spiral into the origin, and the origin is called a *focus*. In Fig. 14.4 the origin is called a *node*.

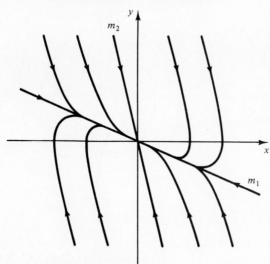

Figure 14.4   Example 14.3.2: node.

The next two examples are concerned with motor position control systems. Relay control of motors is a low-cost form of control that in many applications can provide adequate servo performance. Relay servos are also called *on−off servos* or *bang-bang servos*. The examples illustrate that different equations may apply in different parts of the phase plane, so the phase-plane portrait may be made up of several distinct patterns.

Example 14.3.3   Servo with Piecewise Linear Gain

Figure 14.5(a) shows a simple model of a motor plus load (inertia J, damping B) in a position servo of which the gain of the amplifier varies with system error $E = \theta_i - \theta_o$ as shown in Fig. 14.5(b). For a step input $\dot{E} = -\dot{\theta}_o$, $\ddot{E} = -\ddot{\theta}_o$, and the system equation $J\ddot{\theta}_o + B\dot{\theta}_o = K(E)$ is written as follows in terms of E:

$$J\ddot{E} + B\dot{E} + K(E) = 0 \qquad (14.9)$$

As in (14.1) to (14.4), this yields the phase-plane equation

$$\frac{d\dot{E}}{dE} = \frac{-(B\dot{E} + K(E))}{J\dot{E}} \qquad (14.10)$$

Assuming that for the lower gain K_1 the system (14.9) is overdamped, the phase-plane portrait for $-E_o \leq E \leq E_o$ will be of the form in Fig. 14.4, as shown in Fig. 14.5. If for gain K_2 the system is underdamped, the trajectories for $-E_o > E > E_o$ must be of the spiral form in Fig. 14.3. The foci of these spirals are the points $-E_1$ and E_1 identified in Fig. 14.5(b).

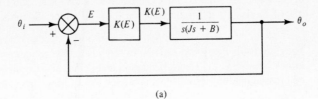

(a)

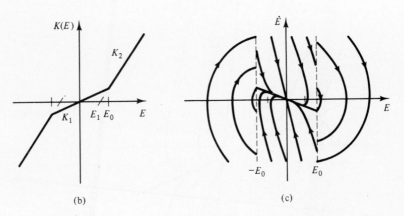

(b)   (c)

Figure 14.5   Example 14.3.3: piecewise linear gain.

Example 14.3.4   Relay Servo with a Ramp Input

In Fig. 14.6(a), the relay is assumed to have a deadband, and the input is a ramp $\theta_i = At$. Hence the derivatives of the error $E = \theta_i - \theta_o$ are

$$\dot{E} = A - \dot{\theta}_o \qquad \ddot{E} = -\ddot{\theta}_o$$

and the system equation $J\ddot{\theta}_o + B\dot{\theta}_o = f(E)$ becomes

$$J\ddot{E} + B\dot{E} + f(E) = BA \qquad f(E) = 0 \qquad \text{for } -E_0 < E < E_0$$
$$= +T \qquad \text{for } E > E_0$$
$$= -T \qquad \text{for } E < -E_0$$

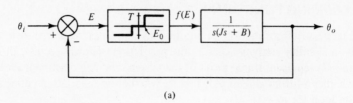

(a)

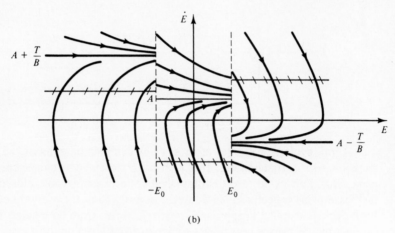

(b)

Figure 14.6   Example 14.3.4: relay servo.

The phase-plane equation is

$$\frac{d\dot{E}}{dE} = \frac{-B\dot{E} - f(E) + BA}{J\dot{E}} = \frac{-B(\dot{E} - A)}{J\dot{E}} \qquad -E_0 < E < E_0$$

$$= \frac{-B(\dot{E} - A + T/B)}{J\dot{E}} \qquad E > E_0 \qquad (14.11)$$

$$= \frac{-B(\dot{E} - A - T/B)}{J\dot{E}} \qquad E < -E_0$$

It is seen that in each range $d\dot{E}/dE$ depends only on $\dot{E}$, so the isoclines are horizontal lines. For any chosen value of $\dot{E}$ the slope $d\dot{E}/dE$ can be calculated from (14.11) and line segments at this slope drawn on the isocline, as shown in Fig. 14.6(b) for an example in each range.

In this manner the phase-plane portrait can be constructed. Note in particular the isoclines identified in Fig. 14.6(b) on which the slope $d\dot{E}/dE$ is zero, so the trajectory is horizontal. These are velocity limits, on which the available torque equals that needed to overcome friction, so that none is available to change $\dot{E}$. The portrait shows that the servo cannot follow a ramp input for which $A \geqslant T/B$. The velocity limit for $E > E_0$ then lies on or above the E-axis, so $\dot{E} \geqslant 0$ and motion cannot be in the direction of decreasing error E. Also verify that, for $A < T/B$, the final steady-state error must equal half the deadband region.

In the case of an ideal relay, the portrait is as in Fig. 14.6(b) with the central region reduced to zero width. For an ideal relay with a step input, one may verify that the pattern is that in Fig. 14.6(b) for $A = 0$.

14.4 DESCRIBING FUNCTIONS

The describing function technique is a frequency response method, and its main use is in stability analysis (that is, the prediction of limit cycles). In Fig. 14.7, where G_1 and G_2 represent linear parts of the system and N a nonlinear element, the question is whether a limit cycle exists, that is, whether an oscillation can maintain itself around the loop for $R = 0$.

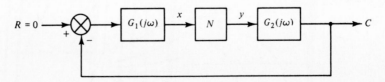

Figure 14.7 System configuration.

Limit cycles for second-order systems can also be constructed by phase-plane methods. As illustrated in Fig. 14.8, they are represented by closed curves in the phase plane. But limit cycles are distinguished from other possible closed curves in that the phase-plane trajectories tend toward or away from them asymptotically. A stable limit cycle is one that is approached by trajectories from both sides. Even the slightest disturbance causes trajectories to depart from unstable limit cycles. As indicated in the first of Fig. 14.8, an initial condition or input outside the limit cycle leads to an unstable transient growth, while transients following a sufficiently small disturbance decay to zero. Thus it is also necessary to determine the type of limit cycle.

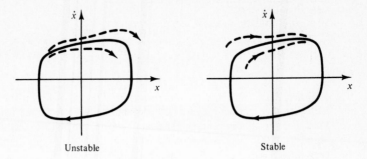

Unstable   Stable

Figure 14.8 Limit cycles.

The model for N in Fig. 14.7 used in this analysis is based on the following assumption:

The input x to the nonlinearity is sinusoidal:

$$x = A \sin \omega t \qquad (14.12)$$

There is an apparent contradiction here. In Fig. 14.9 the square-wave output of an ideal relay for a sinusoidal input is a periodic function, so it can be represented by a *Fourier series* of the general form

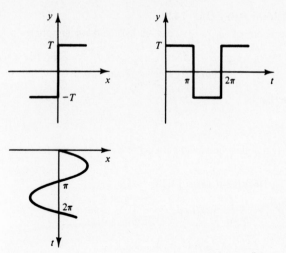

Figure 14.9   Ideal relay.

$$y(t) = b_0 + \sum_{n=1}^{\infty} (a_n \sin n\omega t + b_n \cos n\omega t) \tag{14.13}$$

Thus y contains harmonics ($n > 1$) in addition to the fundamental Fourier component

$$y_f = a_1 \sin \omega t + b_1 \cos \omega t \tag{14.14}$$

The harmonics would pass around the loop via G_2 and G_1 back to x, contradicting the assumption (14.12). But for most practical systems:

1. y_f is considerably larger than the harmonics.
2. G_2 acts as a low-pass filter that attenuates the harmonics much more strongly than y_f.

The combination of these effects usually justifies (14.12) and also implies that only the fundamental component y_f of the output of the nonlinearity needs to be considered. The describing function that models the nonlinearity is therefore defined as follows.

Definition: The *describing function* (DF) N of a nonlinearity is the ratio

$$N = \frac{y_f}{x} \tag{14.15}$$

From the theory of Fourier series, the *Fourier coefficients* a_1 and b_1 for y_f in (14.14) are

$$a_1 = \frac{\omega}{\pi} \int_0^{2\pi/\omega} y \sin \omega t \, dt \qquad b_1 = \frac{\omega}{\pi} \int_0^{2\pi/\omega} y \cos \omega t \, dt \tag{14.16}$$

If y can be extended into an odd function of time, as in Fig. 14.9, then $b_1 = 0$ since $\cos \omega t$ is an even function of time, and the DF becomes

$$N = \frac{a_1}{A} \tag{14.17}$$

The DF is an equivalent linear gain that depends on the amplitude A, and sometimes also the frequency ω, of the input x.

Example 14.4.1 DF of an Ideal Relay (Fig. 14.9)

Since the DF is independent of ω, $\omega = 1$ can be assumed for simplicity so that $x = A \sin t$.

$$a_1 = \frac{1}{\pi} \int_0^{2\pi} y \sin t \, dt = \frac{2}{\pi} \int_0^{\pi} T \sin t \, dt = \frac{4T}{\pi} \tag{14.18}$$

So the describing function is

$$N = \frac{4T}{\pi A} \tag{14.19}$$

As expected, this shows that the equivalent gain decreases as the input A increases, since the output is constant.

Example 14.4.2 Relay with Deadband (Fig. 14.10)

$$a_1 = \frac{1}{\pi} \int_0^{2\pi} y \sin t \, dt = \frac{2}{\pi} \int_\alpha^{\pi - \alpha} T \sin t \, dt$$

$$= \frac{2T}{\pi} (-\cos t) \Big|_\alpha^{\pi - \alpha} = \frac{4T}{\pi} \cos \alpha \tag{14.20}$$

Here α is the value of t at which x equals b, that is,

$$\alpha = \sin^{-1} \frac{b}{A} \quad \text{or} \quad A \sin \alpha = b \tag{14.21}$$

Since $\cos \alpha = \sqrt{1 - (b/A)^2}$, it follows that, for $A > b$ ($N = 0$ for $A \leqslant b$),

$$N = \frac{4T}{\pi A} \sqrt{1 - \left(\frac{b}{A}\right)^2} \tag{14.22}$$

The plot in Fig. 14.10 is as expected, with $N = 0$ for $A < b$, N rising to a maximum close to $A = b$, where the equivalent gain is largest, and then decreasing to zero as A increases while the output is constant.

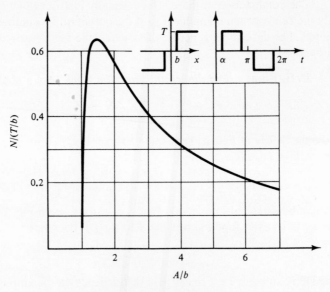

Figure 14.10 Relay with deadband.

Example 14.4.3   Saturation [Fig. 14.11(a)]

$$a_1 = \frac{2}{\pi} \int_0^{\pi} y \sin t \, dt = \frac{4}{\pi} \int_0^{\pi/2} y \sin t \, dt$$

$$= \frac{4}{\pi} \left[\int_0^{\alpha} KA \sin^2 t \, dt + \int_{\alpha}^{\pi/2} (Kb) \sin t \, dt \right]$$

$$= \frac{4KA}{\pi} \left(\frac{\alpha}{2} - \frac{\sin 2\alpha}{4} \right) + \frac{4Kb}{\pi} \cos \alpha$$

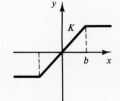

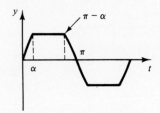

(a)

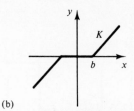

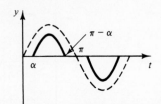

(b)

Figure 14.11   (a) Saturation;
(b) deadband.

Saturation is reached when $\sin \alpha = b/A$, so

$$\cos \alpha = \sqrt{1 - \left(\frac{b}{A} \right)^2} \qquad \sin 2\alpha = \frac{2b}{A} \sqrt{1 - \left(\frac{b}{A} \right)^2}$$

Substitution yields, for $A > b$ ($N = K$ for $A \leq b$),

$$N = \frac{2K}{\pi} \left[\sin^{-1} \frac{b}{A} + \frac{b}{A} \sqrt{1 - \left(\frac{b}{A} \right)^2} \right] \qquad (14.23)$$

This is plotted in Fig. 14.12. As Fig. 14.11(a) suggests, the equivalent gain decreases as A increases beyond the saturation limit.

Example 14.4.4   Deadband [Fig. 14.11(b)]

Instead of deriving N as above, one may prove and use the property that the DF of a sum of nonlinear functions equals the sum of the individual DFs. From this it may be verified that the DF equals K minus the DF (14.23) for saturation. This describing function is plotted in Fig. 14.12. For large A the equivalent gain approaches K.

Extensive tables and plots of DFs are given in books on nonlinear control systems. These include DFs that depend on frequency and DFs where y_f and x are not in phase, as in the following example.

Example 14.4.5   Relay with Hysteresis and Deadband

In Fig. 14.13, half the width of each square pulse is

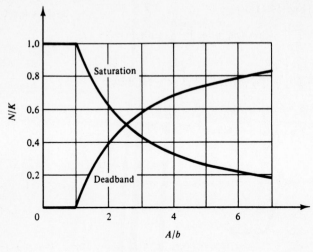

Figure 14.12　Describing functions for saturation and deadband.

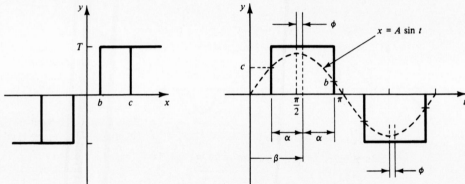

Figure 14.13　Relay with hysteresis and deadband.

$$\alpha = 0.5\left(\pi - \sin^{-1}\frac{c}{A} - \sin^{-1}\frac{b}{A}\right)$$

The center of the first square pulse is located at

$$\beta = \alpha + \sin^{-1}\frac{c}{A}$$

Relative to the center of the first half of the sine wave x, this is at an angle

$$\phi = \beta - \frac{\pi}{2}$$

Hence y_f will lag x by a phase shift $-\phi$, and N is no longer real.

$$-\phi = -0.5\left(\sin^{-1}\frac{c}{A} - \sin^{-1}\frac{b}{A}\right) \tag{14.24}$$

To find the amplitude of y_f, it is noted that if in Example 14.4.2 the width of each square pulse is $(\pi - 2\alpha_1)$, then from (14.20) the DF α_1/A is $N = (4T/\pi A)\cos \alpha_1$. To use this result in Fig. 14.13, let $2\alpha = \pi - 2\alpha_1$. Then $\cos \alpha_1 = \cos[(\pi/2) - \alpha] = \sin \alpha$, so the DF becomes

$$N = \frac{4T \sin \alpha}{\pi A} e^{-j\phi} \qquad (14.25)$$

where $e^{-j\phi}$ shows the phase lag of y_f relative to x.

14.5 STABILITY ANALYSIS USING DESCRIBING FUNCTIONS

Since N in Fig. 14.7 is an equivalent linear gain, and $E = R - C = -C$ for $R = 0$,

$$C = G_2 N G_1 E = -GNC \qquad C(GN + 1) = 0 \qquad (14.26)$$

where

$$G = G_1 G_2 = \text{product of linear elements in loop} \qquad (14.27)$$

From (14.26), the condition for a nonzero solution for C (that is, a limit cycle) is

$$GN + 1 = 0 \qquad G = \frac{-1}{N} \qquad (14.28)$$

Graphical Interpretation.   Limit cycles are identified by the intersections of the polar plot of $G(j\omega)$ and a plot of $-1/N$.

Example 14.5.1   Relay-Controlled Servo (Fig. 14.14)

Physically, there must be a limit cycle since the relay must switch back and forth near $E = 0$. $-1/N = -\pi A/(4T)$, so $-1/N$ for increasing A lies along the entire negative real axis. From Chapter 7, the plot of this $G(j\omega)$ has the form shown, and therefore an intersection, and a limit cycle, will exist.

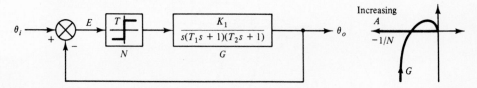

Figure 14.14   Relay-controlled servo.

The frequency of the limit cycle is that corresponding to the intersection along the $G(j\omega)$ curve. Its amplitude A at the input to the nonlinearity can be found from the $-1/N$ plot at the intersection.

To determine the stability of limit cycles, it is recalled that a linear system with loop gain function KG is on the verge of instability when the polar plot of KG passes through the -1 point, that is, when the polar plot of G passes through the critical point $-1/K$. Analogously, for nonlinear systems the critical point is $-1/N$, and the plot of $-1/N$ shows how in this case the point changes with conditions at the input to the nonlinearity. Use is now made of the Nyquist criterion for open-loop stable systems. This criterion states that the system is stable if the critical point $-1/N$ lies to the left of the polar plot of G.

Suppose that a system is in a limit cycle at amplitude A and frequency corresponding to the intersection of G and $-1/N$. Now, due to some disturbance, let A increase slightly, moving the critical point slightly away from the intersection along $-1/N$. If this brings it to the right of G, the system is unstable, causing A to increase, farther away from the limit cycle. If A decreased slightly from the value at the intersection, it would then be to the left of G. This indicates stability and continuing decrease of A. Thus disturbances in either direction from the limit cycle would cause trajectories to depart farther from it, showing it to be unstable.

Example 14.5.1 (continued)

From Fig. 14.14, the frequency of the limit cycle can be found from the condition that the phase angle of $G(j\omega)$ is $-180°$. G is the same as equation (7.21) of Example 7.6.2, where this frequency was found to be $\omega = 1/\sqrt{T_1 T_2}$, and $G = -K_1 T_1 T_2/(T_1 + T_2)$. The amplitude A of the limit cycle is given by $G = -1/N$, and so by

$$\frac{-K_1 T_1 T_2}{T_1 + T_2} = -\frac{\pi A}{4T} \quad \text{or} \quad A = \frac{4TK_1 T_1 T_2}{\pi(T_1 + T_2)}$$

It is also readily verified that the intersection in Fig. 14.14 represents a stable limit cycle. An increase of A brings the critical point to the left of G and causes A to decrease back to the intersection. A further decrease would cause instability and a return to the intersection.

Example 14.5.2   Relay-Controlled Servo with Deadband

In Example 14.5.1 let the ideal relay be replaced by one with deadband, of which Fig. 14.10 gives the DF. $N = 0$ for $A \le b$, rises to a maximum and then decreases back to zero with increasing A. Thus $-1/N$, real and negative, changes as indicated in Fig. 14.15 with increasing A. There are two intersections, P and Q, so two limit cycles. It may be verified that P is unstable and Q stable. Thus the system is stable for an initial condition or disturbance for which A does not reach P, but goes into a limit cycle at Q if excited sufficiently severely, beyond P.

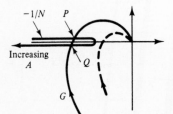

Figure 14.15   Example 14.5.2.

If G is as in Example 14.5.1 with $K_1 = 1$, $T_1 = 1$, $T_2 = 0.5$, the frequency of any limit cycle is $\omega = 1/\sqrt{T_1 T_2} = 1.414$ and its amplitude must satisfy $-1/N = G = -1/3$, so $N = 3$. Using Fig. 14.10, if $T = 1$, $b = 0.1$, then $N/(T/b) = 0.3$ and two limit cycles can exist, satisfying $A/b = 1.04$ and $A/b = 4.1$. Hence the unstable limit cycle at P has an amplitude 0.1, and the stable limit cycle at Q, an amplitude of 0.41.

System design should evidently aim to avoid intersections between G and $-1/N$. In Fig. 14.15, if the gain of G is reduced, as shown by the dashed polar plot, there is no intersection and so no limit cycle, and the system is stable. Phase-lead or phase-lag compensation can also serve to modify G in order to achieve this.

14.6 SECOND, OR DIRECT, METHOD OF LIAPUNOV

Liapunov's method for stability analysis is in principle very general and powerful. The major drawback, which seriously limits its use in practice, is the difficulty often associated with construction of the *Liapunov function* or *V-function* required by the method.

The system dynamics must be described by a state-space model, discussed in detail in Chapter 11. It is a description in terms of a set of first-order differential equations. For example, a nonlinear system might be described by a set of n first-order nonlinear differential equations

$$\dot{x}_i = f_i(x_1, x_2, \ldots, x_n, t) \qquad i = 1, \ldots, n \tag{14.29}$$

Referring to Section 11.2 or to Appendix A, this can be written compactly in the form of a *state-space model* as

$$\dot{\mathbf{x}} = \mathbf{f}(\mathbf{x}, t) \tag{14.30a}$$

where

$$\mathbf{x} = \begin{bmatrix} x_1 \\ \vdots \\ x_n \end{bmatrix} \qquad \dot{\mathbf{x}} = \begin{bmatrix} \dot{x}_1 \\ \vdots \\ \dot{x}_n \end{bmatrix} \qquad \mathbf{f}(\mathbf{x}, t) = \begin{bmatrix} f_1(x_1, \ldots, x_n, t) \\ \vdots \\ f_n(x_1, \ldots, x_n, t) \end{bmatrix} \tag{14.30b}$$

The vector $\mathbf{x}$ is the *state vector,* and its elements are *state variables.* The origin $\mathbf{x} = \mathbf{0}$ ($x_1 = \cdots = x_n = 0$) of the state space will be assumed to be an equilibrium solution, where $f_i = 0$, $i = 1, \ldots, n$. The phase plane in Section 14.3 is in effect a state space for second-order systems. It is recalled that the state-model description of a given system is not unique but depends on which variables are chosen as state variables.

The Liapunov function, $V(x_1, \ldots, x_n)$, is a scalar function of the state variables. To motivate the following and to make the stability theorems plausible, let V be selected to be

$$V(\mathbf{x}) = \|\mathbf{x}\|^2 = \sum_{i=1}^{n} x_i^2 \tag{14.31}$$

Here, $\|\mathbf{x}\|$ is the Euclidean norm of $\mathbf{x}$, the length of the vector $\mathbf{x}$, and the distance to the origin of the state space. V is evidently positive and $V(\mathbf{0}) = 0$. Now let

$$\dot{V} = \frac{dV}{dt} = \frac{\partial V}{\partial x_1}\dot{x}_1 + \cdots + \frac{\partial V}{\partial x_n}\dot{x}_n \tag{14.32}$$

be calculated by substituting (14.29). If $\dot{V}$ were found to be always negative, with $\dot{V}(\mathbf{0}) = 0$, then apparently V decreases continuously, and the state must end up in the origin of the state space, implying asymptotic stability.

It could be that $\dot{V}$ is only negative in a small enough region around the origin. Hence the following distinctions are appropriate:

1. A system is *globally asymptotically stable* if it returns to $\mathbf{x} = \mathbf{0}$ after any size disturbance.

2. It is *locally asymptotically stable* if it does so after a sufficiently small disturbance.

3. It is *stable* if for a given size disturbance the solution stays inside a certain region.

To develop these concepts, the following definitions are used for the sign of V (and $\dot{V}$):

1. V is *positive (negative) definite* in a region containing $\mathbf{x} = \mathbf{0}$ if it is positive (negative) everywhere except that $V(\mathbf{0}) = 0$.

2. *V* is *positive (negative) semidefinite* if it has uniform positive (negative) sign but is zero also at points other than the origin.

3. *V* is *indefinite* if both signs occur in the region.

For example, $V(x_1, x_2) = x_1^2 + x_2^2$ is positive definite, $V(x_1, x_2) = x_1 + x_2$ is indefinite, and $V(x_1, x_2, x_3) = -x_1^2 - x_2^2$ is negative semidefinite, because it is zero along the x_3-axis.

Sylvester's theorem is used to find such properties for a general quadratic form

$$Q = \sum_{i=1}^{n} \sum_{j=1}^{n} a_{ij} x_i x_j \qquad (a_{ij} = a_{ji}) \tag{14.33}$$

From Appendix A, this can also be written as

$$Q = \mathbf{x}'\mathbf{A}\mathbf{x}$$
$$\mathbf{x}' = [x_1 \quad \cdots \quad x_n]$$

$$\mathbf{A} = \begin{bmatrix} a_{11} & a_{12} & \cdots & a_{1n} \\ a_{12} & a_{22} & & \\ \vdots & & & \vdots \\ a_{1n} & & \cdots & a_{nn} \end{bmatrix} \tag{14.34}$$

Here $\mathbf{x}'$ is the transpose of $\mathbf{x}$, and $\mathbf{A}$ is a symmetrical matrix.

Sylvester's theorem. *Q* is positive definite if and only if all principal minors of the determinant $|\mathbf{A}|$ are larger than zero:

$$a_{11} > 0, \qquad \begin{vmatrix} a_{11} & a_{12} \\ a_{12} & a_{22} \end{vmatrix} > 0, \quad \ldots, \quad |\mathbf{A}| > 0 \tag{14.35}$$

If one or more are zero, *Q* is semidefinite. A matrix $\mathbf{A}$ is said to be, say, positive definite if the corresponding quadratic form is positive definite, and $-\mathbf{A}$ is then negative definite.

Example 14.6.1

$$Q = x_1^2 + 2x_2^2 + 9x_3^2 + 2x_1x_2 + 4x_1x_3 + 6x_2x_3$$

From (14.33), the coefficient of $x_i x_j$, $i \neq j$, is $a_{ij} + a_{ji} = 2a_{ij}$. Using this yields

$$a_{11} = 1, \qquad a_{22} = 2, \qquad a_{33} = 9, \qquad a_{12} = 1, \qquad a_{13} = 2, \qquad a_{23} = 3$$

$$a_{11} = 1, \qquad \begin{vmatrix} a_{11} & a_{12} \\ a_{12} & a_{22} \end{vmatrix} = \begin{vmatrix} 1 & 1 \\ 1 & 2 \end{vmatrix} > 0, \qquad \begin{vmatrix} a_{11} & \cdot & a_{13} \\ \cdot & \cdot & \cdot \\ a_{13} & \cdot & a_{33} \end{vmatrix} = \begin{vmatrix} 1 & 1 & 2 \\ 1 & 2 & 3 \\ 2 & 3 & 9 \end{vmatrix} = 4 > 0$$

Hence *Q* is positive definite.

The following theorems can now be stated:

Liapunov Stability Theorem. If there exists a positive-definite *V*, and $V \to \infty$ as $\|\mathbf{x}\| \to \infty$, the system is asymptotically stable in the region in which $\dot{V}$ is negative definite and is stable if $\dot{V}$ is negative semidefinite. The properties are global if the region extends over the entire state space.

Liapunov Instability Theorem.   If there exists a V such that $\dot{V}$ is negative definite, and $\dot{V} \rightarrow -\infty$ as $\|\mathbf{x}\| \rightarrow \infty$, the system is unstable in the region in which V is not positive (semi-) definite.

REMARKS

1. The reason for two theorems is that if the origin is unstable it will be impossible to find a V-function that satisfies the stability theorem. But one satisfying the instability theorem will exist and, if it can be found, will prove instability.
2. The V-function is not unique, and different choices in general will indicate different stability regions. As this implies, a system in general does not become unstable where $\dot{V}$ changes sign. The theorems give only sufficient conditions, and the predicted stability boundaries are usually quite conservative.
3. Asymptotic stability can often be proved even if $\dot{V}$ is only semidefinite: If the curve on which $\dot{V} = 0$ is found not to satisfy the system equations, it is not a trajectory, so the state cannot remain on this curve, and hence the system must be asymptotically stable.

The numerous techniques for deriving Liapunov functions, discussed in books on nonlinear control systems, often apply to certain classes of systems. As noted earlier, the difficulties here can be considerable, and experience is quite important. The quadratic form is often suitable and will be used in the following examples to illustrate the application of the theorems.

Example 14.6.2   Nonlinear Spring–Mass–Damper System

A spring–mass–damper system for which the damping force is proportional to the third power of the velocity is described by the differential equation

$$\ddot{y} + 0.5\dot{y}^3 + y = 0$$

Let mass position y and velocity $\dot{y}$ be chosen as state variables: $x_1 = y$, $x_2 = \dot{y}$. Then $\dot{x}_1 = x_2$ and, from the differential equation, $\dot{x}_2 = \ddot{y} = -0.5\dot{y}^3 - y = -0.5x_2^3 - x_1$ are the state equations. Attempt the positive-definite V-function $V = x_1^2 + x_2^2$ ($V \rightarrow \infty$ as $\|\mathbf{x}\| \rightarrow \infty$).

$$\dot{V} = 2x_1\dot{x}_1 + 2x_2\dot{x}_2 = 2x_1x_2 - x_2^4 - 2x_1x_2 = -x_2^4$$

$\dot{V}$ is negative semidefinite, so the system is globally stable. But since the x_1-axis ($x_2 = 0$) is not a trajectory, the system is also globally asymptotically stable, by remark 3. However, usually $\dot{V}$ for this simple V-function would be indefinite, and would not prove either stability or instability.

Example 14.6.3   Aizerman's Method (Fig. 14.16)

Figure 14.16 shows a motor position servo in which loading and saturation effects on the controller–amplifier combine to produce the nonlinear characteristic $u = f(e)$ indicated.

As in Section 14.3, the system equation is $\ddot{e} + \dot{e} + f(e) = 0$. Choosing the state variables $x_1 = e$, $x_2 = \dot{e}$, the state equations are

$$\dot{x}_1 = x_2 \qquad \dot{x}_2 = -x_2 - f(x_1)$$

Let $f(x_1)/x_1 = 1$ be a linear approximation to the nonlinearity. A V-function will first be sought for the linearized system

$$\dot{x}_1 = x_2 \qquad \dot{x}_2 = -x_2 - x_1$$

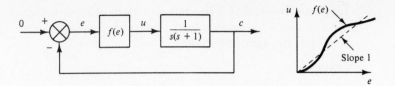

Figure 14.16   Example 14.6.3.

Let $V = a_{11}x_1^2 + 2a_{12}x_1x_2 + a_{22}x_2^2$. Then

$$\dot{V} = -2a_{12}x_1^2 + 2(a_{11} - a_{12} - a_{22})x_1x_2 + 2(a_{12} - a_{22})x_2^2$$

Constrain $\dot{V}$ to, say, $\dot{V} = 2x_1^2 + 2x_2^2$. This requires $a_{11} = -3$, $a_{12} = -1$, $a_{22} = -2$, so

$$V = -3x_1^2 - 2x_1x_2 - 2x_2^2$$

Sylvester's theorem will show this to be negative definite, so the linearized system is globally asymptotically stable.

This negative-definite V-function is now used for the nonlinear system. It is found that

$$\dot{V} = 2x_1 f(x_1) + 4x_2 f(x_1) - 4x_1x_2 + 2x_2^2$$

$$= 2\frac{f(x_1)}{x_1}x_1^2 + 4\left[\frac{f(x_1)}{x_1} - 1\right]x_1x_2 + 2x_2^2$$

Sylvester's theorem yields the following conditions for $\dot{V}$ to be positive definite:

$$k > 0, \qquad k - (k - 1)^2 = -k^2 + 3k - 1 > 0 \qquad \left[\frac{f(x_1)}{x_1} \equiv k\right]$$

From the roots of $k^2 - 3k + 1 = 0$, sufficient conditions for global asymptotic stability are

$$0.38 < \frac{f(x_1)}{x_1} < 2.62$$

V-functions that provide wider bounds on the sector in which the nonlinearity may be located usually exist, and certainly do in Example 14.6.3. But to obtain this improvement other than by trial and error requires more advanced methods. It is noted also that Liapunov functions are not well adapted to nonlinear system design.

14.7 POPOV AND CIRCLE CRITERIA FOR STABILITY

The Popov criterion and the circle criterion give sufficient conditions for stability of nonlinear systems in the frequency domain. They have direct graphical interpretations, similar to that with describing functions, and are convenient for design as well as analysis. The disadvantage is that predicted stability limits are frequently overly conservative. Figure 14.17 shows the system configuration considered. $G(s)$ represents the linear part, and the nonlinearity f is constrained to a sector bounded by slopes $\alpha\ (= \varepsilon)$ and $\beta\ (= k)$. The nonlinearities considered are the following:

(n1)   *Single-valued and time-invariant:* a unique output u for each input e
(n2)   *With hysteresis:* nonlinearities with memory, where u depends on the time

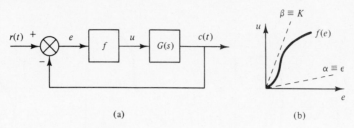

Figure 14.17   System and nonlinearity.

history of e; for example, a relay with hysteresis, or backlash in mechanical connections

(n3)   *General nonlinearities:* time varying and perhaps including hysteresis

   It will be assumed that $r(t)$ and any disturbance inputs are bounded and square integrable; that is, $\int_0^\infty r^2(t)\, dt < \infty$. Note that this implies that $r \to 0$ as $t \to \infty$.

Popov's Method

1. $G(s)$ is assumed to be open-loop stable.
2. The nonlinearity is assumed to satisfy

$$\varepsilon < \frac{u}{e} < K \tag{14.36}$$

where $\varepsilon = 0$ if all poles of G are inside the left-half s-plane, and $\varepsilon > 0$, arbitrarily small, if G has poles on the imaginary axis. This is because for $f = 0$ the closed-loop poles coincide with the open-loop poles, so one or more would be on the imaginary axis and the system could not be asymptotically stable.

   Use is made of a modified frequency response function $G^*(j\omega)$, defined by

$$\text{Re } G^* = \text{Re } G \qquad \text{Im } G^* = \omega\, \text{Im } G \qquad \omega \geqslant 0 \tag{14.37}$$

in the following graphical interpretation of the Popov conditions for global asymptotic stability.

   Popov's Theorem.   For any initial condition, the system output is bounded and tends to zero as $t \to \infty$ if the plot of $G^*(j\omega)$ lies entirely to the right of the *Popov line,* which crosses the real axis at $-1/K$ at a slope $1/q$. Here the restrictions on q and K depend on the nonlinearity:
(n1) $-\infty < q < \infty$ if $0 < K < \infty$; $0 \leqslant q < \infty$ if $K = \infty$.
(n2) $-\infty < q \leqslant 0$ and $0 < K < \infty$.
(n3) $q = 0$ and $0 < K \leqslant \infty$.

Figure 14.18 illustrates the use of these criteria. In Fig. 14.18(a) the polar plot of $G^*(\omega \geqslant 0)$ lies to the right of the Popov line shown, so the system is asymptotically stable. In Fig. 14.18(b), the intersection identified as $-1/K$ gives the maximum K for which the theorem guarantees stability, and then only for the single-valued, time-invariant nonlinearities (n1), for which $q > 0$ is admissible. For a general nonlinearity (n3) the restriction on q is more severe. As indicated in Fig. 14.18(c), with q constrained to be zero, the maximum K is now that which corresponds to the vertical

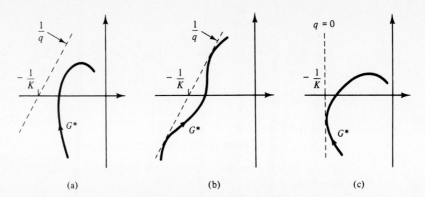

Figure 14.18 Popov's method.

tangent of G^*. If in (c) the nonlinearity were type (n1), the maximum K would be that corresponding to the intersection of G^* with the negative real axis, because the tangent at this point would be an admissible Popov line. It is important to observe that the intersections of G and G^* with the negative real axis are the same, and that this intersection therefore also gives the maximum stable gain of a linear system, by the Nyquist criterion. Hence, for nonlinearities (n1) the *Popov sector* of stability for Fig. 14.18(c), and (a) as well, is the same as the *Hurwitz sector* (that is, the range of stable gains if the system were linear). Since the Hurwitz sector gives the largest permissible gain, an indication of the degree to which the prediction of the Popov sector could be conservative is available by inspection.

The circle criterion for stability is of the same nature as the Popov theorem, with a similar form of graphical interpretation. An advantage is that it uses the Nyquist diagram or polar plot of $G(j\omega)$ and not the modified function G^*.

The Circle Criterion

Circle Theorem.   A system with the nonlinearity in the sector $\alpha < u/e < \beta$ has an output that is bounded and tends to zero as $t \to \infty$ for any initial condition if:

1. $\beta > \alpha > 0$: $G(j\omega)$ does not touch or encircle the circular disk in Fig. 14.19(a).
2. $\beta > 0$, $\alpha = 0$: $G(j\omega)$ lies to the right of the vertical at $-1/\beta$ indicated in Fig. 14.19(b).
3. $\alpha < 0 < \beta$: $G(j\omega)$ lies inside the circle in Fig. 14.19(c).

REMARKS

1. The nonlinearity may be time varying, and in cases 1 and 2 may also include memory. Indeed, case 2 corresponds directly to the Popov line for the nonlinearity (n3), where $q = 0$, since the real parts of G and G^* are the same.
2. Evidently, for case 3, G may not have poles at the origin since then $|G| \to \infty$ as $\omega \to 0$.
3. Unlike the Popov criterion, the circle criterion for case 1 can be restated to apply to open-loop unstable systems:

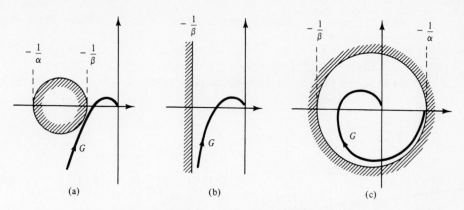

Figure 14.19   The circle criterion.

The Nyquist diagram of $G(j\omega)$, $-\infty < \omega < \infty$, lies outside the circle in Fig. 14.19(a) and encircles it as many times in the counterclockwise direction as there are poles of $G(s)$ with positive real parts.

This extension is exactly what the Nyquist criterion would suggest. The critical point $-1/K$ for a linear system with loop gain KG can be visualized as having expanded into a circle to account for nonlinear effects.

The Popov and circle criteria do not consider the actual shape of the nonlinearity, while the describing function does. Therefore, the stability prediction resulting from DFs can be expected to be in general less conservative and to indicate stability even if G does enter the critical disk.

14.8 CONCLUSION

The behavior of nonlinear systems, types of nonlinearities, and the classical phase-plane and describing function techniques have been discussed. The second method of Liapunov was introduced, as well as the very useful Popov and circle criteria for stability analysis in the frequency domain. Numerous extensions of these frequency domain criteria are available, and the discussion in Section 14.7 should also serve as a suitable introduction to this important area of nonlinear feedback system analysis and design.

PROBLEMS

14.1. Determine linearized models about the given operating points for the following nonlinearities:
   (a) $z = f(x, y) = x^3 + 2y^2$, $(x_0, y_0) = (1, 1)$
   (b) $y = f(x) = 3x^3 + 2x$, $x_0 = 1$

14.2. Determine a linearized gain to represent the nonlinear element $y = f(x) = x + 0.5x^2 - 0.05x^3$ for small variations about each of the following operating points.
   (a) $x_0 = 0$ **(b)** $x_0 = 1$ **(c)** $x_0 = 2$ **(d)** $x_0 = 3$

14.3. Linearize the following nonlinearities for small variations about the operating points indicated.

(a) $z = k_1 x + k_2 x^2 y$, $(x_0, y_0) = (1, 1)$
(b) $z = x^3 + 2x^2 y + 4xy^2 + 4y^3$, $(x_0, y_0) = (1, 1)$

14.4. Sketch the general form of the phase-plane patterns for each of the systems in Fig. P14.4.

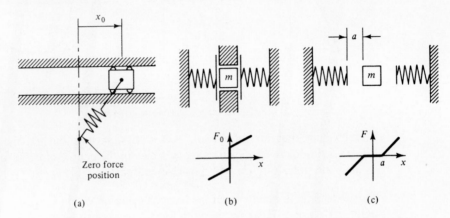

(a)	(b)	(c)

Figure P14.4   (a) Dual equilibrium; (b) preloaded springs; (c) springs and deadband.

14.5. Sketch the phase-plane portrait for step inputs for the relay servo shown in Fig. P14.5.

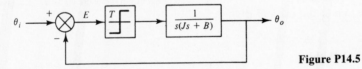

Figure P14.5

14.6. Repeat Problem 14.5 for ramp inputs $\theta_i = At$.

14.7. **(a)** Verify that Fig. P14.7 is a model for a motor position servo with Coulomb friction on the motor shaft.

(b) Obtain the phase-plane and isocline equations for step inputs. What is the nature of the isoclines, and where do they intersect the E-axis of an $\dot{E}$–E phase plane?

(c) Sketch the phase-plane pattern, assuming underdamped system behavior.

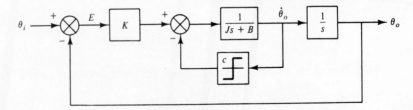

Figure P14.7

14.8. **(a)** Verify that Fig. P14.8 represents a motor position servo with tachometer feedback that is subject to saturation.

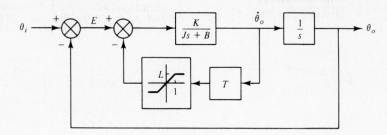

Figure P14.8

(b) Obtain the phase-plane and isocline equations for step inputs. Where do the isoclines intersect the E-axis?

(c) Sketch the phase-plane pattern, taking into account that for large $|\dot{\theta}_o|$ the tachometer feedback will be relatively less effective.

14.9. Repeat Problem 14.8 for ramp inputs.

14.10. A motor position servo with deadband in the error detector is modeled as shown in Fig. P14.10. For step inputs:

(a) Obtain the phase-plane and isocline equations and find the isocline intersections with the E-axis.

(b) Sketch the phase-plane pattern, assuming underdamped behavior for the value of K.

(c) Comment on the steady-state errors.

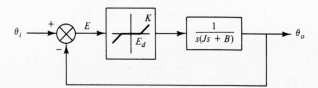

Figure P14.10

14.11. Saturation, shown in a motor position servo in Fig. P14.11, is a very common nonlinearity. Obtain the phase-plane equation and isocline equation for step inputs and sketch the phase-plane pattern.

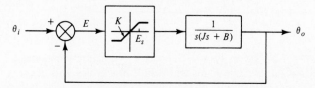

Figure P14.11

14.12. Repeat Problem 14.11 for ramp inputs. Verify the value of the steady-state error from that for small-signal operation. Compare the pattern for large error with that for a relay in Problem 14.6 and comment.

14.13. Sketch the phase-plane pattern of the system of Fig. 14.6 for step inputs.

14.14. Determine the describing function for the nonlinear element in Fig. P14.14, with input x and output y and slopes K_1 and K_2. (Use the describing functions for deadband and/or saturation.)

14.15. In Fig. P14.15, let $G = 10/[s(0.2s + 1)(0.05s + 1)]$. Determine the amplitude at the input of the nonlinearity and the frequency of any limit cycle that may be present for zero input.

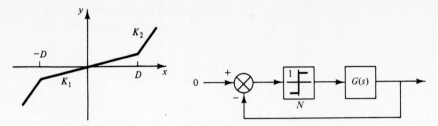

Figure P14.14   **Figure P14.15**

14.16. In Fig. P14.15, calculate and compare the frequency and amplitude of the error signal limit cycle for:
(a) $G = 10/[s(0.1s + 1)(0.01s + 1)]$ (b) $G = 10/[s(0.1s + 1)(0.02s + 1)]$

14.17. In Fig. P14.15, if $G = 5/[s(0.05s + 1)]$, will there be a limit cycle and, if so, what are its amplitude and frequency?

14.18. Let Fig. P14.15 represent a relay servo with

$$G = \frac{100}{s(0.05s + 1)(0.01s + 1)}$$

What are the amplitude and frequency of any limit cycle at the system output?

14.19. If Fig. P14.15 is a relay servo with $G = K/[s(s + 1)(s + 2)]$, find and compare the amplitude and frequency of the limit cycle in the system output for $K = 1$ and $K = 6$.

14.20. In Fig. P14.20, a tachometer feedback loop has been added to a relay servo to eliminate finite-amplitude limit cycles. Show that this is possible, and use Bode plot sketches to determine a reasonable choice for T.

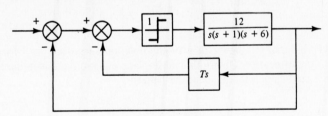

Figure P14.20

14.21. In Fig. P14.21, $G(s) = 12/[(s(s + 1)(s + 5)]$ and N is a saturation element.
(a) Determine the limit on its slope K for which no limit cycles will occur.
(b) If $K = 5$ and the saturation limit of the input to the nonlinearity is $b = 1$, find the amplitude and frequency of any limit cycle and determine its stability.

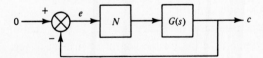

Figure P14.21

14.22. In Problem 14.21, let N be a deadband nonlinearity with a deadband from -1 to $+1$ and a slope $K = 5$.
(a) Find the amplitude and frequency of any limit cycle.

(b) What is the nature of this limit cycle, and for what range of amplitudes at the input to the nonlinearity is the system stable, if any?

14.23. In Problem 14.21, let N be a relay with a deadband $b = 1$ and output $T = 5$. Determine the amplitude and frequency of any limit cycles. Find the nature of these limit cycles and the stability of system behavior for different amplitudes at the input to the nonlinearity.

14.24. In Fig. P14.21 with $N = f(e)$ and $G(s) = 1/[s(s + 1)]$ as in Example 14.6.3, use Liapunov functions to determine the Hurwitz sector of stability, that is, the range of linear gains of N for which the system is stable. Does the result agree with the actual size of this sector as obtained, say, from a root locus sketch?

14.25. In Problem 14.24, if $f(e)/e = K$ is a linear approximation to the nonlinearity, use Aizerman's method based on the Liapunov function for arbitrary K found in Problem 14.24 to determine the stability sector for:
(a) $K = 2$ (b) $K = 4$ (c) $K = 10$
Compare the results also with the stable sector $0.38 < f(e)/e < 2.62$ found in Example 14.6.3 for $K = 1$.

14.26. In Fig. P14.21 with $N = f(e)$ and $G(s) = 5/[s(s + 5)]$, determine the stability sector of the nonlinearity by Aizerman's method. Use $f(e)/e = 1$ as a linear approximation.

14.27. In Fig. P14.27, N is a nonlinearity with output $f(x_1)$.
(a) Obtain a state-space model with x_1 and x_2 as state variables. [*Hint:* For the last block, $-x_1/x_2 = (s - 3)/(s + 4)$, and cross-multiplication and inverse transformation give $\dot{x}_2 = 3x_2 - \dot{x}_1 - 4x_1$, where $\dot{x}_1$ can be eliminated by using the same approach for the preceding block.]
(b) Determine the Hurwitz sector of stability (that is, the range of linear gains in N for which the system is stable).
(c) Determine the stability sector of $f(x_1)/x_1$ by Aizerman's method, using $f(x_1) = 0$ as a linear approximation to the nonlinearity.

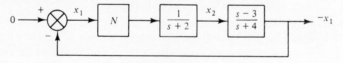

Figure P14.27

14.28. For Problem 14.27, apply the Popov stability criterion to find the Popov sector of stability and the type of nonlinearity permitted. Compare with the Hurwitz sector for an idea of to what extent the Popov sector could be conservative.

14.29. For the system of Fig. P14.21 with $G(s) = 20/[s(s + 2)(s + 4)]$:
(a) Determine the stability sector of the nonlinearity by the Popov method, assuming a general nonlinearity.
(b) Repeat part (a) for a single-valued time-invariant nonlinearity. Compare it with the Hurwitz sector of stability to determine to what extent the prediction may be conservative.

14.30. For the system of Problem 14.29, use the circle criterion to predict the stability sectors of a general nonlinearity, and compare with the Popov predictions in Problem 14.29:
(a) If the lower sector boundary approaches zero.
(b) If the lower sector boundary is $f(x, t)/x = 0.5$.

14.31. For the system of Fig. P14.21 with $G(s) = 15/[s(s + 2)(s + 6)]$:
  (a) Find the Popov stability sector of N for a general nonlinearity.
  (b) Repeat for a single-valued time-invariant nonlinearity.
  (c) Compare both with the Hurwitz stability sector.

14.32. For the system of Problem 14.31, predict the stability sectors for a general nonlinearity by the circle criterion:
  (a) If the lower sector boundary approaches zero.
  (b) If the lower sector boundary is $f(x, t)/x = 0.5$.

14.33. Apply the circle criterion to the system considered in Problems 14.27 and 14.28. Compare the stability sector obtained with the Hurwitz and Popov sectors. Note that the negative gain limit of the Hurwitz sector suggests the use of the form of the circle criterion that allows a negative slope sector boundary.

14.34. For the system in Fig. P14.21 with $G(s) = (s - 1)/(s + 1)^2$:
  (a) What is the Hurwitz stability sector?
  (b) Find the Popov stability sector and the type of nonlinearity permitted.

14.35. Use the circle criterion to determine the stability sector for the system of Problem 14.34, allowing for a negative slope sector boundary, and compare with the Hurwitz and Popov sectors.

14.36. The restriction of the Popov stability sector to zero minimum slope, which is a serious disadvantage in Problems 14.28 and 14.34, where the Hurwitz sector includes negative gains, can be overcome by *pole shifting*. Show that the system of Problem 14.34, with nonlinearity $m = f(x)$ and $G(s) = -x/m$, is equivalent to one with nonlinearity m' and G' as given next, where L is a chosen constant:

$$m' = g(x) = m + Lx$$
$$G' = \frac{-x}{m'} = \frac{s - 1}{(s + 1)^2 - L(s - 1)}$$

Note that the poles of G' differ from those of G; hence the name "pole shifting," and that L can be chosen to change a sector boundary of $f(x)/x$ with negative slope to one with zero or larger slope of $g(x)/x$.

14.37. Use the result of Problem 14.36 to apply pole shifting to Problem 14.34. Choose L such that the lower (negative slope) boundary of the Hurwitz sector becomes a zero slope boundary for the nonlinearity $g(x)$. Determine the Popov stability sector for $g(x)/x$ and, from it, that for $f(x)/x$. Compare the result with the Hurwitz sector. What type of nonlinearity is permitted?

Appendix A

Vectors, Matrices, and Determinants

A.1 VECTORS AND MATRICES

The $m \times n$ matrix A in (A.1) is a rectangular array of mn elements, where a_{ij} is the element in row i and column j. The notation $\{a_{ij}\}$ is often used to identify the matrix. If $m = n$, the matrix is square and n dimensional. For $n = 1$, $\mathbf{A}$ is a column vector, such as $\mathbf{x}$ in (A.1), and for $m = 1$ it is a row vector, such as $\mathbf{y}$ in (A.1).

$$\mathbf{A} = \{a_{ij}\} = \begin{bmatrix} a_{11} & a_{12} & \cdots & a_{1n} \\ a_{21} & a_{22} & & a_{2n} \\ \vdots & & & \vdots \\ a_{m1} & a_{m2} & \cdots & a_{mn} \end{bmatrix} \quad \mathbf{x} = \begin{bmatrix} x_1 \\ x_2 \\ \vdots \\ x_n \end{bmatrix} \quad \text{(A.1)}$$

$$\mathbf{y} = [y_1 \quad y_2 \quad \cdots \quad y_n]$$

Some important definitions are as follows:

1. *Diagonal matrix:* A square matrix of which all elements not on the main diagonal are zero: $a_{ij} = 0$, $i \neq j$.

2. *Unit matrix or identity matrix* $\mathbf{I}$: A diagonal matrix of which all elements on the main diagonal are equal to 1.

3. *Zero matrix or null matrix* $\mathbf{0}$: A matrix of which all elements are zero.

4. *Symmetrical matrix:* $a_{ij} = a_{ji}$ for all $i \neq j$.

5. *Transpose matrix* $\mathbf{A}'$ *of* $\mathbf{A}$: Obtained by interchanging the rows and columns of $\mathbf{A}$. The transpose of the column vector $\mathbf{x}$ in (A.1) is the row vector $\mathbf{x}' = [x_1 \quad x_2 \quad \cdots \quad x_n]$. $(\mathbf{A}')' = \mathbf{A}$, and a matrix is symmetrical if $\mathbf{A}' = \mathbf{A}$.

6. *Trace of a square matrix* $\mathbf{A}$: The sum of the elements on the main diagonal.

$$\text{tr } \mathbf{A} = a_{11} + a_{22} + \cdots + a_{nn} \qquad (A.2)$$

7. *Partitioned matrices:* A matrix can be partitioned into submatrices or vectors. Equations (A.3) show the $m \times n$ matrices $\mathbf{A} = \{a_{ij}\}$ and $\mathbf{B} = \{b_{ij}\}$ partitioned into, respectively, column vectors and row vectors.

$$\mathbf{A} = [\mathbf{a}_1 \ \cdots \ \mathbf{a}_n]$$

$$\mathbf{b}_i' = [b_{i1} \ \cdots \ b_{in}] \qquad \mathbf{B} = \begin{bmatrix} \mathbf{b}_1' \\ \vdots \\ \mathbf{b}_m' \end{bmatrix} \qquad \mathbf{a}_i = \begin{bmatrix} a_{1i} \\ \vdots \\ a_{mi} \end{bmatrix} \qquad (A.3)$$

A.2 VECTOR AND MATRIX OPERATIONS

For matrices $\mathbf{A} = \{a_{ij}\}$, $\mathbf{B} = \{b_{ij}\}$ or vectors $\mathbf{x} = \{x_i\}$, $\mathbf{y} = \{y_i\}$, some important properties and operations are the following.

1. $\mathbf{A} = \mathbf{B}$ ($\mathbf{x} = \mathbf{y}$) if and only if $a_{ij} = b_{ij}$ ($x_i = y_i$) for all i, j. The matrices (vectors) are then equal.
2. Multiplication by a scalar h, $h\mathbf{A} = \{ha_{ij}\}$, $h\mathbf{x} = \{hx_i\}$. Each element is multiplied by the scalar.
3. Addition and subtraction, $\mathbf{C} = \mathbf{A} \pm \mathbf{B}$, $\mathbf{z} = \mathbf{x} \pm \mathbf{y}$. Corresponding elements are added or subtracted: $c_{ij} = a_{ij} \pm b_{ij}$, $z_i = x_i \pm y_i$. Some properties:

$$\mathbf{A} + \mathbf{B} = \mathbf{B} + \mathbf{A} \qquad (\mathbf{A} + \mathbf{B}) + \mathbf{C} = \mathbf{A} + (\mathbf{B} + \mathbf{C})$$

$$(\mathbf{A} + \mathbf{B})' = \mathbf{A}' + \mathbf{B}'$$

$$(h_1 + h_2)\mathbf{x} = h_1\mathbf{x} + h_2\mathbf{x} \qquad \mathbf{A} + \mathbf{0} = \mathbf{A} \qquad \mathbf{A} - \mathbf{A} = \mathbf{0}$$

4. Multiplication, $\mathbf{C} = \mathbf{AB}$. The number of columns of $\mathbf{A}$ must equal the number of rows of $\mathbf{B}$. If $\mathbf{A}$ is $m \times n$ and $\mathbf{B}$ is $n \times p$, then element c_{ij} of the $m \times p$ matrix $\mathbf{C} = \{c_{ij}\}$ is

$$c_{ij} = \sum_{k=1}^{n} a_{ik}b_{kj} \qquad (A.4)$$

That is, element k in row i of $\mathbf{A}$ is multiplied by element k in column j of $\mathbf{B}$, and these products are added for all k in this row i and column j to obtain c_{ij}. Multiplication using partitioned matrices gives the same result:

$$\mathbf{C} = \mathbf{AB} = \begin{bmatrix} \mathbf{a}_1' \\ \vdots \\ \mathbf{a}_m' \end{bmatrix} [\mathbf{b}_1 \ \cdots \ \mathbf{b}_p]$$

$$c_{ij} = \mathbf{a}_i'\mathbf{b}_j = [a_{i1} \ \cdots \ a_{in}] \begin{bmatrix} b_{1j} \\ \vdots \\ b_{nj} \end{bmatrix} = \sum_{k=1}^{n} a_{ik}b_{kj}$$

Some properties: In general, $\mathbf{AB} \neq \mathbf{BA}$ (that is, matrix multiplication is not commutative). If $\mathbf{A}$ and $\mathbf{B}$ commute, then $\mathbf{AB} = \mathbf{BA}$.

$$\mathbf{AI} = \mathbf{IA} = \mathbf{A} \qquad \mathbf{III} = \mathbf{I}^3 = \mathbf{I} \qquad (\mathbf{AB})\mathbf{C} = \mathbf{A}(\mathbf{BC}) = \mathbf{ABC}$$

$$\mathbf{A}(\mathbf{B} + \mathbf{C}) = \mathbf{AB} + \mathbf{AC} \qquad \mathbf{A0} = \mathbf{0A} = \mathbf{0}$$

Note that in the matrix case $\mathbf{AB} = \mathbf{0}$ need not imply that $\mathbf{A}$ or $\mathbf{B}$ is a null matrix. For example,

$$\begin{bmatrix} a & b \\ 0 & 0 \end{bmatrix} \begin{bmatrix} b & 0 \\ -a & 0 \end{bmatrix} = \begin{bmatrix} 0 & 0 \\ 0 & 0 \end{bmatrix}$$

5. $(\mathbf{AB})' = \mathbf{B}'\mathbf{A}'$: The transpose of a product is the product of the transposes in reverse order.

6. *Linear transformations and algebraic equations:* Alternative forms of the same set of simultaneous linear algebraic equations are as follows:

$$a_{11}x_1 + \cdots + a_{1n}x_n = y_1 \qquad y_i = \sum_{j=1}^{n} a_{ij}x_j \qquad i = 1, \ldots, m$$

$$\vdots \qquad\qquad\qquad \vdots \qquad\qquad\qquad\qquad\qquad\qquad\qquad \text{(A.5)}$$

$$a_{m1}x_1 + \cdots + a_{mn}x_n = y_m \qquad \mathbf{y} = \mathbf{Ax} \qquad \mathbf{A} = \{a_{ij}\}$$

The matrix representation may be verified by applying the general multiplication rule to the matrix–vector product, which results in a vector. Equations (A.5) may represent a transformation of variables, used extensively to simplify problem formulations. It can also be a set of equations to be solved for $\mathbf{x}$. If $\mathbf{y} = \mathbf{0}$, the set is homogeneous, and a necessary condition for a nontrivial solution, that is, a solution $\mathbf{x}$ of which not all elements are zero, is known to be that the determinant of the matrix $\mathbf{A}$ of the coefficients be zero.

7. *Quadratic forms $\mathbf{x}'\mathbf{Ax}$:*

$$\mathbf{x}'\mathbf{Ax} = [x_1 \quad \cdots \quad x_n] \begin{bmatrix} a_{11} & \cdots & a_{1n} \\ \vdots & & \vdots \\ a_{n1} & \cdots & a_{nn} \end{bmatrix} \begin{bmatrix} x_1 \\ \vdots \\ x_n \end{bmatrix} = \sum_{i=1}^{n} x_i y_i \qquad \text{(A.6)}$$

where y_i is as defined in (A.5). Note that the product is a scalar number.

8. *Vector products and orthogonality:* The scalar product, or dot product, or inner product of two vectors $\mathbf{x}$ and $\mathbf{y}$ is a scalar number.

$$\mathbf{x}'\mathbf{y} = [x_1 \quad \cdots \quad x_n] \begin{bmatrix} y_1 \\ \vdots \\ y_n \end{bmatrix} = \sum_{i=1}^{n} x_i y_i = \mathbf{y}'\mathbf{x}$$

$$\text{(A.7)}$$

$$\mathbf{x}'\mathbf{x} = \sum_{i=1}^{n} x_i^2 \qquad |\mathbf{x}| = \sqrt{x_1^2 + \cdots + x_n^2} = \sqrt{\mathbf{x}'\mathbf{x}}$$

As an extension of three-dimensional concepts, $|\mathbf{x}|$ identifies the length of vector $\mathbf{x}$. A vector is said to be normalized if $|\mathbf{x}| = 1$ and is then called a unit vector. Two vectors are orthogonal if their inner product is zero: $\mathbf{x}'\mathbf{y} = \mathbf{y}'\mathbf{x} = 0$. This is a generalization of the mutually perpendicular coordinate axes for rectangular coordinate systems in three-dimensional space. The axes are given by the unit vectors $\mathbf{e}_1' = [1 \quad 0 \quad 0]$, $\mathbf{e}_2' = [0 \quad 1 \quad 0]$, and $\mathbf{e}_3' = [0 \quad 0 \quad 1]$, with zero inner products.

A.3 DETERMINANTS, THE INVERSE, AND THE RANK OF A MATRIX

The determinant $|\mathbf{A}|$ or det $\mathbf{A}$ of an $n \times n$ matrix $\mathbf{A}$ is a scalar number or function. It is found via the use of minors and cofactors. The minor m_{ij} of element a_{ij} is the determinant of a matrix of order $n - 1$ obtained from $\mathbf{A}$ by removing the row and column containing a_{ij}. The cofactor c_{ij} of a_{ij} is $c_{ij} = (-1)^{i+j} m_{ij}$. The determinant of $\mathbf{A}$ is then

$$|\mathbf{A}| = \sum_{j=1}^{n} a_{ij} c_{ij} = \sum_{i=1}^{n} a_{ij} c_{ij} \qquad (A.8)$$

As this implies, $|\mathbf{A}|$ may be found by expansion in this manner along any row or column of $\mathbf{A}$. For example, if $\mathbf{A}$ is 3 by 3, expansion along the first row gives

$$|\mathbf{A}| = a_{11}(a_{22}a_{33} - a_{23}a_{32}) - a_{12}(a_{21}a_{33} - a_{23}a_{31})$$
$$+ a_{13}(a_{21}a_{32} - a_{22}a_{31}) \qquad (A.9)$$

Some properties are that $|\mathbf{A}'| = |\mathbf{A}|$, $|h\mathbf{A}| = h^n|\mathbf{A}|$ for scalars h, and $|\mathbf{AB}| = |\mathbf{A}|\,|\mathbf{B}|$. Interchanging two columns or two rows of $\mathbf{A}$ changes the sign of $|\mathbf{A}|$, and $|\mathbf{A}| = 0$ if $\mathbf{A}$ has two equal rows or columns.

The inverse matrix $\mathbf{A}^{-1}$ of $\mathbf{A}$ is now found by first defining the adjoint matrix adj($\mathbf{A}$) of $\mathbf{A}$. It is the transpose of the matrix obtained by replacing each element of $\mathbf{A}$ by its cofactor. From (A.8), it is readily verified that $\mathbf{A}$ adj($\mathbf{A}$) = adj($\mathbf{A}$)$\mathbf{A}$ = $|\mathbf{A}|\mathbf{I}$. The inverse $\mathbf{A}^{-1}$ of $\mathbf{A}$ is defined by the relations $\mathbf{A}^{-1}\mathbf{A} = \mathbf{AA}^{-1} = \mathbf{I}$ and is therefore equal to

$$\mathbf{A}^{-1} = \frac{\text{adj}(\mathbf{A})}{|\mathbf{A}|} \qquad (A.10)$$

If $|\mathbf{A}| = 0$, the matrix $\mathbf{A}$ is said to be singular, and the inverse does not exist. Some properties are that $(\mathbf{A}^{-1})^{-1} = \mathbf{A}$, $(\mathbf{A}^{-1})' = (\mathbf{A}')^{-1}$, and $(\mathbf{AB})^{-1} = \mathbf{B}^{-1}\mathbf{A}^{-1}$. The last equation says that the inverse of a product is equal to the product of the inverses in reverse order.

The rank $R(\mathbf{A})$ of $\mathbf{A}$ is defined to be the dimension of the largest nonsingular matrix, that is, square and with a nonzero determinant, contained in $\mathbf{A}$. Equivalently, it is the maximum number of linearly independent rows or columns of $\mathbf{A}$. A set of vectors $\mathbf{x}_1, \ldots, \mathbf{x}_n$ is linearly dependent if there are constants $c_1, \ldots, c_n$ that are not all zero such that $c_1\mathbf{x}_1 + c_2\mathbf{x}_2 + \cdots + c_n\mathbf{x}_n = \mathbf{0}$. This is equivalent to $\mathbf{Ac} = \mathbf{0}$, where $\mathbf{A} = [\mathbf{x}_1 \; \cdots \; \mathbf{x}_n]$ and $\mathbf{c}' = [c_1 \; c_2 \; \cdots \; c_n]$. $|\mathbf{A}| = 0$ is the necessary condition for a nontrivial solution, so $\mathbf{A}$ must be nonsingular for linear independence. Some rank properties are that $R(\mathbf{A}') = R(\mathbf{A})$ and $R(\mathbf{A}) = R(\mathbf{A}'\mathbf{A}) = R(\mathbf{AA}')$.

A.4 MATRIX CALCULUS

The derivative or integral of a vector or a matrix is the vector or matrix consisting of the derivatives or integrals of the elements. For example, if $\mathbf{A}(t) = \{a_{ij}(t)\}$, then

$$\frac{d}{dt}(\mathbf{A}(t)) = \left\{\frac{d}{dt}(a_{ij}(t))\right\}$$

Some properties are the following:

$$\frac{d}{dt}(\mathbf{AB}) = \mathbf{A}\frac{d}{dt}(\mathbf{B}) + \frac{d}{dt}(\mathbf{A})\mathbf{B}$$

$$\frac{d}{dt}(h\mathbf{A}) = h\frac{d}{dt}(\mathbf{A}) + \frac{dh}{dt}\mathbf{A} \qquad h \text{ scalar}$$

$$(A.11)$$

Appendix B

Computer Aids for Analysis and Design

B.1 PACKAGE FEATURES AND SYSTEM REQUIREMENTS

A disk is provided with this book that gives all source code and all executable programs for the following:

1. Plotting the transient response for unit step inputs
2. Plotting root loci
3. Constructing polar poles
4. Constructing asymptotic and/or actual Bode plots

In addition to graphical output on the screen, the package supports hardcopy output on:

- IBM Graphics-compatible dot-matrix printers
- Hewlett-Packard-compatible laser printers

An IBM-compatible PC with a math coprocessor is required, and the system must be configured with version 5.0 of MS-DOS. This is because no special graphics package is used, but plotting is accomplished entirely by using the limited version of QBasic that comes with DOS 5.0.

The package and its individual programs are discussed later with examples of their use. First, however, it is appropriate to point out that several general program packages are available for the computer-aided analysis and design of control systems, generally with capabilities considerably beyond those provided on the present disk. A good example is outlined next.

B.2 THE STUDENT EDITION OF MATLAB

General program packages that can be used for the problems in this book include the following:

1. For VAX-compatible computers:
   CTRLC (System Control Technology)
   MATRIX-X (Integrated Systems, Inc.)
2. For IBM PC-compatible computers:
   PC-MATLAB (Mathworks, Inc.)
   CC (Prof. P. Thompson, Caltech)

Particularly suitable in the present context is the version of MATLAB available in **The Student Edition of MATLAB** from Prentice Hall, Englewood Cliffs, New Jersey, 1992. This book is accompanied by two 3.5-inch disks or by 5.25-inch disks and includes the *Systems and Signals Toolbox,* one of numerous application tool boxes available with the professional version. The *Student Edition* is otherwise the same as the professional version except for the following:

- Limitations on the sizes of vectors and matrices
- No graphics postprocessor feature for hardcopy output
- A math coprocessor is not required but is used if it is available

   The Systems portion of the *Toolbox* includes a number of important functions:

1. Step response plots for continuous state-space models and transfer functions
2. Root locus plots for continuous and discrete-time state-space models and transfer functions
3. Polar plots and Nyquist plots for continuous state models and transfer functions
4. Bode plots for continuous state-space models and transfer functions
5. Optimal regulator and observer design in state space
6. Transformations between state space, transfer functions, and pole–zero representations, both continuous and discrete-time
7. Conversions from continuous to discrete-time models

B.3 GENERAL DESCRIPTION

The package is interactive in nature and is design oriented in that controller transfer functions can be entered separately from the plant transfer function. The latter may be entered via the keyboard or from a file. The numerator and denominator of the transfer function may each consist of a product of several polynomials. The programs can also be used to find the roots of a polynomial by entering it as the denominator of a transfer function and choosing numerical output on the screen or printer.

   The computations are performed by FORTRAN programs available from the preceding edition, modified to write x–y data points to output files:

trans.exe   rloc.exe   polr.exe   bodepl.exe

The FORTRAN source code, including all subroutines, is given on the disk. Table B.1 is a listing of these programs.

TABLE B.1   LISTING OF FORTRAN MAIN PROGRAMS AND SUBROUTINES

Main Programs:
Unit step response	trans.f
Polar plots	polr.f
Root loci	rloc.f
Bode diagrams	bodepl.f

Subroutines:
Transient response	eqroot, ing, proot, respon
Polar plots	calcpx, ing, mutpol
Root loci	ing, mutpol, proot, reduce, spcadd
Bode plots	checkn, ing, mgplt, mutpol, phplt, proot, rcheck, reduce, writem

Calls by subroutines:
mgplt	amplt, bdplt	ing	mutpol
phplt	aplt	proot	quad
bdplt	bode	respon	diff
bode	sort	sort	fmax

The executable versions of the FORTRAN programs are on the disk, but for the QBasic programs only the source code (.BAS files) is needed. This is because, in order to limit the need for QBasic to the facilities available with MS-DOS 5.0, the RUN instruction of the QBasic environment is used, and it requires only .BAS files. OBasic is used because it is well suited to produce the desired plots.

The operation is coordinated by DOS batch files, with a separate set for the two types of printers:

- Dot matrix printers: go1.bat, i1.bat, i = 1...4
- Laser printers: go2.bat, i2.bat, i = 1...4
- Exit from the package: 5.bat

The go files present a menu of five choices on the screen. The choice entered, for example, 21, starts the corresponding batch file for computation and plotting, and at the end of this file a choice is presented to start the go file again or to exit via batch file 5.

B.4 SOME ASPECTS OF OPERATION AND PLOTTING

Typing go1 or go2, depending on the type of printer used, starts the program and presents the choices. For a dot-matrix printer the DOS files GRAPHICS.COM and GRAPHICS.PRO must be in the path. The QBasic screen will show briefly when the plotting section of the program begins. Prompts for axes scaling then appear. Note that the commas indicated in the prompts should be inserted when entering the data. Depressing PrintScreen when the plot is on the screen will give hardcopy output.

Then, or immediately if no hardcopy is needed, press any key to remove the plot from the screen and see the next prompt appear.

The aspect ratio, asp, is available to allow a scale adjustment on the vertical axis to ensure that the scales on both axes will be the same for root locus plots and polar plots. Its value is 1 for a VGA screen, but for an Epson LX800 a value asp = 1.2 is needed to ensure that a circle on the screen will also appear as a circle on the printer.

Since the QBasic .BAS files used for plotting do not need to be compiled, they can readily be edited to accommodate screens other than the VGA screen for which they are written.

For an EGA screen, screen 12 in the programs must be replaced by screen 2, and in the PMAP instruction used for axes labeling, the parameter $w = 16$ must be replaced by $w = 8$. For a VGA screen with 480 pixels per column and 30 rows, the number of pixels per row is $w = 480/30 = 16$. For an EGA screen with 200 pixels per column and 25 rows, the number per row is $w = 200/25 = 8$.

Also, the aspect ratio asp must be changed to have a circle appear as a circle on the screen. For a VGA screen the vertical to horizontal pixel ratio, 480/640 equals the ratio of the screen dimensions, so the pixels are square, and the aspect ratio is 1. For an EGA screen the pixel ratio is 200/640, so 200 pixels cover a distance analogous to 480 on a VGA screen. Therefore, the aspect ratio should be changed to asp = 200/480 = 0.41667. These asp values are used in the WINDOW statements that define the ranges of the variables to be covered by the screen window.

B.5 STEP RESPONSE OF A TRANSFER FUNCTION MODEL

This program is based on the partial fraction expansion technique [2]. The transform $C(s)$ of the output is a ratio of polynomials $P(s)$ and $Q(s)$. It is assumed that $Q(s)$ is of higher degree than $P(s)$ and that the roots R_n of $Q(s)$ are all distinct.

The subroutine PROOT is used to determine the roots of a polynomial with real coefficients and occurs in many of the programs. It is based on the Lin–Bairstow method [1], in which an iterative routine is used to find and divide out successive quadratic factors of the polynomial. Their roots, plus that of any remaining first-order factor, constitute the solution.

For distinct roots, the residues K_n in the partial fraction expansion

$$C(s) = \frac{P(s)}{Q(s)} = \sum_{n=1}^{m} \frac{K_n}{(s - R_n)}$$

are known to be given by

$$K_n = \lim_{s \to R_n} (s - R_n) \frac{P(s)}{Q(s)}$$

Since $Q(s) = (s - R_1) \cdots (s - R_n) \cdots (s - R_m)$, it may be seen that (L'Hopital's rule)

$$\lim_{s \to R_n} \frac{s - R_n}{Q(s)} = \lim_{s \to R_n} \frac{1}{Q'(s)} = \frac{1}{Q'(R_n)}$$

where $Q'(s)$ is the derivative of $Q(s)$. Hence

$$K_n = \frac{P(R_n)}{Q'(R_n)}$$

and inverse transformation gives the response

$$c(t) = \sum_{n=1}^{m} \frac{P(R_n)}{Q'(R_n)} e^{R_n t}$$

This is valid for both real and complex distinct roots. A subroutine ING, common to this and the following programs, is used to enter the coefficients of $P(s)$ and $Q(s)$. $P(s)$, in particular, is frequently available in the form of a product of several polynomials. Therefore, ING asks first how many polynomial factors are contained in each of $P(s)$ and $Q(s)$. For each in turn it will then repeat as often as necessary the prompts to enter the order plus one of the polynomial factor, and its coefficients in order of ascending power, separated by spaces or carriage returns. Note that only the transfer function data are entered. The pole at the origin due to the step input is added by the program, which also checks to ensure that all poles are distinct. The screen prompts the user to choose the numerical values to be shown on the screen or printed on paper.

Example B.5.1

Figure B.1 shows the step response of

$$G(s) = \frac{1}{s^2 + s + 1}$$

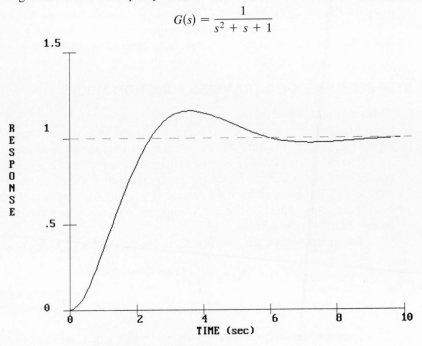

Figure B.1   Example of transient response plot.

B.6 ROOT LOCUS PLOTS

This program plots the loci of the closed-loop poles of a system with loop gain function $G(s) = KN(s)/D(s)$ for varying K. $N(s)$ and $D(s)$ are polynomials and the poles are the roots of the polynomial

$$D(s) + KN(s)$$

which are calculated for specified values of K. The polynomials $N(s)$ and $D(s)$ can again be entered in terms of their factors as described for the step response program. The closed-loop poles are calculated a specified number of times with a specified increment of gain K. The screen prompts the user to choose whether the numerical values of the roots are to be shown and, if so, whether on the screen or on paper. Manual scaling permits a part of the plot to be enlarged. The axes can be moved as desired for a more efficient plot. The program allows for the separate entry of a dynamic compensator by the same routine ING as that used for input of the plant transfer function.

Example B.6.1

Figure B.2 shows the root locus plot for the loop gain function

$$G(s) = \frac{K(s + 3)}{(s + 1)(s + 2)}$$

plotted on a laser printer with aspect ratio asp $= 1$.

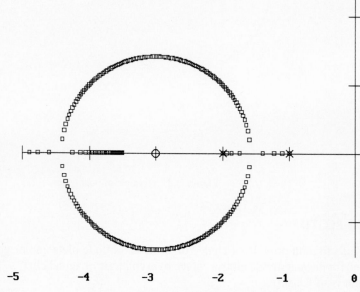

| -5 | -4 | -3 | -2 | -1 | 0 |

Figure B.2   Example of root locus plot.

B.7 POLAR PLOTS

This program constructs a polar plot of the transfer function

$$G(s) = \frac{KN(s)}{D(s)}$$

where K is a constant and $N(s)$ and $D(s)$ are polynomials. The program is design oriented in that it allows for separate entry of a dynamic compensator and provides a polar plot of the product. The plot is obtained by calculating the magnitude and phase angle of the transfer function for $s = j\omega$ a specified number of times with a specified increment of ω. $N(s)$ and $D(s)$ can again be entered in terms of their polynomial factors. A warning is given if at the starting frequency the magnitude is infinite, with a request to enter another starting frequency. The program allows the origin of the axes system to be positioned as desired.

Example B.7.1

Figure B.3 shows the polar plot of the transfer function

$$G(s) = \frac{1}{s^2 + s + 1}$$

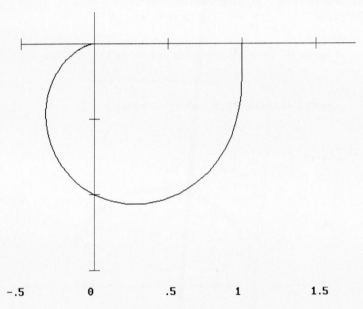

−.5	0	.5	1	1.5

Figure B.3   Example of polar plot.

B.8 BODE PLOTS

This program provides a choice of Bode magnitude plots and/or phase-angle curves and for the magnitude plots a choice of asymptotic or actual curves. The transfer function is of the form

$$G(s) = \frac{A(s)}{s^n B(s)}$$

where $A(s)$ and $B(s)$ are polynomials with nonzero constant terms and n is a, possibly zero, integer constant. $A(s)$ and $B(s)$ may again be entered in terms of their polynomial factors, and the program is oriented to design in that a dynamic compensator can be entered separately by the same subroutine ING as that used for input of the plant.

To plot $G(s)$, the roots of $A(s)$ and $B(s)$ are found, and these roots are stored in a vector X in the order of increasing frequency. Here the frequency corresponding to a complex root $c + jd$ is calculated as $\sqrt{c^2 + d^2}$. If there are N such break frequencies, the vector X is of dimension $N + 2$, with $X(1)$ and $X(N + 2)$ identifying the lowest and highest frequencies for which the plot is to be made.

The asymptotic magnitude plot is constructed by flagging the roots of $A(s)$ to cause a $+20$ dB/dec change of slope and those of $B(s)$ for a -20 dB/dec change of slope. These flags are stored in a vector Flag of dimension $N + 1$, where Flag(1) gives the slope of the low-frequency asymptote; Flag(1) $= -20n$. The asymptotic

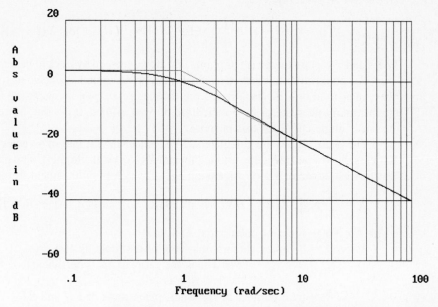

Figure B.4   Example of Bode magnitude plot.

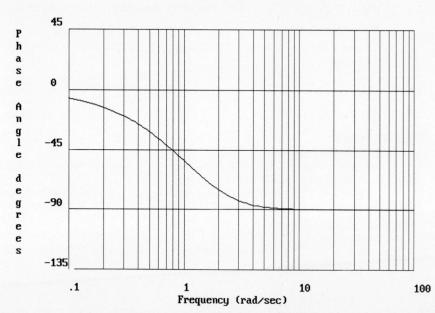

Figure B.5   Example of phase-angle curve.

magnitudes for the frequencies stored in X can now be calculated and are stored in a vector Y of dimension $N + 2$. The starting magnitude is

$$Y(1) = 20 \log K + \text{Flag}(1) \log X(1)$$

where K is the ratio of the constant terms of $A(s)$ and $B(s)$. The other elements of Y are, with $i = 2, \ldots, N + 2$,

$$Y(i) = Y(i - 1) + \left[\sum_{j=1}^{i-1} \text{Flag}(j)\right][\log X(i) - \log X(i - 1)]$$

The X and Y vectors form a set of points that are connected by lines to generate the plot.

For nonminimum-phase transfer functions an option is included to allow the phase margin to be still visualized from the magnitude plot by changing the slope at nonminimum-phase zeros by -20 instead of $+20$ dB/dec. It is safer, however, to use the phase-angle curve for this purpose.

The phase angles are calculated at Np frequencies, stored in a vector $X1$, where $X1(1)$ is the lowest frequency for the plot and $X1(Np)$ is the highest. The phase angles are stored in a vector $Y1$ of dimension Np, of which the ith element is found from

$$Y1(i) = \sum_{j=1}^{Na} \arg[X1(i) - RN(j)] - 180SK - \sum_{j=1}^{Nb} \arg[X1(i) - RN(j)] - 90n$$

where $Na, Nb =$ order of polynomial $A(s)$, $B(s)$

$RN, RD =$ complex vector consisting of the roots of $A(s)$, $B(s)$

$SK = 0, 1$ if $(-1)^{NPR}$ times the ratio of the coefficients of the highest powers of $A(s)$ and $B(s)$ is >0, <0

$NPR =$ total number of right-half-plane roots of RN and RD

The $X1$ and $Y1$ vectors form a set of points that are connected by lines to generate the phase-angle curve. If its smoothness is inadequate, Np should be increased.

Example B.6.1

Figures B.4 and B.5 show Bode magnitude and phase-angle curves for the transfer function

$$G(s) = \frac{s + 3}{(s + 1)(s + 2)}$$

B.9 REFERENCES

1. Curtis, F. G., *Applied Numerical Analysis*, 2nd ed. Reading, Mass.: Addison-Wesley Publishing Company, Inc., 1980.
2. Churchill, R. V., *Modern Operational Mathematics in Engineering*. New York: McGraw-Hill Book Company, 1944, pp. 44–45.

References

ANDERSON, B.D.O., and J.B. MOORE, *Linear Optimal Control*. Englewood Cliffs, N.J.: Prentice Hall, 1971.

ANDERSON, B.W., *The Analysis and Design of Pneumatic Systems*. New York: John Wiley & Sons, Inc., 1967.

ASHLEY, H., *Engineering Analysis of Flight Vehicles*. Reading, Mass.: Addison-Wesley Publishing Company, Inc., 1974.

ÅSTRÖM, K.J., and B. WITTENMARK, *Computer-Controlled Systems—Theory and Design*, 2nd ed. Englewood Cliffs, N.J.: Prentice Hall, 1990.

ATHERTON, D.P., *Nonlinear Control Engineering*. New York: Van Nostrand Reinhold Company, 1975.

AUSLANDER, D.M., Y. TAKAHASHI, and M.J. RABINS, *Introducing Systems and Control*. New York: McGraw-Hill Book Company, 1974.

BELSTERLING, C.A., *Fluidic Systems Design*. New York: Wiley-Interscience, 1971.

BLACKBURN, J.F., G. REETHOF, and J.L. SHEARER, *Fluid Power Control*. Cambridge, Mass.: The MIT Press, 1960.

BLAKELOCK, J.H., *Automatic Control of Aircraft and Missiles*. New York: John Wiley & Sons, Inc., 1965.

BRYSON, A.E., JR., and Y.C. HO, *Applied Optimal Control*. Waltham, Mass.: Ginn and Company, 1969.

CANFIELD, E.B., *Electromechanical Control Systems and Devices*. New York: John Wiley & Sons, Inc., 1965.

CANNON R.H., JR., *Dynamics of Physical Systems*. New York: McGraw-Hill Book Company, 1967.

CASSELL, D.A., *Microcomputers and Modern Control Engineering*. Reston, Va.: Reston Publishing Co., Inc., 1983.

CHEN, C.T., *Analysis and Synthesis of Linear Control Systems*. New York: Holt, Rinehart and Winston, 1975.

————, *Introduction to Linear System Theory*. New York: Holt, Rinehart and Winston, 1970.

————, *Linear System Theory and Design*. New York: Holt, Rinehart and Winston, 1984.

CROSSLEY, T.R., and B. PORTER, "Synthesis of Aircraft Modal Control Systems Having Real or Complex Eigenvalues," *Aeronautical Journal of the Royal Aeronautical Society*, Vol. 73, pp. 138–142, Feb. 1969.

DAVISON, E.J., and H.W. SMITH, "A Note on the Design of Industrial Regulators: Integral Feedback and Feedforward Controllers," *Automatica*, Vol. 10, pp. 329–332, May 1974.

————, and S.H. WANG, "On Pole Assignment in Linear Multivariable Systems Using Output Feedback," *IEEE Transactions on Automatic Control*, Vol. AC-20, pp. 516–518, Aug. 1975.

D'AZZO, J.J., and C.H. HOUPIS, *Linear Control System Analysis and Design*, 3rd ed. New York: McGraw-Hill Book Company, 1988.

DeRUSSO, P.M., R.J. ROY, and C.M. CLOSE, *State Variables for Engineers*. New York: John Wiley & Sons, Inc., 1965.

DESHPANDE, P.B., and R.H. ASH, *Elements of Computer Process Control—With Advanced Control Applications*. Englewood Cliffs, N.J.: Prentice Hall, 1981.

DiSTEFANO, J.J., III, A.R. STUBBERUD, and I.J. WILLIAMS, *Feedback and Control Systems* (Schaum's Outline Series). New York: Schaum Publishing Co., 1967.

DOEBELIN, E.O., *System Modeling and Response*. New York: John Wiley & Sons, Inc., 1980.

DORF, R.C., *Modern Control Systems*, 6th ed. Reading, Mass.: Addison-Wesley Publishing Company, Inc., 1992.

DRANSFIELD, P., *Engineering Systems and Automatic Control*. Englewood Cliffs, N.J.: Prentice Hall, 1968.

D'SOUZA, A.F., *Design of Control Systems*. Englewood Cliffs, N.J.: Prentice Hall, 1988.

ELGERD, O.I., *Control Systems Theory*. New York: McGraw-Hill Book Company, 1967.

EVELEIGH, V.W., *Introduction to Control Systems Design*. New York: McGraw-Hill Book Company, 1972.

FALLSIDE, F., *Control System Design by Pole–Zero Assignment*. New York: Academic Press, Inc., 1977.

FORTMAN, T.E., and K.L. HITZ, *An Introduction to Linear Control Systems*. New York: Marcel Dekker, Inc., 1977.

FOSTER, K., and G.A. PARKER, *Fluidics—Components and Circuits*. New York: Wiley-Interscience, 1970.

FRANKLIN, G.F., and J.D. POWELL, *Digital Control of Dynamic Systems*, 2nd ed. Reading, Mass.: Addison-Wesley Publishing Company, Inc., 1989.

————, J.D. POWELL, and ABBAS EMAMI-NAEINI, *Feedback Control of Dynamic Systems*. Reading, Mass.: Addison-Wesley Publishing Company, Inc., 1986.

GELB, A., and W.E. VAN DER VELDE, *Multiple-Input Describing Functions and Nonlinear System Design*. New York: McGraw-Hill Book Company, 1968.

GIBSON, J.E., *Nonlinear Automatic Control*. New York: McGraw-Hill Book Company, 1963.

GRAHAM, D., and R.C. LATHROP, "The Synthesis of Optimum Response: Criteria and Standard Forms," *Transactions of the AIEE*, Vol. 72, pp. 273–288, 1953.

————, and D. McRuer, *Analysis of Nonlinear Control Systems*. New York: John Wiley & Sons, Inc., 1961.

Gupta, S.C., and L. Hasdorff, *Fundamentals of Automatic Control*. New York: John Wiley & Sons, Inc., 1970.

Hale, F.J., *Introduction to Control System Analysis and Design*, 2nd ed. Englewood Cliffs, N.J.: Prentice Hall, 1988.

Harrison, H.L., and J.G. Bollinger, *Introduction to Automatic Controls*, 2nd ed. Scranton, Pa.: International Textbook Company, 1969.

Hasdorff, L., *Gradient Optimization and Nonlinear Control*. New York: John Wiley & Sons, Inc., 1976.

Hawkins, D.J., "Pseudodiagonalization and the Inverse Nyquist Array Method," *Proceedings of the IEE*, Vol. 119, pp. 337–342, March 1972.

Hsu, J.C., and A.U. Meyer, *Modern Control Principles and Applications*. New York: McGraw-Hill Book Company, 1968.

Hung, Y.S., and A.G.J. MacFarlane, *Multivariable Feedback: A Quasi-Classical Approach*. New York: Springer-Verlag, 1982.

Jacquot, R.G., *Modern Digital Control Systems*. New York: Marcel Dekker, Inc., 1981.

Johnson, M.A., "Diagonal Dominance and the Method of Pseudodiagonalizaton," *Proceedings of the IEE*, Vol. 126, pp. 1011–1017, Oct. 1979.

Kailath, T., *Linear Systems*. Englewood Cliffs, N.J.: Prentice Hall, 1980.

Keller, R.E., *Statics and Dynamics of Components and Systems*. New York: John Wiley & Sons, Inc., 1971.

Kirk, D.E., *Optimal Control Theory—An Introduction*. Englewood Cliffs, N.J.: Prentice Hall, 1970.

Kosut, R.L., "Suboptimal Control of Linear Time-Invariant Systems Subject to Control Structure Constraints," *IEEE Transactions on Automatic Control*, Vol. AC-15, pp. 557–563, Oct. 1970.

Kuo, B.C., *Automatic Control Systems*, 6th ed. Englewood Cliffs, N.J.: Prentice Hall, 1991.

Kuo, D.C., *Digital Control Systems*. New York: Holt, Rinehart and Winston, 1980.

Kwakernaak, H., and R. Sivan, *Linear Optimal Control Systems*. New York: Wiley-Interscience, 1972.

Lapidus, L., and R. Luus, *Optimal Control of Engineering Processes*. Waltham, Mass.: Blaisdell Publishing Company, 1967.

Lee, T.H., G.E. Adams, and W.M. Gaines, *Computer Process Control—Modeling and Optimization*. New York: John Wiley & Sons, Inc., 1968.

Leininger, G.G., "Diagonal Dominance Using Function Minimization Algorithms," *Proceedings of the IFAC Symposium on Multivariable Technological Systems*, pp. 105–112, Fredericton, Canada, July 1977.

————, "Multivariable Compensator Design Using Bode Diagrams and Nichols Charts," *Proceedings of the IFAC Symposium on Computer Aided Design*, pp. 127–132, Zürich, 1979.

Leitman, G., *An Introduction to Optimal Control*. New York: McGraw-Hill Book Company, 1966.

Levine, W.S., and M. Athans, "On the Determination of the Optimal Constant Output Feedback Gains for Linear Multivariable Systems," *IEEE Transactions on Automatic Control*, Vol. AC-15, pp. 44–50, Feb. 1970.

————, T.L. Johnson, and M. Athans, "Optimal Limited State Variable Feedback Controllers for Linear Systems," *IEEE Transactions on Automatic Control.* Vol. AC-16, pp. 785–792, Dec. 1971.

Lewis, E., and H. Stern, *Design of Hydraulic Control Systems.* New York: McGraw-Hill Book Company, 1962.

MacFarlane, A.G.J., "Return-Difference and Return-Ratio Matrices and Their Use in Analysis and Design of Multivariable Feedback Control Systems, *Proceedings of the IEE,* Vol. 117, pp. 2037–2049, Oct. 1970.

————, and J.J. Belletrutti, "The Characteristic Locus Design Method," *Automatica,* Vol. 9, pp. 575–588, Sept. 1973.

————, and B. Kouvaritakis, "A Design Technique for Linear Multivariable Feedback Systems," *International Journal of Control,* Vol. 25, No. 6, pp. 837–874, 1977.

————, and I. Postlethwaite, "The Generalized Nyquist Stability Criterion and Multivariable Root Loci," *International Journal of Control,* Vol. 25, pp. 81–127, Jan. 1977.

McCausland, I., *Introduction to Optimal Control.* New York: John Wiley & Sons, Inc., 1969.

McGillem, C.D., and G.R. Cooper, *Continuous and Discrete Signal and System Analysis,* 2nd ed. New York: Holt, Rinehart and Winston, 1984.

McRuer, D.T., *Aircraft Dynamics and Automatic Control.* Princeton, N.J.: Princeton University Press, 1973.

Melsa, J.L., and D.G. Schulz, *Linear Control Systems.* New York: McGraw-Hill Book Company, 1969.

Merritt, H.E., *Hydraulic Control Systems.* New York: John Wiley & Sons, Inc., 1967.

Minorsky, N., *Theory of Nonlinear Control Systems.* New York: McGraw-Hill Book Company, 1969.

Munro, N., *Modern Approaches to Control System Design.* Stevenage, U.K.: Peter Peregrinus Ltd., 1979.

Nise, N.S., *Control Systems Engineering.* Redwood City, Calif.: Benjamin/Cummings Publishing Company, Inc., 1992.

Ogata, K., *Modern Control Engineering,* 2nd ed. Englewood Cliffs, N.J.: Prentice Hall, 1990.

————, *State Space Analysis of Control Systems.* Englewood Cliffs, N.J.: Prentice Hall, 1967.

Ower, J.C., and J. Van de Vegte, "Classical Control Design for a Flexible Manipulator: Modeling and Control System Design," *IEEE Journal of Robotics and Automation,* Vol. RA-3, No. 5, pp. 485–489, 1987.

Palm, W.J., III, *Modeling, Analysis and Control of Dynamic Systems.* New York: John Wiley & Sons, Inc., 1983.

Patel, R.V., and N. Munro, *Multivariable System Theory and Design.* Elmsford, N.Y.: Pergamon Press, 1982.

Phillips, C.L., and R.D. Harbor, *Feedback Control Systems,* 2nd ed. Englewood Cliffs, N.J.: Prentice Hall, 1991.

————, and H.T. Nagle, Jr., *Digital Control System Analysis and Design,* 2nd ed. Englewood Cliffs, N.J.: Prentice Hall, 1990.

Porter, B., and R. Crossley, *Modal Control—Theory and Applications.* London: Taylor and Francis Ltd., 1972.

PRIME, H.A., *Modern Concepts in Control Theory.* New York: McGraw-Hill Publishing Company, 1969.

RAVEN, F.H., *Automatic Control Engineering,* 3rd ed. New York: McGraw-Hill Book Company, 1978.

ROSENBROCK, H.H., "Design of Multivariable Control Systems Using the Inverse Nyquist Array," *Proceedings of the IEE,* Vol. 116, pp. 1929–1936, Nov. 1969.

————, *State-Space and Multivariable Theory.* New York: Thomas Nelson and Sons, 1970.

————, "Progress in the Design of Multivariable Control Systems," *Measurement and Control,* Vol. 4, pp 9–11, Jan. 1971.

————, *Computer-Aided Control System Design.* New York: Academic Press, Inc., 1974.

SAGE, A.P., *Linear Systems Control.* Champaign, Ill.: Matrix Publishers, Inc., 1978.

————, *Optimum Systems Control.* Englewood Cliffs, N.J.: Prentice Hall, 1968.

SCHULZ, D.G., and J.L. MELSA, *State Functions and Linear Control Systems.* New York: McGraw-Hill Book Company, 1967.

SHINNERS, S.M., *Modern Control System Theory and Application.* Reading, Mass.: Addison-Wesley Publishing Company, Inc., 1972.

SHINSKEY, F.G., *Process Control Systems,* 2nd ed. New York: McGraw-Hill Book Company, 1979.

SMITH, H.W., and E.J. DAVISON, "Design of Industrial Regulators," *Proceedings of the IEE,* Vol. 119, pp. 1210–1215, Aug. 1972.

STRINGER, J.D., *Hydraulic Systems Analysis.* London: The Macmillan Press Ltd., 1976.

TAKAHASHI, T., *Mathematics of Automatic Control.* New York: Holt, Rinehart and Winston, 1966.

TAKAHASHI, Y., M.J. RABINS, and D.M. AUSLANDER, *Control and Dynamic Systems.* Reading, Mass.: Addison-Wesley Publishing Company, Inc., 1970.

THALER, G.J., and M.P. PASTEL, *Analysis and Design of Nonlinear Feedback Control Systems.* New York: McGraw-Hill Book Company, 1962.

VAN DE VEGTE, J., "Classical Design of Two-by-Two Systems with Severe Interaction," *International Journal of Control,* Vol. 44, No. 4, pp. 1017–1028, 1986.

————, "Classical Design, with application to a 3×3 Turbofan Engine Model," *International Journal of Control,* Vol. 45, No. 1 pp. 1–16, 1987.

WEST, J.C., *Analytical Techniques for Nonlinear Control Systems.* London: English Universities Press, 1960.

WIBERG, D.M., *State Space and Linear Systems* (Schaum's Outline Series). New York: McGraw-Hill Book Company, 1971.

WOLOVICH, W.A., *Linear Multivariable Systems.* New York: Springer-Verlag, Inc., 1974.

ZIEGLER, J.G., and N.B. NICHOLS, "Optimum Settings for Automatic Controllers," *Transactions of the ASME,* Vol. 64, No. 8, p. 759, 1942.

Index

Accelerometer, 53
Accuracy, Bode plots, 230
  digital control systems, 308
  dynamic, 231
  steady-state, 103
AC servomotor, 40
Active dynamic compensation, 75
Actuator, 2
  hydraulic, 50
  pneumatic, 2, 48
Adaptive control, 98
A/D converter, 266
Adiabatic processes, 46
Adjoint matrix, 428
Aizerman's method, 415
Algebraic equations, 427
Algorithms. *See* Control algorithms
Aliasing, 274
Analog computer simulation, 77
Analog prefilters, 275
Angle condition, root loci, 159
Armature controlled motor, 40
Asymptotic approximations, 211
Asymptotic decoupling, 368
Asymptotic stability, *See* Stability
Attitude control:
  Bode plot design, 242
  rockets, 70
  root locus design, 171, 173
  satellites, 68
Attitude control with structural
    resonance:
  Bode plot design, 249
  root locus design, 183
  spacecraft, 69
Auxiliary equation, 118

Backlash, 400
Backward difference, 276
Backward rectangular rule, 277
Bandwidth, 225
  on Bode plot, 231
  versus crossover frequency, 228
  on Nichols chart, 234
Bang-bang servos, 403
Bilinear transformation, 302, 310
Block diagram modeling, 35, 60
  of physical systems, 60

  of pneumatic controllers, 70
Block diagram reduction, 13, 79
  for digital control systems, 289
  for parallel loops, 82
  reduction rules, 81
Bode plots, 204
  actual magnitude curves, 213
  asymptotic magnitude plots, 211
  computer-aided analysis and design, 436
  design, 235
  digital control systems, 310
  gain margin, 211
  performance measures, 230
  phase margin, 210
  quadratic lag, 207
  simple lag, 207
Brake, mechanical, 57
Break frequency, 207
Bridged-T network, 38
Bulk modulus, 46
Bumpless transfer, 267

Capacitance, 36
  fluids, 46
  thermal, 42
Cascade, control, 267
Central band, 272
Central difference, 276
Characteristic curves, nonlinear, 401
Characteristic equation, 14
  for discrete state models, 342
  for state space models, 329
Characteristic polynomial, 14, 390
  desired, 355
Circle criterion, 418
Closed-loop control, 4
Closed-loop frequency response, polar
    plots, 215
  Nichols chart, 232
Closed-loop transfer function, 13, 80
  transfer function matrix, 387
Cofactor matrix, 428
Column dominance ratio, 388
Column vector, 425
Companion matrix, 323
Compensation, 134. *See also* Controllers
Complementary strips, 272
Compressibility, 46